Biological Investigations

Sixth Edition

Form, Function, Diversity, and Process

Warren D. Dolphin

Iowa State University

Boston   Burr Ridge, IL   Dubuque, IA   Madison, WI   New York   San Francisco   St. Louis
Bangkok   Bogotá   Caracas   Kuala Lumpur   Lisbon   London   Madrid   Mexico City
Milan   Montreal   New Delhi   Santiago   Seoul   Singapore   Sydney   Taipei   Toronto

McGraw-Hill Higher Education

A Division of The **McGraw-Hill** *Companies*

BIOLOGICAL INVESTIGATIONS: FORM, FUNCTION, DIVERSITY, AND PROCESS
SIXTH EDITION

Published by McGraw-Hill, a business unit of The McGraw-Hill Companies, Inc., 1221 Avenue of the Americas, New York, NY 10020. Copyright © 2002, 1999, 1997 by The McGraw-Hill Companies, Inc. All rights reserved. No part of this publication may be reproduced or distributed in any form or by any means, or stored in a database or retrieval system, without the prior written consent of The McGraw-Hill Companies, Inc., including, but not limited to, in any network or other electronic storage or transmission, or broadcast for distance learning.

Some ancillaries, including electronic and print components, may not be available to customers outside the United States.

This book is printed on recycled, acid-free paper containing 10% postconsumer waste.

2 3 4 5 6 7 8 9 0 QPD/QPD 0 9 8 7 6 5 4 3 2 1

ISBN 0–07–303141–0

Sponsoring editor: *Patrick E. Reidy*
Developmental editor: *Margaret B. Horn*
Senior marketing manager: *Lisa L. Gottschalk*
Project manager: *Joyce Watters*
Production supervisor: *Sherry L. Kane*
Designer: *K. Wayne Harms*
Cover image: *Telegraph Colour Library/FPG International*
Photo research coordinator: *John C. Leland*
Photo research: *LouAnn K. Wilson*
Senior supplement producer: *Stacy A. Patch*
Executive producer: *Linda Meehan Avenarius*
Compositor: *Shepherd, Inc.*
Typeface: *10/12 Times Roman*
Printer: *Quebecor World Dubuque, IA*

The credits section for this book begins on page 447 and is considered an extension of the copyright page.

Some of the laboratory experiments included in this text may be hazardous if materials are handled improperly or if procedures are conducted incorrectly. Safety precautions are necessary when you are working with chemicals, glass test tubes, hot water baths, sharp instruments, and the like, or for any procedures that generally require caution. Your school may have set regulations regarding safety procedures that your instructor will explain to you. Should you have any problems with materials or procedures, please ask your instructor for help.

www.mhhe.com

BRIEF CONTENTS

CONTENTS

PREFACE

This lab manual is dedicated to the many students and colleagues who have been my patient teachers. I hope that it returns some of what has been learned so that a new generation of biologists may soon add to our wonder of nature's ways while advancing our understanding of life's diverse forms and processes.

As reflected in the subtitle, this lab manual reflects fundamental biological principles based on the common thread of evolution: form reflects function; unity despite diversity; and the adaptive processes of life. The manual was written for use in a two-semester introductory biology course serving life science majors. I have emphasized investigatory, quantitative, and comparative approaches to studying the life sciences and have integrated physical sciences principles where appropriate. In choosing topics for inclusion, I sought to achieve a balance between experimental, observational, and comparative activities. The comments of several expert reviewers were incorporated into this revision, clarifying many points from previous editions. The activities included in each lab topic have been tested in multisection lab courses and are known to work well in the hands of students.

Throughout the manual, the concept of hypothesis testing as the basic method of inquiry has been emphasized. Starting with lab topic 1 on the scientific method, and reiterated in experimental topics throughout the manual, students are asked to form hypotheses to be tested during their lab work and then are asked to reach a conclusion to accept or reject their hypotheses. Hypothesis testing and a comparative trend analysis also have been added into the more traditional labs dealing with diversity so that students are guided to look across several labs in reaching conclusions. Labs investigating physiological systems and morphology emphasize the concept of form reflects function. Comparative activities are included to demonstrate the adaptations found in several organisms.

Nature of the Revisions

Several major changes were made in this edition. The plant section was thoroughly revised. The old plant phylogeny lab topic is now divided into two topics, the seedless and seed plants, to better reflect the time needed to study plant phylogeny, and alternation of generations is given greater emphasis. The section on the functional biology of angiosperms was also extensively revised. The

old transport lab topic was divided into two lab topics, one emphasizing plant tissue systems and primary root structure, and the other emphasizing primary and secondary growth in stems. In addition some experiments were changed in other labs. In Lab Topic 1 about the scientific method, the experiment was changed from one testing physical fitness to one that emphasizes reaction time so that less athletic students will feel included and the results are not as predictable before the experiment. A new fruit fly experiment has been added which has more of an investigative theme requiring students to determine the genotypes of unknowns they are given. It can be completed in two weeks rather than the four required for the old experiments. The microevolution lab topic was rewritten and now includes student activities and computer simulations to teach the Hardy-Weinberg Principle instead of drawing beads from a container to illustrate statistical sampling. The taxonomic classifications for bacteria and protists were updated to reflect current thinking and the information in textbooks. In several of the exercises, the student activities were streamlined deleting experiments that usually were not performed for lack of time. All exercises were edited to improve clarity based on experience with students at Iowa State University.

New teaching elements were added as well. Each lab topic now starts with a Pre-lab Preparation section. In this section key vocabulary terms are listed and key concepts are named. The expectation is that students will realize that they must study vocabulary and concepts before coming to lab. Lab instructors can reinforce this realization by giving short quizzes before starting lab work. At the end of each lab topic, there is a section entitled "Learning Biology by Writing." For those departments that have strong writing-across-the-curriculum emphases, the suggested assignments will complement their goals. Several new Critical Thinking and Lab Summary Questions have also been added at the end of each lab topic.

Organization of Lab Topics

The lab topics have a standard format. All start with the Pre-lab Preparation section. This is followed by a list of equipment, organisms, and solutions to be used during the lab, informing students about what they will encounter in the lab. A brief introduction explains the biological principles to be investigated. These introductions are not meant to replace a textbook. They are included to summarize ideas that students will have had in lecture and to discuss how they apply to the lab. The lab instructions are detailed and allow students to proceed at their own pace through either experimental or observational lab work. Dangers are noted and explained. Data tables help students organize their lab observations. Questions are interspersed to avoid a cookbook approach to science and spaces are provided for answers and sketches. New terms are in boldface the first time used and are followed by a definition. At the end of each lab topic, several alternative suggestions are given for summarizing the lab work. A Learning Biology by writing section usually describes a writing assignment or lab report. Critical thinking questions emphasize applications. A lab summary based on several questions organizes the reporting of lab activities in a more stepwise approach. An Internet sources section points the students toward information sources on the WWW. Appendices include discussions of the use of significant figures, directions on making graphs, a description of elementary statistics, and instructions of how to write a lab report.

WWW Site

Under the sponsorship of McGraw-Hill, a WWW site has been established for this manual at http//www.mhhe.com/dolphin/

There you will find a preparator's manual giving recipes of chemical solutions and sources of supplies for each of the exercises. Also included is a list of links to other WWW sites which have materials relevant to the topics that students are investigating in the labs. If you know of links that should be included, please send them to me by E-mail (wdolphin@iastate.edu).

Acknowledgments

I would especially like to thank James Colbert, Associate Professor of Botany at Iowa State University, for his helpful comments and his patience in explaining plant biology. I also wish to thank the critical reviewers who made constructive suggestions throughout the writing of this manual: William Barstow, University of Georgia; Daryl Sweeney, University of Illinois; Gerald Gates, University of Redlands; Marvin Druger, Syracuse University; Thomas Mertens, Ball State University; Cynthia M. Handler, University of Delaware; Stan Eisen, Christian Brothers College; Paul Biebel, Dickinson College; Stephen G. Saupe, St. Johns University (Minnesota); Sidney S. Herman, Lehigh University; Margaret Krawiec, Lehigh University; Charles Lycan, Tarrant County Junior College; Olukemi Adewusi, Ferris State University; Karel Rogers, Adams State College; Peter A. Lauzetta, Kingsborough Community College (CUNY); Maria Begonia, Jackson State University; Thomas Clark Bowman, Citadel Military College; Gary A. Smith, Tarrant County Junior College;
Timothy A. Stabler, Indiana University Northwest; William J. Zimmerman, University of Michigan-Dearborn; and Nancy Segsworth, Capilano College (British Columbia).

Reviewers

Naomi D'Alessio, *Nova Southeastern University*
Carolyn Alia, *Sarah Lawrence College*
Linda L. Allen, *Lon Morris College*
Gordon Atkins, *Andrews University*
E. Rena Bacon, *Ramapo College of New Jersey*
Nina Caris, *Texas A & M University*
James T. Colbert, *Iowa State University*
Angela Cunningham, *Baylor University*
Carolyn Dodson, *Chattanooga State Technical Community College*
Frank J. Dye, *Western Connecticut State University*
Phyllis C. Hirsch, *East Los Angeles College*
Cathleen M. Jenkins, *Cuyahoga Community College*
Shelley Jones, *Florida Community College at Jacksonville*
Elaine King, *Environmental Biologist, Consultant*
Sonya Michaud Lawrence, *Michigan State University*
Raymond Lewis, *Wheaton College*
Brian T. Livingston, *University of Missouri—Kansas City*
Charles Lycan, *Tarrant County Junior College Northwest Campus*
Jacqueline S. McLaughlin, *Penn State Berks-Lehigh Valley College*
Susan Petro, *Ramapo College of New Jersey*
Gary Shields, *Kirkwood Community College*
Gary A. Smith, *Tarrant County Junior College*
Joan F. Sozio, *Stonehill College*
David Steen, *Andrews University*
Geraldine W. Twitty, *Howard University*
Carl Vaughan, *University of New Hampshire*
Lise Wilson, *Siena College*
Ming Y. Zheng, *Houghton College*

Margaret Horn, editor at McGraw-Hill Publishers, was most helpful during the preparation of the revisions, and I thank her for her patience and support. Special thanks goes to my friend and illustrator Dean Biechler who operates Chichaqua Bend Studios and to students of the Biological/Pre-Medical Illustration Program at Iowa State University. They prepared the illustrations for this and several of the earlier editions of the lab manual. By working directly with them, I have clarified many of my understandings of biology and have truly developed an appreciation of how form reflects function in biological systems. Last, but certainly not least, I thank my family—Judy, Jenny, Garth, Shannon and Lara—for their support throughout the preparation of this and earlier editions.

If you have questions or comments, please contact me by E-mail (wdolphin@iastate.edu.).

How lab topics correlate with chapters in major textbooks

Lab Topic	Audesirk & Audesirk & Biology, 5th ed.	Campbell, Reece & Mitchell Biology, 5th ed.	Lewis et al. Life, 4th ed.	Mader Biology, 7th ed.	Purves, Sadava, Orianes & Heller Life, 6th ed.	Raven & Johnson Biology, 6th ed.	Solomon, Berg, & Martin Biology, 5th ed.
1. Science: A Way of Gathering Knowledge	1	1	1	1	1	1	1
2. Techniques in Microscopy	6	7	3	4	4	5	4
3. Cellular Structure Reflects Function	6	7	3	4	4	5	4
4. Determining How Materials Enter Cells	5	8	4	5	5	6	5
5. Quantitative NA Techniques and Statistics		NA	NA	NA	NA	NA	NA
6. Determining the Properties of an Enzyme	4	6	5	6	6	8	6
7. Measuring Cellular Respiration	8	9	6	6	7	9	7
8. Determining Chromosome Number in Mitotic Cells	11	12	8	9	9	11	9
9. Observing Meiosis and Determining Crossover Frequency	11	13	9	10	9	13	10
10. Using Mendelian Principles to Determine the Genotypes of Fruit Flies	12	14, 15	10, 11	11, 12	10	13	10
11. Isolating DNA and Working with Plasmids	9, 13	16, 20	12	14, 17	11, 17	14, 19	11, 14
12. Testing Assumptions in Microevolution and Inducing Mutations	15	23	13, 15	16, 19	21	20, 21	18
13. Using Bacteria as Experimental Organisms	19	27	20	29	26	34	23
14. Diversity Among Protists	19	28	21	30	27	35	24
15. Plant Phylogeny: Seedless Plants	21	29	22	32	28	37	26
16. Plant Phylogeny: Seed Plants	21	30	22	32	29	37	27
17. Fungal Diversity and Symbiotic Relationships	20	31	23	31	30	36	25
18. Early Events in Animal Development	36	32, 47		51	16, 43	60	49

Lab Topic							
19. Animal Phylogeny: Evolution of Body Plan	22	33	24	33	31	44	28
20. Protostomes I: Evolutionary Development of Complexity	22	33	24	34	31	45	29
21. Protostomes II: A Body Plan Allowing Great Diversity	22	33	24	34	32	46	29
22. Deuterstomes: Origins of the Vertebrates	22	37	25	35	33	47, 48	30
23. Investigating Plant Tissues and Root Structure	23	35	26, 27	36	34	38, 39	31
24. Investigating Stem Structure, Growth, and Function	23	36	27	36, 37	35	39	33
25. Investigating Leaf Structure and Photosynthesis	7	10	6	7	8	10	8, 32
26. Investigating Angiosperm Reproduction and Development	24	38, 39	28, 29	39	37, 38	40, 42, 43	35, 36
27. Investigating Digestive and Gas Exchange Systems	28, 29	41, 42	36, 37	43, 44	48, 50	51, 53	44, 45
28. Investigating Circulatory Systems	27	42	35	41	49	52	42
29. Investigating the Excretory and Reproductive Systems	30, 35	44, 46	38	50	40, 42, 51	58, 59	46, 48
30. Investigating Form and Function in Muscle and Skeletal Systems	34	49	34	48	47	50	38
31. Investigating the Nervous and Sensory Systems	33	48, 49	31	46, 47	44, 45, 46	54, 55	39, 40, 41
32. Statistically Analyzing Simple Behaviors	37	51	41	22	52	27	50
33. Estimating Population Size and Growth	38	52	43	23	54	24	51
34. Standard Assays of Water Quality	40	54	44	25	56	29, 30	54, 55

LAB TOPIC 1

Science: A Way of Gathering Knowledge

Supplies

Preparator's guide available at
  http://www.mhhe.com/dolphin

Materials

Meter sticks
Photo copies of newspaper, magazine, and journal
  articles about biology (AIDS, rainforests, or cloning
  would be good examples, especially if articles
  were coordinated so students see same material
  intended for different audiences.)

Prelab Preparation

Before doing this lab, you should read the introduction
and sections of the lab topic that have been scheduled
by the instructor.

You should use your textbook to review the
definitions of the following terms:

  Dependent variable
  Hypothesis
  Independent variable
  Scientific literature

You should be able to describe in your own words
the following concepts:

  Critical reading
  Experimental design
  Reaction time
  Scientific method

As a result of this review, you most likely have
questions about terms, concepts, or how you will do
the experiments included in this lab. Write these
questions in the space below or in the margins of the
pages of this lab topic. The lab experiments should
help you answer these questions, or you can ask your
instructor for help during the lab.

Objectives

1. To understand the central role of hypothesis testing
   in the modern scientific method
2. To design and conduct an experiment using the
   scientific method
3. To summarize sample data as charts and graphs;
   to learn to draw conclusions from data
4. To evaluate writing for its science content and style

Background

Many dictionaries define science as a body of knowledge
dealing with facts or truths concerning nature. The emphasis is on facts, and there is an implication that absolute truth is involved. Ask scientists whether this is a
reasonable definition and few will agree. To them, science is a process. It involves gathering information in a
certain way to increase humankind's understanding of
the facts, relationships, and laws of nature. At the same
time, they would add that this understanding is always
considered tentative and subject to revision in light of
new discoveries.

Science is based on three fundamental principles:

The *principle of unification* indicates that any explanation
  of complex observations should invoke a simplicity of
  causes such that the simplest explanation with the least
  modifying statements is considered the best; also known
  as the law of parsimony.

The second principle is that *causality is universal;* when
  experimental conditions are replicated, identical results
  will be obtained regardless of when or where the work is
  repeated. This principle allows science to be self-
  analytical and self-correcting, but it requires a standard
  of measurement and calibration to make results
  comparable.

The third principle is that of the *uniformity of nature;*
  it states that the future will resemble the past so that
  what we learned yesterday applies tomorrow.

For many, science is just a refined way of using common sense in finding answers to questions. During our
everyday lives, we try to determine cause and effect relationships and presume that what happened in the past has a
high probability of happening in the future. We look for relationships in the activities that we engage in, and in the
phenomena that we observe. We ask ourselves questions
about these daily experiences and often propose tentative
explanations that we seek to confirm through additional
observations. We interpret new information in light of

previous proposals and are always making decisions about whether our hunches are right or wrong. In this way, we build experience from the past and apply it to the future. The process of science is similar.

The origin of today's scientific method can be found in the logical methods of Aristotle. He advocated that three principles should be applied to any study of nature:

1. One should carefully collect observations about the natural phenomenon.

2. These observations should be studied to determine the similarities and differences; i.e., a compare and contrast approach should be used to summarize the observations.

3. A summarizing principle should be developed.

While scientists do not always follow the strict order of steps to be outlined, the modern scientific method starts, as did Aristotle, with careful observations of nature or with a reading of the works of others who have reported their observations of nature. A scientist then asks questions based on this preliminary information-gathering phase. The questions may deal with how something is similar to or different from something else or how two or more observations relate to each other. The quality of the questions relates to the quality of the preliminary observations because it is difficult to ask good questions without first having an understanding of the subject.

After spending some time in considering the questions, a scientist will state a research **hypothesis,** a general answer to a key question. This process consists of studying events until one feels safe in deciding that future events will follow a certain pattern so that a prediction can be made. In forming a hypothesis, the assumptions are stated and a tentative explanation proposed that links possible cause and effect. A key aspect of a hypothesis, and indeed of the modern scientific method, is that the hypothesis must be falsifiable; i.e., if a critical experiment were performed and yielded certain information, the hypothesis would be declared false and would be discarded, because it was not useful in predicting any natural phenomenon. If a hypothesis cannot be proven false by additional experiments, it is considered to be tentatively true and useful, but it is not considered absolute truth. Possibly another experiment could prove it false, even though scientists cannot think of one at the moment. Thus, recognize that science does not deal with absolute truths but with a sequence of probabilistic explanations that when added together give a tentative understanding of nature. Science advances as a result of the rejection of false ideas expressed as hypotheses and tested through experiments. Hypotheses that over the years are not falsified and which are useful in predicting natural phenomena are called theories or principles—for example, the principles of Mendelian genetics.

Hypotheses are made in mutually exclusive couplets called the **null hypothesis** (H_o) and the **alternative hypothesis** (H_a). The null hypothesis is stated as a negative and the alternative as a positive. For example, when crossing fruit flies a null hypothesis might be that the principles of Mendelian genetics do not predict the outcomes of the experiment. The alternative hypothesis would be that Mendelian principles do predict the outcome of the experiment. As you can see, rival hypotheses constitute alternative, mutually exclusive statements: both cannot be true.

The purpose in proposing a null hypothesis is to make a statement that could be proven false if data were available. Experiments or reviews of previously conducted experiments provide the data and are therefore the means for testing hypotheses. In designing experiments to test a hypothesis, predictions are made. If the hypothesis is accurate, predictions based on it should be true. In converting a research hypothesis into a prediction, a deductive reasoning approach is employed using if-then statements: if the hypothesis is true, then this will happen when an experimental variable is changed. The experiment is then conducted and as certain variables are changed, the response is observed. If the response corresponds to the prediction, the hypothesis is supported and accepted; if not, the hypothesis is falsified and rejected.

The design of experiments to test hypotheses requires considerable thought! The variables must be identified, appropriate measures developed, and extraneous influences must be controlled. The **independent variable** is that which will be varied during the experiment; it is the cause. The **dependent variable** is the effect; it should change as a result of varying the independent variable. **Control variables** are also identified and are kept constant throughout the experiment. Their influence on the dependent variable is not known, but it is reasoned that if kept constant they cannot cause changes in the dependent variable and confuse the interpretation of the experiment.

Once the variables are defined, decisions must be made regarding how to measure the effect of the variables. Measures may be quantitative (numerical) or qualitative (categorical) and imply the use of a standard. The metric system has been adopted as the international standard for science. If the independent variables are to be varied, a decision must be made concerning the scale or level of the treatments. For example, if something is to be warmed, what will be the range of temperatures used? Most biological material stops functioning (dies) at temperatures above 40°C and it would not be productive to test at temperatures every 10°C throughout the range 0° to 100°C. Another aspect of experimental design is the idea of replication: how many times should the experiment be repeated in order to have confidence in the results and to develop an appreciation in the variability of the response.

Once collected, experimental data are reviewed and summarized to answer the question: does the data falsify or support the null hypothesis? The research conclusions then state the decision regarding the acceptability of the null hypothesis and discuss the implications of the decision.

If the experimental data are consistent with the predictions from the null hypothesis, the hypothesis is supported, but not proven absolutely true. It is considered true only on a trial basis. If the hypothesis is in a popular area of research, others may independently devise experiments to test the same hypothesis. A hypothesis that cannot be falsified, despite repeated attempts, will gradually be accepted by others as a description that is probably true and worthy of being considered as suitable background material when making new hypotheses. If, on the other hand, the data do not conform to the prediction based on the null hypothesis, the hypothesis is rejected and the alternative hypothesis is supported.

Modern science is a collaborative activity with people working together in a number of ways. When a scientist reviews the work of others in journals or when scientists work in lab teams, they help one another with interpretation of data and in the design of experiments. When a hypothesis has been tested in a lab and the results are judged to be significant, she or he then prepares to share this information with others. This is done by preparing a presentation for a scientific meeting or a written article for a journal. In both forms of communication, the author shares the preliminary observations that led to the forming of the hypothesis, the data from the experiments that tested the hypothesis, and the conclusions based on the data. Thus, the information becomes public and is carefully scrutinized by peers who may find a flaw in the logic or who may accept it as a valuable contribution to the field. Thus, the scientific discussion fostered by presentation and publication creates an evaluation function that makes science self-correcting. Only robust hypotheses survive this careful scrutiny and become the common knowledge of science.

LAB INSTRUCTIONS

You will create a research hypothesis, design an experiment to test it, conduct the experiment, summarize the data, and come to a conclusion about the acceptability of the hypothesis. You will also practice evaluating scientific information from various published sources.

Using the Scientific Method

Description of the Problem

Working in groups of four, you are to develop a scientific hypothesis and test it. The topic will be neuromuscular reaction time. This can be easily measured in the lab by measuring how quickly a person can grasp a falling meter stick. The person whose reaction time is being measured sits at a table with her or his forearm on the top and the hand extended over the edge, palm to the side and the thumb and forefinger partially extended. A second person holds a meter stick just above the extended fingers and drops it. The subject tries to catch it. The distance the meter stick drops is a measure of reaction time.

Your assignment is to create a scientifically answerable question regarding reaction time in individuals with different characteristics and to express this as testable hypotheses. You will then design an experiment to test the hypotheses, collect the data, analyze, and come to a decision to reject or accept your hypothesis. For example, you might investigate the differences between those who play musical instruments and those who do not or try a more complex design that investigates gender differences in reaction time for students who are in some type of athletic training versus those who are not. The design will depend on the hypotheses that you decide to test as a group in your lab section. Continuing the example, you might propose a null hypothesis that there will be no significant differences in reaction time between musicians and nonmusicians. An alternative hypothesis would be that there is a significant difference in the reaction times between the two types.

Summarizing Observations

Start your discussion of this assignment by summarizing the collective knowledge of your group about neuromuscular response time. Are these responses the same for all people or might they vary by athletic history, gender, body size, age, hobbies requiring manual dexterity, left versus right hand, or other factors? Be sure to consider these factors in both a qualitative and quantitative light. You might expect differences in the physiological responses of those who exercise. What other factors might influence the response time? As your group discussion proceeds, make notes below that summarize the group's knowledge and observations about what characteristics influence reaction time.

Asking Questions

Research starts by asking questions which are then refined into hypotheses. Review the group observations that you listed and write down scientifically answerable questions that your group has about reaction time in people with different characteristics. Be prepared to present your group's best questions to the class and to record the best questions from the class on a piece of paper.

Forming Hypotheses

With your group, review the questions posed in the class discussion. Examine the questions for their answerability. Do some lack focus? Are they too broad? Are others too simple, with obvious answers? By what criteria would you judge a good question?

As a group take what you think is the best question and state it as a prediction. For example, because piano players constantly train their neuromuscular units you might expect that they would have short reaction times. Use this prediction as a basis for forming a testable couplet of hypotheses. Continuing with the example, you might propose for a null hypothesis that there would be no significant difference between piano players and nonmusicians in reaction time. The alternate hypothesis would be that there is a significant difference. Remember that hypotheses are proposed in mutually exclusive couplets and they must be testable through experimentation or further data gathering such that one will be proven false. State your null hypothesis and an alternative hypothesis.

H_o

H_a

Be prepared to describe your group's couplet of hypotheses to the class and to indicate why you think they are significant and will add to the body of knowledge that the class has expressed through its earlier observations. Describe how your hypotheses are testable.

Designing an Experiment

To test the null hypothesis, a controlled experiment must be devised. It should be designed to collect evidence that would prove the null hypothesis false. Within your group, discuss what the experiment should be. Your discussion should address the variables in the experiment.

Which of the variables is (are) the independent variable(s), the one(s) that will be varied to invoke a response?

Which of the variables is (are) the dependent variable(s), the one(s) that are the effects? How will the measurements be made and over what time?

What variables will be controlled and how will they be controlled?

Having decided which variables fit into these categories, you must now decide on a level of treatment and how it will be administered. How will you standardize measurements across groups?

Recognizing that the subject may anticipate the dropping of the meter stick or be momentarily distracted when it is dropped, how many observations should be made and over what time period? How many times will you repeat the experiment to have confidence in your results?

TABLE 1.1 Reaction times measured as millimeters free fall.

Description of Subject (age, gender, musician?, athlete?, other?)	Replication 1	Replication 2	Replication 3	Average
Subject 1				
Subject 2				
Subject 3				

Note that in your design, all groups in the lab section do not have to conduct exactly the same experiment. Continuing an earlier example, half of the groups could work with males, half with females. These could be further subdivided into musicians and nonmusicians with the gender categories. The results could be pooled at the end to determine if there were any differences.

Procedure

After answering these questions as a group, write a set of instructions on how the experiment should be performed. Your group should then perform the experiment. One person should be the subject, chosen according to the procedures. The others should each take different jobs. One can be the director of the experiment. Another can be the person who drops the stick, and another can record the data after each try.

Data Recording

Look over table 1.1 and fill in the information required in the title. Begin your experiment and record the data in the table. If you are doing more than three replications, you can write additional numbers in the extra space.

Data Summarization

Different groups should now record the average reaction time and subject descriptions on the blackboard. This data can be analyzed in a number of different ways. What is the average reaction in the lab section? _____ What is the average reaction time for females? _____ Males? _____ For right hand? _____ Left hand? _____ Musicians? _____ Nonmusicians? _____ Other factors investigated?

Data Interpretation

Write a few sentences that summarize the trends that you see in the data and the differences between groups.

Conclusion

Return to the hypotheses that you made at the beginning of the experiment. Compare them to the experimental results. Must you accept or reject the null hypothesis? Why? Cite the data used in making the decision. If you determine that there is a difference in reaction time between categories of people, how can you decide if it is a significant difference?

Discussion

Discuss with your partners how the experiments added to the knowledge base of the class which was outlined before the experiment began. Do you see any significance to the knowledge gained? Explain.

As you conducted this experiment and analyzed the results, additional questions probably came into your mind. As a result of this thinking and the results of this experiment, what do you think would be a significant hypothesis to test if another experiment were to be done?

Evaluate the design of your experiment. Be as critical as you can. Were any variables not controlled that should have been? Is there any source of error that you now see but did not before?

Students might find it interesting to check their reaction time at an interactive WWW site: http://netra. exploratorium.edu/baseball/reactiontime.html. Can you correlate this independent measurement with your experimental results? How?

Scientific Method Assignment

Your instructor may ask you to write up this experiment as a scientific report and to hand it in at the next lab meeting. Refer to appendix D for instructions on how to write such a report.

Evaluating Published Information

(adapted from notes prepared by Chuck Kugler at Radford University and Chris Minor at Iowa State University)

We daily are exposed to scientific information in newspapers, magazines, over the World Wide Web, and through scholarly reports in journals and books. How do you evaluate such information? Is a newspaper best because it is available daily or is the WWW better because no editors have changed words to fit a story in the column space? In classes throughout your undergraduate years, in your future jobs, and in everyday life, you will be asked to evaluate what you read and make decisions about the quality of information.

In this section you will learn how to evaluate a written report. Your instructor will pass out photocopies of a newspaper, magazine, and journal article reporting on the same scientific discovery. Read the articles quickly so that you have a rough idea of what is in them. When you are finished with the articles, read the following material in the lab manual. Refer back to the photocopies as you read and try to find examples of the writing styles mentioned in the lab manual.

Evaluate Format

First, be suspicious of any scientific report that is not written in a style that parallels the scientific method where the hypotheses are clearly identified, data are presented, and the reasoning leading to the conclusions is explained. The formal elements of a scientific paper are discussed in appendix D. If a report lacks these elements, it is not a scientific report and should not be used as a source of observations upon which to create a hypothesis or to test one. On the other hand, reading about a discovery in the newspaper can alert you to locate the actual report in a journal that the reporter read before writing the story.

Evaluate the Source

Several thousand journals publish information of interest to biologists. The journals range from magazines such as *National Geographic* and *Scientific American* to scholarly journals, published by professional associations, such as the *American Journal of Botany, Journal of Cell Biology, Genetics, Ecology, Science,* etc. Magazine articles are usually written by science journalists and not by scientists who did the research. They can be quite helpful in developing a general appreciation for a topic, but they are not ultimate sources of scientific information. Scholarly journals are considered the most reliable sources of information and even these will vary in the quality of the work that is published. What makes these journals so reliable is the use of a peer review system. Articles are written by scientists and sent to the journal editor, who is usually a scientist. When she receives the article, she sends it to three other scientists who are working on similar problems and asks them to make comments about the work. Often these reviews can be harsh and may criticize writing style and content. The reviewers' comments are returned to the author who then revises the paper before it is published. It is this peer review system that maintains the quality of the information appearing in journals. Popular magazines such as *Time,* or television shows (even those on the Discovery channel), or movies have been created for entertainment purposes and are not sources of evidence.

Evaluate Writing Style

Good scientific writing is factual and concise. It is not overly argumentative, nor should it be an appeal to the emotions. As you read any scientific report, watch for the following:

1. Forceful statements: made to build the reader's confidence;
2. Repetition: some authors believe that the more they say something, the more likely you are to believe it;
3. Dichotomous simplification: expressing a complex situation as if there were only two alternatives;
4. Exaggeration: often identifiable by the use of the words "all" or "never";
5. Emotionally charged words: the author is attempting to get you to agree based on "feelings," not reason.

Evaluate the Arguments

Examine how the author seeks to convince you that what is reported is true, significant, and applicable to science. Be on guard for the following types of rhetorical arguments:

1. Appeals to authority: citing a well-known person or organization to make a point, e.g., "the American Dental Association recommends. . . ." You should ask what is the basis of their recommendation and are they experts in the field under discussion. Authorities can be biased, be experts in fields other than the one under consideration, and be wrong.
2. Appeals to the democratic process: using the phrase "most people" believe, use, or do. Remember they could be wrong. Only 200 years ago, most people believed in the spontaneous generation of life.
3. Use of personal incredulity: implying that you could not possibly believe something, e.g., "how could something as complex as the human just evolve, didn't it need a designer?"
4. Use of irrelevant arguments: statements that might be true but which are not relevant, e.g., "suggesting that complex animals could have resulted by chance is like saying that a clock could result from putting gears in a box and shaking it."
5. Using straw arguments: presenting information incorrectly and then criticizing the information because it is wrong, e.g., "the evolution of a wing requires 20 simultaneous mutations—an impossibility." There is no basis for saying that the evolution of a wing requires 20 mutations; it could be fewer but most likely many more.
6. Arguing by analogy: using an analogy to suggest that an idea is correct or incorrect., e.g., "intricate watches are made by careful designers, so complex organisms must have had a designer."

Science: A Way of Gathering Knowledge **7**

A 60-pound bushel of soybeans contains
about 48 pounds of meal and 11 pounds of oil.

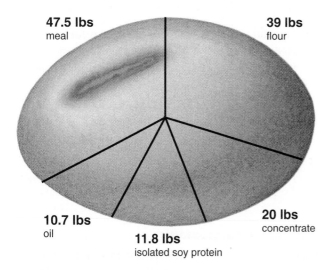

47.5 lbs
meal

39 lbs
flour

10.7 lbs
oil

11.8 lbs
isolated soy protein

20 lbs
concentrate

What does this chart tell you?

Do the numbers add up?

Would you use the information in this chart to make a decision?

Do you trust this data?

SOURCE: Iowa Soybean Review, 1995/1996 Soya Bluebook

Evaluate the Evidence

Before getting too involved in interpreting trends in the data, spend a few moments thinking about the type of evidence that is presented. Was the evidence collected using the scientific method and is there a hypothesis that is being tested? Be especially skeptical of reports that have the following flaws in their evidence:

1. Distinguish between evidence and speculation: evidence includes data, whereas speculation is simply a statement based on an educated guess.

2. Use of anecdotal evidence: anecdotes are stories usually involving single events and are not the results of carefully designed experiments, e.g., "bee stings are lethal; my uncle died when he was stung."

3. Correlation used to imply cause and effect: correlation is a probability of two events occurring together. While it is interesting to speculate that one might cause the other, this is not necessarily so; e.g., at the instant that a major earthquake has struck a major city, there is a high probability that someone was slamming a car door. Did the slam cause the earthquake?

4. Sample size and selection: in statistical studies, a large number of situations should be examined and the procedures used to select the situations should be free of bias. You do not choose to report only the experiments that support your beliefs.

5. Misrepresentation of source: source material can be quoted out of context or badly paraphrased; e.g., an actual statement "Moderate drinking of alcohol may benefit the consumer" could be misrepresented as "Drinking is good for you."

Check the Data

When data are presented, get in the habit of doing routine checks. If percentages are involved, do you know the sample size? It is an impressive statement to say that 75% of the people surveyed preferred brand X, but it is less impressive to find out that this calculation is based on a sample size of 4 rather than 400 or more. When percentages are reported, be sure to check that they add up to 100. If on the eve of the election 42% of the voters are for Gore and 41% are for Bush, it would seem that Gore has won, except that 17% of the voters are unaccounted for and could swing the election one way or the other.

Continue the habit of doing simple arithmetic checks when examining data in tables. If totals are given for columns of numbers, do some quick math to see if things check out. If they do not, you might not want to base major decisions on the report. Besides you do not know what other kinds of errors went undetected!

With the advent of computer graphics, it is now rather easy to use computer programs to produce interesting looking and appealing graphics. However, one should not accept data based on its beauty of presentation. To illustrate this point, see figures 1.1 and 1.2 for some interesting graphics that appeared in newspapers or trade publications.

Evaluate the Conclusions

In a scientific paper, the conclusions should come near the end of the article. Conclusions are not a summary of the data. Conclusions deal with the decision that is to be made about the hypothesis that was being tested. You should ask, "Are the data thoroughly reviewed to test the hypothesis?" Ask yourself whether there is another explanation to what

Vital statistics

These numbers may help you decide where you stand on the issues.

Married mothers in the workplace

Percentage of married working women with children under the age of six.

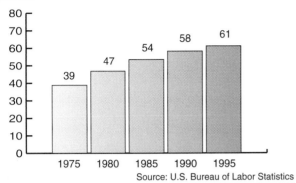

Source: U.S. Bureau of Labor Statistics

Child poverty

Percentage of children living in poverty.

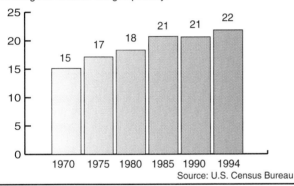

Source: U.S. Census Bureau

Juvenile violent crime

Violent crime arrests per 100,000 juveniles.

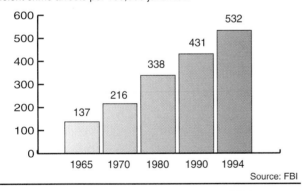

Source: FBI

Most non-custodial fathers don't pay child support

Percentage of non-custodial fathers who paid child support 1989

63% - Paid nothing
26% - Paid full amount
12% - Paid partial amount

Source: U.S. Census Bureau

Figure 1.2 This graphic appeared in a major newspaper. Focus on the trends in the data. Has the number of working mothers increased significantly in the past five years? Do significantly more children live in poverty in 1994 compared to 1985? What is the sample size? Were the same populations of people compared? What is the message of this collection of graphics when considered as a whole? Why is this not an acceptable scientific report? What is the difference between correlation and cause and effect when considering two or more trends?

the author is telling you. Once a decision is made to accept or reject the hypothesis, the implications of the decision are discussed. In some cases, the implications are then extrapolated to new situations, but overextrapolation can result in problems. For example, raising a frog's body temperature from 10°to 20°C may increase the frog's metabolic rate twofold, but this does not mean by extrapolation that raising it to 100°C will increase metabolic rate tenfold. In fact, the frog will die when its body temperature approaches 40°C.

As you look back through the newspaper, magazine, and journal articles, which one of these forms of publication used more of the nonscientific forms of writing and arguing?

Evaluating Scientific Literature Assignment

Go to the library and choose a science-related article from a periodical of your choice, such as a newspaper, popular magazine, or a science journal. Your instructor or a librarian can suggest some journals to skim through to locate an article that is of interest to you. Photocopy the article because you will be writing on it. Once photocopied, write at the top of the first page, the name of the journal from which it was copied, the volume, and the number (or month) of the issue. Your assignment is to analyze the article using the information given on the next page. You will be marking all over this article as you analyze it. You should turn in the marked article along with the following form next week.

Journal Analysis Form

Evaluate the Source

Who is (are) the author(s)?
Where does (do) the author(s) work?
Consider situations where the author(s) could have vested interests?
What type of source is this?
Are articles peer reviewed in this source?

What Is the Hypothesis?

State the hypothesis tested in the work reported. If none, so indicate.

Examine the Writing Style

Use a light-colored marker to highlight on the photocopy any passages that seem to deviate from a factual and concise style. Write a number next to the highlighted area indicating the type of writing style used according to the following key:

1. Forceful statement
2. Use of repetition
3. Dichotomous simplification
4. Exaggeration
5. Use of emotionally charged words

Examine the Arguments

Use a light-colored marker to highlight on the photocopy any arguments used in the article. Write a number next to the highlighted area indicating the type of argument that is used according to the following key:

1. Appeals to authority
2. Appeals to the democratic process
3. Uses personal incredulity
4. Uses irrelevant arguments
5. Uses straw arguments
6. Argues by analogy

Analyze the Evidence

Underline the sections of the photocopied article that present the arguments of the author. Write a letter next to the arguments according to the following key:

A. Speculation
B. Evidence collected using the scientific method
C. Anecdotal evidence
D. Correlation, not cause and effect
E. Description of sample size and selection method
F. Possible place to check for misrepresentation of source

Check the Data

Do all percentages given add up to 100%? If not, circle where the omission is located in the text.
Do all numbers in columns or charts add up to the indicated totals or are there math mistakes?
Circle the mistakes.
Are flashy graphics used to catch your attention?
Do they add to your understanding or simply emotionally excite you? Write comments next to the questionable graphics.

Examine the Conclusions

Are the conclusions easy to find and clearly stated?

Are the conclusions based on a review of the data and a test of the hypotheses presented in the introduction?

Are the conclusions supported by evidence collected using the scientific method?

Has the author extrapolated beyond the range of data collected?

Attach this evaluation form to your photocopied article and turn it in for grading.

LAB TOPIC 2

Techniques in Microscopy

Supplies

Preparator's guide available at
http://www.mhhe.com/dolphin

Equipment

Compound microscope
Dissecting microscope

[handwritten notes:]
$1m = 10^9 nm$
$= 10^3 mm$
$= 10^6 Mm$ micrometers
$1mm = 10^3 Mm$
$= 10^2 nm$

Materials

Ocular micrometer
Stage micrometer
Small colored letters from printed page
Slides and coverslips
Dropper bottles with water
Dissecting needles and scissors

Prelab Preparation

Before doing this lab, you should read the introduction and sections of the lab topic that have been scheduled by the instructor.

You should use your textbook to review the definitions of the following terms:

Brightness
Calibration
Contrast
Magnification
Resolution

You should be able to describe in your own words the following concepts:

Light path through parts of a microscope
How to make wet-mount slide
How to calculate an ocular micrometer

As a result of this review, you most likely have questions about terms, concepts, or how you will do the experiments included in this lab. Write these questions in the space below or in the margins of the pages of this lab topic. The lab experiments should help you answer these questions, or you can ask your instructor for help during the lab.

Objectives

1. To learn the parts of a microscope and their functions
2. To investigate the optical properties of a microscope, including image orientation, plane of focus, and measuring objects
3. To understand the importance of magnification, resolution, and contrast in microscopy

Background

Since an unaided eye cannot detect anything smaller than 0.1mm (10^{-4} meters) in diameter, cells, tissues, and many small organisms are beyond our visual capability. A light microscope extends our vision a thousand times, so that objects as small as 0.2 micrometers (2×10^{-7} meters) in diameter can be seen. The electron microscope further extends our viewing capability down to 1 nanometer (10^{-9} meters). At this level, it is possible to see the outlines of individual protein or nucleic acid molecules. Needless to say, microscopy has greatly improved our understanding of the normal and pathological functions of organisms.

Although 300 years have passed since its invention, the standard light microscope of today is based on the same principles of optics as microscopes of the past. However, manufacturing technology has developed to a point that quality instruments for classroom use are now mass produced. Your microscope is as good as those used by Schleiden, Schwann, and Virchow, the biologists who founded cell theory in the mid-nineteenth century, and is far superior to the one used by Robert Hooke, the first person to use the word "cell" in describing biological materials.

Microscope quality depends upon the capacity to **resolve,** not magnify, objects. The distinction between microscopic resolution and magnification can best be illustrated by an analogy. If a photograph of a newspaper is taken from across a room, the photograph would be small, and it would be impossible to read the words. If the photograph were enlarged, or magnified, the image would be larger, but the print would still be unreadable. Regardless of the magnification used, the photograph would never make a fine enough distinction between the points on the printed page. Therefore, without **resolving power,** or the ability to distinguish detail, magnification is worthless.

Modern microscopes increase both magnification and resolution by matching the properties of the light source and precision lens components. Today's light microscopes are limited to practical magnifications in the range of 1000 to

Figure 2.1   Correct (a) and incorrect (b) ways to carry a microscope.

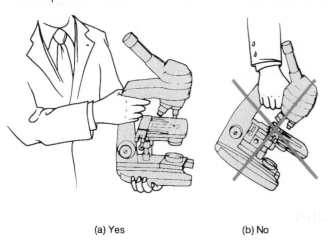

(a) Yes   (b) No

2000× and to resolving powers of 0.2 micrometers. Most student microscopes have magnification powers to 450×, or possibly to 980×, and resolving properties of about 0.5 micrometers. These limits are imposed by the expense of higher power objectives and the accurate alignment of the lens elements and light sources.

The theoretical limit for the resolving power of a microscope depends on the **wavelength** of light (the color) and a value called the **numerical aperture** of the lens system, times a constant (0.61). The numerical aperture is derived from a mathematical expression that relates the light delivered to the specimen by the condenser to the light gathered by the objective lens. If all other factors are equal, resolving power is increased by reducing the wavelength of light used. Microscopes are often equipped with blue filters because blue light has the shortest wavelength in the visible spectrum. Therefore,

minimum distance that can be resolved

$$= \frac{\text{wavelength}}{\text{numeric aperature}} \times 0.61$$

For example, if green light with a wavelength of 500 nanometers is used and the numerical aperture is 2, the theoretical resolving power is 153 nanometers, or 0.153 micrometers.

Even with sufficient magnification and resolution, a specimen can be seen on a microscope slide only if there is sufficient **contrast** between parts of the specimen. Contrast is based on the differential absorption of light by parts of the specimen. Often a specimen will consist of opaque parts or will contain natural pigments, such as chlorophyll, but how is it possible to view the majority of biological materials that consist of highly translucent structures?

Microscopists improve contrast by using stains that bind to cellular structures and absorb light to provide contrast. Some stains are specific for certain chemicals and bind only to structures composed of those chemicals. Others are nonspecific and stain all structures.

To summarize, good microscopy involves three factors: resolution, magnification, and contrast. A beginning biologist must learn to manipulate a microscope with these factors in mind to gain access to the world that exists beyond the unaided eye.

LAB INSTRUCTIONS

AVOIDING HAZARDS IN MICROSCOPY

Use care in handling your microscope. The following list contains common problems, their causes, and how they can be avoided.

1. Microscope dropped or ocular falls out
   a. Carry microscope in upright position using both hands, as shown in figure 2.1.
   b. When placing the microscope on a table or in a cabinet, hold it close to the body; do not swing it at arm's length or set it down roughly.
   c. Position electric cords so that the microscope cannot be pulled off the table.

2. Objective lens smashes coverslip and slide
   a. Always examine a slide first with the low- or medium-power objective.
   b. Never use the high-power objective to view thick specimens.
   c. Never focus downward with the coarse adjustment when using high-power objective.

3. Image blurred
   a. High-power objective was pushed through the coverslip (see number 2) and lens is scratched.
   b. Slide was removed when high-powered objective was in place, scratching lens. Remove slide only when low-power objective is in place.
   c. Use of paper towels, facial tissue, or handkerchiefs to clean objectives or oculars scratched the glass and ruined the lens. Use only *lens tissue* folded over at least twice to prevent skin oils from getting on the lens. Use distilled water to remove stubborn dirt.
   d. Clean microscope lenses before and after use. Oils from eyelashes adhere to oculars, and wet-mount slides often encrust the objectives or substage condenser lens with salts.

4. Mechanical failure of focus mechanism
   a. Never force an adjustment knob; this may strip gears.
   b. Never try to take a microscope apart; you need a repair manual and proper tools.

Figure 2.2 A binocular compound microscope.

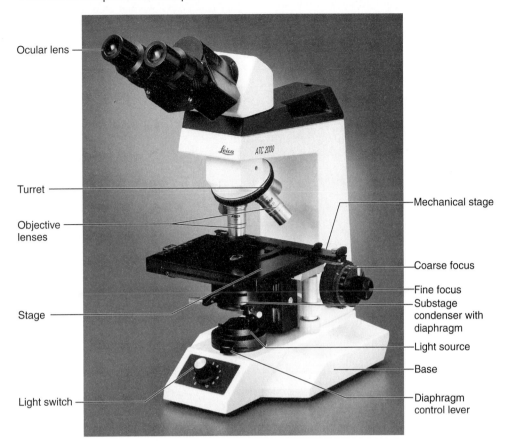

Labels: Ocular lens, Turret, Objective lenses, Stage, Light switch, Mechanical stage, Coarse focus, Fine focus, Substage condenser with diaphragm, Light source, Base, Diaphragm control lever

The Compound Microscope

▶ Get your microscope from its storage place, using the precautions just mentioned. Depending on its age, manufacturer, and cost, your compound microscope may have only some of the features discussed in this section. Look over your microscope and find the parts described, referring to figure 2.2.

Parts of a Microscope

Ocular Lens
The **ocular lens** is the lens you look through. If your microscope has one ocular, it is a **monocular** microscope. If it has two, it is **binocular.** In binocular microscopes, one ocular is adjustable to compensate for the differences between your eyes. Ocular lenses can be made with different magnifications. What magnification is stamped on your ocular lens housing?

The ocular lens is actually a system of several lenses that may include a pointer and a measuring scale called an **ocular micrometer.** Never attempt to take an ocular lens apart.

Body Tube
The body tube is the hollow housing through which light travels to the ocular. If the microscope has inclined oculars, as in figure 2.2, the body tube contains a prism to bend the light rays around the corner.

Objective Lenses
The **objective lenses** are a set of three to four lenses mounted on a rotating **turret** at the bottom of the body tube. Rotate the turret and note the click as each objective comes into position. The objective gathers light from the specimen and projects it into the body tube. Magnification ability is stamped on each lens. What are the magnification abilities of your objectives?

Scanning (small) Lens	_____
Low-power (medium) Lens	_____
High-power (large) Lens	_____
Oil Immersion (largest) Lens	_____ (optional)

Stage
The horizontal surface on which the slide is placed is called the **stage.** It may be equipped with simple clips for holding the slide in place or with a **mechanical stage,** a geared device for precisely moving the slide. Two knobs, either on top of or under the stage, move the mechanical stage.

Substage Condenser Lens
The substage **condenser** lens system, located immediately under the stage, focuses light on the specimen. An older microscope may have a mirror instead.

Diaphragm Control

The **diaphragm** is an adjustable light barrier built into the condenser. It may be either an **annular** or an **iris** type. With an annular control, a plate under the stage is rotated, placing open circles of different diameters in the light path to regulate the amount of light that passes to the specimen. With the iris control, a lever projecting from one side of the condenser opens and closes the diaphragm. Which type of diaphragm does your microscope have?

Use the smallest opening that does not interfere with the field of view. The condenser and diaphragm assembly may be adjusted vertically with a knob projecting to one side. Proper adjustment often yields a greatly improved view of the specimen.

Light Source

The light source has an off/on switch and may have adjustable lamp intensities and color filters. To prolong lamp life, use medium to low voltages whenever possible. A second diaphragm may be found in the light source. If present, experiment with it to get the best image.

Base and Body Arm

The base and body arm are the heavy cast metal parts.

Coarse Focus Adjustment

Depending on the type of microscope, the **coarse adjustment** device either raises and lowers the body tube or the stage to focus the optics on the specimen. Use the coarse adjustment only with the scanning (4×) and low-power (10×) objectives. Never use it with the high-power (40×) objective. (The reasons for this will be explained later.)

The Focus Adjustment

The **fine adjustment** changes the specimen-to-objective distance very slightly with each turn of the knob and is used for all focusing of the 40× objective. It has no noticeable effect on the focus of the scanning objective (4×), and little effect when using the 10× objective.

The Microscope and Your Eyes

Students often wonder if they should remove their glasses when using a microscope. If you are nearsighted or farsighted, there is no need to wear your glasses. The focus adjustments will compensate. If you have an astigmatism, however, you should wear your glasses because microscope lenses do not correct for this problem.

If your microscope is monocular, you will probably tend to use it with one eye closed. Eyestrain will develop if this is continued for long. Learn to keep both eyes open as you look through the microscope and ignore what you see with the other eye. This will be hard at first. Remove all light-colored papers from your field of view or try covering your eye with your hand.

Figure 2.3 Procedure for making a wet-mount slide. (*a*) Place a drop of water on a clean slide. (*b*) Place specimen in water. (*c*) Place edge of coverslip against the water drop and lower coverslip onto slide.

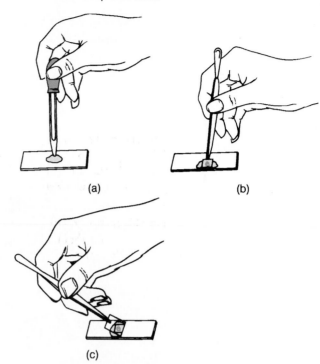

(a)　(b)

(c)

Making Slides and Using a Microscope

Figure 2.3 shows how to prepare a specimen as a wet mount on a microscope slide. Take a magazine or an old printed page and cut out a colored lowercase letter *e* or *a* or the number *3, 4,* or *5.* Clean a microscope slide with a tissue, add a drop of water to the center, and place the letter in the drop. Add a coverslip and place the slide in its normal orientation on the microscope stage with the scanning objective in place.

Now you are ready to view the slide. Follow the steps listed in the box on the next page.

The seven steps listed are the usual procedures for using the microscope. Always start with a clean scanning objective and proceed in sequence to high power, making minor adjustments to the focus and light source. Using a microscope is similar to changing the channels on a television set and adjusting the picture at each new setting. Your skill in using and tuning your microscope will determine what you will see on microscope slides throughout this course.

The following activities are designed to familiarize you with your microscope. Use the wet-mount slide you just made to carry out these activities.

The Compound Microscope Image

A compound microscope image has several properties, including image orientation, magnification, field of view, brightness, focal plane, and contrast.

STEPS USED IN VIEWING A SLIDE

1. Check that the ocular and all objective lenses as well as the slide are clean.

2. Turn the illuminator on and open the diaphragm. Center the specimen over the stage opening.

3. Look through the ocular. Starting with the scanning objective as close to the slide as possible while looking through the oculars, back off with the coarse adjustment knob until the specimen is in sharp focus.

4. Readjust the light intensity and center the specimen in the field of view by moving the slide. Close down the iris diaphragm and, if possible, adjust the substage condenser until the edges of the diaphragm are in focus.

5. Switch from the scanning objective to the low-power (10×) objective. The lens stop should click when the objective is in place. Sharpen the focus, adjust the centering of the specimen, and readjust the condenser height and diaphragm opening.

6. Switch to the high-power (40×) objective. Adjust the focus with the *fine focus adjustment only*. If you use the coarse adjustment, you may hit the slide and damage the high-power objective.

7. If you have a binocular microscope, adjust the ocular lenses for the differences between your eyes. Determine which ocular is adjustable. Close the eye over that lens and bring the specimen into sharp focus for the open eye. Open the other eye, and close the first. If the specimen still is not in sharp focus, turn the adjustable ocular until the specimen is in focus. You need not repeat this procedure when you look at other specimens, but should do it each time you get the microscope from the cabinet because other students may also be using your microscope and adjusting it for their eyes.

[handwritten: Center Specimen]

Image Orientation

With the scanning objective in place, observe the letter on the slide through the microscope and then with the naked eye. Is there a difference in the orientation of the images? While looking through the microscope, try to move the slide so that the image moves to the left. Which way did you have to move the slide? Try to move the image down. Which way did you have to move the slide?

When showing someone an interesting specimen, you can describe the location of the specimen by referring to the field of view as a clock face. (Thus, the point of interest might be described as being at one o'clock or seven o'clock.)

Some microscopes have pointers built into the ocular. In such cases, the structure of interest can simply be moved to the end of the pointer.

Magnification

Compound microscopes consist of two lens systems: the objective lens, which magnifies and projects a "virtual image" into the body tube, and the ocular lens, which magnifies that image further and projects the enlarged image into the eye.

The ocular lens only increases the magnification of the image and does not enhance the resolution. The objective lens magnifies and resolves. The total magnification of a microscope is the product of the magnification of the objective and the ocular. If the objective lens has a magnification of 5× and the ocular 12×, then the image produced by these two lenses is 60 times larger than the specimen.

What magnifications are possible with your microscope?

Scanning power =
Low power   =
High power   =
Oil immersion =

[handwritten: working distance becomes less with higher magnification]

Field of View and Brightness

Observe your microscope slide with the scanning, low-, and high-power objectives. Note that as magnification increases, the diameter of the field of view decreases and the brightness of the field is reduced. Note also that the **working distance,** the distance between the slide and the objective, decreases as the magnification is increased. (This is the reason you never focus on thick specimens with a high-power objective.) These relationships are summarized in figure 2.4.

Focal Plane and Optical Sectioning

The concept of the **focal plane** is important in microscopy. Like the eye, a microscope lens has a limited depth of focus; therefore, only part of a thick specimen is in focus at any one setting. The higher the magnification, the thinner the focal plane. In practical terms, this means that you should make constant use of the fine adjustment knob when viewing a slide with the high-power objective. If you turn the knob a quarter turn back and forth as you view a specimen, you will get an idea of the specimen's three-dimensional form. It would be possible, for example, to reconstruct the three-dimensional structure in figure 2.5 from sections (1), (2), and (3).

Image Contrast

The contrast of the image can be changed by closing the diaphragm, although this usually results in poorer resolution. Light rays are deflected from the edges of the diaphragm and enter the slide at oblique angles. Scattered light makes materials appear darker because some rays of light take longer to reach the eye than others. This can be an advantage when looking at unstained specimens. Thus, the benefits of greater contrast sometimes outweigh the loss of resolving power. Contrast is also improved by reducing light intensity or brightness.

Techniques in Microscopy

Figure 2.4

Comparison of the relative diameters of fields of view, light intensities, and working distances at three different objective magnifications.

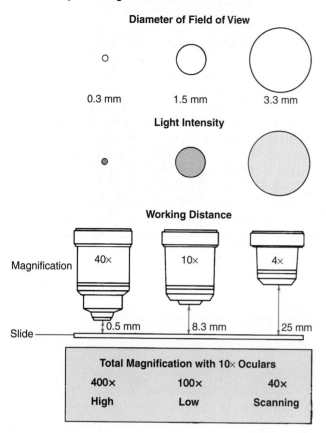

Diameter of Field of View

0.3 mm   1.5 mm   3.3 mm

Light Intensity

Working Distance

Magnification   40×   10×   4×

Slide   0.5 mm   8.3 mm   25 mm

Total Magnification with 10× Oculars		
400×	100×	40×
High	Low	Scanning

Figure 2.5

(a) Sequentially focusing at depths (1), (2), and (3) yields (b) three different images that can be used to reconstruct the original three-dimensional structure.

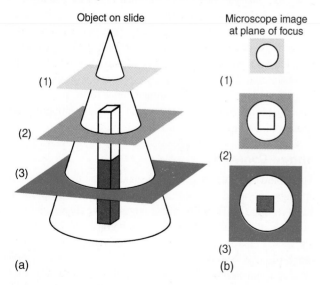

Object on slide   Microscope image at plane of focus

(1)

(2)

(3)

(a)   (b)

Measurement of Microscopic Structures

Measuring microscopic structures requires a standardized **ocular micrometer.** It is a small glass disc on which are etched uniformly spaced lines in arbitrary units. The disc is inserted into an ocular of the microscope, and the etched scale is superimposed on the image of the specimen when you look through a microscope. Does your microscope have an ocular micrometer? _____ The spacing between the lines on the disc must be calibrated with a very accurate standard ruler called a **stage micrometer.**

The ocular micrometer must be calibrated for each objective. Any object can then be measured by superimposing the ocular scale on it and measuring its size in ocular units. These units can then be multiplied by the calibration factor to obtain the actual size of the object.

To calibrate an ocular micrometer, obtain a stage micrometer from the supply area. Look at it with the scanning objective. What are the units? What is the smallest space equal to in these units? Follow the steps given in figure 2.6.

Determine how many spaces on the stage micrometer are equal to 100 spaces on the ocular micrometer at the following powers and record in the table below. Divide the number of stage units in millimeters by 100 to determine the calibration for one ocular unit when using the scanning objective. Record below. Repeat for each objective. *Be careful not to push the high-power objective through the stage micrometer. (They are expensive!)*

	Stage Units (mm)	Ocular Units	mm per Ocular Unit	Converted to μm
Scanning	____	____	____	____
Low	____	____	____	____
High	____	____	____	____

Stereoscopic Dissecting Microscopes

The stereoscopic microscope (fig. 2.7), usually called a **dissecting microscope,** differs from the compound microscope in that it has two (rather than one) objective lenses for each magnification. This type of microscope always has two oculars. The stereoscopic microscope is essentially two microscopes in one. The great advantage of this instrument is that

Figure 2.6 A stage micrometer is used to calibrate an ocular micrometer.

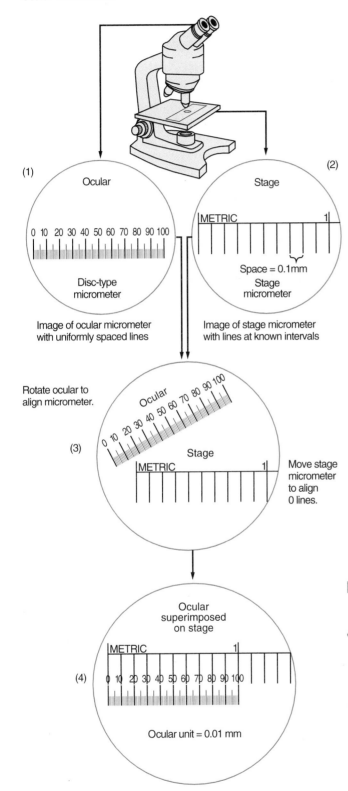

(1) Ocular

0 10 20 30 40 50 60 70 80 90 100

Disc-type micrometer

Image of ocular micrometer with uniformly spaced lines

(2) Stage

METRIC   1

Space = 0.1mm
Stage micrometer

Image of stage micrometer with lines at known intervals

Rotate ocular to align micrometer.

(3) Ocular

0 10 20 30 40 50 60 70 80 90 100

Stage

METRIC   1

Move stage micrometer to align 0 lines.

(4) Ocular superimposed on stage

METRIC   1

0 10 20 30 40 50 60 70 80 90 100

Ocular unit = 0.01 mm

Figure 2.7 Stereoscopic microscope.

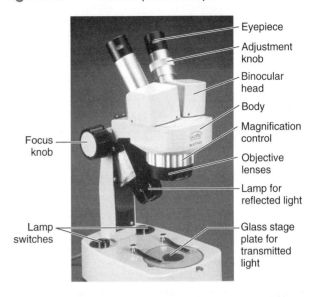

Eyepiece

Adjustment knob

Binocular head

Body

Magnification control

Objective lenses

Lamp for reflected light

Glass stage plate for transmitted light

Focus knob

Lamp switches

objects can be observed in three dimensions. Because the alignment of the two microscopes is critical, the resolution and magnification capabilities of a stereoscopic microscope are less than in a compound microscope. Magnifications on this type of microscope usually range from 4× to 50×. The oculars can be adjusted for individual eye spacing and for focus, as in the compound binocular microscope.

Stereoscopic microscopes are often used for the microscopic dissection of specimens. The light source may come from above the specimen and be reflected back into the microscope, or it may come from underneath and be transmitted through the specimen into the objectives. The stage may be clear glass or an opaque plate, white on one side and black on the other. The choice of illumination source depends on the task to be performed and on whether the specimen is opaque or translucent.

Set up your dissecting microscope with reflected light. Place your hand on the stage and observe the nail on your index finger. Move your hand so the image travels to the right and down. How does image movement correspond to actual movement?

Change the illumination to transmitted light. Place the previously prepared slide of a printed letter on the stage and focus on it using the highest magnification. Determine which ocular is adjustable. Close the eye over the adjustable ocular and focus the microscope sharply on the edge of the letter. Now close the other eye and open the first. Is the edge still in sharp focus? If not, turn the adjustable ocular until it is. This procedure should be followed whenever a stereoscopic microscope is used for long periods to avoid eyestrain.

Your instructor may have a supply of flowers, seeds, or dead insects to examine with the stereoscopic microscope.

Learning Biology by Writing

Write a short essay (about 200 words) describing how magnification, resolution, and contrast are important considerations in microscopy. Indicate how microscopists can increase contrast in viewing specimens.

As an alternative assignment, your instructor may ask you to complete some or all of the lab summary and critical thinking questions.

Lab Summary Questions

1. Define magnification and resolution. How do these properties of a microscope differ?
2. In the table below, enter one of the words "increase," "decrease," or "no change" to describe how the properties of the image change as you use different objectives on your microscope.
3. Describe how you should calibrate an ocular micrometer in a microscope.
4. When you calibrated your microscope, what were the sizes of one ocular unit at: 40×_____; 100×_____; and 400×_____?

Critical Thinking Questions

1. When looking through the oculars of a binocular compound light microscope, you see two circles of light instead of one. How would you correct this problem? If you saw no light at all, just a dark field, what correction would you make? Now, you finally have an interesting structure in view using your 10× objective lens, but, when you switch to the 40× objective lens, the structure is not in the field of view. What happened? How would you correct this?
2. What type of microscope would you use to observe the tube feet of a sea star? What type of microscope would you use to determine the sex of a live fruit fly? If you wanted to look at the chromosomes of the fruit fly, what type of microscope would you use?

Image Properties	Objectives		
	Scanning	Low	High
Magnification			
Field of view			
Brightness of field			
Resolving power			

LAB TOPIC 3

Cellular Structure Reflects Function

Supplies

Preparator's guide available at
http://www.mhhe.com/dolphin

Equipment

Compound microscopes with ocular micrometers and
oil immersion objectives, if available
Optional: microtone and wax-embedded
specimens for sectioning demonstration

Materials

Living organisms
Blue-green algae cultures of *Anabaena* and
Microcystis
Mixed culture of algae amoebas, flagellates, and
ciliates
Onion
Elodea
Yogurt
Prepared slides of
Gram-stained slide containing mixed cocci, bacilli,
and spirilla
Human skin
Frog skin
Areolar connective tissue
Neurons from cow's spinal-cord smear
Pine stem, tangential section or macerate
Coverslips and slides
Razor blades and forceps

Solutions

Methyl cellulose or *Protoslo*
Neutral red stain
India ink

Prelab Preparation

Before doing this lab, you should read the introduction
and sections of the lab topic that have been scheduled
by the instructor.
You should use your textbook to review the
definitions of the following terms:

Bacteria
Cyanobacteria
Epidermis
Epithelium
Eukaryotic

Plant vascular tissues
Prokaryotic
Protists

You should be able to describe in your own words
the following concepts:

Cell theory
Structure reflects function
Cell compared to tissue

As a result of this review, you most likely have
questions about terms, concepts, or how you will do
the experiments included in this lab. Write these
questions in the space below or in the margins of the
pages of this lab topic. The lab experiments should
help you answer these questions, or you can ask your
instructor for help during the lab.

Objectives

1. To learn the differences between prokaryotic and
eukaryotic cell types
2. To observe living cells
3. To introduce students to staining methods
4. To observe representative tissue types in plants and
animals
5. To identify an unknown tissue
6. To collect evidence that cellular structure reflects
function

Background

In 1665, Robert Hooke first used the word **cell** to refer to the
basic units of life. One hundred and seventy-three years later,
after other scientists had observed cells and the many variations that occur in cell structure, two German biologists,
M. Schleiden and T. Schwann, published what is called the
cell theory. This theory states that the cell is the basic unit of
life and that all living organisms are composed of one or more
cells or the products of cells. The cell theory is not the result of
one person's work but is based on the observations of many

microscopists. Today the cell theory is accepted as fact. All living organisms are constructed of cells and the products of cells. Only viruses defy inclusion in this generalization.

If cells are the basic units of life, then the study of basic life processes is the study of cells. Today cell biologists strive to understand how cells function by using tools such as microscopes, centrifuges, and biochemical analyses. This quest for knowledge is driven by a logical relationship; if normal organismal function is dependent on cell function, then disease and abnormal functioning can also be understood at the cellular level.

Biologists recognize two organizational plans for cells. **Prokaryotic cells** lack a nuclear envelope, chromosomal proteins, and membranous cytoplasmic organelles. Bacteria and blue-green algae are prokaryotic cells. **Eukaryotic cells** have the structural features that prokaryotes lack. Protozoan, algal fungal, plant, and animal cells are eukaryotic.

Although these two types of cells are distinctly different, they share many characteristics. A plasma membrane always surrounds a cell and regulates the passage of materials into and out of the cell. Both types of cells have similar types of enzymes, depend on DNA as the hereditary material, and have ribosomes that function in protein synthesis. The eukaryotic types evolved after the prokaryotic cells and are more complex.

LAB INSTRUCTIONS

You will observe the differences between prokaryotic and eukaryotic cells, as well as variations within these groups. This will introduce you to the paradox that biologists constantly face: the unity and the diversity of living forms. Moreover, you should come to appreciate a maxim in biology: form reflects function.

Prokaryotic Cells

Prokaryotic cells are found in all members of the Kingdoms Archaebacteria and Eubacteria.

Bacteria

In 1884, the Danish bacteriologist Christian Gram developed a diagnostic staining technique, which is used to separate bacteria into two groups: Gram positive and Gram negative. Dead Gram-positive bacteria retain crystal violet dye while being washed in alcohol, but Gram-negative bacteria do not.

Modern microscopists know this is due to chemical differences in the composition of the bacterial cell walls. Gram-negative bacteria have more lipid material in the cell wall. When washed with alcohol, the lipids are extracted and the crystal violet stain no longer binds to the cell. Gram-negative cells are colorless after the Gram staining, but then a second stain (safranin) makes them visible as pink cells. The identification of thousands of different types of bacteria is based on this diagnostic test in combination with other traits.

Figure 3.1   Scanning electron micrographs of the three types of bacterial cells: (*a*) cocci, (*b*) bacilli, (*c*) spirilla.

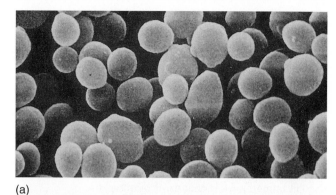

(a)

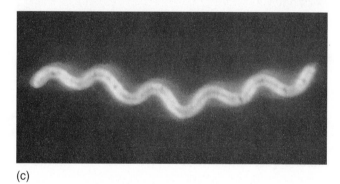

(b)

(c)

Obtain a prepared slide of mixed types of bacteria. Observe with the 40× objective or with an oil immersion objective, if available. (Your instructor will explain how to use an oil immersion objective.) The slide should contain both Gram-positive and Gram-negative bacteria and three shapes of bacterial cells **Cocci** are sphere-shaped bacteria, **bacilli** are rod-shaped bacteria, and **spirilla** are comma- or corkscrew-shaped bacteria (fig. 3.1).

Indicate whether both Gram-positive (violet) and Gram-negative (pink) forms are found for each shape. If your microscope has a calibrated ocular micrometer, measure the bacterial cell sizes and record below:

Sizes of Bacteria

Cocci _____
Bacilli _____
Spirilla _____

Figure 3.2   Transmission electron micrograph of bacterium *Pseudomonas aeroginosa*. Magnification, ×67,200.

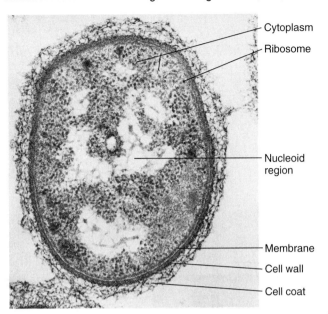

Figure 3.3   Transmission electron micrograph of the cyanobacterium *Oscillatoria* sp. Magnification, ×80,000.

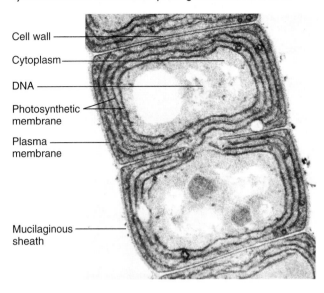

Because of their small size, it is impossible to see detail inside bacterial cells with the light microscope. Figure 3.2 is a transmission electron micrograph of a section of a bacterial cell. Note the **cell wall, cell membrane, cytoplasm, ribosomes,** and **nucleoid region** containing DNA. Notably absent is any evidence of a nuclear envelope, chromosomes, or any internal organelles.

Yogurt is made by adding the bacteria *Lactobacillus* sp. and *Streptococcus thermophilus* to milk and allowing the bacteria to anaerobically metabolize milk sugar. Lactic acid is produced and excreted by the bacteria. It curdles the milk, producing the semisolid yogurt.

Take a very small amount of yogurt and mix it on a slide with a drop of water. Add a coverslip and observe the slide through your microscope. What are the shapes of the bacteria in yogurt? What are their approximate sizes? Record your observations below.

Cyanobacteria (Blue-Green Algae)

The common name "blue-green algae" characterizes the predominant feature of about half the organisms found in this group: they are blue because of the presence of a pigment called phycocyanin and green because they contain chlorophyll. However, some species may be brown or olive because of other pigments. All cyanobacteria are prokaryotes and most are surrounded by a gelatinous matrix. They live in soils, on moist surfaces, and in water.

Two cyanobacteria are available for study in this lab— *Anabaena* (see fig. 14.3) and *Microcystis*. Make a wetmount slide by placing small drops of each culture on a slide. Take a dissecting needle and dip it in India ink and touch the wet needle to the drop of blue-green algae culture. Some of the the ink will transfer and improve the viewing. Press a coverslip down and blot away excess liquid.

Which species is surrounded by an extensive gelatinous matrix? Can you see structures inside the individual cells? Below make a sketch of each organism.

Anabaena cells form filaments composed of three cell types: small spherical **vegetative cells;** elongate spores called **akinetes;** and large spherical **heterocyst** cells, which function in nitrogen fixation. Label these cells in your drawing above.

Observe the internal structure of the cyanobacteria cell in figure 3.3. Note the characteristic **mucilaginous sheath** outside the cell wall. Note the **cell wall, plasma membrane,**

Figure 3.4 Three types of protozoa.

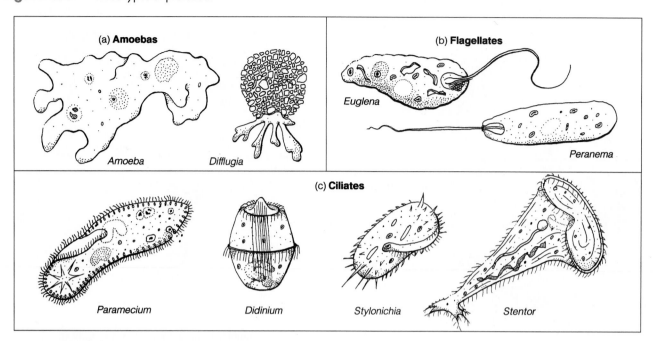

(a) **Amoebas**
Amoeba   Difflugia

(b) **Flagellates**
Euglena   Peranema

(c) **Ciliates**
Paramecium   Didinium   Stylonichia   Stentor

photosynthetic membrane, and **clear DNA regions.** The staining procedure used in preparing this specimen does not allow you to see the ribosomes in this picture, but all prokaryotic cells contain thousands of these important cell organelles.

Eukaryotic Cells

Eukaryotic cells include protist, fungal, plant, and animal cells.

Protists

Protozoa and algae are single-celled and colonial eukaryotic organisms that some scientists include in a single kingdom, Protista, while others separate into several kingdoms too numerous to discuss here. The term "protozoa" means first animals and at one time was used as a phylum name by zoologists. Now it is a term of convenience including organisms from several kingdoms as does the term "algae." What structure not found in bacteria should you be able to see in protozoa and algae?

The answer you just gave to this question is a hypothesis that you can test by observation. A culture of mixed protozoa and algae is available in the lab. Make a wet-mount slide with some of the culture debris from the bottom of the container. Look at it first with the scanning objective and then with high power to identify the three protozoan types (fig. 3.4) and algae. To observe internal cellular structure, you may have to add methyl cellulose, a thickening agent, or neutral red stain.

Sketch examples of three types of protozoa below and describe the distinguishing characteristics.

Do your observations support or contradict your hypothesis? What evidence do you have that protists are eukaryotes?

Plant Cells

The cells of plants differ from those of animals in several characteristics. Plant cells are always surrounded by a **cell wall** composed of cellulose and resinous materials. The living part of a cell surrounded by the cell wall is called the

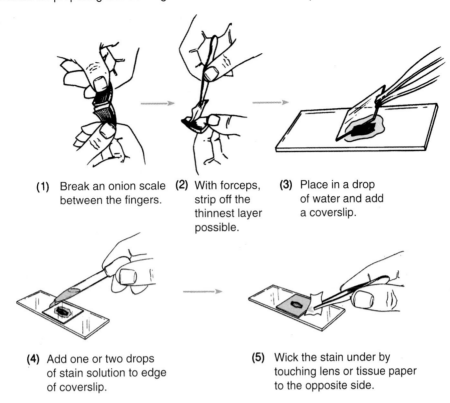

(1) Break an onion scale between the fingers.

(2) With forceps, strip off the thinnest layer possible.

(3) Place in a drop of water and add a coverslip.

(4) Add one or two drops of stain solution to edge of coverslip.

(5) Wick the stain under by touching lens or tissue paper to the opposite side.

protoplast. In the cytoplasm of some protoplasts, unique organelles called **chloroplasts** are found. They carry out the complex chemical reactions of photosynthesis. Protoplasts also usually have a large central **vacuole** filled with water and dissolved materials.

As in animals, the cells of plants are organized into **tissues,** cells that are similar in structure and function. In this part of the lab exercise, you will look at three types of plant cells.

Epidermal Cells

Epidermal cells are found on the surfaces of plants and function as a protective barrier. What shape would you hypothesize best suits the function of epidermal cells? Do you think they would be plate-like, cuboidal, or tubular? Why?

Figure 3.5 demonstrates how to prepare a wet-mount slide of onion epidermis. After making your slide, observe it under a 10× objective and test the hypothesis that you made concerning the shape of epidermal cells. Once the specimen is in focus, adjust the light intensity and condenser so that the cells are clearly visible.

Note the individual cells outlined by thin cell walls composed of cellulose. The **plasma membrane** lies just inside the cell wall but cannot be seen because its thickness is less than the resolution of the light microscope. A large, fluid-filled vacuole is in the center of some cells. Another membrane, the **tonoplast,** surrounds the vacuole and regulates what passes in and out.

Switch to the high-power objective and locate the **nucleus** in the periphery of the cell. Sketch a typical cell and label it, including dimensions, in the space below. Measure the length and width of one cell. Focus up and down through a single cell. Do your observations support or disprove your hypothesis about cell shape?

Cellular Structure Reflects Function   **23**

How does the three-dimensional shape of an epidermal cell relate to its function?

Elodea sp. is an aquatic plant whose leaves are only two cells thick. Break off a leaf and mount it in a drop of water on a slide. Add a coverslip and observe.

 Besides the nucleus, what two other structures characteristic of plant cells can you see?

 What evidence have you observed that plants have eukaryotic cells?

Plant Vascular Tissues

Plants have vascular tissues that transport water, minerals, and other materials from the roots to the leaves, and also transport the products of photosynthesis from the leaves to other regions of the plant. Two basic tissues make up the plant vascular system: **xylem** transports water and minerals and **phloem** transports photosynthetic products (fig. 3.6). These cells also provide structural support for a plant. Their elongate, tubular form with thick cell walls reflects their function. You will look at xylem.

Xylem: Obtain a slide of macerated pine wood from the supply area. The trunks of pine trees, except for a narrow ring just beneath the bark, are composed primarily of **tracheids.**

Examine your slide first under scanning and then in detail with the high-power objective. The tracheids are elongated cells with long, tapering end walls. The side walls of the cell are perforated by pits. Water passes to adjacent tracheids through the pits so that water moving from the roots to needles follows a zigzag pathway.

Figure 3.6 Two types of cells found in xylem.

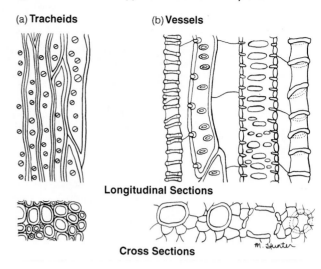

(a) **Tracheids** (b) **Vessels**

Longitudinal Sections

Cross Sections

Note two important aspects of the tracheids: (1) the cell walls are thickened so that the tracheids not only transport but also structurally support the plant, and (2) no **protoplast** (living cell) is visible in the tracheid. The protoplast of a tracheid functions only to make the cell wall and then dies, leaving the tubelike wall to function in water transport and support.

Sketch a few tracheids below. Be sure to show the pit structure in the side walls. Use your ocular micrometer to measure a tracheid and add dimensions to your illustration.

In some plants, a second type of xylem cell is found. Called **vessels,** these cylindrical cells are not tapered at the ends and have cell walls that are reinforced by ringed and spiral thickenings.

? If form reflects function, do your observations of xylem cells support the idea that they conduct fluids and support the plant? Explain.

Figure 3.7 How plane of sectioning affects shape seen on slide. (*a*) Sections through a bent tube. (*b*) The effects of slicing through an egg in different directions. Clearly, preparation affects what is seen.

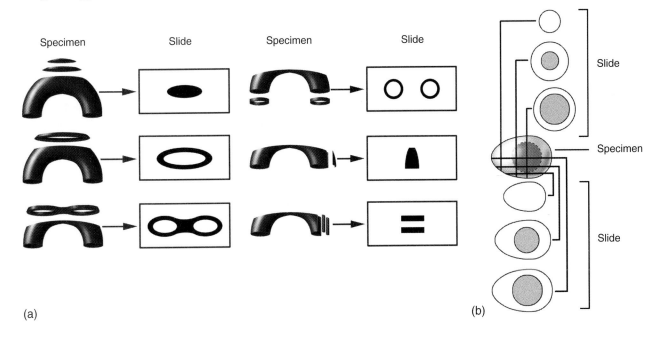

(a)   (b)

A redwood tree is 100 m tall. Use your measurements of tracheid cell length to calculate how many tracheid cells must be stacked on end to form a conduit from the ground to the top of the tree.

Animal Cells

Animal tissues are difficult, if not impossible, to hand section because the absence of supporting cell walls allows the cells to be crushed during sectioning. To prepare animal tissues (and many plant tissues) for microscopic observation, it is necessary to instantly kill and fix the cells with a chemical, then either freeze them or infiltrate them with plastic or wax to make the tissue rigid. Thin sections then can be cut from this rigid block using a special cutting machine called a **microtome.** If a microtome is available in the lab, your instructor will demonstrate its use.

Tangential, longitudinal, or cross-sectional cuts may be made on embedded tissue. As figure 3.7 indicates, the same basic structure may look different, depending on the plane of the section. After they are cut, the sections are attached to slides and stained to illustrate various structures. (Staining reactions depend on the chemical composition of the structure to be illustrated.) Because the same tissue can be stained by different dyes, it is a good practice not to "memorize" tissues by color. For example, skin could be stained blue on one slide and red on another.

Permanent slides are made by mounting coverslips over the tissue sections using a resin, such as balsam, instead of water. This procedure takes several hours and is beyond the scope of a simple laboratory. Throughout the rest of the course, you will use slides that have been commercially prepared in this manner.

In most animals, cells are specialized to carry out particular functions. Four basic tissue types are found: epithelial, connective, nerve and muscle. In this section of the exercise, you will look at examples of the first three on stained, prepared slides. At the end of this portion of the exercise, you should be able to look at an unknown tissue and identify it as one of these basic tissue types.

Epithelial Tissue

Tissues found on the body surfaces, lining cavities, or forming glands are **epithelial** tissues (fig. 3.8). They are characterized by (1) having one surface not in contact with other cells, (2) having another surface in contact with a basement membrane, and (3) having no materials between adjacent cells.

Obtain two slides: one a cross section of a frog's skin and the other a cross section of human skin. Look at each under the low power of your compound microscope and determine which surface was outermost in the animal. As you study both, consider how the skin of both animals acts as an

Figure 3.8   Epithelial tissue: (*a*) simple epithelium; (*b*) artist's interpretation.

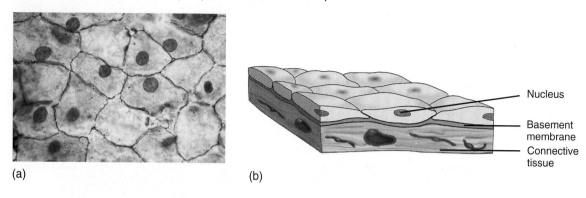

(a)

(b)

- Nucleus
- Basement membrane
- Connective tissue

Figure 3.9   Cross section of human skin showing dried, compressed cells of outer layers (stratum corneum), the epidermal layer that divides to produce outer cells, melanized cells that give color to skin, and connective tissue of layer beneath the skin (dermis).

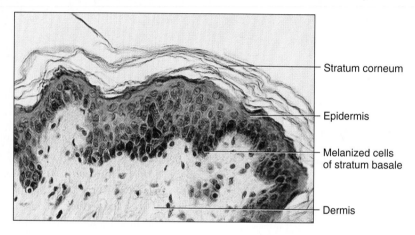

- Stratum corneum
- Epidermis
- Melanized cells of stratum basale
- Dermis

outer barrier that prevents water loss and invasion by microorganisms. (See figure 3.9.)

Human skin is a more effective barrier because it is **stratified,** squamous epithelium, meaning that it consists of many layers of flattened cells, while the frog's squamous epithelium is simply a single cell layer. Which organism, frog or human, is more fully adapted to a terrestrial environment? Why?

Below, make two comparison sketches of the outer layers of frog and human skin.

Epithelial cells lining the digestive system have a columnar shape rather than being flattened, and those lining the trachea and bronchi of the respiratory system are ciliated. Demonstration slides of these tissues are available in the lab and should be studied. Does each of these tissues demonstrate the three characteristics of epithelial tissues listed earlier? What are they?

Connective Tissue

This tissue is characterized by a nonliving, extracellular **matrix,** which is secreted by the basic connective tissue cell type, the **fibroblast** (fig. 3.10). The matrix contains mucopolysaccharide gel material and many protein fibers. A common fiber in connective tissue is made of the protein collagen. Cartilage and bone are examples of **supportive** connective tissues. In bone, the matrix contains calcium salts as well as collagen fibers and the cells are confined to small chambers. (See lab topic 28.)

Obtain a slide of areolar connective tissue (fig. 3.10) and observe it under low power with your compound microscope. Note the many fibers and scattered fibroblasts. This tissue attaches the skin to the body and strengthens the walls of blood vessels and organs in the body. Below, sketch a small section of the slide and label the diagnostic features.

Figure 3.10   Connective tissue: (*a*) loose, fibrous, connective tissue (areolar); (*b*) artist's interpretation.

(a)

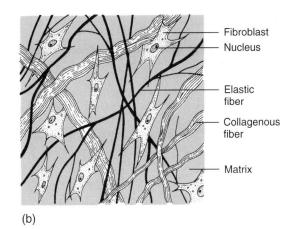

Fibroblast
Nucleus

Elastic fiber

Collagenous fiber

Matrix

(b)

Use your ocular micrometer to measure the size of ten fibroblasts. Calculate the average size from your measurements.

Cell #	Longest Dimension in μm
1	————
2	————
3	————
4	————
5	————
6	————
7	————
8	————
9	————
10	————
Average length =	————

How is the extracellular matrix related to the function of connective tissues?

Cellular Structure Reflects Function

Figure 3.11   Nerve cell.

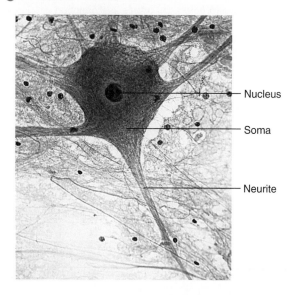

— Nucleus

— Soma

— Neurite

Use your ocular micrometer to measure the size of the soma of a neuron. Add the dimensions to your drawing.

 Is a neuron larger, smaller, or the same size as fibroblast that you measured when you looked at connective tissue?

 How does the shape of a neuron differ from that of other cells you have studied?

Nerve Tissue

Nerve cells are specialized to transmit messages from one part of the body to another as nerve impulses. In mammals, most nerve cell bodies reside in the spinal cord or brain and cytoplasmic extensions pass out to muscles or to sensory receptors (fig. 3.11).

▶ Obtain a slide of a neuron prepared by smearing a section of a cow's spinal cord on the slide. Observe it under low power with your compound microscope. Note the cell body (**soma**) containing the nucleus. Extending from the soma are cytoplasmic extensions called **neurites.** Those that conduct impulses away from the cell body are called **axons** and those that conduct impulses toward the cell body are called **dendrites.** Sketch a neuron below.

 How does the form of a neuron reflect its function?

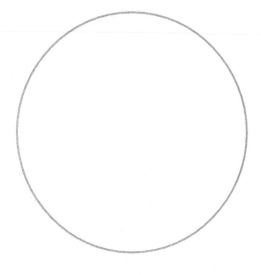

Unknowns

▶ On a table in the laboratory, your instructor may have set up five microscopes. Each has a slide of an unknown plant or animal tissue in focus at the end of the pointer. Identify the basic tissue types and list your reasons for naming each type in the following table.

Tissue	**Reason**
1.	
2.	
3.	
4.	
5.	

Learning Biology by Writing

Based on your observations during this exercise, write a short essay (about 200 words) on the theme "form reflects function" at the cellular level in plants and animals. Cite four examples that you looked at during this lab session.

As an alternative assignment, your instructor may ask you to complete the Lab 3 summary and critical thinking questions.

Lab Summary Questions

1. List the structural differences found between prokaryotic and eukaryotic cells.
2. How do plant cells differ from animal cells?
3. Epidermal cells in both plants and animals are often flattened, that is, they are thin and wide. How is this example of form reflecting function?
4. Using your observations of connective and nerve tissues, explain how form reflects function at the cellular level.
5. Explain how the shapes of xylem tracheids are related to their functions in plants.
6. You made several measurements in this lab. Indicate the size of the following:

Bacillus _____
Onion epidermal cell _____
Xylem tracheid _____
Fibroblast _____
Soma of neuron _____

Cellular Structure Reflects Function

Critical Thinking Questions

1. During your observation of cells, what similarities did you notice between prokaryotic cells and organelles of eukaryotic cells?

2. Describe the extracellular matrix of the following tissues: bone, cartilage, and loose connective tissue. Why is blood considered a connective tissue? Why is muscle not considered a connective tissue?

3. Give two examples of both plant and animal cells where form reflects function. Explain how this principle applies to your examples.

4. Describe what a chair would look like when viewed at three different depths of field. Describe what it would look like from three different sides. Could you describe the whole structure if you saw only one cross section and one longitudinal section?

LAB TOPIC 4

Determining How Materials Enter Cells

Supplies

Preparator's guide available at
http://www.mhhe.com/dolphin

Equipment

Compound microscopes
Balance (0.1 g sensitivity)

Materials

Slides and coverslips
Small glass rods
Dropper bottles
Markers
Thistle tube with dialysis membrane over end
Dialysis tubing precut to 15 cm
1 ml pipettes
Test tubes and rack
Albustix (drugstore)
Millimeter rulers, 5 mm
Alcohol lamp
Cork borer
250 ml beaker
Paper towels
Petri plates containing 5 mm of 2% plain agar
Live organisms
    Elodea
    Paramecium caudatum

Solutions

India ink
3% NaCl
2×10^{-3}M sodium azide
1% albumin solution in 3% NaCl
0.25% soluble starch in 1% Na_2SO_4
I_2KI solution (5 g I_2: 10 g KI: 100 ml H_2O)
1 M $AgNO_3$
2% $BaCl_2$
0.1 M Na_2SO_4
Concentrated HCl
Concentrated NH_4OH
30% sucrose with Congo Red

Prelab Preparation

Before doing this lab, you should read the introduction and sections of the lab topic that have been scheduled by the instructor.

You should use your textbook to review the definitions of the following terms:

Brownian movement
Diffusion
Hypertonic
Hypotonic
Isotonic
Kinetic energy
Osmoregulation
Osmosis
Plasmolysis
Turgor
Water expulsion vesicle

You should be able to describe in your own words the following concepts:

Facilitated diffusion
Osmotic pressure
Differentially permeable membrane

As a result of this review, you most likely have questions about terms, concepts, or how you will do the experiments included in this lab. Write these questions in the space below or in the margins of the pages of this lab topic. The lab experiments should help you answer these questions, or you can ask your instructor for help during the lab.

Objectives

1. To observe Brownian movement
2. To determine if diffusion and osmosis both occur through differentially permeable membranes
3. To observe the effects of turgor and plasmolysis in plant cells
4. To test experimental hypotheses about the effects of hypotonic, isotonic, and hypertonic solutions on the water-pumping rates of *Paramecium*.

Figure 4.1   The fluid mosaic model of a cell membrane.

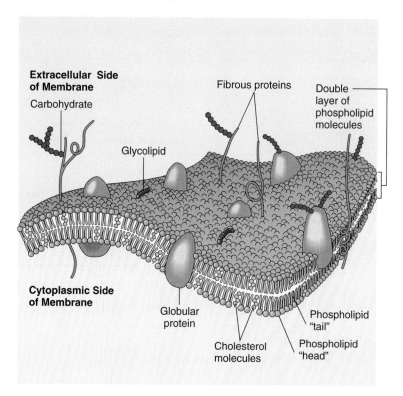

Background

The maintenance of a constant internal environment in a cell or organism is called **homeostasis.** In a constant environment, enzymes and other cellular systems are able to function at optimum efficiency. One component of a cell's homeostatic mechanisms is the ability to exchange materials selectively with the environment. Ions and organic compounds, such as sugars, amino acids, and nucleotides, must enter a cell, whereas waste products must leave a cell. Regardless of the direction of movement, the common interface for these processes is the plasma membrane. The cell walls of plants and bacteria offer little, if any, resistance to the exchange of molecules.

The plasma membrane is a mobile mosaic of lipids and proteins (fig. 4.1). Materials cross this outer cell boundary by several processes. Large particles are engulfed in membrane, forming a vesicle or vacuole that can pass into or out of the cell. Some small molecules diffuse through the spaces between lipid molecules in the membrane. Others bind with proteins in the membrane and are transported into or out of the cell.

To understand cellular transport, you should recognize that atoms, ions, and molecules in solution are in constant motion, continuously colliding with one another because of their **kinetic energy.** As the temperature of any phase is raised, the speed of movement increases so that molecules collide more frequently with greater force. A directly observable consequence of this constant motion is **Brownian movement,** an erratic, vibratory movement of small parti-

cles in aqueous suspension caused by collisions of water molecules with the particles.

Diffusion also results from the kinetic energy of molecules. For example, when a few crystals of a soluble substance are added to water, molecules break away from the crystal surface and enter solution, some traveling to the remotest regions of the solution. This process continues until the substance is equally distributed throughout the solvent. To generalize this example, in any localized region of high concentration, the movement of molecules is, on the average, away from the region of highest concentration and toward the region of lowest concentration. The gradual difference in concentration over the distance between high and low regions is called the **concentration gradient.** The steeper the concentration gradient, the more rapid the rate of diffusion. The rate of diffusion is also directly proportional to temperature and inversely proportional to the molecular weight of the substance involved. (All molecules move rapidly at high temperatures, but larger molecules move more slowly than small molecules at the same temperature.)

Substances diffuse into and out of cells by passing through the spaces between membrane molecules or dissolving in the lipid or protein portions of the membrane. However, due to size or charge, some substances cannot pass through membranes. Membranes that block or otherwise slow passage of certain substances are described as being **differentially permeable.** Differential permeability accounts for the phenomenon of **osmosis,** or the diffusion of water through a membrane under special conditions.

Figure 4.2 Model of osmosis through a differentially permeable membrane. Small water molecules can pass through pores in membrane but larger protein molecules cannot.

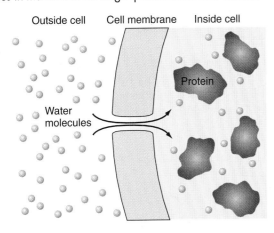

Outside cell   Cell membrane   Inside cell

Protein

Water molecules

The conditions for osmosis are shown in figure 4.2, where a porous membrane is pictured with water on one side and protein in water on the other. The special condition is that the small water molecules can pass through the pores of the membrane, but the large proteins cannot. Hence, water at a greater concentration outside (because it is not diluted by protein) will tend to diffuse into the cell. If the cell were encased in a rigid box, the increasing water pressure would cause the water to flow back out of the cell to the low-pressure area. Eventually, an equilibrium would be reached when the flow of water into the cell, due to concentration differences, balances the flow out of the cell, caused by pressure differences. The pressure at equilibrium is called the **osmotic pressure** of the solution.

Since all cells contain molecules in solution that cannot pass through the membrane, osmosis always occurs when cells are placed in dilute aqueous solutions. In bacteria and plants, the cell wall prevents the cell from bursting by providing a rigid casing that helps regulate osmotic pressure within the cell. In animals, an osmoregulatory organ is found, such as the kidney, which adjusts the concentration of substances in the body fluids that bathe the cells.

Many ions and organic molecules important to cell metabolism are taken into cells by specific transport proteins found in the cell membranes. **Facilitated diffusion** occurs when such a protein simply serves as a binding and entry port for the substrate. In essence, the protein is a pipeline for a specific substance. The direction of flow is always from a region of high concentration to one of low concentration, but gradients are maintained because many molecules, upon entering the cell, are metabolically converted to other types of molecules.

For many other materials, favorable diffusion gradients do not exist. For example, sodium ions are found at higher concentrations outside mammalian cells, yet the net movement of sodium is from inside to outside the cell. For such materials, cellular energy must be expended to transport the molecules across the cell membrane. **Active transport** oc-

LAB INSTRUCTIONS

You will observe Brownian movement, diffusion, and the properties of differentially permeable structures. Start the experiment "Simultaneous Osmosis and Diffusion" first so that sufficient time can elapse to see results. You will formulate and test experimental hypotheses regarding the activity of the water-expulsion vesicle in *Paramecium*.

curs when proteins in the cell membrane bind with the substrate and with a source of energy to drive the "pumping" of a material into or out of a cell.

Simultaneous Osmosis and Diffusion

Diffusion and osmosis can be demonstrated simultaneously in one setup. Dialysis tubing is an artificial membrane material with pore sizes that allow small molecules to pass through it but not large molecules.

Water, NaCl, and Na_2SO_4 have molecular weights of 18, 58.5, and 146 respectively. Starch and proteins have molecular weights greater than 100,000. If dialysis tubing is a differentially permeable membrane, which molecules would you hypothesize can pass through the membrane?

Obtain a 15 cm section of dialysis tubing that has been soaked in distilled water. Tie or fold and clip one end of the tubing to form a leakproof bag. Half fill the bag with a solution of 1% protein (albumin) dissolved in 3% NaCl. Also add a 3 ml sample of the same solution into each of four test tubes labeled **"Inside Start."**

Now tie the bag closed with a leakproof seal. Wash the bag with distilled water, blot it on a paper towel, weigh it to the nearest 0.1 g, and record the weight in table 4.1.

Place the bag in a 250 ml beaker containing a solution of 0.25% soluble starch dissolved in 1% Na_2SO_4. Place 3 ml samples of the fluid from the beaker in each of four test tubes labeled **"Outside Start."** The starting conditions are summarized in figure 4.3. This experiment will run for approximately two hours. Go on to the other experiments while this experiment runs in the background.

At 15-minute intervals, swirl the beaker containing the bag or place the beaker on a slowly turning magnetic stirrer.

Figure 4.3 Starting conditions for osmosis and diffusion experiment, showing composition of solutions inside and outside dialysis bag.

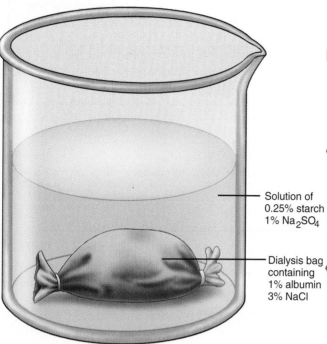

Solution of 0.25% starch 1% Na₂SO₄

Dialysis bag containing 1% albumin 3% NaCl

Test for Protein

▶ Dip Albustix reagent strips (usually used in urinalysis) into a start and end sample for both inside and outside solutions. The paper will turn green to blue-green if albumin is present.

Test for Starch

▶ Add a few drops of I₂KI to each remaining tube. If a blue color appears before mixing, it indicates the presence of starch. If no color develops, add a few crystals of KI without mixing, then add I₂ crystals. If a blue color develops as the iodine dissolves but then disappears, this is still a positive test for starch.

❓ Which set of test tubes served as a control in this experiment?

❓ Describe which ions were able to move through the dialysis membrane. Which direction did they move in relation to their concentration gradient? What are the molecular weights of these ions?

Analysis

▶ After two hours or longer, take four 3 ml samples from the beaker and place them in four test tubes labeled **"Outside End."** Now remove the bag, rinse it with distilled water, blot it on a paper towel, and weigh it to the nearest 0.1 g. Record the weight in table 4.1.

Empty the contents into a beaker, take four 3 ml samples, and place them in four test tubes labeled **"Inside End."**

▶ Now assay the inside and outside samples from the start and end for the presence of the compounds added at the beginning of the experiment. Record the results of your analysis in table 4.1, using plus and minus symbols to indicate the presence or absence of material both before and after incubation. The following are specific, easy-to-perform indicator tests:

Test for Chloride Ion

▶ Add a few drops of 1 M AgNO₃ to one inside and one outside tube for both start and end samples. A milky white precipitate of AgCl indicates the presence of Cl⁻.

Test for Sulfate Ion

▶ Add a few drops of 2% BaCl₂ solution to one inside and one outside tube for both start and end samples. If SO₄⁻ is present, a white precipitate of BaSO₄ will form.

❓ Did starch and protein move through the dialysis membrane? What are their typical molecular weights?

❓ What evidence do you have that water moved through the dialysis membrane?

TABLE 4.1   Results of osmosis/diffusion experiment with dialysis tubing

| | Outside Bag | | Inside Bag | |
	Start	End	Start	End
NaCl				
Na$_2$SO$_4$				
Protein				
Starch				
H$_2$O (weight)	XXX	XXX		

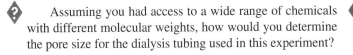

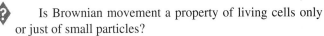

Assuming you had access to a wide range of chemicals with different molecular weights, how would you determine the pore size for the dialysis tubing used in this experiment?

Is Brownian movement a property of living cells only or just of small particles?

Brownian Movement

The vibratory movement exhibited by small particles in suspension in a fluid was first observed in 1827 by Robert Brown, a Scottish botanist. Brown erroneously concluded that living activity caused this motion, but scientists now know that Brownian movement results from the constant collision of water molecules with particles. Small particles 10 micrometers or less in size are noticeably displaced by the collision, whereas larger particles are not.

To illustrate Brownian movement, place a drop of water to one side of the center of a microscope slide. Add about one-half of an *Elodea* leaf to the drop and tap on the leaf with a polished glass rod to pulverize the leaf. Cover the preparation with a coverslip. Place a second drop of water on the slide. Dip a dissecting needle in India ink and then touch the needle tip to the new water drop (Note: India ink consists of small carbon particles suspended in a fluid.) Add a coverslip so that you now have two coverslips side by side. Observe the slide with a high-power objective and dim lighting.

Briefly record your impressions of the movement of each set of particles, pointing out any differences due to size. If you gently warm the slide over a bulb or alcohol lamp, what happens? (Do not boil!)

Osmosis

The rate of water movement in osmosis can be observed in an osmometer (fig. 4.4). A sugar solution in a thistle tube is separated from a beaker containing water by a dialysis membrane that allows water to pass through though not sugar molecules. What do you expect will happen over time in a setup such as in figure 4.4?

Early in the lab period, observe the height of the column of fluid in the thistle tube. At approximately 20-minute intervals during the lab, repeat. Record the time and height in table 4.2.

Describe what is happening to both sugar and water molecules in the osmometer.

Figure 4.4   Osmometer demonstrates passage of water across a differentially permeable membrane.

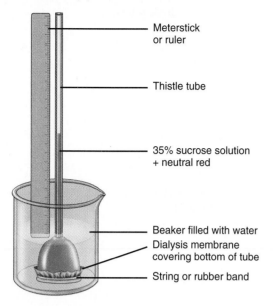

- Meterstick or ruler
- Thistle tube
- 35% sucrose solution + neutral red
- Beaker filled with water
- Dialysis membrane covering bottom of tube
- String or rubber band

TABLE 4.2	Fluid height in osmometer	
Time	**Elapsed time**	**Height**

Over time do you expect that the rate of water movement will increase, decrease, or remain the same? Why?

Figure 4.5   Setup for diffusion of gases. Use glass tube 1 inch in diameter and 16 inches long with cotton-tipped wooden applicator sticks inserted in rubber stoppers at ends. Remove stoppers and simultaneously dip one in concentrated HCl and the other in concentrated NH_4OH. Simultaneously reinsert and observe for formation of white ring of NH_4Cl.

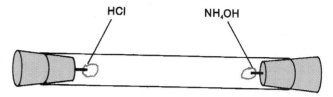

HCl NH_4OH

Diffusion in Gels

Diffusion in highly viscous solutions can be demonstrated using agar as a gelling agent. Obtain a petri plate containing 1% agar poured to a depth of 5 mm. Use a 5 mm cork borer to make two wells about 2 cm apart. Add a solution of sodium sulfate to one and barium chloride to the other. Observe at intervals during the lab by placing the plate on a black paper. When these two compounds diffuse through the water in the agar and meet, the white precipitate barium sulfate will be formed. Record your results as a time diary below.

Diffusion in Gases (Optional Demonstration)

Figure 4.5 shows a simple apparatus for determining the effect of molecular weight on diffusion in gases. Your instructor may do this as a demonstration. If concentrated HCl is introduced at one end and NH_4OH at the other, the gases HCl and NH_3 diffuse toward the center. HCl has a molecular weight of 36 and NH_3 of 17. Which gas should diffuse faster? State this as a hypothesis to be tested.

When HCl and NH_3 meet, they form NH_4Cl, a white salt, not a gas. It precipitates in the tube. Measure the distance each gas traveled in the tube.

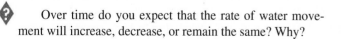

CAUTION

The HCl is a concentrated acid and can harm you and your clothes. The NH_4OH is very irritating to the eyes and nose. For these reasons, your instructor may choose to do this part of the exercise as a demonstration.

Record the results of your experiment below. What is the ratio of these distances? What is the ratio of the molecular weights? What is the ratio of the square roots of the molecular weights? Is diffusion rate directly or inversely proportional to molecular weight or the square root of molecular weight? Refer back to your hypothesis regarding the rate of diffusion. Do you accept or reject this hypothesis?

 What do you think would happen if you now placed these cells in plain water? Formulate this prediction as a hypothesis. Test the hypothesis by wicking distilled water under the coverslip. Record your results below. A cell in which the central vacuole is expanded, pressing the cytoplasm against the cell wall, is said to be turgid. Based on this observation must you accept or reject your hypothesis?

 Use the concept of osmosis to explain why the cytoplasm pulled away from the cell wall and then expanded again under the different conditions tested.

Osmosis in Living Cells

Elodea Plasmolysis

Plant cells are surrounded by rigid cell walls, and under normal conditions, the cytoplasm of the cell is closely pressed against these cell walls. Cells in this condition are said to be **turgid.**

▶ Remove a leaf from an *Elodea* plant, cut a very small piece from the tip, and mount it in water on a microscope slide. Using your compound microscope, identify individual cells and note the position of the cytoplasm in relation to the cell walls. Sketch a few cells. While looking at these cells, you may see cytoplasmic streaming (cyclosis). If you think in terms of molecular mixing, can you propose how this might benefit the cells in *Elodea?*

▶ Now treat the cells with a concentrated solution of 35% sucrose by placing a drop or two of the solution next to the coverslip and wicking it under the coverslip. (See fig. 3.5.) Observe the cells looking at the position of the cytoplasm relative to the cell wall. When the cytoplasm pulls away from the cell wall due to water loss from the central vacuole, the cell is said to **plasmolysed.**

Osmoregulation in Paramecium

Aquatic animals and plants can be exposed to environmental media that differ significantly in their osmotic concentrations. An **isosmotic** medium has the same concentration of osmotically active particles and, therefore, the same osmotic pressure as a cell does. This is sometimes also referred to as isotonic medium. A cell in an isosmotic medium will neither gain nor lose water since there is no net "pulling" force on the water in either the internal or external environments.

If a cell is placed in a **hyperosmotic** (hypertonic) medium, or a medium that has a higher concentration of osmotically active particles and thus a higher osmotic pressure than the cell, the cell tends to lose water to the medium. A **hypoosmotic** (hypotonic) medium contains a lower number of osmotically active particles than a cell and thus has a lower osmotic pressure. In a hypoosmotic medium, a cell gains water by osmosis. You just finished experimenting with the effects of hypoosmotic and hyperosmotic solutions on *Elodea* leaf cells.

Paramecium is a freshwater, single-celled organism. Its habitat is a hypoosmotic environment and *Paramecium* continually gains water. A mechanism is required for the *Paramecium* to get rid of this excess water. The **water-expulsion vesicle** (or contractile vacuole) functions in this capacity (fig. 4.6). The osmotic regulation function of the water-expulsion vacuole can be observed on a microscope slide. The preparation is difficult and requires patience and careful observation.

Figure 4.6 Photomicrograph of *Paramecium caudatum* showing one of its two water expulsion vesicles (WEV). In life, these collapse and fill in a regular cycle.

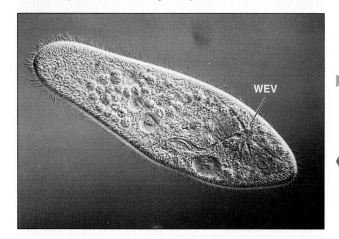

WEV

CAUTION

Sodium azide is a deadly poison that blocks cellular respiration by binding to the cytochromes. Do not ingest.

Baseline Observations

Obtain a drop of culture fluid containing paramecia that have been kept in a boiled wheat grain culture. Pick up some culture debris to aid in trapping the organisms. Place the sample on a clean glass slide and add a coverslip. Using medium power, look for the water-expulsion vacuole. (*Note:* It will be necessary to slow the animal down. Draw some of the fluid off with a paper towel. Try to trap protozoans in culture debris under the coverslip. If this does not work, shred a small piece of lens paper or cotton and put it on a new slide. Add a drop of culture and a coverslip. Look for paramecia caught in the fiber matrix.)

Observe a trapped *Paramecium* with high power, looking for one of the two water expulsion vesicles. Are the water-expulsion vacuoles temporary or permanent? Do the anterior and posterior vesicle contract at the same time? At the same rate? The normal rates of pulsation vary somewhat with culture conditions. You may expect a rate of six to ten pulsations per minute in the culture fluid. Draw a *Paramecium* below and record your observations as notes. Record the number of pulsations per minute on the first row of table 4.3.

Experimental Observations

Using your baseline observations for comparison, you should now determine the rate of pulsation (pulsations/minute) of the vesicle when the organisms are in the following solutions:

1. distilled water
2. 3% NaCl
3. **Optional:** culture fluid + 2×10^{-3}M sodium azide

In each case, mix one drop of the culture fluid containing paramecia with one drop of the desired solution. Counts can easily be made at the magnification provided by a $10\times$ objective and $10\times$ ocular once the *vesicles* and their pulsations have been observed.

Would you predict that the pulsations per minute in distilled water would be higher or lower than the number when in culture medium? Formulate this prediction as a testable null hypothesis (H_o) and alternative hypothesis (H_a) and record below:

H_o

H_a

Now count the number of pulsations per minute in distilled water and record in table 4.3.

Review your hypotheses and make a decision to accept or reject your null hypothesis.

Formulate a new set of testable hypotheses that relates expected vacuole pulsation rate in 3% NaCl to that seen in normal culture media. State the hypotheses below:

H_o

H_a

Add 3% NaCl to your slide and "wick" it under the coverslip. Count the number of pulsations and record in table 4.3.

Review the hypotheses you made for this experiment and accept or reject the null hypothesis.

TABLE 4.3   Water expulsion vesicle pumping rates in *Paramecium*

	Pulsations/Min	Vacuolar Volume (μ^3)	Volume Pumped (μ^3/Min)
Culture fluid (control)			
Distilled water			
3% NaCl			
2 mM sodium azide (optional)			

Optional: Sodium azide is a metabolic poison that stops the ability of a cell to make ATP during aerobic respiration. If energy is required for the water-expulsion vesicle to operate, it will stop functioning in cells treated with sodium azide.

Formulate a testable null hypothesis (H_o) and alternative (H_a) that relates azide treatment to normal vesicle pulsation rate. State the hypotheses below:

H_o

H_a

Make a fresh slide of *Paramecium* and wick a solution of sodium azide under the coverslip. Let sit for a few minutes and then measure the rate of vacuole pulsation. Record in table 4.3.

Must you accept or reject your null hypothesis? Why?

Analysis

If the volume of the water-expulsion vesicle at maximum expansion were known, it would be possible to estimate the amount of water expelled per minute. To calculate the vesicular volume, it is necessary to know the vesicle's diameter. This can be measured with an ocular micrometer. Since the use of the ocular micrometer is somewhat involved (especially in measuring the diameter of the water-

expulsion vesicle in a *Paramecium*), your lab instructor may give you a value for vacuole diameter. However, keep in mind that vesicle diameters vary from individual to individual and that the diameter of a vesicle within an individual may change in response to the osmotic pressure of the environmental medium.

Using the following formula, calculate the vesicle volume. This vesicular radius should be in micrometers (10^{-6} meters).

$$\text{volume (for a sphere)} = \frac{4\pi(\text{radius})^3}{3}$$

$$= 4.19r^3$$

Since you now know the volume of the water-expulsion vesicle and the pumping rates per minute of the vesicle, you can calculate the amount of water pumped per minute by a *Paramecium*. Enter the values in table 4.3. The units will be in cubic micrometers. A milliliter is a cubic centimeter.

 How many cubic micrometers are in a cubic centimeter? (Note: a cubic centimeter equals a milliliter.) How many milliliters is the *Paramecium* pumping per minute? Per day?

Why do you think the pumping rate was greatest in distilled water and least in the 3% NaCl solution? What does the sodium azide treatment demonstrate?

Learning Biology by Writing

Prepare a brief lab report about the water-expulsion vesicle experiments in this exercise. In the introduction to the report, state the hypotheses that were tested by your experiments. Next, describe the methods used to test each hypothesis. Finally, report the results of the tests, including the tables you completed. Discuss what these experiments demonstrated about the osmotic challenges that freshwater organisms face. Do marine organisms face the same osmotic challenges?

As an alternative assignment your instructor may ask you to complete the following lab summary and critical thinking questions.

Internet Sources

Reverse osmosis is a recently developed process to purify seawater for drinking water or to clean up polluted waters. Current information on this process can be found on the World Wide Web. Use an Internet browser to locate this information.

When in your browser, click on the **Search** button at the top of the screen. Choose one of the search engines by clicking on its name. When you get the dialog box, enter the words *"reverse osmosis"* and submit the search.

Look over the results from your search. Answer the following questions. What is reverse osmosis? Describe the process. List three applications of reverse osmosis. List two URLs that would be good resources for learning about this process.

Lab Summary Questions

1. What data do you have to support the conclusion that diffusion is inversely related to molecular weight?
2. What data do you have to support the conclusions that (1) dialysis membranes are differentially permeable and that (2) osmosis and diffusion can both occur at the same time through a dialysis membrane?
3. Explain how osmosis can be considered a special case of diffusion.
4. Explain why *Paramecium* continually gains water from its environment. How does it get rid of excess water?

Critical Thinking Questions

1. If a person's blood volume drops due to injury or severe dehydration, why do doctors administer isotonic saline intravenously instead of pure water?
2. How would an increase in temperature affect the rate of diffusion of gases? How might it affect osmosis? Explain.
3. How would increasing the concentration of albumin in your simultaneous osmosis and diffusion experiment affect the results?
4. When preparing sauerkraut, shredded cabbage is layered with coarse salt crystals. What affect would this have on the cabbage?
5. What osmotic regulatory challenges would a fish living in freshwater have versus a fish living in salt water?
6. We've all seen pictures from Third World countries of malnourished children with spindly limbs and huge distended bellies. This condition is called *kwashiorkor*. It is the result of a subsistence diet of corn meal, which supplies caloric requirements but lacks an essential amino acid. In effect, this diet is almost completely lacking in useable proteins. Based on what you've learned about osmosis, how could low blood protein levels lead to the distended bellies so characteristic of the disorder?
7. If fruits are crushed, cooked, and then stored at room temperature, they spoil. If high concentrations of sugar are added as in making jams, their shelf life is greatly extended. Explain this in terms of the osmotic challenges facing invading bacteria and fungi.

LAB TOPIC 5

Quantitative Techniques and Statistics

Supplies

Equipment

Spectrophotometer and tubes
Balances (sensitivity minimum 0.1 g)

Materials

5 ml pipettes or micropipetters
Suction devices for pipettes
Test tubes and racks
50 ml beakers

Solutions

Distilled water
Bromophenol blue standard solution (0.02 mg/ml)
Bromophenol blue solution for unknown

Prelab Preparation

Before doing this lab, you should read the introduction and sections of the lab topic that have been scheduled by the instructor.

You should use your textbook to review the definitions of the following terms:

Absorbance
Balance
Gram
Histogram
Mean
Micropipet
Milliliter
Spectrophotometer
Standard deviation
Variance

You should be able to describe in your own words the following concepts:

Beer's law
Absorption spectrum
Standard curve
Review Appendix C, pages 437–443

As a result of this review, you most likely have questions about terms, concepts, or how you will do

the experiments included in this lab. Write these questions in the space below or in the margins of the pages of this lab topic. The lab experiments should help you answer these questions, or you can ask your instructor for help during the lab.

Objectives

1. To learn to use pipettes and balances
2. To use a spectrophotometer to measure absorbance by colored solutions
3. To use a standard curve to determine the concentration of dye in an unknown solution
4. To analyze data using simple statistical techniques

In several forthcoming exercises, you will measure the volumes of liquid chemical solutions, weigh various samples, use a spectrophotometer, and obtain results that must be graphed or statistically analyzed. This exercise will give you the opportunity to practice these techniques before you have to apply them to a biological problem.

Lab Instructions

Pipette Technique

Your instructor will demonstrate the use of common volumetric glassware, including graduated cylinders, volumetric flasks, and beakers with approximate markings. In chemical tests, it is frequently necessary to measure a small volume of liquid (between 0 and 10 milliliters) quickly and accurately. Various calibrated automatic dispensers or syringes may be used to do so. For years, however, the standard measuring device used in research laboratories has been the glass **pipette** (fig. 5.1) because of handling ease and low cost. Recently, pipetting devices called micropipetters (fig. 5.2) that have disposable plastic tips have replaced the standard glass pipette in most laboratories.

A glass pipette is filled by using a syringe or valved rubber bulb attached to the end of the pipette (fig. 5.3). To use a

Figure 5.1 One type of glass pipette used in laboratories. The serological pipette must drain dry to dispense the calibrated volume.

Serological pipette

Figure 5.2 A micropipetter has a dial that allows one to set the volume to be measured, and a disposable tip that is discarded after use.

pipette, immerse just the tip in the appropriate fluid and draw it up beyond the zero mark using a suction device. Look at the surface of the fluid in the pipette and note the **meniscus,** the concave upper surface caused by surface tension. Hold the pipette vertically and allow the fluid to escape until the bottom of the meniscus touches the zero line. Any drops hanging from the tip of the pipette should be removed by touching the tip to the inside of the beaker from which the solution was drawn. When releasing fluid from the pipette, be careful not to allow the liquid to run freely because droplets will cling to the inner walls. The flow should be regulated, aiming for a rate of less than one-half the free flow.

C A U T I O N
If you use glass pipettes, never draw chemical or bio-hazardous solutions into a pipette by mouth suction, as you would do with a straw, because you may accidentally ingest a poison or pathogen. Use a suction device (fig 5.3.)

If you have to use an automatic micropipetter, using it is a bit more involved and proper technique will be demonstrated by your instructor, usually the following steps are involved.

1. Select a pipetter that dispenses volumes in the range you want.

2. Note the units on the dial and dial in the volume you wish to dispense.

3. Put the correct colortip on the pipette with a twisting motion.

4. Depress the button on the top to the first stop and, while holding it in that position, submerge it in the fluid to be drawn up. Slowly release the button to draw the fluid up.

5. Without blotting, put the tip into the tube to receive the fluid. Press the button to dispense the fluid, continuing to press it past the first step to the second, thus blowing out the fluid. Touch the tip to the side wall to remove any drops clinging to the tip.

6. If a new solution is to be pipetted, press the tip ejection lever and put on a new tip. The same tip can be used over again when repetively dispensing the same fluid.

To practice your pipetting technique, obtain a 5 ml pipette and a 50 ml beaker.

Weigh the empty beaker to the nearest tenth of a gram. Your laboratory may be equipped with electronic or triple-beam balances. Your instructor will give you directions on how to use the available equipment. Be sure that you know how to zero the balance and then to weigh a specimen. Record the beaker's weight below.

Now use the pipette to add the following volumes of water in milliliters to the beaker:

5.0	4.0	3.0
4.5	3.5	2.5
4.3	3.3	1.5
4.2	3.2	1.0

Theoretically a total of 40 ml should have been added.

Verifying Techniques

To check the accuracy of your pipetting, you will weigh the amount of water added to the beaker and then use the density of water to convert this weight to a volume. The density of a substance is its weight per unit volume. The density of water at room temperature (23°C) is 0.998 g/ml.

Weigh the beaker containing the 40 ml of water and record the weight below. Subtract the weight of the empty beaker from the filled beaker weight to obtain the actual weight of water in the beaker. This operation—subtracting the weight of the container from the weight of the material plus the weight of the container—is called **taring.**

Weight of beaker containing water _____ g

Minus weight of empty beaker _____ g

= Weight of water _____ g

Figure 5.3 Filler devices used with glass pipettes: (a) rubber bulb with valves, (b) syringe with rubber connector. Other suction devices may be used and, if available, will be demonstrated by your instructor.

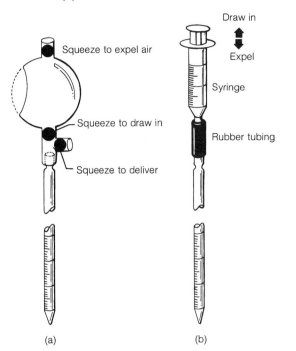

(a)　　　　　　(b)

Did you leave out a pipetting? Did you repeat one? Did you correctly read the pipette and meniscus? Did you properly use the balance? Is the balance accurate? Check it with a standard, known weight. What would be the effect on your data if the balance was improperly calibrated and consistently gave readings that were heavier than they were supposed to be? Would your % error be positive or negative?

List below your explanations for the error in your measurement.

If you did a similar experiment with larger pipettes in which you added two 20 ml portions, would you expect your error to change? Why?

Divide the weight of water in the beaker by the density of water to calculate the volume of water in the beaker. How much water, in milliliters, is in your beaker?

Calculate the experimental error as a percentage (called percentage error) using the following equation:

$$\% \text{ error} = \frac{\begin{array}{c}\text{actual} - \text{theoretical}\\ \text{value} \quad\quad \text{value}\end{array}}{\text{theoretical value}} \times 100$$

$$= \frac{\text{actual volume of } H_2O - 40}{40} \times 100$$

% error = _____

In recording the results of the above calculation, you undoubtedly had to make a decision about the number of significant figures to write down. The section on significant figures and rounding at the beginning of appendix A contains the rules to use in such situations. Read it!

If your percentage error is not zero, think about how you did this experiment. What does a positive percent error mean? A negative? Honestly evaluate your work.

Simple Statistics

Different students probably obtained slightly different values for their water volume measurements at the beginning of the exercise. Even if gross errors (such as wrong calculations, failure to follow directions, or incorrect readings) are ruled out, there would still be some variation due to minor experimental and chance errors. Therefore, the measurements you have made are not true values but are simply estimations of a true value. Some estimations obtained by the class are lower than the true value, while others are higher. When it is important for scientists to obtain a true measurement, they repeat the measurement several times and calculate a **mean** (average) value.

Chance errors in data sets cancel each other out when means are calculated—that is, a value that is too high due to chance error is balanced by a value that is low for the same reason. The **range** of observations gives some sense of the variability in measurement or the precision of the analysis. However, range is not a very good estimator of variability because it can be artificially inflated by one or two outlying values. Consider the following two sets of hypothetical data:

	Set A	Set B
	30	30
	29	40
	31	20
	28	32
	32	31
	30	30
	29	31
	31	26
Σ (= sum)	240	240
N (= number of observations)	8	8
Mean	30	30
Range	28 – 32(±2)	20 – 40(±10)

One pair of values in set B created an extremely broad range even though the means were the same. Because of this problem and the need to convey information about the amount of variability in a set of measurements, scientists use prescribed calculations to obtain variability estimators called **variance** and **standard deviation.** (Appendix C contains a more thorough discussion of statistics.) These estimators are obtained by expressing all values in a data set as plus or minus variations from the mean, that is, as *measured value – mean value*. Variance and standard deviation are calculated as follows:

$$\text{variance} =$$

$$\frac{\Sigma\ (\text{measured value for each sample – mean})^2}{(N-1)\ \text{one less than number of observations}}$$

$$\text{standard deviation} = \pm\sqrt{\text{variance}}$$

Use the following work table to calculate the standard deviation of the hypothetical set of measured values given in the first column.

Measured value	Measured minus mean	(Measured minus mean)²
15		
20		
25		
16		
20		
24		
10		
20		
30		

Σ	_____		Σ	_____
N	_____		N – 1	_____
Mean	_____		Variance	_____
			SD	_____

Do your answers agree with those of other students? Check your arithmetic if they do not.

Analysis of Class Data

Having practiced how to calculate a standard deviation, you will now apply this concept to the class data on the volume of water pipetted at the beginning of this exercise.

To create a class data set, all students should go to the blackboard and write the values they calculated from the density data for the volume of water pipetted into the beaker at the beginning of this exercise. Copy the data from the blackboard into the first two columns of table 5.1.

Scan the values that you entered in table 5.1 and answer the following questions:

1. Are some values in the class data grossly different from all the others? If so, check for errors in calculation or procedure that will allow you to objectively eliminate the data. If any data are rejected, indicate why.

2. Now, calculate the mean value for the water volume and enter the range of values. Is it possible to determine if variation is due to measurement of weight or of water volume? Compare the mean values for water volume to the theoretical value of 40 ml. Is the mean closer to the real value than are many individual measurements? Why? Is the mean volume for water volume skewed significantly away from 40 ml? If so,

TABLE 5.1 Class data work table

Calculated Water Volume (ml)	(Measured Vol – Mean Vol)	Measured Vol – Mean Vol)²

Sum (Σ) ＿＿＿ Σ ＿＿＿

Range ＿＿＿ Σ/(N – 1) ＿＿＿

Number of observations (N) ＿＿＿

Mean (Σ/N) ＿＿＿ $\sqrt{\Sigma/N-1}$ ＿＿＿

check to see if the skewing is due to a few values or whether all values show bias. In the latter case, check the balance. Balances that weigh consistently heavy or light would cause such a bias.

3. Do the range values give you an estimate of the variability in a measurement? If good lab technique is used, should the range be large or small?

4. Calculate the standard deviations for the volume of water from the class data. Use the last two columns in table 5.1 to organize your calculations. Record the mean and standard deviation below.

   Mean =

   SD = ±

Read the first part of appendix C and explain, in your own words, what it signifies when a scientist writes that the mean for a set of measurements was 40 plus or minus a standard deviation of 1.5. Write your explanation below.

Students who have experience with the use of spreadsheets on microcomputers should realize that such software often contains functions for simple statistical analyses. You need only enter the data set into a spreadsheet and call up the functions. In this lab manual, you will often collect data that consist of repeated measurements. You should learn how to use a spreadsheet program such as *Excel* to calculate a mean and standard deviation. Your instructor may have handouts to help you use these programs.

Histograms

The class data on the volume of water can also be used to make a histogram, sometimes called a bar graph (see appendix B). Histograms are useful in that they visually convey to the reader the amount of variability in a given set of measurements. Many spreadsheet programs have built-in graph functions and can also be used to create histograms or line graphs directly from data entered into the program.

To create a histogram for the class data, look at table 5.1 and determine the range of water volumes. Create four categories by dividing four into the difference between the highest and lowest water volumes. Add this value to the lowest water volume to obtain the interval of

Quantitative Techniques and Statistics

Figure 5.4   Schematic drawing of the path of light through a spectrophotometer.

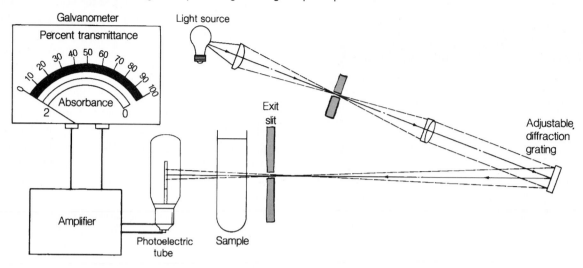

the first category. Sequentially add this value to the upper limit of each category to define the intervals for all categories. Record these values below.

Intervals   **Frequencies**

Now count how many values in table 5.1 fall into each category and record the frequencies above.

Use this data to construct a bar graph in which the x-axis is the category and the y-axis is the frequency of occurrence in the data set. Label all axes and indicate the scale.

Add an arrow above the histogram to indicate where the mean of the data set lies and add a cross bar perpendicular to the arrow that spans plus or minus one standard deviation.

Many spreadsheet programs have built-in graph functions and can also be used to create histograms or line graphs directly from data entered into the program. If you are to use such a program, your instructor will give you directions.

Spectrophotometry

Many organic molecules absorb radiant energy because of the nature of their chemical bonds. (Light-absorbing organic molecules have a system of single and double bonds between adjacent carbons or carbon and nitrogen.) For example, proteins and nucleic acids absorb ultraviolet light in the wavelength interval 240 to 300 nanometers (nm), pigments and dyes absorb visible light (about 400 to 770 nm), and other organic compounds absorb infrared energy (above 770 nm). Our perception of color is related to the ability of pigment molecules in cone cells to absorb light energy. If an object appears to be red, it contains molecules whose chemical bonds absorb blue or green **photons** of light while reflecting red light back to the red sensitive cones of one eye.

A **spectrophotometer** is an instrument designed to detect the amount of radiant energy absorbed by molecules in a solution. Spectrophotometers have five basic components: a **light source,** a **diffraction grating,** an **aperture** or **slit,** a detector (a **photoelectric tube**), and a **meter** or digital readout to display the output of the phototube. The arrangement of these parts is shown in figure 5.4.

When light passes through the diffraction grating, it is split into its component colors or wavelengths, which then diverge. Sections of the projecting spectrum can be either blocked or allowed to pass through the slit, so that only one color will pass to the other sections of the spectrophotometer.

Light that passes through the slit travels to the phototube, where it creates an electric current proportional to the number of photons striking the phototube. If a **galvanometer** or its digital meter equivalent is attached to the phototube, the electric current output—which represents the quantity of light striking the phototube—can be measured. The meter scale is usually calibrated in two ways: **percent transmittance,** which runs on a scale from 0 to 100, and **absorbance,** which runs from 0 to 2 in most practical applications.

Zero transmittance control   100% transmittance control   Wavelength control

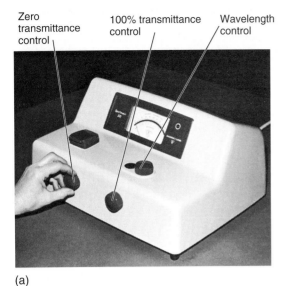

(a)

Sample compartment

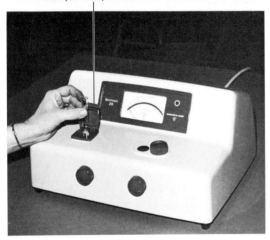

(b)

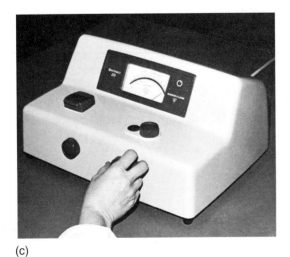

(c)

Figure 5.5 Steps for using a Spectronic 20 spectrophotometer. About twenty minutes before the measurements are to be made, turn on the instrument by rotating the zero control knob clockwise. (*a*) The wavelength knob (on top of the spectrophotometer) should be adjusted to a wavelength of 425 nanometers. Adjust the meter needle to zero transmittance by rotating the zero control knob. (*b*) Insert a tube containing 8 ml of solvent into the sample holder. Keep the index line of the tube aligned with the index line on the sample holder. Close the tube chamber cover. (*c*) Adjust the meter to read 100% transmittance by rotating the 100% transmittance control knob. (This knob regulates the amount of light passing through the spectrophotometer slit, thus controlling the amount of light that reaches the sample and the phototube.) Remove the tube and close the lid. If the galvanometer needle does not return to zero, readjust accordingly. Reinsert the tube to see if the instrument still registers 100% transmittance; if not, readjust with the 100% transmittance control knob. If the spectrophotometer is used for any length of time, recheck these readings now and then. Once the meter is adjusted for 100% transmittance, measurements can be made on experimental samples at that wavelength only. You must recalibrate at each new wavelength.

Before the light-absorbing properties of a solution can be measured, three adjustments on the spectrophotometer are necessary. First, the diffraction grating must be adjusted so that the desired color of light passes through the slit. This is usually the color of light that is best absorbed by the compound under consideration. Second, the output of the phototube must be adjusted or calibrated to correct for drift in the electronic circuit. Third, a compensation must be made for dirt or contaminating colored material in the light path between the source and the detector.

To calibrate the electrical circuits, an adjustment knob is turned when there is no sample in the instrument and the meter is set to read 0% transmission. To adjust for materials in the light path, fill a clean sample tube with the same solvent (usually water) to be used in dissolving the dye. Place the tube containing the solvent in the sample compartment of the spectrophotometer (see fig. 5.5). This tube is called a **blank** and serves as a control. The galvanometer needle will deflect to about 100% transmittance, or 0 absorbance, depending on which scale you are using. The needle should be set to exactly 100 or 0 by adjusting the current output adjustment knob. This standardizes the spectrophotometer.

If a colored solution is put in the tube in place of the pure solvent, some of the light coming from the slit will be absorbed by the dye molecules, and some will be transmitted to the phototube. The amount absorbed will be proportional to the number of the dye molecules per unit volume (or **concentration**) of the solution.

The meter needle will reflect the phototube current output, which will be less than that seen in the standardizing procedure when there were no dye molecules. If the transmittance scale is used, the amount of light transmitted by the solution is measured as a percentage by the spectrophotometer. This measurement is described by the following equation:

Percent transmittance (T)

$$= \frac{\text{intensity of light through sample}}{\text{intensity of light through blank}} \times 100$$

$$= \frac{I_s \times 100}{I_b}$$

If the **absorbance** scale is used, the measurement equation is somewhat more involved. Instead of measuring the amount of light transmitted by the solution, the spectrophotometer reads the amount of light absorbed and converts this measurement into absorbance (A) units described by the following equation:

$$A(\text{absorbance}) = \log_{10}\left(\frac{1}{T}\right)$$

An example may help to clarify the relationship between transmittance and absorbance. If a dye solution is placed in the spectrophotometer and is found to transmit 10% of the light, its absorbance can be calculated thus:

Since T = 0.10

And $A = \log_{10}\left(\frac{1}{T}\right)$

$A = \log_{10}(10) = 1$

Scientists prefer to work with absorbance units because they are directly and linearly related to the concentration of the dye in solution (fig. 5.6). This relationship is described by the **Lambert-Beer Law,** which states that for a given concentration range and a sample tube of constant diameter, the absorbance is directly proportional to the concentration of solute molecules.

This relationship will be true regardless of the dye used, providing: (1) monochromatic (one-color) light is used, and (2) this color is the wavelength of light best absorbed by the dye. The wavelength of maximum absorption will always be given in the directions for any experiment.

Absorbance Curves

To learn how to use the spectrophotometer, you will explore the Lambert-Beer relationship using bromophenol blue dye. Your instructor will supply you with a 0.02mg/ml solution of bromophenol blue in water.

First, determine the wavelength of maximum absorption by following these steps:

1. Turn on the spectrophotometer and familiarize yourself with the position of the control knobs. (Refer to fig. 5.5 for their functions on a Spectronic 20.)

2. Adjust the meter to 0% *transmittance,* using the zero control knob.

Figure 5.6 Standard curve of the absorbances of known dye concentrations. First, the absorbances of four known concentrations of dye are measured and plotted to make a graph. Then the absorbance of an unknown concentration is measured. The graph is used to find the concentration of dye that would give the measured absorbance.

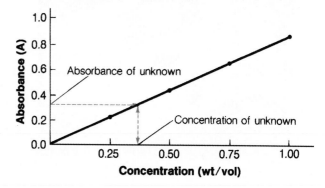

TABLE 5.2 Absorbance readings for bromophenol blue

Wavelength (nm)	Absorbance Units
425	_____
450	_____
475	_____
500	_____
525	_____
550	_____
575	_____
600	_____
625	_____

3. Obtain two matched sample tubes and clean them, if necessary, with tissue. Fill one tube with distilled water (solvent) and use it as a blank. Fill the other with bromophenol blue solution. This will be your experimental tube.

4. Set the wavelength control at 425 nm, insert the blank tube containing water, and adjust the reference knob, so that the meter reads 0 *absorbance.*

5. Remove the blank tube and recheck the 0% *transmittance.* If you must readjust, repeat step 4. Insert the tube containing 0.02 mg/ml bromophenol blue.

6. Read the meter and record the *absorbance* in table 5.2. Be sure to read the correct scale. Do not read the % transmittance scale.

TABLE 5.3 Concentration and absorbance for eight dye dilutions at _____ nanometers

Tube	ml of Dye	ml of H_2O	Concentration	A
1	0	8	0.00 mg/ml	___
2	1	7	___	___
3	2	6	___	___
4	3	5	___	___
5	4	4	___	___
6	5	3	___	___
7	6	2	___	___
8	7	1	___	___
9	8	0	0.02 mg/ml	___
10	unknown	0	___	___

7. Change the wavelength to 450 nm and repeat the adjustments in steps 4, 5, and 6. Continue these measurements at 25 nm intervals up to 625 nm. Remember to always recalibrate (steps 4 and 5) at each new wavelength.

At which wavelength in the range 425–625 nm does bromophenol blue absorb the most light?

Plot the measured *absorbance* as a function of wavelength on the graph paper at the end of this exercise. Wavelength, the independent variable, should be on the x-axis and absorbance on the y-axis. Instructions for drawing graphs are in appendix B.

Standard Curves

In this section you will construct a *standard curve,* which demonstrates the linear relationship between absorbance and concentration. You will then use this standard curve to determine the concentration of dye in an unknown.

Obtain a stock solution of bromophenol blue containing 0.02 mg/ml. Prepare a series of dilutions in eight test tubes, using the proportions of dye and water listed in table 5.3. After adding the solutions, mix by holding the test tube in the left hand between your thumb and forefinger. Gently strike the bottom of the tube several times with the forefinger of your right hand to create a swirling motion of the fluid in the tube. Do not strike the tube so hard that the dye splashes out.

Set the spectrophotometer at the maximum absorption wavelength for bromophenol blue. Record the wavelength in the heading of table 5.3. After calibrating the spec-

trophotometer with water as a blank, read the *absorbance* for all eight tubes in the dilution series at this wavelength. Because you are not changing wavelength, you do not need to blank the spectrophotometer between readings. Record your results in the last column of table 5.3.

To calculate the actual concentration of bromophenol blue in each tube in units of mg/ml, do the following:

1. Determine the total mg of dye added to each tube by multiplying the number of ml of added dye by the dye concentration, which was 0.02 mg/ml.

2. Then divide the value by 8 ml, the total volume of fluid present after adding water. Record the concentrations in the fourth column of table 5.3.

These calculations are summarized in the work table that follows:

Tube	ml dye	total mg dye	Dye concentration
1	___	___	___
2	___	___	___
3	___	___	___
4	___	___	___
5	___	___	___
6	___	___	___
7	___	___	___
8	___	___	___

Enter values from last column in table 5.3 under concentration.

Use the graph paper at the end of this exercise to plot your data with absorbance as a function of dye concentration. Instructions for drawing graphs are given in appendix B. Remember that the x-axis, or abscissa, is always the independent variable and the y-axis, or ordinate, is always the dependent variable. In this experiment, what is the independent variable?

 structed? Briefly outline the steps below. Perform the procedure and record the result below.

Label both axes. After plotting your data points, draw the line that, on the average, best fits all points of the data. Do not connect the points and create bumpy curves; draw a straight line that best fit the points. This technique compensates for some of the random variability in the data.

The plot of absorbance as a function of dye concentration is called a **standard curve.** By reading the graph, you can determine the absorbance of any dye concentration within the range of concentrations tested. The line on the graph may be extrapolated to predict the absorbance of concentrations beyond the highest tested, but there is always a danger that the Lambert-Beer Law does not apply at very high concentrations.

▶ Your instructor will now give you a solution of bromophenol blue that contains an unknown amount of dye.

How can you determine the dye concentration using the spectrophotometer and the standard curve you just con-

Unknown concentration = _____ mg/ml

Your instructor will now tell you the actual concentration of dye in the unknown. Calculate your percentage error.

Learning Biology by Writing

Prepare a lab report in which you present your data on determining the dye concentration in the unknown (include your graph). Be sure you state the problem and then discuss how you solved it. In your discussion, describe in general terms some sources of experimental error in lab work and the value of working with repetitive measurements.

As an alternative assignment, your instructor may ask you to complete the following summary and critical thinking questions.

Lab Summary Questions

1. Explain how you would use a standard curve to determine the amount of orange dye that had been added to a can of orange soda.
2. Describe the differences between mean, range, and standard deviation.

3. Hand in table 5.1 showing the mean and standard deviation for the class data on the volume of water.
4. Hand in the histogram for the class data on volume of water.
5. Turn in your standard curve of absorbance versus dye concentration.

Critical Thinking Questions

1. Explain why the sample cuvettes (sample tubes) used with the spectrophotometer are "matched." Why were the outsides cleaned each time before placing them in the spectrophotometer? What effect does the sample volume have on readings? What effect does tube placement have on readings?
2. Describe a "normal distribution" curve. What percentage of the data is encompassed by one standard deviation?
3. What is meant by "grading on the curve" or "curving" the grades?

LAB TOPIC 6

Determining the Properties of an Enzyme

Supplies

Preparator's guide available at
    http://www.mhhe.com/dolphin

Equipment

Constant temperature water baths or large trays to
    serve as water baths
Spectrophotometers at 500 nm

Materials

Blender or mortar and pestle
Fresh white turnip, horseradish root, or potato
Tissues and markers
Optical surface tubes for the spectrophotometer
15 ml test tubes and rack
50 ml beakers
5 ml pipettes graduated in 0.1 ml units with suction
    devices or automatic pipetters
Thermometer (alcohol)

Solutions

10 mM H_2O_2
25 mM guaiacol (Sigma Chemical Co.)
2% Hydroxylamine (Sigma Chemical Co.)
Citrate-phosphate buffers at pHs 3, 5, 7, and 9

Prelab Preparation

Before doing this lab, you should read the introduction
and sections of the lab topic that have been scheduled
by the instructor.

You should use your textbook to review the
definitions of the following terms:

Enzyme
Inhibitor
pH
Peroxidase
Product
Spectrophotometer
Substrate

You should be able to describe in your own words
the following concepts:

Structure of an enzyme
Effect of pH on enzyme structure
Effect of temperature on enzyme structure

Effect of competitive inhibitors on enzymes
Review Appendix B, pages 435–436

As a result of this review, you most likely have
questions about terms, concepts, or how you will do
the experiments included in this lab. Write these
questions in the space below or in the margins of the
pages of this lab topic. The lab experiments should
help you answer these questions, or you can ask your
instructor for help during the lab.

Objectives

1. To perform a quantitative assay of the activity of
   an enzyme in a tissue extract using a
   spectrophotometer
2. To organize the data as concise tables and graphs
   for inclusion in a lab report describing the
   properties of the enzyme peroxidase
3. To test the following null hypotheses:
   a. The amount of enzyme does not influence the
      rate of reaction.
   b. The temperature of the solution does not
      influence the activity of an enzyme.
   c. The pH of the solution does not influence the
      activity of an enzyme.
   d. Boiling an enzyme before a reaction does not
      influence its activity.
   e. Other molecules with shapes similar to an
      enzyme's substrate have no effect on the activity
      of an enzyme.

Background

The thousands of chemical reactions occurring in a cell
each minute are not random events but are highly controlled by biological catalysts called **enzymes.** Like all catalysts, enzymes lower the **activation energy** of a reaction,
the amount of energy necessary to trigger a reaction.

Figure 6.1 Enzymes are large proteins that bind substrates at their active sites. The substrate is converted into products, which are released, allowing the enzyme to bind to another substrate molecule. In this example, sucrase converts sucrose into glucose and fructose.

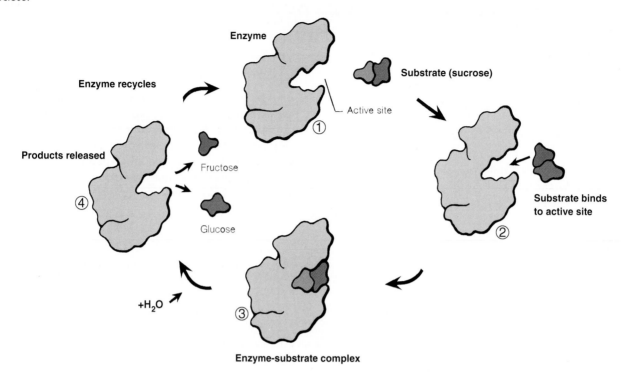

Most enzymes are proteins with individual shapes determined by their unique amino acid sequences. Since these sequences are spelled out by specific genes, the chemical activities of a cell are under genetic control. The shape of an enzyme, especially in its **active site,** determines its catalytic effects (fig. 6.1). The active site of each type of enzyme will bind only with certain kinds of molecules—for example, some enzymes bind with glucose but not with ribose because the former is a six-carbon sugar while the latter has only five carbons.

A molecule that binds with an enzyme and undergoes chemical modification is called the **substrate** of that enzyme. Often metallic ions, such as Fe^{+++}, Mg^{++}, Ca^{++}, or Mn^{++}, aid in the binding process, as do vitamins or other small molecules called **co-factors** or **coenzymes.**

The binding between enzyme and substrate consists of weak, noncovalent chemical bonds, forming an **enzyme-substrate complex** that exists for only a few milliseconds. During this instant, the covalent bonds of the substrate either come under stress or are oriented in such a manner that they can be attacked by other molecules, for example, by water in a hydrolysis reaction.

The result is a chemical change in the substrate that converts it to a new type of molecule called the **product** of the reaction. The product leaves the enzyme's active site and is used by the cell. The enzyme is unchanged by the reaction and will enter the catalytic cycle again, provided other substrate molecules are available.

Individual enzyme molecules may enter the catalytic cycle several thousand times per second; thus, a small amount of enzyme can convert large quantities of substrate to product. Eventually enzymes wear out; they break apart and lose their catalytic capacity. Cellular proteinases degrade inactive enzymes to amino acids, which are recycled by the cell to make other structural and functional proteins.

The amount of a particular enzyme found in a cell is determined by the *balance* between the processes that *degrade* the enzyme and those that *synthesize* it. When no enzyme is present, the chemical reaction catalyzed by the enzyme does not occur at an appreciable rate. Conversely, if enzyme concentration increases, the rate of the catalytic reaction associated with that enzyme will also increase.

The pH or salt concentrations of a solution affect the shape of enzymes by altering the distribution of + and − changes in the enzyme molecules which, in turn, alters their substrate-binding efficiency. Temperature, within the physiological limits of 0° to 40°C, affects the frequency with which the enzyme and its substrates collide and, hence, also affects binding. All factors that influence binding obviously affect the rate of enzyme-catalyzed reactions. Some of these factors will be investigated during this laboratory.

Peroxidase

During this lab, you will study an enzyme called **peroxidase.** It is a large protein containing several hundred amino acids and has an iron ion located at its active site. Peroxidase makes an ideal experimental material because it is easily prepared and assayed. Turnips, horseradish roots, and potatoes are rich sources of this enzyme.

The normal function of peroxidase is to convert toxic hydrogen peroxide (H_2O_2), which can be produced in certain metabolic reactions, into harmless water (H_2O) and oxygen (O_2).

$$2\,H_2O_2 \xrightarrow{\text{peroxidase}} 2\,H_2O_2 + 2\,O$$

The oxygen often reacts with other compounds in the cell to form secondary products.

The peroxidase reaction can be measured by following the formation of oxygen. The amount of oxygen present after the reaction can be measured in two ways: by the accumulation of gas in a closed system connected to a manometer or by the appearance of chemically active oxygen.

Many dyes will react with active oxygen by changing from a colorless to a colored state, and dye techniques are easier to perform than volumetric tests of gases in teaching laboratories. Such tests are called **dye-coupled reactions.** The enzyme-catalyzed reaction produces a product that enters into a secondary reaction with the dye. The enzyme, itself, does not bind with or affect the dye.

You will use the dye **guaiacol,** which turns brown when oxidized. The entire peroxidase reaction, including the measure of active oxygen through guaiacol, is as follows:

$$2\,H_2O_2 \xrightarrow{\text{peroxidase}} 2\,H_2O + 2\,O$$

$$O + \text{guaiacol} \rightarrow \text{oxidized guaiacol}$$
$$\quad\ \ \ \text{(colorless)} \qquad\qquad \text{(brown)}$$

To quantitatively measure the amount of brown color in the final product, the enzyme, substrate, and dye can be mixed in a tube and immediately placed in a spectrophotometer. As color accumulates, the absorbance at 500 nm will increase. The procedure for using a spectrophotometer was explained in lab topic 5. You should review those instructions before proceeding. (See fig. 5.5.)

LAB INSTRUCTIONS

In this exercise, you will determine the effects of several factors on the activity of peroxidase.

Preparing an Extract Containing Peroxidase

These steps will be done by the instructor before class to save time:

1. Weigh 1 to 10 g of peeled turnip, horseradish, or potato tissue on a double- or triple-beam balance.

2. Homogenize the tissue by adding it to 100 ml of cold (4°C) 0.1 M phosphate buffer at pH 7. Grind the mixture in a cold mortar and pestle with sand or blend it for 15 seconds at high speed in a cold blender. The extract will keep overnight in a refrigerator.

Standardizing the Amount of Enzyme

The extract contains hundreds of different types of enzymes, including peroxidase. The activity of each enzyme will vary, depending on the size and age of the turnip, horseradish, or potato; the extent of the tissue homogenization; and the age of the extract. Only peroxidase, however, will react with H_2O_2.

To demonstrate that the amount of enzyme influences the rate of the reaction and to determine the correct amount of extract to use in future experiments, a trial run should be performed in which the amount of enzyme added is the only variable. Your instructor may do this section as a demonstration to show you how to best organize the procedures used in assaying an enzyme.

Before starting an experiment, a null (H_o) and an alternative (H_a) hypothesis should be stated. These should relate the rate of reaction to the amount of enzyme added. An example will be given here, but you will make your own hypotheses in the other experiments done in this lab.

H_o: *The amount of enzyme added to the reaction will have no effect on the rate of reaction.*

H_a: *The amount of enzyme added to the reaction changes the rate of the reaction.*

To test your H_o, you should use the following directions to set up the chemical reactions and to conduct the experiment:

1. Label four 50 ml beakers as follows: *turnip (or horseradish or potato) extract; buffer, pH 5; 10 mM H_2O_2; and 25 mM guaiacol.* Fill each about half full with the appropriate stock solution. Label four pipettes to correspond with the beakers. Alternatively, your lab instructor may have these reagents available in burettes or dispensers. If that is the case, you will be given verbal directions on how to add solutions to your test tubes.

2. Number seven test tubes from 1 to 7. The contents of the tubes will be:

    1: Control with no extract to be used in calibrating the spectrophotometer
    2: Substrate and indicator dye
    3: Dilute extract
    4: Substrate and indicator dye (same as 2)
    5: Medium concentration of extract
    6: Substrate and indicator dye (same as 2)
    7: Concentrated extract

    Pairs 2 and 3, 4 and 5, or 6 and 7 will be mixed together when it is time to measure a reaction. Mix a pair only when you are ready to measure that reaction in the spectrophotometer. The exact quantities to be added to each tube are listed in table 6.1.

3. Add stock solutions to each tube using the corresponding graduated 5 ml pipette or dispensing device. Use of the wrong pipette or dispenser will cross contaminate your reagents and introduce errors into your subsequent experiments.

Determining the Properties of an Enzyme   **55**

TABLE 6.1 Mixing table for trial run to determine extract concentration (all values in milliliters)

Tube	Buffer (pH 5)	H_2O_2	Extract	Guaiacol (Dye)	Total Volume
1 Control	5.0	2.0	0	1.0	8
2	0	2.0	0	1.0	3
3	4.5	0	0.5	0	5
4	0	2.0	0	1.0	3
5	4.0	0	1.0	0	5
6	0	2.0	0	1.0	3
7	3.0	0	2.0	0	5

TABLE 6.2 Results from trial run of enzyme activity (entries are absorbance units at 500 nm)

Time (Sec)	Tubes 2 and 3 0.5 ml Extract	Tubes 4 and 5 1.0 ml Extract	Tubes 6 and 7 2.0 ml Extract
20			
40			
60			
80			
100			
120			

4. Use the directions in figure 5.5 to adjust the spectrophotometer to zero absorbance at 500 nm. Pour the contents of test tube 1 into a cuvette (the special spectrophotometer test tube made of optical glass). This tube is used to "blank" the spectrophotometer, so that any color caused by contaminants in the reagents will not influence subsequent measurements.

5. If you are working as teams of students, one person can be a timer, another a spectrophotometer reader, and another a data recorder. Note the time to the nearest second and mix the contents of tubes 2 and 3 by pouring them back and forth twice. Mixing should be completed within ten seconds.

6. Add the reaction mixture to a cuvette by pouring or using an eye dropper, wipe the outside, and place the cuvette in the spectrophotometer. Read the absorbance at 20-second intervals from the start of mixing. If you are a little late in reading the meter, record the absorbance and change the table to show the actual time of the reading. Record your measurements in table 6.2. After two minutes (six readings) remove the tube from the spectrophotometer and visually note the color change. Discard the solution.

7. Mix the contents of tubes 4 and 5, transfer to a cuvette, and repeat your measurements for two minutes at 20-second intervals. Record the results in table 6.2.

8. Mix the contents of tubes 6 and 7, transfer to a cuvette, and record the absorbance measurements in table 6.2.

Analysis of Amount of Enzyme Data

Now plot the values in table 6.2 on one panel of graph paper provided at the end of this exercise. The abscissa should be the independent variable (time in seconds) and the ordinate the dependent variable (absorbance units). Explain why absorbance is considered the dependent variable.

Plot all three tests on the same coordinates using different plotting symbols. Using a clear plastic ruler, draw the single straight line that best fits the points for each of the conditions. (Curves may plateau at the end.) Appendix B discusses how to make graphs.

In mathematical terminology, what characteristic of the graphed line is a measure of enzyme activity?

Temperature Effects

To determine the effects of temperature on peroxidase activity, you will repeat the enzyme assay in water baths at four temperatures:

1. In a refrigerator at approximately 4°C
2. At room temperature (about 23°C)
3. At 32°C
4. At 48°C

 What are the units of activity in this experiment?

State a null (H_o) hypothesis that relates change in enzyme activity to the temperature of the solutions used.

H_o

 Which amount of enzyme gave a linear absorbance change from 0 to 1 in approximately 120 seconds?

Use this amount in all subsequent experiments in this exercise.

State an alternative hypothesis.

H_a

 Change the amount of enzyme called for in mixing tables 6.3, 6.5, 6.7, and 6.9 as necessary based on your first experiment. The standardizing procedure must be repeated for each batch of extract. Why?

To test your H_o, you should use the following directions to set up the reactions and conduct the experiment:

If constant temperature baths are not available, improvise with plastic containers, adding hot and cold water to adjust the temperature. Number nine test tubes in sequence 1 through 9. Refer to table 6.3 for the volumes of reagents to be added to each tube.

Preincubate all the solutions at the appropriate temperatures for at least 15 minutes before mixing. After reaching temperature equilibrium and adjusting the spectrophotometer with the contents of test tube 1, mix pairs of tubes (2 and 3, 4 and 5, 6 and 7, and 8 and 9) one pair at a time. After mixing one pair, measure the change in absorbance for two minutes at 20-second intervals for each temperature. The temperatures will not remain exact, but the effects can be overlooked. After measuring the absorbance changes, mix the second pair and measure the absorbance change, and so on.

Note: The room-temperature experiment can be performed immediately while the other tubes temperature-equilibrate.

Record changes in absorbance for the reaction mixture at each temperature in table 6.4.

Do you accept or reject your H_o (null hypothesis) regarding rate of reaction and amount of enzyme? Why?

Analysis of Temperature Data

These results should be graphed at the end of the laboratory period on one of the panels of graph paper at the end of the exercise. The slopes of the linear portions of these curves

Factors Affecting Enzyme Activity

If the scheduled laboratory period is short, your instructor may divide you into teams, each of which will test for the effects of one or more of the following experimental variables. The results will be shared at the end of the lab period and may be included in your report.

Temperature	Tube	Buffer (pH 5)	H₂O₂	Extract	Guaiacol (Dye)	Total Volume
	1 Control	5.0	2.0	0	1.0	8
4°C	2	0	2.0	0	1.0	3
	3	4.0	0	1.0	0	5
23°C	4	0	2.0	0	1.0	3
	5	4.0	0	1.0	0	5
32°C	6	0	2.0	0	1.0	3
	7	4.0	0	1.0	0	5
48°C	8	0	2.0	0	1.0	3
	9	4.0	0	1.0	0	5

TABLE 6.4   Temperature effects on peroxidase activity (entries are absorbance units at 500 nm)

Time (Sec)	Tubes 2 and 3 4°C	Tubes 4 and 5 23°C	Tubes 6 and 7 32°C	Tubes 8 and 9 48°C
20				
40				
60				
80				
100				
120				

are a measure of enzyme activity. Does activity vary with temperature? What is the optimum temperature?

_____ _____ °C

Do you accept or reject the H₀ stated earlier? Why?

To show clearly this relationship, you should prepare a derivative graph after lab. Determine the absorbance change per minute (slope) at each temperature treatment from your linear graphs. Record below.

Temperature   **Activity (ΔA/min)**

4°

23°

32°

48°

On graph paper, plot the activity values as functions of temperature. Fit a curve to the plotted points. Your curve should be bell shaped. The top of the curve indicates the **temperature optimum,** the temperature at which the maximum rate is observed.

pH Effects

Begin by stating null (H₀) and alternative (Hₐ) hypotheses that relate change in enzyme activity to the pH of the solutions used.

H₀

Hₐ

TABLE 6.5   Mixing table for pH experiment (all values in milliliters)

pH	Tube	Buffer	H_2O_2	Extract	Guaiacol (Dye)	Total Volume
5	1 Control	5.0 (pH 5)	2.0	0	1.0	8
3	2	0	2.0	0	1.0	3
	3	4.0 (pH 3)	0	1.0	0	5
5	4	0	2.0	0	1.0	3
	5	4.0 (pH 5)	0	1.0	0	5
7	6	0	2.0	0	1.0	3
	7	4.0 (pH 7)	0	1.0	0	5
9	8	0	2.0	0	1.0	3
	9	4.0 (pH 9)	0	1.0	0	5

TABLE 6.6   Effects of pH on peroxidase activity (entries are absorbance units at 500 nm)

Time (Sec)	Tubes 2 and 3 pH 3	Tubes 4 and 5 pH 5	Tubes 6 and 7 pH 7	Tubes 8 and 9 pH 9
20				
40				
60				
80				
100				
120				

To determine the effect of pH on peroxidase, perform the following experiment.

Your instructor will supply buffers at pHs of 3, 5, 7, and 9. Number nine test tubes 1 through 9. Set up pH-effect tests by adding the reagents described in table 6.5.

After adjusting the spectrophotometer with the contents of test tube 1, mix pairs of tubes one at a time (2 and 3, 4 and 5, 6 and 7, 8 and 9. Measure absorbance changes at 20-second intervals for two minutes for each pair before mixing the next pair. Record the results in table 6.6.

Analysis of pH Data

The values in this table should be graphed at the end of the laboratory period on one panel of graph paper at the end of the exercise. The slopes of the linear portions of these curves are a measure of enzyme activity. Does activity vary with pH? What are the units of activity? What is the optimum pH?

Do you accept or reject the H_o regarding pH effects on enzymes? Why?

To show clearly this relationship, you should prepare a derivative graph after lab. Determine the absorbance change per minute from your plots of absorbance versus time at each pH. Record below.

pH	Activity (ΔA/min)
3	
5	
7	
9	

Plot the activity values as functions of pH on graph paper. The resulting curve should be bell shaped with the top (peak) indicating the **pH optimum,** the pH at which the maximum rate is observed.

Determining the Properties of an Enzyme

TABLE 6.7 — Mixing table for boiling extract (all values in milliliters)

Temperature	Buffer (pH 5)	H_2O_2	Boiled Extract	Guaiacol	Total Volume
1 Control	5.0	2.0	0	1.0	8
2	0	2.0	0	1.0	3
3	4.0	0	1.0	0	5

Effect of Boiling on Peroxidase Activity

Most proteins are denatured when they are heated to temperatures above 70°C. **Denaturation** is a nonreversible change in a protein's three-dimensional structure.

If heating a protein to 100°C irreversibly alters its shape (denaturation), what do you predict will happen to measured enzyme activity?

State null (H_o) and alternative (H_a) hypotheses that relate heat treatment of an enzyme (boiling) to the expected effect on that enzyme's activity.

H_o

H_a

TABLE 6.8 — Results from using boiled extract (entries are absorbance units at 500 nm)

Time (Sec)	Tubes 2 and 3 Boiler Extract	
20		
40		
60		
80		
100		
120		

Do you accept or reject the H_o made at the beginning of this experiment? Why?

Consult your text to see what kinds of chemical bonds are disrupted in proteins heated to 100°C. Describe these bonds and the effect on protein shape below.

To perform the experiment that tests your null hypothesis, follow these directions:

Add 3 ml of extract to a test tube and place it in a boiling water bath. After five minutes, remove the tube and let it cool to room temperature. Number three test tubes and add reagents as called for in mixing table 6.7.

Use the contents of tube 1 to blank the spectrophotometer. Mix the contents of tubes 2 and 3, pour the mixture into a cuvette, and read the absorbance at 20-second intervals for two minutes. Record the results in table 6.8.

Compare the activity of peroxidase after boiling to the activity of peroxidase kept at room temperature and pH 5 (see table 6.4 or 6.6). How did boiling affect the activity?

TABLE 6.9 Mixing table for inhibitor experiments (all values in milliliters)

Tube	Buffer (pH 5)	H_2O_2	Extract	Hydroxylamine Treated Extract	Guaiacol	Total Volume
1 Control	5.0	2.0	0	0	1.0	8
2	0	2.0	0	0	1.0	3
3	4.0	0	1.0	0	0	5
4	0	2.0	0	0	1.0	3
5	4.0	0	0	1.0	0	5

In this experiment, there was no control. What should have been in another pair of tubes to act as an experimental control?

TABLE 6.10 Enzyme inhibition result (entries are absorbance units at 500 nm)

Time (Sec)	Tubes 2 and 3 Normal Extract	Tubes 4 and 5 Hydroxylamine-treated Extract
20		
40		
60		
80		
100		
120		

Optional: The Effects of Inhibitors

Hydroxylamine (HONH$_2$) has a structure similar to hydrogen peroxide (HOOH). Calculate the molecular weight of hydrogen peroxide: _____. Calculate the molecular weight of hydroxylamine: _____. Hydroxylamine binds with the iron atom at the active site of peroxidase and prevents hydrogen peroxide from entering the site. What do you predict will be the effect on enzyme activity? State your predicted effects as null (H_o) and alternative (H_a) hypotheses.

H_o

H_a

treated enzyme preparation to that of the enzyme without the inhibitor. Table 6.9 lists the proportions of each solution to be used for the tests.

After adjusting the spectrophotometer with the contents of test tube 1, mix pairs one at a time (2 and 3, 4 and 5) and measure the changes at 20-second intervals for two minutes. Record your measurements in table 6.10.

These data should be graphed at the end of the laboratory period. The slopes of the linear portions of the curves are a measure of enzyme activity. What are the units? Explain why the slopes differ.

To test these hypotheses, mix five drops of 2% hydroxylamine (neutralize to pH 7) and 2 ml of enzyme extract, letting the mixture stand for at least ten minutes. Then measure the peroxidase activity, comparing the activity in this

Do you accept or reject the H_o made at the beginning of this experiment? Why?

Analysis

You have collected quantitative information about enzymes and how they work, using the enzyme peroxidase as an example. Your instructor will discuss the results in class and will indicate how to share the data from your experiments with the class. Directions for making graphs are given in appendix B.

 In all these experiments, tube 1 has been listed as a control. What is it controlling for? No reaction happens in this tube. Why was it included?

Learning Biology by Writing

This exercise provides material well suited for a written scientific report. You have tested five hypotheses and the collected results describe the properties of the enzyme peroxidase. Appendix D contains general directions on report writing. Your instructor may also have specific instructions for you to follow in writing your lab report. A suggested form follows.

Purpose

A good lab report begins with a statement of purpose, which summarizes the hypotheses tested in the experiments. Write the purpose in the third person impersonal, concentrating on the scientific questions involved and not the teaching and learning objectives.

Techniques

Describe in about two paragraphs the techniques you used to extract the peroxidase and to measure enzyme activity.

Results

Use graphs to report your data. (See appendix B.) Time should be recorded on the abscissa and absorbance on the ordinate. A separate graph should be made for each factor—pH, temperature, and inhibitors. All curves for any

one factor should be on the same set of coordinates. Indicate the slope values for each curve, labeling curves and coordinates, and writing a descriptive legend (for example, "Peroxidase activity at different pH levels. Symbols used are . . .").

The results from the temperature or the pH experiments should be summarized in separate graphs in which you calculate the slope of your absorbance versus time plots at each temperature (or pH) and plot the slope as a function of temperature (or pH). These derivative plots make it easier to see what the temperature (or pH) optimum is for peroxidase.

Discussion

Read the section in your textbook about enzymes. Discuss how the experiments you performed can be interpreted in terms of the structural properties of a protein subjected to different chemical and physical treatments. What are the limitations of these experiments? Will all enzymes have the same pH and temperature optima? Will all denature at the same temperature or be inhibited by hydroxylamine? How would you do the experiments differently, if you were to repeat them? What are the sources of error?

As an alternative assignment, your instructor may ask you to complete the following lab summary and critical thinking questions.

LAB TOPIC 7

Measuring Cellular Respiration

Supplies

Preparator's guide available at
    http://www.mhhe.com/dolphin

Equipment

Barometer
Hot plates
pH meter

Materials

Distillation apparatus (see fig. 7.2)
Yeast packets
Four- to six-day germinating peas
Test tubes and racks
One-hole stoppers with 1 ml × 0.01 ml pipettes
    inserted (see fig. 7.3)
Tuberculin syringes with 1 1/2-inch #18 needles
Aluminum foil
Test tube rack
Nonabsorbent cotton
Absorbent cotton
Cheesecloth
Mortar and pestle
Miscellaneous beakers
Weighing pans
Pasteur pipettes

Solutions

Karo syrup
0.1 M $Ba(OH)_2$
I_2KI (5 g I_2:10 g KI:100 ml H_2O)
1.5 M NaOH (*Caution:* caustic)
15% KOH in dropper bottles
95% ethanol in dropper bottles

Prelab Preparation

Before doing this lab, you should read the introduction and sections of the lab topic that have been scheduled by the instructor.

You should use your textbook to review the definitions of the following terms:

Aerobic
Anaerobic
Electron Transport System

Ethanol
Fermentation
Glycolysis
Krebs cycle

You should be able to describe in your own words the following concepts:

How ethanol is produced by yeast
How CO_2 is produced in respiration
Where O_2 is used in respiration
Review Appendix B.

As a result of this review, you most likely have questions about terms, concepts, or how you will do the experiments included in this lab. Write these questions in the space below or in the margins of the pages of this lab topic. The lab experiments should help you answer these questions, or you can ask your instructor for help during the lab.

Objectives

1. To identify the end products of aerobic and anaerobic respiration in yeast
2. To test a null hypothesis regarding the effect of freezing on the aerobic respiration of peas
3. To measure the rate of oxygen consumption in germinating pea seedlings

Background

All organisms, whether plant or animal, bacteria, protists or fungi, perform **cellular respiration.** Respiration involves several enzyme-catalyzed reactions that break down organic molecules and yield energy that a cell can use to perform the work of growth, maintenance, and function. Biologists often use the metabolism of glucose as a model for respiration because the metabolic pathways are well understood and occur in most organisms.

After entering a cell, glucose can be broken down into two molecules of pyruvic acid by a series of approximately ten enzyme-catalyzed reactions known collectively as **glycolysis.** The enzymes for these reactions float free in the aqueous portion of the cytoplasm. As the chemical reactions of glycolysis occur, energy contained in the covalent bonds of the glucose molecule is released. Some of this energy passes from the organism as heat, and some is used in the synthesis of the energy storage compound **ATP.** It is subsequently used by the cell to fuel various work functions.

To obtain energy from glucose, hydrogen atoms are removed from the glucose molecule as it is metabolized. These hydrogen atoms can be removed only by hydrogen (electron) carriers, such as the compound **NAD$^+$** (nicotinamide-adenine dinucleotide), which acts as a coenzyme in the chemical reactions. Since a finite amount of NAD$^+$ occurs in the cell and since each NAD$^+$ molecule can combine with only two hydrogens, there must be a mechanism for removing hydrogen from NAD•2H complexes so that glycolysis can continue. Without such a mechanism, glycolysis would cease when all NAD$^+$ molecules were saturated with hydrogen.

In many organisms, respiration can occur under anaerobic conditions where no oxygen is required to breakdown glucose. Many bacteria, yeast, and some animals, ferment glucose, producing lactic acid or ethanol. In these fermentations, hydrogens are removed from glucose, passed to the electron carrier NAD, and then on to pyruvic acid, the end product of glycolysis, converting it to lactic acid or ethanol which is excreted from the cell. Fermentation allows cells to make ATP in the absence of oxygen.

Cells with the enzymes for forming the end products of anaerobic metabolism can live in oxygen-deficient environments and still gain energy from glucose. Such cells do not, however, harvest all of the energy contained in the chemical bonds of glucose, since they excrete large organic molecules that contain substantial amounts of energy. Cells metabolizing glucose by fermentations harvest only about 5% of the available energy.

Many cells, when in an environment containing oxygen, are capable of metabolizing glucose by **aerobic respiration.** In this series of approximately 30 enzyme-catalyzed reactions, the glucose molecule is completely disassembled to yield CO_2 and H_2O. Of the usable energy contained in the chemical bonds of glucose, about half escapes as heat and about half is trapped in ATP during aerobic respiration.

In aerobic respiration, the initial series of reactions is the same as in anaerobic respiration; both begin with glycolysis. The end products of glycolysis are also the same and include pyruvic acid, NAD•2H, and ATP. The difference between the two types of respiration lies in how the NAD•2H is regenerated and in the fate of pyruvic acid. In anaerobic respiration, the hydrogens are passed to pyruvic acid; in aerobic respiration, they are passed to oxygen through a series of compounds located in the cristae of the mitochondria or on the mesosomes of bacteria.

As the hydrogens pass through this series, known as the **electron transport system (ETS),** or **cytochrome system,** they lose energy. Some of this energy escapes as heat, but a large portion is trapped through the mechanism of **chemiosmosis** as usable energy in ATP. When the hydrogen finally combines with oxygen, water is formed. Thus in organisms that gain energy through aerobic respiration, the oxygen they consume leaves the body as the oxygen atom in a water molecules, not as the oxygen atoms in carbon dioxide.

Pyruvic acid, also produced in aerobic respiration, can be further metabolized. The covalent bonds of pyruvic acid are broken by the enzymes of the **Krebs cycle** found in the mitochondria, yielding CO_2 and additional hydrogen that combines with electron carriers like **NAD$^+$.** These carriers are regenerated by means of the electron transport system, yielding more ATP by the mechanism of chemiosmosis.

Overall, 18 times more ATP is produced by aerobic respiration than is produced by anaerobic respiration. However, for aerobic respiration to occur, molecular oxygen is absolutely necessary. Organisms with aerobic metabolism are more efficient but are constrained to environments containing oxygen. Figure 7.1 summarizes the differences between these two types of respiration.

LAB INSTRUCTIONS

You will observe some of the properties of aerobic and anaerobic respiration. Because of the number of experiments included here, your instructor may omit some experiments, or do some as demonstrations.

Respiration in Yeast

Yeast can break down glucose and obtain energy under both aerobic and anaerobic conditions. When oxygen is present, yeasts will break down glucose aerobically, using the metabolic sequence of glycolysis, Krebs cycle, and electron transport. Water and carbon dioxide are the end products. When oxygen is not available, the Krebs cycle and electron transport system shut down. Glucose is metabolized to form pyruvate, which, in turn, is converted to two waste products, carbon dioxide and ethanol.

To test this metabolic duplicity in yeast, your instructor will set up two yeast cultures about 24 hours before class. This was done by adding 140 ml of white corn syrup and 280 ml of H_2O to each of two 1-liter flasks. A tablespoon of baker's yeast was added to each flask. A third flask will contain the same as the other two but no yeast will have been added. It is the control.

A two-hole stopper with a long aeration tube and short escape tube was added to one flask, which was labeled **aerobic.** The tube was connected to an air pump and the culture vigorously aerated for the 24 hours (fig. 7.2).

Figure 7.1 Summary of reactions involved in anaerobic and aerobic metabolism of glucose.

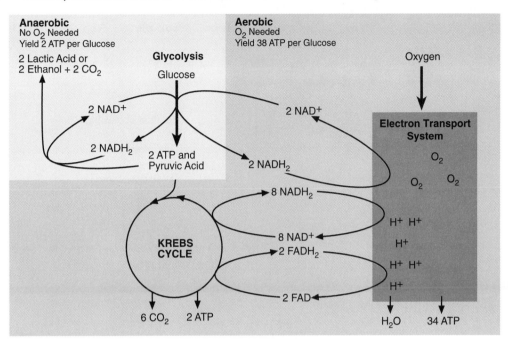

Figure 7.2 Culture setup for growing yeast aerobically and anaerobically.

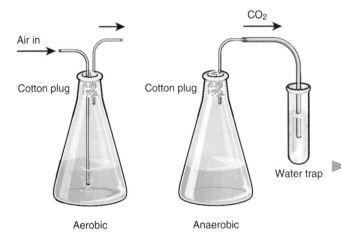

A one-hole stopper was added to the remaining flask. A short piece of glass tubing was passed through the stopper's hole and about 50 cm of rubber tubing was connected to the glass tubing. The open end of the rubber tubing was placed in a test tube two-thirds full of water. This water valve allows gases to escape but prevents air from entering the culture. This flask was labeled **anaerobic** (fig.7.2).

When you come to lab, your instructor will perform a simple classic chemical test to determine whether both cul-

tures are producing carbon dioxide. If a gas is bubbled through 0.1 M $Ba(OH)_2$, any carbon dioxide present reacts to form barium carbonate, a white precipitate, according to the following equations:

$$CO_2 + H_2O \rightarrow H_2CO_3$$

$$H_2CO_3 + Ba(OH)_2 \rightarrow BaCO_3 \downarrow + 2H_2O$$

What are the results when the gases emitted by the cultures are bubbled through $Ba(OH)_2$?

Aerobic _____

Anaerobic _____

The other products of respiration should be water or ethanol. Since you cannot test for water in an aqueous culture, you will test for ethanol. Ethanol can be removed from the culture by distillation because it boils at 78°C, 22°C less than the boiling point of water.

Three distillation setups should be assembled (fig. 7.3). About 100 ml of **aerobic** culture should be added to one, 100 ml of **anaerobic** to another, and 100 ml from the control to the last. Turn on the hot plates and collect about 30 ml of distillate from each still. Remember to turn off the hot plates! What you have just done is a scaled-down version of the process used in the liquor and gasohol industries.

In which of the three flasks do you hypothesize you will find ethanol? _____

Ethanol in the distillates can be detected by a simple procedure called to iodoform test. Take five test tubes,

Figure 7.3   A reflux air-cooled distillation setup. The glass tube should measure 30 cm from the flask to the curve to allow reflux action.

TABLE 7.1   Distillate tests from yeast cultures

Tube	Sample	Iodoform Test Results
1	Water	_____
2	Ethanol and water	_____
3	Distillate of aerobic culture	_____
4	Distillate of anaerobic culture	_____
5	Distillate of control	_____

◆   Which tube(s) gave a positive test for ethanol? Is your hypothesis supported or falsified?

number each one, and add 1 ml of strong I_2KI and 1.5 ml of 1.5 M NaOH to each. These are the test reagents. Now add the following samples to each tube:

Tube

1	2.5 ml distilled water
2	1.25 ml distilled water and 1.25 ml 95% ethanol
3	2.5 ml of distillate from the **aerobic** culture
4	2.5 ml of distillate from the **anaerobic** culture
5	2.5 ml of distillate from the **control**

Mix all solutions and let stand for five minutes. If ethanol is present, it will react with the iodine in the presence of NaOH to form **iodoform,** which will settle out as a yellow precipitate. If no reaction occurs in tube 2 (the known sample of ethanol), add 1 ml more of I_2KI to *all* tubes and mix. Record your results in table 7.1. Use *NR* to record where there is no reaction, and + to indicate where a precipitate is found.

◆   Why were tubes 1, 2 and 5 included in the iodoform test procedure? What purposes do they serve?

Aerobic Respiration in Peas

If living tissues or small organisms are placed in a closed chamber, they will consume oxygen and produce carbon dioxide. If the CO_2 is chemically removed as it is produced, the pressure in the chamber will drop in proportion to the O_2 consumed. The simple device shown in figure 7.4 can be constructed to measure this change.

Potassium hydroxide serves as an efficient CO_2 trap, as shown by the following reactions:

$$H_2O + CO_2 \text{ (gas)} \rightarrow H_2CO_3 \text{ (solution)}$$

$$H_2CO_3 + 2\,KOH \rightarrow K_2CO_3 \text{ (crystalline)} + 2H_2O$$

If a small drop of dye is placed in the open end of the pipette in the apparatus shown in figure 7.4, it will move inward as the pressure in the chamber decreases due to oxygen consumption and to fluctuations in the classroom temperature and atmospheric pressure. The markings on the pipette allow direct readings as apparent volume units.

To convert apparent volume to actual volume of oxygen consumed, two corrections will be necessary. One correction will offset the effects of fluctuations in temperature and pressure during the experiment. The other will adjust the **apparent volume** to **standard volume** at a standard pressure of 760 Torr (mm of Hg) and standard temperature of 273 Kelvin. The first correction is made by using an experimental control called a **thermobar** and the second by a calculation using the **combined gas laws.**

Figure 7.4 Apparatus for measuring oxygen consumption in pea seedlings.

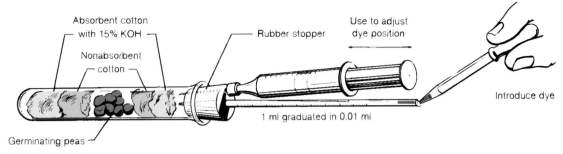

Respirometer

Forming the Hypothesis

You will use the apparatus in figure 7.4 to determine the effect of freezing and thawing on pea seedlings. Peas are often planted very early in the spring and germinating peas may experience very cold frosts. You will investigate whether freezing has any effect on germination by measuring oxygen consumption as a measure of aerobic respiration capability.

Convert this question into a testable null hypothesis (H_o) and alternative (H_a). State your hypotheses:

H_o

H_a

How will you set up a control for this experiment to eliminate the effects of temperature and pressure variations?

Experimental Setup

Before lab, your instructor will have frozen and thawed some germinating peas, and will also have on hand normal, untreated germinating peas all at room temperature. These are your experimental organisms.

To begin, label and weigh two empty weighing pans. If you are using an electronic balance, adjust it to zero (tare) using the weighing pans. Now add about 8 to 12 four-to-six-day-old normal germinating peas to one pan and the same number of freeze/thaw-treated peas to the other. After weighing the pans with peas, subtract the empty pan weight to obtain the tissue weight.

	Normal	Freeze/Thaw-Treated
Weight of peas and pans	_____	_____
Weight of empty pans	_____	_____
Weight of peas	_____	_____

Record the pea weights in table 7.4.

Take three tubes and add a small ball of *absorbent* cotton 2 cm in diameter to the bottom of each tube. Hold the tube vertically and drop in four or five drops of 15% KOH, so that the drops fall directly on the cotton and do not run down the sidewalls. Cover the moistened cotton with a layer of *nonabsorbent* cotton. Add normal peas to one of the tubes, freeze/thaw-treated peas to another, and no peas to the third. This third tube serves as the *thermobar*. Now add a second ball of *nonabsorbent* cotton to all three tubes and cover it with a ball of *absorbent* cotton. Moisten the absorbent cotton with two drops of 15% KOH.

In all of these procedures, care should be taken not to wet the sides of the tubes with KOH, since this will injure plant tissue. Add dry stoppers with pipettes and syringes to all three tubes. The setups should look like that in figure 7.4.

Measurement

Place the tubes on the table and let them equilibrate for five minutes. (Temperature equilibration after holding the tubes in your warm hands is absolutely essential for accurate results.)

Is your body temperature above or below room temperature? What happens to the pressure exerted by a gas when it cools?

TABLE 7.2 Raw oxygen consumption in milliliters

Tube	Contents	Reading Time (minutes)					
		0	3	6	9	12	15
1	Normal peas						
2	Freeze/thaw-treated peas						
3	Thermobar						

TABLE 7.3   Thermabar-corrected data in milliliters

Tube	Contents	Reading Time (minutes)					
		0	3	6	9	12	15
1	Normal peas						
2	Freeze/thaw-treated peas						

After 5 minutes, add a drop of dye into the end of each pipette by means of an eye dropper and adjust the front surface of the dye drop to a position of 0 ml in all three by pulling out on the syringe plunger. Record zero in table 7.2, along with the starting time.

Read the position of the front surface of the dye drop at two- to five-minute intervals (three minutes suggested but should be modified according to rates observed), depending on the amount of activity. Record the readings in table 7.2. If fluid moves out from the 0 mark at the end of the pipette, estimate the volume change and use a minus sign to indicate readings less than zero.

Do not touch the tubes during the experiment because temperature increases will cause the gases in the tubes to expand and give false readings.

Analysis

▶ Correct the raw data by subtracting the value of the empty thermobar. This tube corrects for any apparent changes during the experiment due to changes in atmospheric pressure or temperature. Thus, changes in the thermobar tube reflect environmental variation, whereas those in the other tubes reflect changes due to metabolism plus environmental variation. Subtraction of the thermobar value, which may be either plus or minus, corrects for the effects of environmental variation. (Remember that if you subtract a negative number, you must actually add.) Enter the corrected values in table 7.3.

Now plot both sets of the apparent-volume-of-oxygen-consumed figures as a function of time on one piece of graph paper at the end of the exercise. Directions for making graphs are given in appendix B. Use different plotting symbols for each treatment and label all axes. Draw the straight line that best fits each data set.

Calculate the slope of each line and enter the slopes in table 7.4. Divide the slope values by the pea weights to arrive at a *specific rate of oxygen consumption per gram* of tissue for each treatment.

If you are going to compare rates of oxygen consumption between the samples, why is it necessary to calculate oxygen consumption on a per-gram-of-tissue basis?

These values are only apparent rates of consumption and must be corrected to standard conditions before they can be reported and compared to published data. By convention, all gas volumes in the literature are reported as volumes at 760 Torr (mm of Hg) pressure and 273 absolute temperature in Kelvin units.

The gas laws, mentioned earlier, are used to make this correction, as follows:

If V_1 equals apparent volume at P_1 (atmospheric pressure during the experiment) and at T_1 (experimental temperature in Kelvins, $°C + 273$), and if V_2 equals corrected volume at standard pressure P_2 and at standard temperature T_2, then by the gas laws:

$$\frac{P_1 V_1}{T_1} = \frac{P_2 V_2}{T_2}$$

To calculate the actual rate of oxygen consumption, then, rearrange:

$$V_2 = \frac{P_1 T_2 \times V_1}{T_1 P_2}$$

LAB TOPIC 8

Determining Chromosome Number in Mitotic Cells

Supplies

Preparator's guide available at
http://www.mhhe.com/dolphin

Equipment

Compound microscopes
Water bath or heating block at 60°C

Materials

Prepared slides of whitefish blastula
Prepared slide of sectioned onion root tip
Onion root tips two to four days old. Immerse green
    onions from grocery in water with aeration.
Glass pestles, 4 inches × ⅛ inch; round end in flame
    and then file flat area
Alcohol lamps
Small vials (15 to 20 ml) with caps
Small watch glass
Razor blades
Forceps
Slides and coverslips

Solutions

Fixative solution of one part glacial acetic acid to three
    parts 100% methanol made at start of lab
1 M HCl (Caution)
45% acetic acid (Caution)
Fresh Feulgen stain (Caution)

Prelab Preparation

Before doing this lab, you should read the introduction
and sections of the lab topic that have been scheduled
by the instructor.
    You should use your textbook to review the
definitions of the following terms:

   Anaphase
   Cell cycle
   Centromere
   Centrosome
   Centriole
   Chromatin
   Chromasome
   Chromatid
   Cytokinesis
   Daughter Cells

   Metaphase
   Mitosis
   Prophase
   Spindle
   Telophase

You should be able to describe in your own words
the following concepts:

   Structure of a chromosome
   How and when chromosomes replicate
   Positions of chromosomes at stages of mitosis
   Significance of mitosis
   Review Appendix C, pages 437–443

   As a result of this review, you most likely have
questions about terms, concepts, or how you will do
the experiments included in this lab. Write these
questions in the space below or in the margins of the
pages of this lab topic. The lab experiments should
help you answer these questions, or you can ask your
instructor for help during the lab.

Objectives

1. To identify stages of mitosis and cytokinesis on
   prepared slides of plant and animal cells
2. To stain chromosomes in dividing plant tissues
3. To determine the number of chromosomes in
   cultivars of onions
4. To apply chromosome counting techniques and
   descriptive statistics in order to test a hypothesis

Background

An important characteristic of living cells is their ability to
divide, producing two daughter cells, which are genetically
identical. Cell division in eukaryotes involves two
processes; **Karyokinesis** is the division of the nucleus by
either mitosis or meiosis and **cytokinesis** is division of the
cytoplasm. Prior to division, cells undergo a growth process

Figure 8.1　How DNA is organized in a chromosome. A chromosome consists of two sister chromatids joined at centromeres during metaphase.

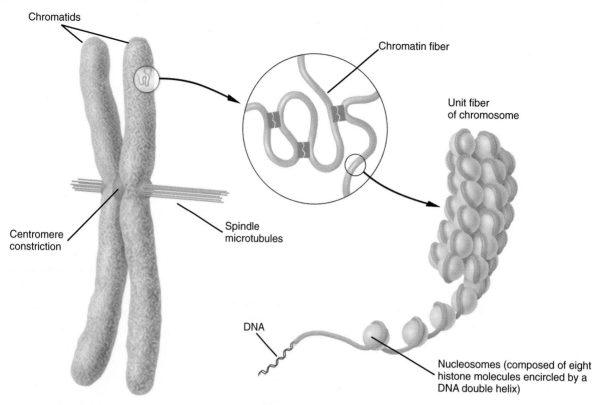

in which molecules, such as fats, proteins, and nucleic acids, are synthesized from food molecules using energy derived from respiration. Molecular synthesis alone, however, is not sufficient to ensure proper growth. Molecules must assemble or be assembled into eukaryotic cellular components, such as plasma membranes, ribosomes, mitochondria, and chromosomes. Before dividing, a cell contains hundreds of mitochondria, thousands of ribosomes, and literally trillions of small molecules, such as amino acids and sugars. Cells in balanced, continuous growth double their components and then divide these components in half, producing two equal daughter cells.

Most eukaryotic cells have only one nucleus and its division involves a special mechanism. The important contents of the nucleus are the **chromosomes,** the carriers of hereditary information. (Different organisms have different numbers of chromosomes in their nuclei: for example, the donkey has 66, humans have 46, and fruit flies, 8.)

During the growth period, these chromosomes make copies of themselves. Each duplicated chromosome consists of two strands of genetic information called **sister chromatids** (fig. 8.1). Each sister chromatid consists of a single long DNA molecule that is coiled, folded, and wrapped around structures called nucleosomes. These are composed of proteins called histones. Each microscopic chromosome in an onion cell contains a highly folded DNA molecule that is about a meter long!

Mitosis is a karyokinesis process by which the nucleus equally divides its contents, including the copies of chromosomes, to form two daughter nuclei. As a cell enters mitosis, its nuclear envelope breaks down, a spindle forms, and the chromosomes line up at the center, or **equator,** of the cell. Spindle fibers radiating from opposite poles of the cell attach to each of the two chromatids. The fibers attach to each chromatid at the **centromere,** a locally constricted region of a chromosome where the chromatids are held together.

As a cell progresses through mitosis, the centromeres split, and the spindle fibers pull the separated sister chromatids to opposite poles of the cell. The cell then divides along the centerline, producing two daughter cells that each have a copy of all the chromosomes that were in the mother cell prior to growth and division (fig. 8.2).

Following karyokinesis, the cytoplasm divides. Fundamentally different cytokinesis mechanisms are found in animal and plant cells. In animal cells, the cytoplasm divides by constricting inward in a process called **furrowing.** In the furrow region, the protein actin, the same one involved in muscle contraction, encircles the cell. This contractile ring gradually pinches the cell in half, forming the daughter cells.

In plant cells, there is no constriction process. Instead, membrane vesicles containing cell wall components and derived from the Golgi apparatus migrate to the center of the cell and form a plate (phragmoplast) across the center

Figure 8.2   Chromosome positions during stages of mitosis in a hypothetical animal cell.

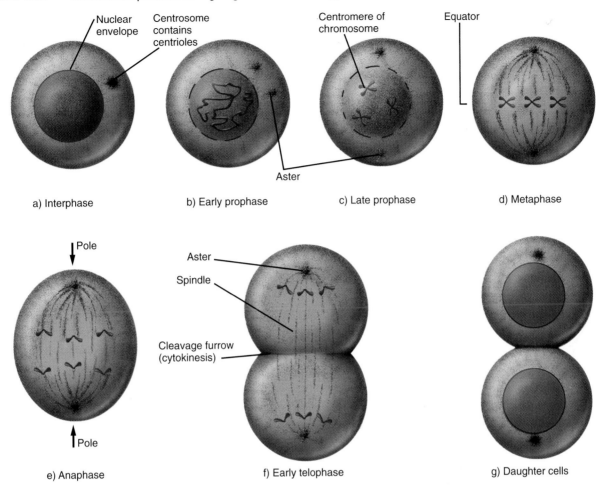

a) Interphase

b) Early prophase

c) Late prophase

d) Metaphase

e) Anaphase

f) Early telophase

g) Daughter cells

of the mother cell. These vesicles fuse with each other and the plasma membrane to form the end membranes of two new daughter cells. Hence, the phrase cytokinesis by **cell plate** formation is used.

The alternating periods of growth (interphase) and division (phases of mitosis and cytokinesis) are called the **cell cycle.** Interphase can be divided into three subphases known as G_1, **S,** and G_2. During the S subphase, the DNA of the chromosome duplicates.

LAB INSTRUCTIONS

You will identify different mitotic stages in animal and plant cells. You will prepare slides for microscopic observation using selective staining techniques that stain only DNA in chromosomes. Using the slide that you make, you will determine the number of chromosomes that are characteristic for the species used in the experiment. It is best to start the staining procedure first and then look at the stages of mitosis in prepared slides during lulls in the staining procedure.

Mitosis in Animal Cells

Mitosis is most easily observed in growing, embryonic tissues, which have many cells dividing at the same time. The whitefish **blastula** is an early embryonic stage in the development of the whitefish from a fertilized egg. At this stage, the embryo is essentially a disc of dividing cells on top of a globe of yolk.

Biological supply houses "fix" the cells of the blastula; that is, they rapidly kill the cells and preserve them chemically. The disc is then embedded in wax or plastic to make it rigid and sectioned into thin slices. The sections are mounted on microscope slides. The tissue is stained with a dye to make the chromosomes visible. A resin is then placed on top of the stained specimen with a coverslip to make a permanent slide.

Obtain a prepared slide of a whitefish blastula and observe it under scanning power with your compound microscope. Note that several sections of the blastula are on the slide. Each contains many cells, some of which contain darkly stained chromosomes.

Center a cell containing chromosomes in the field of view and observe it first under medium power and then

Mitosis in Plant Cells

▶ If time permits, obtain a slide of a longitudinal section of an onion root tip. Cells just behind the tip divide by mitosis as the root elongates. Study this area on your slide and identify cells in interphase, prophase, metaphase, anaphase, and telophase (fig. 8.3). Sketch cells in these stages below.

with the high-power objective. Sketch the cell, indicating such features as the **spindle, chromosomes, centromeres,** and **asters.** Though not visible through the light microscope, centrioles are found at the center of the asters in animal cells. The clear area in the center of the aster is called the **centrosome.** Centrioles do not occur in plant cells.

▶ Now, locate another cell in which the chromosomes are visible. Chances are that the chromosomes are not aligned in the same patterns as in the previous cell. During mitosis, the chromosomes undergo choreographed movements, ensuring that each daughter cell will obtain a full chromosome complement. In a blastula, the cell divisions are not synchronized, so different cells are in different stages of mitosis or may not have been dividing at the time the tissue was fixed and prepared. Furthermore, since you are looking at sectioned material, there may be cells in which neither chromosomes nor nuclei are visible because the section was taken from the edge of a cell and did not include any nuclear material. The absence of nuclear material in these cells is called an **artifact** of slide preparation. An artifact is a phenomenon that results from technique and not from some biological mechanism.

▶ Over the next 15 minutes, look at various cells on the blastula slide and identify the mitotic stages. You should find examples of **interphase, prophase, metaphase, anaphase,** and **telophase.** Use figure 8.2 to help you identify the stages. Sketch each stage and label the structures.

? Did you see asters in any of the onion root tip cells?

Look at a cell in late telophase. Can you see a cell plate forming? The line of vesicles forming across the center of the long axis of the spindle is called the **cell plate.**

Mitosis lasts for about 90 minutes in onion root tip cells. Each of the four phases takes a different amount of **?** time. The phase lasting the longest will be the most commonly observed. Why do you think this is so?

▶ Look at 50 *dividing* onion root tip cells, and tally the frequency of occurrence of the phases in table 8.1.

Because frequency of occurrence is directly proportional to the length of a phase, multiply the percentage of the cells in a phase times the duration of mitosis to obtain the time required for a phase in mitosis. Enter the time values in the last column of table 8.1.

? Which phase of mitosis is longest?

Figure 8.3 Stages of mitosis in an onion root tip as seen in two types of preparations: (a) root tip sectioned and stained so cells remain intact, and (b) root tip stained with Feulgen reagent and then squashed.

  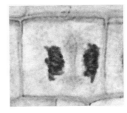 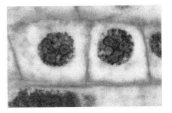

(a)

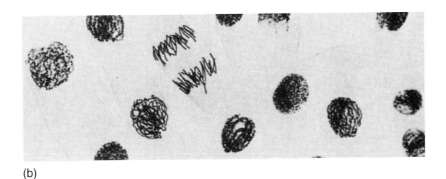

(b)

TABLE 8.1 Determining duration of mitotic phases

Phase	Number Seen	% of Total	Length in Minutes
Prophase	——	——	——
Metaphase	——	——	——
Anaphase	——	——	——
Telophase	——	——	——
Total	——	——	——

Staining Dividing Cells

Actively growing root tips, young leaves, flower buds, or other **meristematic** (dividing) plant tissues have a high percentage of cells in mitosis. These tissues can be prepared for study by the sectioning technique described earlier or by a squashing technique in which the tissue is softened, stained, flattened, and observed (fig. 8.4).

The squashing method has two advantages over sectioning. First, intact cells are easy to identify in squashes, whereas in sectioned material, the plane of the sectioning may be tangential to the nucleus and may thus exclude some chromosomes. Second, with care, cells may be flattened without bursting. This allows easy observation of the chromosomes, although it does distort the original three-dimensional organization of the cell.

The use of onion root tips is recommended, though it would be possible to use any other plant, bulb, or germinating seed that has vigorously growing roots.

Tissue Preparation Steps

1. Pour 3 ml of *freshly* prepared methanol-acetic acid **fixative** into a small vial. Remove the tips of three roots and place them immediately in the fixative. The fixative rapidly kills the cells by denaturing proteins and extracting lipids. The root tips should be fixed for 15 minutes at 60°C.

2. To stain and squash the tissue, it is necessary to **hydrolyze** it, that is, to partially break down the cells and their components. After fixation, work in a ventilated area or fume hood and slowly pour off the fixative into a watch glass, being careful not to lose the root tips. Discard the fixative in a waste container.

   Add a few milliliters of 1 M HCl to the vial and incubate the vial in a water bath for ten minutes at 60°C to soften the plant cell walls and partially hydrolyze the DNA in the chromosomes. *The*

Figure 8.4   Steps for preparing tissue by the Feulgen stain/squash method.

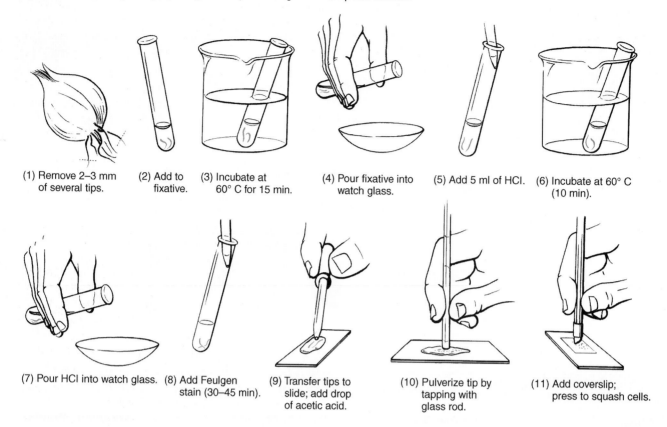

(1) Remove 2–3 mm of several tips.

(2) Add to fixative.

(3) Incubate at 60° C for 15 min.

(4) Pour fixative into watch glass.

(5) Add 5 ml of HCl.

(6) Incubate at 60° C (10 min).

(7) Pour HCl into watch glass.

(8) Add Feulgen stain (30–45 min).

(9) Transfer tips to slide; add drop of acetic acid.

(10) Pulverize tip by tapping with glass rod.

(11) Add coverslip; press to squash cells.

temperature and time in this step are critical. Too short or too long a hydrolysis may result in poor staining in the next step.

> **C A U T I O N**
> HCl is hydrochloric acid which, if splashed, will damage your eyes (wear goggles) or clothing (wear apron). Feulgen stain appears colorless, but if it gets on your hands or clothing, it will stain them vivid pink. Be careful! Household bleach will remove stains.

3. After hydrolysis, pour off the acid into a watch glass, taking care not to lose the tips. Add a milliliter or so of **Feulgen stain** to the vial. Maximum stain intensity will be reached after 45 minutes to one hour at room temperature. The slide may be prepared after 30 minutes.

4. To make a slide, transfer a root tip to a very small drop of 45% acetic acid on a slide. Using a razor blade, cut off 1 to 2 mm of the root tip and discard the remaining older portion. Look at the tip carefully. It should have a prominent purple to pink band of staining which is less than a millimeter wide. This is where the dividing cells are located.

5. Pulverize the root tip into a fine pulp on the slide by tapping on it with a polished glass rod about 100 times. The difference between making a superior and a mediocre slide lies in the pulping. It is important to use only a small amount of liquid so that the pulp is easily made. Now place a clean coverslip over the preparation. The amount of liquid on the slide should be just enough to flow out to the edges of the coverslip (fig. 8.4).

6. Lay the slide on the table and cover the slide with two thicknesses of paper towels. Put your thumb on the towel over the coverslip and press directly downward as hard as you can. Use care to prevent the coverslip from moving sideways—this will roll cells on top of one another and destroy the material. If the tissue is especially tough, extreme pressure may be necessary.

7. Examine the slide using low power to locate pink-stained chromosomes on the slide. You should reduce the light intensity to see the chromosomes. Small quantities of 45% acetic acid may be added to the edge of the coverslip to prevent the preparation from drying. If the cells are not sufficiently spread, they may be squashed further by repeating the pressing procedure. The Feulgen stain will gradually fade (in about two hours).

Study your slide and identify cells in the four mitotic phases and in interphase. Compare what you see to figure 8.3. How many chromosomes are in these cells? How many chromatids make up the chromosomes at different phases? Sketch some of the cells and label the stages.

cells using the same techniques presented in this exercise and then statistically analyze their data. In this portion of the exercise, you will estimate the diploid (2N) number of chromosomes for the species used to make your slide. The following describes what could be a real-life application of this procedure.

Imagine that you work in a biological laboratory that performs chromosome analyses for hospitals and anyone else who has an interest and is willing to pay the going rate.

A major horticultural firm has come to your lab with a problem it would like you to work on. The firm, Seeds Unlimited, has been selling a new strain of onion seed, which has been very successful. However, another horticultural firm, Gardeners Incorporated, claims that Seeds Unlimited stole the variety from them. Furthermore, Gardeners Incorporated wants damages because they spent years and thousands of dollars developing the new variety, which is protected under a horticultural patent. Your job is to prove in a court of law that the two varieties in question are genetically different and thus do not represent a patent infringement.

A literature search of horticultural patents yielded the basic karyotype information about the Gardeners Incorporated variety. It is a registered variety with a diploid number of chromosomes equal to 18. You must now determine the number of chromosomes in the Seeds Unlimited variety.

A null hypothesis (H_o) to be tested by collecting data on number of chromosomes would be: *The number of chromosomes in the Seeds Unlimited variety (the onions you have in lab) is equal to 18.*

If your microscope is equipped with an ocular micrometer and you calibrated it in lab topic 2, use it to measure the length of one of the chromosomes in a squashed onion cell. How long is an onion chromosome in millimeters?

State an alternative hypothesis (H_a).

Testing a Hypothesis

You have now learned to recognize the stages of mitosis and to carry out a classical chromosome-staining technique. You will now apply this knowledge in a simulation of real world situations.

A technique often used by taxonomists to identify closely related species is chromosome counting, also known as **karyotyping.** These scientists will often prepare

As a head technician in the laboratory, you give the following directions to all other technicians:

Procedure

To count chromosomes, look only at mi cells. The number of chromosomes m other pole of the cell represents the di

Determining Chromosome Number in Mitoti

TABLE 8.2 Number of chromosomes observed by you in onion root cells

Cell #		Cell #	
1.	_____	11.	_____
2.	_____	12.	_____
3.	_____	13.	_____
4.	_____	14.	_____
5.	_____	15.	_____
6.	_____	16.	_____
7.	_____	17.	_____
8.	_____	18.	_____
9.	_____	19.	_____
10.	_____	20.	_____

species. Select 10 to 20 cells in which the chromosomes are well spread by squashing, and count the number of chromosomes at one pole. Record the count in table 8.2. Now count those at the other pole and record. Find another cell and repeat until you have 20 counts.

When counting, be sure to use the high-power objective and to focus up and down as you count so as to include chromosomes that may lie above or below the plane of focus for the objective. It may be helpful to count the visible ends of chromosomes and then to divide by two to estimate the chromosome number. If this method is used, always round up to the next even number when an odd number of ends is observed, because each chromosome has two ends and one may have been missed in counting.

A facsimile of table 8.3 will be on the blackboard. Record your observations on the board, using a tic mark for every observation and a diagonal through a group of four for every fifth observation. After everyone has recorded their observations, count the total number of cells seen by the group having eight chromosomes, nine chromosomes, and so on. Record these numbers in column B.

If the number of chromosomes is constant for a species, why were different numbers of chromosomes seen in different cells?

Analysis

You will now statistically analyze the group data in table 8.3, using the following three steps.

1. Plot a frequency **histogram** of the data in the first two columns of table 8.3 on the graph paper at the end of the exercise. Appendix B contains graphing instructions. This histogram will give you a visual representation of the data variability.

2. Calculate the **average** number of chromosomes per cell from the class data. Rather than adding all of the observations in the class data to obtain the average, a shortcut method can be used as described below:

$$\text{Avg. no. chromosomes per cell} = \frac{\Sigma(A \times B)}{\Sigma B}$$

$$\frac{\Sigma(\text{no. of chromosomes} \times \text{no. of cells})}{\Sigma(\text{no. of cells})}$$

(*Note:* Σ (Sigma) is a Greek letter meaning "sum of.")

The necessary information can be easily derived from table 8.3 by (*a*) summing column **B**, (*b*) multiplying the numbers on each row for columns **A** and **B**, (*c*) summing the column labeled **A** $\times$ **B**, and (*d*) dividing the sum of column **A** $\times$ **B** by the sum of column **B**. What is the average of the group data?

3. Calculate the **standard deviation** of the class data. The calculation of standard deviation is discussed at the end of lab topic 5 and in the beginning of appendix C. However, a shortcut method can again be employed for the data in table 8.3. Remember that:

$$\text{Standard deviation} = \pm\sqrt{\frac{\Sigma(\text{observation-mean})^2}{n-1}}$$

Because the data in table 8.3 are already summarized into categories, the above equation can be translated into a series of arithmetic operations by rows and columns in the table.

To calculate the expression Σ (observations-mean)^2, perform the operation indicated in the fourth column (to the right of the double vertical line): for each row, subtract the average from the value in column **A,** entering the square of the difference in the fourth column. Multiply this value by the value in column **B** and enter the result in the fifth column. Now sum all values in the fifth column.

A No. chromosomes	B No. cells	A × B	(A-Average)²	B(A-Average)²
6				
7				
8				
9				
10				
11				
12				
13				
14				
15				
16				
17				
18				
19				
20				
21				
22				
	Σ _____	Σ _____		Σ _____

To calculate the standard deviation, divide the sum of the fifth column by one less than the sum of column **B.** Take the square root of the quotient so obtained. What is the standard deviation for your sample?

Write the values for the average and standard deviation on your histogram.

Based on the class data summarized in the average, standard deviation, and histogram, what is your conclusion regarding the null hypothesis? Is the number of chromosomes in your class sample (obtained from Seeds Unlim-ited) the same or different from 18 (the number of chromosomes in the Gardeners Incorporated variety)? Must you accept or reject your null hypothesis? Why?

What is your conclusion regarding the horticultural patent suit?

Learning Biology by Writing

Continuing with the situation presented under Testing a Hypothesis, imagine that you must write a report summarizing your laboratory's studies of chromosome number in onion varieties. Your report will be entered as court evidence and must be concise, yet informative. Read appendix D for directions about writing lab reports.

The report should have the following components:

Descriptive title

Purpose of study including hypothesis

Methods
  Brief overview with lab manual cited for specifics

Results
  Histogram of lab data
  Average and standard deviation
    Give formula, substituted values, and answer

Discussion
  Accept or reject null hypothesis and cite evidence
  State potential sources of error

Literature cited

Internet Sources

Much genetic research is directed at building databases that contain what traits are located on what chromosomes. More detailed studies are actually trying to determine what is the nucleotide sequence in the single DNA molecule found in each chromosome. Such projects are now underway for a number of organisms, including the onion. To look for information on the chromosomes of onions use the search engine Google at http://www.google.com. Whe connected enter *"Plant Genome IV Abstracts."* These are abstracts of papers given at a conference in California in 1996. Submit the query.

When the listing of abstracts appears on the screen, several of them will be about the genetics of other plants besides onions. Use the Find feature of your browser (a button or under Edit at top of screen). When the dialog box appears, type *"onion."* This will take you to the abstract that has onion in the title. Click on the abstract and read it. Close the window and repeat to find another abstract on onions. There should be at least four abstracts on onions.

In your own words, summarize the type of research that was reported at this conference. You can find more recent research by entering Plant Genome V in Google (or VI or VII etc.).

As an alternative assignment, your instructor may ask you to complete the following lab summary and critical thinking questions.

Lab Summary Questions

1. Describe the positions of the chromosomes in the following stages of mitosis:

   Prophase
   Metaphase
   Anaphase
   Telophase

2. What is the average number of chromosomes observed by your class in onion root tip cells? What is the range of observations? What is the standard deviation? List two good reasons why there is such variability in the observations.

3. Review the null hypothesis that you made regarding the number of chromosomes in the tested variety of onions. Based on your data, must you accept or reject the hypothesis? Does the company, Seeds Unlimited, have a valid claim?

4. Plot the data in table 8.3 as a frequency histogram on the graph paper at the end of this lab topic and turn it in with this summary. Indicate with arrows the average and standard deviation for the distribution.

5. Why are whitefish blastula and onion root tips used to study mitosis?

6. For a species with a diploid number of chromosomes equal to six, draw a cell as it would look at metaphase.

Critical Thinking Questions

1. Can you think of any reasons why organisms rarely, if ever, have 2N chromosome numbers greater than 100?

2. If two organisms have the same 2N number of chromosomes, are they always members of the same species? Justify your answer.

3. Why do biologists say that mitosis produces genetically identical daughter cells? How can this be true?

4. Many protists, fungi, and plants have haploid cells that divide by mitosis. How many chromosomes would you expect to find in the cells resulting from such a division?

LAB TOPIC 9

Observing Meiosis and Determining Cross-Over Frequency

Supplies

Preparator's guide available at
http://www.mhhe.com/dolphin

Equipment

Dissecting microscopes
Compound microscopes

Materials

Preserved female *Ascaris,* dissected (demonstration)
Prepared slides of *Ascaris* ovary-oviduct-uterus
Sordaria fimicola cultures on petri plates
    (Carolina/Biological Supply Biokit) and fresh
    plates for subculture
Slides and coverslips
Toothpicks or probes

Prelab Preparation

Before doing this lab, you should read the introduction
and sections of the lab topic that have been scheduled
by the instructor.

You should use your textbook to review the
definitions of the following terms:

    Ascus
    Crossing-over
    Diploid (2N)
    Haploid (N)
    Homologous chromosomes
    Spore
    Zygote

You should be able to describe in your own words
the following concepts:

    Chromosome positions during phases of meiosis I
    and II
    Genetic significance of crossing over
    How segregation occurs in meiosis I
    How independent assortment occurs in meiosis I
    General life cycle of a fungus
    Review Chi square, Appendix C pages 437–443

As a result of this review, you most likely have
questions about terms, concepts, or how you will do
the experiments included in this lab. Write these

questions in the space below or in the margins of the
pages of this lab topic. The lab experiments should
help you answer these questions, or you can ask your
instructor for help during the lab.

Objectives

1. To identify several stages of meiosis in *Ascaris* eggs
2. To determine experimentally the frequency of
   crossing-over in *Sordaria,* an ascomycete fungus
3. To perform a chi-square statistical test of a
   hypothesis

Background

Meiosis is a form of nuclear division, which produces
daughter cells having half the number of chromosomes
found in the parent cell. In sexually reproducing species if
meiosis did not occur, the number of chromosomes would
double with each fusion of egg and sperm. For example,
humans have 46 chromosomes in their cells. If meiosis did
not occur in humans, a sperm and egg would each contribute 46 chromosomes to a fertilized egg, so that it would
have 92. Since mitosis always produces daughter cells with
the same number of chromosomes as the parent cell, all
cells in the new individual would also have 92 chromosomes. If this individual mated with another of the same
generation, the third generation of fertilized eggs would
have 184 chromosomes. Obviously, if this exponential progression continued, there soon would not be a cell large
enough to contain the increasing number of chromosomes
in each generation.

Meiosis prevents this problem in sexually reproducing
species by reducing the chromosome number by one-half at
some point in the life cycle. A **diploid** nucleus contains
pairs of chromosomes; a **haploid** nucleus has only one

Figure 9.1   Schematic representation of meiosis. Note a double crossing-over during prophase I in the large pair of chromosomes. The positions of chromosomes at metaphase I and II are the key to understanding how reduction in chromosome number (segregation) occurs during meiosis. Chromosomes are color coded to represent their origins: white are maternal and black, paternal. Random alignment of maternal and paternal chromosomes at metaphase I is the basis for independent assortment.

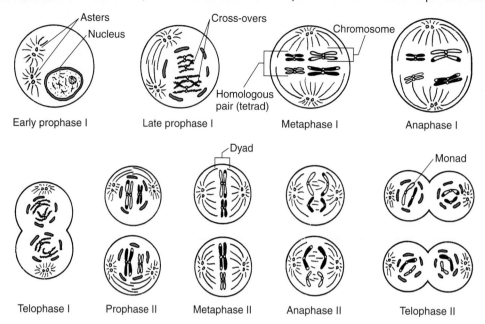

Early prophase I   Late prophase I   Metaphase I   Anaphase I

Telophase I   Prophase II   Metaphase II   Anaphase II   Telophase II

chromosome from each pair, or half as many. In mammals and other higher animals, meiosis always results in either haploid egg or sperm production, but this is not the case in all organisms. In some fungi, for example, meiosis occurs in cells soon after fusion or fertilization, forming haploid spores from diploid fusion cells. The haploid spores are then dispersed by wind and, if they settle on a suitable substrate, germinate and divide by mitosis, producing multicellular mycelia composed of haploid cells. During sexual reproduction in the next generation, cells of one mycelium fuse with those of another to produce the diploid **zygote,** and the cycle repeats.

As a result of fertilization, a zygote contains two similar sets of chromosomes: one inherited from the mother and one from the father. In humans, for example, if both parents are blue eyed, both the egg and the sperm will carry a gene for blue eyes on one of their 23 chromosomes, and the zygote will contain both of these genes on two different chromosomes. However, if one parent has brown eyes and the other blue, the zygote may carry one gene for brown eyes and one for blue. (The child that develops from this egg would have brown eyes, since brown is a physiologically dominant gene.)

These genetic examples illustrate that the chromosomes are in pairs in diploid cells, with one member traceable to each of its parents. Members of such pairs are said to be **homologous;** that is, they can be matched on the basis of the genes they carry. In the example just given, the eye-color trait in general, not the specific color, provides the basis for the homology.

Meiosis reduces the chromosome number by separating the members of each homologous pair. In a human with 46 chromosomes, each egg or sperm produced by meiosis contains one member of each pair, or a total of 23 single chromosomes. However, although a single gamete contains *only one member of each pair,* it usually contains a mixture of the chromosomes derived from an individual's father and mother. There is no division mechanism that separates all maternally derived chromosomes from all paternally derived chromosomes. Therefore, meiosis produces gametes that each contain a mixture of maternal and paternal chromosomes. As you will see in lab topic 10, this leads to variation in future generations.

A stylized representation of meiosis for a hypothetical organism having four chromosomes is shown in figure 9.1. Note that there are two parts to meiosis. In **meiosis I,** the homologous chromosomes line up side by side during metaphase, whereas in **meiosis II,** there are half the number of chromosomes and they line up singly at metaphase. Understanding these differences in alignment is the key to understanding the differences between meiosis I and II.

A cell that begins meiotic cell division has previously undergone a period of growth and **replicated** its chromosomes, so that each chromosome consists of two chromatids joined at the centromere. As this cell enters meiotic **prophase I,** the homologous chromosomes pair off and lie side by side as a **tetrad** of four chromatids, in much the same way that one might hold four skeins of yarn clenched in one hand.

Figure 9.2 Three homologous chromosome pairs showing cross-over points in late prophase I chromosome pairs from a grasshopper (*Chorthippus parallelus*). Count the number of chromatids in each of the three pairs.

LAB INSTRUCTIONS

You will identify cells in several stages of meiosis and determine experimentally the frequency of crossing-over during meiosis.

This pairing process, called **synapsis,** involves forming a connection between adjacent chromatids in tetrads. During this pairing process, the homologous chromosomes reciprocally exchange parts in a process called **crossing-over.** This means that parts of the maternally derived chromosome actually pass and bind to the paternally derived chromosome and vice versa. The net result is that each member of the homologous pair becomes a mosaic of parts derived from both the maternal and paternal chromosomes. Consequently, new gene combinations, or chromosomes, are created. Crossing-over has profound implications as an additional source of evolutionary variation in sexually reproducing species.

As prophase I ends, the recombined chromosomes unravel except at points of attachment called the **chiasmata** (fig. 9.2). Eventually the chromosomes separate, lining up side by side with their homolog at the center of the cell in **metaphase I.** While these events are occurring, the nuclear membrane breaks down and the spindle forms.

At the end of metaphase I, the homologous chromosomes in each pair separate from each other and move to opposite poles of the cell during **anaphase I.** In most animals and many plants, the end of meiosis I is marked by the occurrence of cytokinesis, the division of cytoplasm. Each daughter cell thus produced has half as many chromosomes as the starting cell, but each of these chromosomes is a combination of maternally and paternally derived genes because of crossing-over and each chromosome consists of two chromatids.

In the second part of meiosis, these chromosomes again line up at the center of the cell in **metaphase II,** and then proceed through **anaphase II** and **telophase II.** The result is the separation of the sister chromatids and the production of a total of four cells, each with the haploid number of chromosomes. Each chromosome at this point consists of one chromatid, in which the genes have been recombined as a result of crossing-over.

Meiosis in *Ascaris*

Experimental Organism

Ascaris is a genus of parasitic roundworms (nematodes) found in the intestines of swine. These worms reproduce sexually and have separate sexes, and internal fertilization. *Ascaris* is used to demonstrate meiosis because the diploid number of chromosomes is only four. This means that the chromosomes can be counted during the normal meiotic stages, an ideal characteristic for learning the stages of meiosis. The chromosomes are small, however, and careful observation is required.

Look at the demonstration dissection of a female *Ascaris* and refer to lab topic 19, page 217, for an explanation.

In *Ascaris,* as in most animals, meiosis occurs in the **gonads,** a term that refers to ovaries and testes. Diploid cells in these organs undergo meiosis, producing haploid **gametes,** or egg and sperm. (Because the chromosomes in *Ascaris* are small, crossing-over is not easily seen and will be studied separately.) In *Ascaris,* the production of **oocytes,** through a process called **gametogenesis,** or **oogenesis,** starts in a tubular ovary (fig. 9.3).

As the developing oocytes descend from the ovary to the oviduct, meiosis occurs; different stages can be observed by looking at different regions of the reproductive system. As the developing oocytes enter the lower portions of the female reproductive tract or uterus, meiosis is completed and a sperm nucleus, which had penetrated during fertilization, fuses with the egg nucleus in an event called syngamy. A tough shell forms around the fertilized egg, or **zygote,** as these events occur. Inside the shell, the zygote divides by mitosis, producing the **embryo.**

Procedure

▶ Obtain a prepared slide containing sections of the ovary, oviduct, and uterus of *Ascaris.* Each section can be identified by its diameter. The ovary is the smaller and should be examined first under medium power. You should see a mixture of round, maturing eggs produced by the ovary and triangular sperm introduced when the female mated (fig. 9.4a).

Switch to high power and identify a primary oocyte, a cell destined to undergo meiosis. How many chromosomes should it have?

Figure 9.3 Developmental stages of eggs taken from different parts of female reproductive tract in *Ascaris*. When egg completes meiosis, the sperm and egg nucleus fuse to form zygote. Zygote starts to develop into a worm embryo by mitosis.

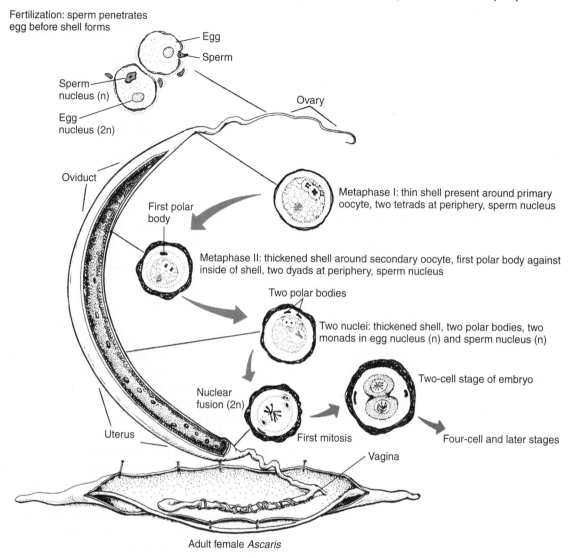

Fertilization: sperm penetrates egg before shell forms

Egg

Sperm

Sperm nucleus (n)

Egg nucleus (2n)

Ovary

Oviduct

First polar body

Metaphase I: thin shell present around primary oocyte, two tetrads at periphery, sperm nucleus

Metaphase II: thickened shell around secondary oocyte, first polar body against inside of shell, two dyads at periphery, sperm nucleus

Two polar bodies

Two nuclei: thickened shell, two polar bodies, two monads in egg nucleus (n) and sperm nucleus (n)

Two-cell stage of embryo

Nuclear fusion (2n)

First mitosis

Four-cell and later stages

Uterus

Vagina

Adult female *Ascaris*

After a sperm fertilizes a primary oocyte, the haploid sperm nucleus rests in the oocyte cytoplasm as the sperm pronucleus, while the oocyte starts to secrete a shell, and its nucleus begins meiosis. The fertilized egg now moves into the oviduct.

While a fertilized egg is in the oviduct, the chromosomes condense, the nuclear envelope disappears, and the two pairs of homologous chromosomes each pair in synapsis, forming two tetrads. What important genetic recombination process occurs in tetrads?

As meiosis continues into metaphase I, the paired, homologous chromosomes (tetrads) move to the center of a spindle that forms off center, or eccentrically (fig. 9.4b). The paired chromosomes now separate during anaphase I, moving to opposite poles. Since the centromeres do not separate, the sister chromatids remain together.

After the homologous chromosomes have separated, unequal division of the cytoplasm (cytokinesis) occurs. This results in one large, functional cell (the secondary oocyte) and one very small, nonfunctional cell called the **first polar body** (fig. 9.4c). How many chromosomes are found in a secondary oocyte? In a first polar body?

Figure 9.4 Photomicrographs of *Ascaris* eggs in stages of meiosis and the first mitotic division: (*a*) sperm penetration with egg nucleus during prophase I, (*b*) metaphase I, (*c*) cytokinesis I, (*d*) metaphase II, (*e*) fusion of egg and sperm nuclei, (*f*) metaphase of mitosis as fertilized egg divides to produce two cells, (*g*) two-cell stage in *Ascaris* development.

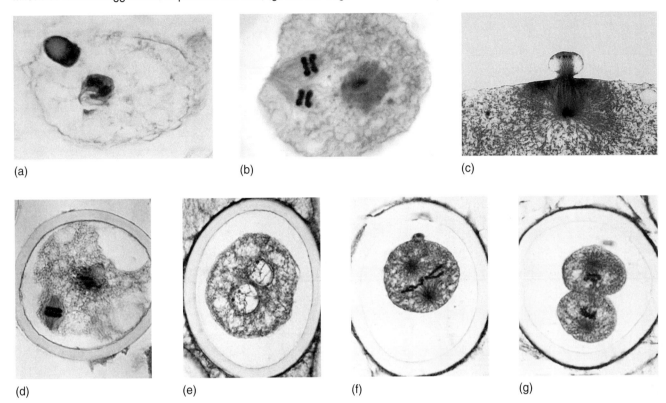

(a)　　　　　　　　　　(b)　　　　　　　　　　(c)

(d)　　　　　　　　(e)　　　　　　　(f)　　　　　　(g)

After an interphase period, the second part of meiosis begins as the secondary oocyte descends into the uterus. The chromosomes, each consisting of two chromatids, line up at the center of the spindle, which again forms eccentrically (fig. 9.4*d*). How many chromosomes are there at this point? _____ Next the centromeres split, and the paired chromatids separate and move to opposite poles as they are pulled by the fibers of a second eccentric spindle. When the chromosomes reach the poles, the cytoplasm divides, resulting in a second polar body and a mature **ovum.**

The sperm nucleus and the newly formed egg nucleus, called pronuclei, in the ovum are both haploid. How many chromosomes should each contain? _____

These nuclei will fuse into a single diploid nucleus (fig. 9.4*e*). How many chromosomes will it have? _____

If present on your slide, these stages will be found in the uterus.

The cell containing the diploid nucleus is large and contains large amounts of stored energy. It is encapsulated in an eggshell and is thus a closed system. As this cell divides to form the embryo, it will draw on energy stored in the cytoplasm as yolk material. The meiotic production of eggs not only reduces the number of chromosomes in the cell but also ensures that a large amount of cytoplasm goes into the ovum for use by the embryo. The sperm essentially contributes only its nuclear material. The polar bodies, which you can see as dark bodies against the inside of the eggshell, are nonfunctional and eventually disintegrate, contributing to the metabolism of the embryo.

Make a series of drawings showing the stages of meiosis. Include eggs in the following stages: sperm penetration, metaphase I, and metaphase II. Label such structures as the eggshell, plasmalemma, spindle, tetrads, sperm nucleus, polar bodies I and II, and chromosomes. Refer to figure 9.4 to help identify the stages.

Figure 9.5

Stages in the life cycle of the fungus *Sordaria*: (a) scanning electron microscope photo of *Sordaria* mycelium; (b) diagram of sexual phases of life cycle.

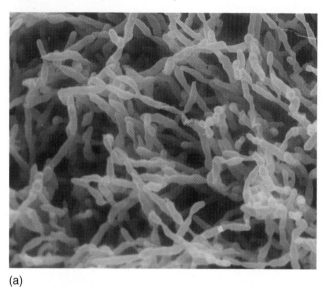

(a)

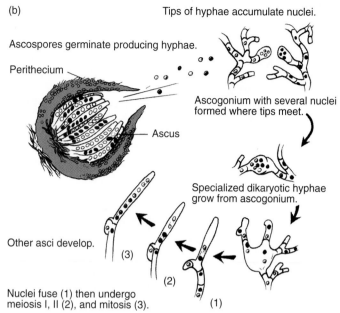

(b)

Tips of hyphae accumulate nuclei.

Ascospores germinate producing hyphae.

Perithecium

Ascus

Ascogonium with several nuclei formed where tips meet.

Specialized dikaryotic hyphae grow from ascogonium.

Other asci develop.

(3)

(2)

(1)

Nuclei fuse (1) then undergo meiosis I, II (2), and mitosis (3).

Measuring the Frequency of Crossing-Over

Background about Experimental Organism

The effects of crossing-over can be demonstrated in *Sordaria fimicola,* an ascomycete fungus. The life cycle of *Sordaria* is shown in figure 9.5. In nature, the fungus grows in the dung piles of herbivorous animals. When mature, a fruiting body called a **perithecium** develops. It contains **ascospores,** small single haploid cells surrounded by a tough outer wall. The ascospores are forcibly ejected and stick to the leaves of grasses or other low-growing plants. Grazing herbivores ingest the spores that pass through their digestive systems and are defecated. The ascospores germinate in the dung to produce thin haploid filaments called **hyphae.** The hyphae grow in length and branch by mitosis, eventually producing a network of filaments called a **mycelium** (fig. 9.5). Cross walls in the hyphae separate individual cells.

Under appropriate conditions, not completely understood, an **ascogonium** having several nuclei is formed by the fusion of two hyphae. Special hyphae, called ascogenous hyphae with two nuclei per cell, are produced by mitosis from this structure. The ascogenous hyphae give rise to **asci,** saclike structures in which ascospores are produced. This process starts with the two nuclei in the cell fusing. The diploid fusion nucleus then undergoes meiosis with crossing-over, forming four haploid nuclei. Each of these nuclei then undergoes mitosis, giving rise to eight ascospores in the ascus.

Several hundred asci will develop in the area of hyphal fusion. As they develop, the asci are surrounded by a growing mass of hyphae, forming the fruiting body or **perithecium** (fig. 9.6) with an opening at its apical end.

Figure 9.6

Perithecium of *Sordaria* has burst, releasing several asci, each containing eight ascospores. The perithecium forms from hyphal tissue that surrounds the dikaryotic cells.

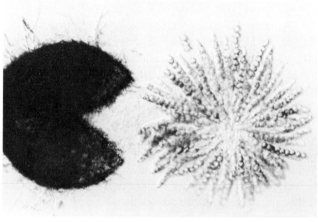

Open perithecium   Many asci each containing 8 ascospores

As individual asci mature, they elongate and forcibly eject their ascospores through the apical opening of the perithecium. Gelatinous materials on the surface of the ascospores attach them to the surfaces of leaves. When these ascospores are ingested by a herbivore, the cycle repeats.

The color of the ascospores is genetically controlled. Normal or wild strains of *Sordaria* produce black ascospores, while mutant forms of *Sordaria* produce tan ascospores. Consequently, in a mixed population of *Sordaria,* three types of fusions (matings) are possible: tan and tan, black and black, and tan and black. A conceptual model of

Figure 9.7 Alternative models of chromosome behavior during *Sordaria*'s sexual reproduction. Models (*a*) and (*b*) will be common. You must find areas where hyphae from tan and black strains have fused to form perithecia before you will observe asci of types (*c*) and (*d*).

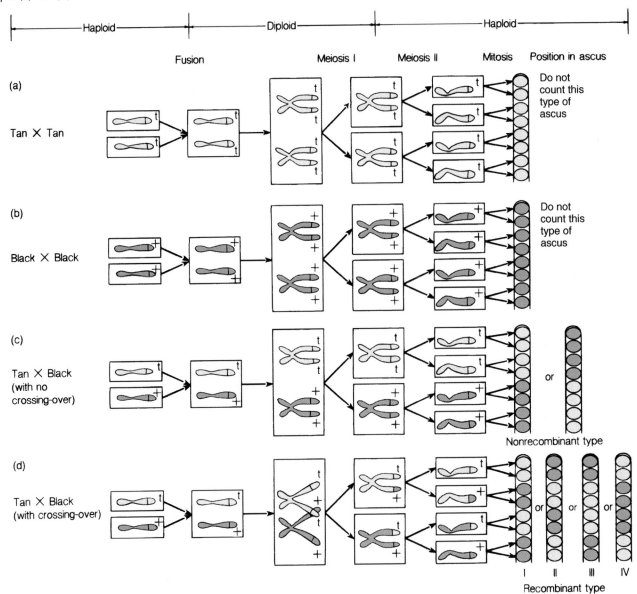

the chromosome behavior in these matings would appear something like figure 9.7.

This figure is based on two experimentally verified assumptions: that the cells in a single ascus result from meiosis followed by a mitosis in each of the meiotic daughter nuclei; and that the eight ascospores in an ascus are arranged in linear order, reflecting the order of the steps in the meiotic and mitotic divisions that occur in the ascus. When the diploid stage is homozygous (both fusion members are genetically the same—for example, black and black), all haploid ascospores are the same. (See fig. 9.7*a* and *b*.) When the diploid stage is heterozygous (for example, black and tan), six possible results can occur. Two of these results are obtained only when

crossing-over fails to occur and are called **nonrecombinants.** The two different chromosomes duplicate, segregate, and undergo an additional mitosis forming the ascospore pattern shown in figure 9.7*c*. However, when crossing-over does occur, it produces the **recombinant** ascospore patterns shown in figure 9.7*d*. Because the ascospores cannot pass by each other in the ascus, they remain aligned in the order that they were produced. Thus, only the process of crossing-over can account for the patterns shown in figure 9.7*d*.

Though all of this may seem complex, it is really a simple biological system. The orderly patterns of colored ascospores in an ascus result from meiotic divisions in which crossing-over did or did not occur.

TABLE 9.1   Results from *Sordaria* ascospore counts

Type	Cross	Ascospore Pattern	Number Counted
(c) Nonrecombinant	Tan × black	4.4	_____
(d) Recombinant	Tan × black	2:2:2:2 or 2:4:2	_____

If you accept the chromosome model shown in figure 9.7 for fusion and ascus development in *Sordaria,* you can determine whether or not crossing-over has taken place by counting the number of asci having different arrangements of tan and black spores.

Procedure

Obtain a petri dish in which a cross was set up between two *Sordaria* strains about two weeks before your laboratory. One strain produced only black spores and the other only tan. Three types of perithecia will be present, corresponding to the three matings previously described.

Use a toothpick or dissecting needle to scrape a number of perithecia from an area where the mycelia have merged, and make a wet mount on a microscope slide. Add a coverslip and, using a pencil or probe, very gently press on the coverslip to squash the perithecia. When a perithecium bursts, it will release many asci, each containing eight ascospores (fig. 9.6). If you press too hard and rupture the asci, you will need to make another slide.

Observe the slide with a microscope on medium or high power. If the perithecia contain only asci with all black or all tan spores, make a new slide using material from a different region of the culture dish. Continue making slides until you find perithecia having asci with both types of spores. These are hybrid asci resulting from black by tan crosses.

The possible spore arrangements in the ascus are shown in figure 9.7.

Count all asci of types (c) or (d) that you can see on your slide. You should count at least 100 asci for statistical accuracy. Record the counts in table 9.1.

Analysis

The relative frequency of recombinant asci equals the number of recombinant hybrid asci observed divided by the total number of hybrid asci observed, or:

$$\% \text{ recombinants} = \frac{\# \text{ recomb.}}{\# \text{ recomb.} + \# \text{ nonrecomb.}} \times 100$$

Calculate the % recombinants found among the asci you observed.

Go to the chalkboard and write the % recombinants you observed. When all students have listed their values, calculate a class average. What is the average relative frequency of recombinants observed by the class? What is the standard deviation of the class data? (Refer to appendix C for method of calculation.)

The relative distance in map units of the spore color gene from the centromere of its chromosome can be calculated from the percentage of recombinants (fig. 9.8). Geneticists apply a logical relationship to calculate map distances—the greater the distance between a gene locus and the centromere, the greater is the frequency of crossing-over. Thus, when the gene locus is at the centromere, no crossing-over will occur.

Why do you think there will be little or no crossing-over when a gene is located very close to the centromere or when two genes are very close together?

Map units are obtained by multiplying the percentage of recombinants by one-half because each recombinant ascus counted represents a tetrad of chromatids in which there has been an exchange involving only two of the four chromatids. So, if you find that 56% of the asci are the recombinant type, the gene for spore color would be said to be 28 units from the centromere.

Calculate the distance of the spore color gene locus from the centromere using the class average for percentage of recombinants.

usual, their occurrence greatly aids genetic analysis, because it is possible to see chromosomal regions that nearly correspond to genes.

In order to view these chromosomes, you must dissect out the salivary glands, stain them, and observe them through your compound microscope (fig. 10.2).

► Locate a third instar larva crawling up the jar wall or the larger ones in the growth medium. Transfer one to a drop of saline on a microscope slide and observe under a dissecting microscope. Push a dissecting needle through the head just behind the black mouthparts and push a second needle through the midbody or hold with forceps. Pull the needles apart, and with a little luck, the mouthparts will separate from the body, trailing the salivary glands. The glands will be two elongated, grapelike, transparent clusters of tissue as shown in figure 10.2. Be careful not to confuse the salivary glands with the glistening fat bodies that stick to them.

► Clean excess tissue off the glands with your needles and move the glands to a clean area of the slide that is devoid of saline. Cover the glands with a drop of acetocarmine or acetoorcein stain and let sit for five minutes. Gently heating the slide over a hot plate or lamp without boiling will speed staining.

► When ready to observe, add a coverslip to the preparation, cover it with a piece of tissue, and press firmly and directly down, to squash the cells and nuclei. Do not let the coverslip move sideways, or a poor slide will result. Observe with your compound microscope using the medium power to find the tissue and the high power to observe chromosomes. If nuclei are not broken, remove the slide from the microscope and press again on the coverslip.

How many chromosomes do you see? _____

Can you see banding in the chromosomes? _____

Sketch part of a chromosome below.

Theoretical Background for Crosses

We will now review some theoretical background before performing experimental genetic crosses with fruit flies.

Genes are located on chromosomes and chromosomes are found as homologous pairs in diploid organisms. The location on a chromosome where a gene is found is said to be the **locus** for that gene. Thousands of gene loci are found on each chromosome. If we could look at a single locus in a population of fruit flies, ignoring all other loci, we might find that some flies have exactly the same gene on both chromosomes in the homologous pair. These flies would be described as being **homozygous** for that gene. We could write this down in the following way. Let "A" equal the symbol for a specific gene. A homozygous individual would have two "A" genes present and this would be written as **AA.** When you write down the genes that are found in an individual, you are said to be writing down the **genotype.** However, as we studied a population of fruit flies, we might find some flies that had a variation of the gene, one that was not exactly the same. This would be called an **allele** of gene "A" and could be written as "a." Hypothetically, some flies could be homozygous for this allele and would have a genotype that was written **aa.** Some, however, could have the "A" form of the gene on one chromosome and the "a" form of the gene on the other chromosome in the homologous pair. The genotype of this individual would be written as **Aa,** and it would be described as being **heterozygous** at this genetic locus.

While the preceding discussion of genes was instructive, it is not possible to took at genes through a microscope and tell what kind are on a chromosome. The only way to determine what genes are present—short of doing DNA analysis, which is not easy—is to do genetic crosses and then to reason what genes must have been present to give the results. When you do genetic crosses, you deal with **phenotypes**—what an organism looks like, not with genotypes—what genes are present. Phenotypes are the result of gene expression (along with environmental influences) and the occurrence of a phenotype is usually a good indicator that a gene is present.

Fruit flies have eight chromosomes: six **autosomes** and two **sex chromosomes.** Females have two X sex chromosomes, so they have four homologous pairs. Males have X and Y sex chromosomes, so they have three homologous pairs plus an X and Y.

Because of these chromosomal differences between the sexes, one would expect to see differences in the inheritance of traits, depending on whether the traits were carried on autosomes or sex chromosomes. These differences are theoretically explained in the next two sections.

Autosome Model

If a gene locus for a trait with two different alleles is on one of the three pairs of autosomes, then a cross between two homozygotes for contrasting traits (alleles *A* and *a*) can be diagramed as follows:

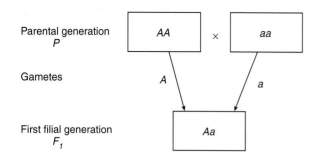

Obviously, one parent must be male and one female, but it makes no difference which parent is homozygous for one allele *(A)* and which is homozygous for the other *(a)*. The result will always be the same: progeny will always have two different alleles (heterozygous) regardless of whether they are male or female.

When Mendel did experiments like this, he had to use the concepts of **dominant** and **recessive** to explain his results. Because the parents are homozygous for different alleles, they look different; *i.e.,* they have different phenotypes. Hypothetically, we could say AA is blue because the "A" allele causes blueness and aa is white because the "a" alleles produce no color development. All progeny are Aa: they have both alleles. What will be their color? The only way to tell is to do an actual genetic cross. When Mendel did similar experiments, he found all progeny were the equivalent of blue. He explained it by saying the allele for blue dominates over white which is recessive.

If the heterozygous offspring from the first filial generation (F_1) are allowed to mate, the results will demonstrate that the recessive allele is present in the heterozygote because the recessive allele segregates from the dominant allele and reappears in the next generation. This can be demonstrated by using a **Punnett square,** a paper and pencil method of keeping track of the kinds of gametes that can be produced and the combination of gametes at fertilization as follows:

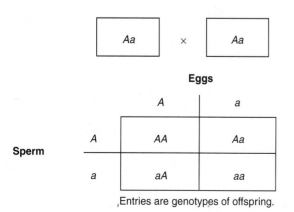

Entries are genotypes of offspring.

The genotypic results from a cross between two heterozygotes would be 1 *AA:* 2 *Aa:* 1 *aa,* or a phenotypic ratio of 3 dominant to 1 recessive if *A* were dominant to *a*. Because the parents have the same genotype, it makes no difference which one is male or which is female.

Sex Chromosome Model

Contrast the theoretical results of the autosomal model to those of a model in which the genes for the traits being studied are at loci on the sex chromosomes. Remember that a female has two X chromosomes but a male has only one plus a Y.

If you start with a gene that has two alleles (+ and *m*), then two types of matings can occur, as diagramed here in Cross A or B. (Note that the alleles are designated as superscripts to X, indicating that they are found at loci on the X chromosome.) In Cross A, the female is homozygous for X^+, but in Cross B, she is homozygous for X^m.

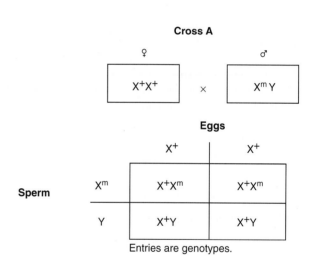

Entries are genotypes.

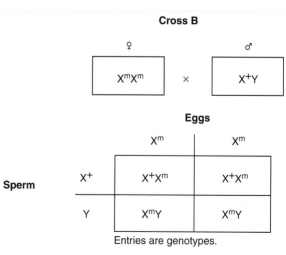

Entries are genotypes.

Summarize the phenotypic ratios, including sex, from the two crosses diagramed, assuming that X^+ is dominant and causes redness and X^m, recessive causing whiteness.

Hypothetical phenotypes from Cross A
Males

Females

Hypothetical phenotypes from Cross B
Males

Females

Obviously in the sex chromosome model, the genotypic and phenotypic ratios will vary according to which parent is homozygous for which trait. This contrasts with the autosomal model in which the sex of the individual is irrelevant to the result. Realize, however, that it is not sex per se that influences the result. It is the inheritance of the sex chromosomes that is important because males have only one X and females have two.

Applying Chromosomal Models to Crosses

In the lab there are several stocks of fruit flies. These flies have been specially bred with attention to four gene loci. These loci determine:

Body color (normal = B^+ and yellow body = b)

Development of eyes (normal = E^+ and eyeless = e)

Wing development (normal = D^+ and dumpy = d)

Normal antenna development
(normal = A^+ and aristapedia = a)

Note: The notation for alleles indicates a dominant allele with a capital letter and a recessive with a lower case. The letter that has been chosen to represent the alleles reflects the trait. The + superscript is to indicate that it is a wild-type allele, *i.e.*, found in natural populations of fruit flies.

Problem to be solved: All of the flies look alike. They have the dominant phenotype. However, some of the flies are homozygous at one or more of the loci and heterozygous for the others. Your job will be to perform a genetic cross to determine the genotypes at each of the loci for the unknown flies that are assigned to you. These will be called the **unknowns.** You will also have to determine whether any of the loci are carried on the X chromosome and are sex linked.

Method to be used: You will do this by performing a **test cross.** A test cross consists of taking an individual that has a dominant, but unknown genotype, and crossing it with one that is known to be homozygous for the recessive alleles at the loci of interest. If the unknown was homozygous for the dominant allele, all of the offspring from the test cross will show the dominant trait. If the unknown is heterozygous, half of the offspring will show the dominant trait and half will show the recessive. The appearance of the recessive phenotype among the offspring is the key observation for determining if the unknown was heterozygous. In the lab, there are male and female flies that have the recessive phenotype. They could be yellow bodied, eyeless, dumpy winged, with abnormal antennae, showing the recessive trait at all four loci, or they could be homozygous recessive at only one locus. These flies will be called the **testers.** You will use these to determine the genotype of the unknown that is assigned to you and your partner.

Before you actually perform the cross, you should engage in a little what-if thinking where you try to anticipate the results based on the autosomal and sex chromosome models described earlier. You will do a single locus by locus analysis. Do this by using the genetic notation given above, writing the genotypes for the parents, and then creating a Punnett square for each hypothetical cross. After you have filled in the Punnett squares, you will perform the actual cross and then compare the results to the what-if scenarios to determine the genotype of the unknown fly that you were given.

Determining Genotypes of Fruit Flies **107**

Body Color Locus

What if the locus for yellow body is on an autosome?

What genotypic and phenotypic ratios would you expect among the offspring from your test cross if the unknown parent is homozygous at this locus?

Note that for this scenario, the required information has been printed to show you what to do. For the other scenarios, you will have to supply all of the information.

What if the unknown is homozygous at this locus?

Hypothetical parental genotypes: Unknown B^+B^+ × Tester bb

Note: *Tester flies will always be homozygous recessive.*

Possible gametes from unknown

	B⁺	B⁺
b	B^+b	B^+b
b	B^+b	B^+b

(left label: Gametes from Tester)

Predicted phenotypic ratios: *All will have normal-colored bodies.*

What if the unknown is heterozygous at this locus?

Hypothetical parental genotypes: Unknown B^+b × Tester bb

	B⁺	b
b		
b		

Predicted phenotypic ratios:

What if the locus for yellow body is on a sex chromosome?

What genotypic and phenotypic ratios would you expect among the offspring from your test cross if the unknown female parent is homozygous at this locus?

Hypothetical parental genotypes: Unknown $X^{B^+}X^{B^+}$ × Tester $X^b y$

	X^{B⁺}	X^{B⁺}
Xᵇ		
Y		

Predicted phenotypic ratios including sex:

But remember that you must also consider the reciprocal cross where the genotypes are reversed for the sexes.

Hypothetical parental genotypes: Unknown $X^{B^+}y$ × Tester X^bX^b

	X^{B⁺}	Y
Xᵇ		
Xᵇ		

Predicted phenotypic ratios including sex:

What if the unknown female is heterozygous at this sex-linked locus?

Hypothetical parental genotypes: Unknown $X^{B^+}X^b$ × Tester X^bY

	X^{B⁺}	Xᵇ
Xᵇ		
Y		

Predicted phenotypic ratios including sex:

Note: Males cannot be heterozygous for a gene on the sex chromosome. Why?

Eye Development Locus

What if the locus for eyelessness is on an autosome?

What genotypic and phenotypic ratios would you expect among the offspring from your test cross if the unknown parent is homozygous at this locus?

Hypothetical parental genotypes: _____ × _____

Note: *Refer to page 107 for allele notation to use.*

Predicted phenotypic ratios:

? **What if the unknown is heterozygous at this locus?**

Hypothetical parental genotypes: _____ × _____

Predicted phenotypic ratios:

? **What if the locus for eyelessness is on a sex chromosome?**

What genotypic and phenotypic ratios would you expect among the offspring from your test cross if the unknown female parent is homozygous at this locus?

Hypothetical parental genotypes: _____ × _____

Predicted phenotypic ratios including sex:

But remember that you must also consider the reciprocal cross where the genotypes are reversed for the sexes.

Hypothetical parental genotypes: _____ × _____

Predicted phenotypic ratios including sex:

? **What if the unknown female is heterozygous at this sex-linked locus?**

Hypothetical parental genotypes: _____ × _____

Predicted phenotypic ratios including sex:

Remember males cannot be heterozygous for traits on X chromosome.

? **Wing Development Locus**

What if the locus for dumpy wings is on an autosome?

What genotypic and phenotypic ratios would you expect among the offspring from your test cross if the unknown parent is homozygous at this locus?

Hypothetical parental genotypes: _____ × _____

Note: *Refer to page 107 for allele notation to use.*

Predicted phenotypic ratios:

? **What if the unknown is heterozygous at this locus?**

Hypothetical parental genotypes: _____ × _____

Predicted phenotypic ratios:

? **What if the locus for dumpy wings is on a sex chromosome?**

What genotypic and phenotypic ratios would you expect among the offspring from your test cross if the unknown female parent is homozygous at this locus?

Hypothetical parental genotypes: _____ × _____

Predicted phenotypic ratios including sex:

But remember that you must also consider the reciprocal cross where the genotypes are reversed for the sexes.

Hypothetical parental genotypes: _____ × _____

Predicted phenotypic ratios including sex:

? **What if the unknown female is heterozygous at this sex-linked locus?**

Hypothetical parental genotypes: _____ × _____

Predicted phenotypic ratios including sex:

Remember males cannot be heterozygous for traits on the X chromosome.

Aristapedia Locus

What if the locus for aristapedia is on an autosome?
What genotypic and phenotypic ratios would you expect among the offspring from your test cross if the unknown parent is homozygous at this locus?

Hypothetical parental genotypes: _____ × _____

Note: *Refer to page 107 for allele notation to use.*

Predicted phenotypic ratios:

What if the unknown is heterozygous at this locus?

Hypothetical parental genotypes: _____ × _____

Predicted phenotypic ratios:

What if the locus for aristapedia is on a sex chromosome?
What genotypic and phenotypic ratios would you expect among the offspring from your test cross if the unknown female parent is homozygous at this locus?

Hypothetical parental genotypes: _____ × _____

Predicted phenotypic ratios including sex:

But remember that you must also consider the reciprocal cross where the genotypes are reversed for the sexes.

Hypothetical parental genotypes: _____ × _____

Predicted phenotypic ratios including sex:

❓ **What if the unknown female is heterozygous at this sex-linked locus?**

Hypothetical parental genotypes: _____ × _____

Predicted phenotypic ratios including sex:

Remember males cannot be heterozygous for traits on X chromosome.

Setting Crosses to Determine Genotypes

▶ Get a culture chamber from the supply area. Following the directions given by your instructor, add flakes of growth medium, distilled water, and yeast to the vial. Let the medium hydrate for about 5 minutes. Label the vial with the code number you are given to identify the unknown and with your name.

Female fruit flies mate once and store sperm in a seminal receptacle. They will lay eggs for several days after mating. Each time eggs are produced some sperm are released from the seminal receptacle and fertilize the eggs. When male and female flies are separated within four hours of emergence from their pupal casing, they are virgins and can be used in controlled mating experiments. Prepared in this manner, all flies you will use are virgins.

Fruit flies can be anesthetized by treatment with carbon dioxide, with a proprietary chemical called Fly Nap, or with ether (*not recommended because of explosion hazard)*. Your instructor will have two or more unknown types of anesthetized flies, separated by sex, available in the lab. Homozygous recessive tester strains will also be available for the four loci you are investigating.

Your instructor will dispense the unknown genotype flies first. Note the sex of the *unknown* you receive and add a *tester* fly of the opposite sex to your vial. Record the code number for the unknown strain below. Your instructor will compare your results later to a master list to determine if you correctly identified the genotype of your unknown.

Code number _____ Gender _____

Once the flies are added, add a sponge stopper, but do not turn the vial upright. Doing so will cause the flies to fall on the surface of the wet growth medium and they will stick there. Keep the vials horizontal until the anesthesia wears off and the flies become active. After the flies have recovered, they may be incubated at room temperature or in a 24°C incubator. New flies will start appearing in the culture vials in about 10 days. You will harvest and count them in two weeks. Some care will be necessary in the interim.

Tending the flies

▶ During next week's lab, you should do the following.

1. Remove the parents, the flies you mated, and put them in a small vial containing a few drops of alcohol to kill them. Label the vial with your name and store it until the next lab.

2. Examine the growth medium. If it seems dry and shows flakes rather than being a homogeneous semi-solid, add a squirt of distilled water from a squeeze bottle. If, on the other hand, it seems wet, take a Chemwipe or piece of tissue and push this into the medium to wick away moisture. Leave the tissue in the vial.

3. Return the vials to the incubator. If pupae are not seen on the side of the vial by the ninth day, raise the temperature to 25°C, but not higher. High temperatures can cause abnormal development.

Killing and counting the offspring

▶ Two weeks after setting the crosses, you will be able to collect your data to test your what-if models. Following instructions from your instructor, over-anesthetize your flies so that you kill them. Pour them out of the vial onto a white index card and examine them, using a dissecting microscope. Use a fine brush to separate them into two piles, males and females. Within each gender, separate the flies by phenotype for body color, eye development, wing development, and antenna development. Count the flies of each type and record in Table 10.1.

Experimental Results

TABLE 10.1	Phenotypes of flies from back cross with unknown: code # _____ gender _____

Phenotype	# of Males	# of Females	Total
Wild-type body			
Yellow body			
Normal eyes			
Eyeless			
Normal wings			
Dumpy wings			
Normal antennae			
Aristopedia antennae			
Totals			

Analysis

▶ Calculate the ratios of the phenotypes among the offspring from your test cross.

▶ Using the notation given at the beginning of the experiment, give the genotype of your unknown at each of the four loci studied.

Code number _____

Genotype at four loci _____

Phenotype for four traits _____

Which of the traits is carried on the X chromosome? _____

Can you think of any other chromosome models that would give you these same experimental results? Explain if necessary.

▶ Compare the ratios from your results to the hypothetical chromosome models for each of the what-if scenarios. Which one of the models best matches your results?

Your instructor may ask you to write your results on a piece of paper and hand them in for grading.

Learning Biology by Writing

Instead of simply turning in your results, you may be asked by your instructor to write a lab report describing how you determined the genotype of the unknown. Do this by first clearly stating the problem you are trying to solve. Discuss the alternative (what-if) scenarios very briefly, giving an example of how you did this for an autosomal and sex-linked trait alternative. Briefly summarize the methods used, citing the lab manual. Your results should be a facsimile of Table 10.1. Your discussion should then explain how you determined the genotype of the unknown.

As an alternative assignment, your instructor may ask you to answer the following Critical Thinking or Lab Summary Questions.

Internet Sources

FlyBase is a large www database that compiles information on the genetics of *Drosophila*. It is located at http://flybase.harvard.edu:7081/

You can use this database to find additional information about the gene loci you studied in this lab. When connected to **FlyBase**, scroll to the query box at the bottom of the page. Enter the name of one of the phenotypes you studied (yellow body, dumpy, aristapedia, or eyeless). Click on *Search Everything*.

Look for information to answer the following questions:
a) On what chromosome is the locus located?
b) How many alleles have been described at this locus?
c) What does the gene do?

Lab Summary Questions

1. Describe the life cycle of *Drosophila* in detail and describe how to distinguish between male and female flies.
2. Describe a backcross and explain why it is a useful tool in genetics studies.
3. Write a paragraph describing the theory behind the experiment that you performed to determine the genotype of the unknown flies you were given.
4. Give the code number, phenotype, and the genotype for the unknown fly assigned to you.
5. The genes for which trait studied in this lab were on the X chromosome? Explain how you arrived at this answer.
6. Define the terms *allele, homozygous, heterozygous, dominant,* and *recessive.*

Critical Thinking Questions

1. In humans, brown eyes are dominant to blue and are autosomal. A child who is brown eyed wonders whether she is homozygous or heterozygous. She has a sister who is blue eyed. What is the "unknown's" genotype?
2. In humans, the ability to metabolize the sugar galactose is genetically determined. The gene locus is on an autosome. The allele for normal metabolism is dominant, and the recessive allele in the homozygous condition cause galactosemia—an uncomfortable but not deadly disease, when recognized. A woman who is galactosemic marries a normal man and they have one child who is normal. Is this convincing proof that the man is homozygous for the normal gene?
3. Hemophilia, a hereditary disease where the blood does not clot well, is determined by a recessive allele that occurs at an X chromosome locus. A man with this condition marries a normal woman who had no history of the disease in her family over 5 generations. He is concerned about having children because he thinks all of their sons will have the disease—though none of their daughters will. Is he correct in his thinking? Explain.

LAB TOPIC 11

Isolating DNA and Working with Plasmids

Supplies

Preparator's guide available on WWW at
http://www.mhhe.com/dolphin

Equipment

Water baths at 42°C and 60°C
Incubator at 37°C
Refrigerator
Spectrophotometer

Materials

Carolina Biological Supply Colony Transformation Kit
(materials marked with asterisk [*] below come in this kit)

*Plasmid-free culture of a competent strain of *E. coli*
    (DH5 α) Gibco 18265017
*Petri plates
*Pipettes and suction device
*Bacteriological loops
*Glass hockey sticks
*Test tubes
Hot plate and 800-ml beaker with boiling water
Beakers, 30-ml
Marker pens
Bunsen burner
Glass rods
Capped plastic tubes, 9.5 cm ×1.5 cm for
    transformation
Test tubes
Cuvettes
Aluminum foil
Autoclave bags for waste
Medium-sized onion

Solutions

10% household bleach
Detergent/enzyme/salt solution—100 ml of
    commercial Woolite in 900 ml of 1.5% NaCl
95% ethanol (stored in freezer)
Ice
DNA dissolved in 4% NaCl
4% sodium chloride
Diphenylamine solution
*Luria broth
*Luria agar, plain and containing 0.1 mg of ampicillin
    per ml

*50 mM $CaCl_2$
*Plasmid DNA (0.01 μg/μl)

Prelab Preparation

Before doing this lab, you should read the introduction and sections of the lab topic that have been scheduled by the instructor.

You should use your textbook to review the definitions of the following terms:

competent cells
genomic DNA
plasmid
spectrophotometer
transformation

You should be able to describe in your own words the following concepts:

Solubility of DNA in polar and nonpolar solvents
How you could "genetically engineer" a bacterium
    to produce a protein normally found only in
    humans
The difference between genonic and plasmid genes
How a spectrophotometer is used to construct a
    standard curve (Lab Topic 5)

As a result of this review, you most likely have questions about terms, concepts, or how you will do the experiments included in this lab. Write these questions in the space below or in the margins of the pages of this lab topic. The lab experiments should help you answer these questions or you can ask your instructor for help during the lab.

Objectives

1. To isolate genomic DNA from an onion
2. To measure the amount of DNA in a sample
3. To conduct a plasmid transformation for ampicillin

resistance in *E. coli* and interpret the results to test the following null hypothesis: There will be no significant difference between plasmid-treated and untreated bacteria in their ability to grow on a growth medium containing ampicillin.

Background

Genetic engineering is one of the newest applications of basic understandings in biology. It involves the isolation of specific genes from one organism and the insertion of the genes into a second organism of the same, or a different species. The development of these techniques has raised much excitement and many questions. Scientists see genetic engineering as a way to explore how genes are regulated, to cure diseases caused by genetic deficiencies, and to add desirable traits to agricultural crops, such as nitrogen fixation to corn, thus eliminating the need for fertilizers.

Private companies have spent millions of dollars to build factories for growing bacteria that have been modified by these techniques to produce human hormones, various antigens/antibodies to diseases, and other useful proteins. Because the courts have ruled that new strains produced by these techniques are patentable, stockbrokers, lawyers, and business executives have had to learn biology as a new industry develops. What discoveries led to this tremendous interest?

For almost 100 years it has been possible to isolate DNA from cells, but it has been only for the last 40 years that its function was understood. The history of molecular biology is discussed well in most textbooks and will not be repeated here. However, the work of one set of investigators is worthy of mention. In classic experiments performed in the 1940s, Oswald Avery, improving on earlier experiments of Fred Griffith, demonstrated that if DNA was isolated from a pneumonia-causing strain of bacteria and added to cultures of nonvirulent bacteria, the recipient cells were transformed into virulent types that caused pneumonia. This **transformation experiment** was the first experiment to show that genes could be artificially transferred in the laboratory.

Since the time of those experiments, our knowledge of gene structure, function, and control has increased astronomically. Modern genetic engineering techniques depend upon four critical factors: (1) techniques that allow scientists to isolate single, whole genes, (2) a suitable host organism in which to insert the foreign gene, (3) a vector (transmission agent) to carry foreign genes into the host, and (4) a means of isolating the host cells that have taken up the foreign gene.

The bacterium *Escherichia coli* is widely used in genetic engineering as a host organism. It has a single large circular DNA molecule, called the **genomic DNA,** as a chromosome containing about two and a half billion base pairs (fig. 11.1). Even though every cell does not take up DNA when exposed to it, *E. coli* has another useful characteristic. It grows rapidly, dividing every 20 minutes, so that a single cell can give rise to a billion descendants in a mat-

Figure 11.1 Electron micrograph of a ruptured *E. coli* cell showing the genomic DNA spewing from cell.

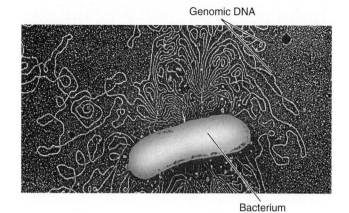

Genomic DNA

Bacterium

Figure 11.2 Plasmid DNA is small in relation to the host cell: see size of genomic DNA in figure 11.1.

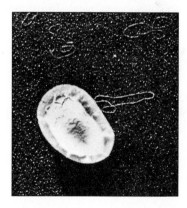

ter of ten hours. Thus, a few transformed cells can provide a huge number within a day.

A **vector** in molecular biology is a substance used as a vehicle to carry genes from one organism to another. It may be a virus, a naked DNA molecule, or a process involving heat or electrical shock. DNA molecules may be modified by gene-splicing techniques so that they carry genes of interest.

A **plasmid** is a small circular DNA molecule containing 1000 to 200,000 base pairs that naturally exists in the cytoplasm of many strains of *E. coli* (fig. 11.2). Independent of the genomic DNA, plasmids replicate as the cell grows and pass to each of the daughter cells during cell division. Plasmids that carry genetic information that is beneficial to the host cell are maintained in a given population. Plasmids frequently carry genes that confer antibiotic resistance, an obvious benefit,

When *E. coli* are exposed to a plasmid, many cells will take up the DNA; *i.e.,* the plasmid somehow crosses the cell membrane. Cells that take up DNA this way are said to be **competent cells.** Competency can be induced by treating cells with divalent cations, such as calcium ions, followed by a heat shock. Although the exact mechanism is not understood, it is often described as making membranes leaky so that large molecules that would not normally pass through the plasmalemma can enter the cell. This treatment does not make all cells competent, but it greatly increases the chances that plasmids will enter the treated cells. Once in a cell, the plasmid replicates and is passed to the daughter cells as they divide.

E. coli carrying a plasmid can be selected from the population by using a **selection medium.** If the plasmid carries a gene for resistance to the antibiotic ampicillin, only those cells that have taken in the plasmid will grow (be selected) when placed in a medium containing ampicillin, and in a matter of a day, billions of ampicillin-resistant cells can be grown.

Genetic engineers use a modification of this technique to insert a foreign gene into *E. coli*. Although the following description is a gross simplification, it outlines the basic strategy used by many groups. The plasmid is modified by inserting a specific foreign gene into the plasmid DNA, perhaps the gene for human insulin. The modified plasmid is then mixed with competent *E. coli* and the *E. coli* are placed in a medium containing ampicillin. Only those cells having the plasmid grow in the medium. They also are the cells that carry the gene (in their new plasmid) for insulin. These cells can then manufacture insulin, which can be isolated from the bacterial cultures and used for insulin therapy of diabetics.

LAB INSTRUCTIONS

You will explore some basic techniques in molecular biology. You will isolate genomic DNA from an onion and estimate the amount of DNA obtained. You will also do a plasmid transformation of *E. coli* so that the cells are changed from ampicillin sensitive to ampicillin resistant.

Isolation of Genomic DNA

For most students, DNA is an abstract substance, one that they have never seen or handled. In this portion of the exercise, you will isolate DNA from an onion and be able to observe it.

The isolation procedure is fairly direct. You will rupture onion cells, releasing their cell contents: proteins, DNA, RNA, lipids, ribosomes, and various small molecules. You will then precipitate the DNA from the suspension by treatment with ethyl alcohol.

Rupturing Onion Cells

The following recipe should make enough DNA for 6 to 8 student groups.

1. Take a medium-sized onion and chop it into ten or so coarse pieces. Add the pieces to a blender along with 100 ml of a detergent/enzyme/salt solution made up by diluting the commercial product Woolite in 1.5% sodium chloride solution. This solution contains

enzymes that degrade proteins, a detergent that emulsifies fats, and salt that creates a polar environment to dissolve the DNA.

2. Briefly blend (20 to 30 seconds) until the onion is completely disintegrated. If the homogenate is too foamy, add another 50 ml of the detergent/enzyme/salt solution.

3. Pour the homogenate into a beaker and incubate in a 60°C water bath for exactly 15 minutes. The time and temperature are critical in this step. Higher temperatures or longer incubations will denature the DNA. What happens when DNA denatures?

4. Cool the mixture in an ice bath for five minutes and then filter the cool homogenate through four layers of cheesecloth to remove pieces of the onion.

5. Each student group should take 6 ml of the filtrate in a test tube.

DNA Precipitation

DNA is not soluble in ethyl alcohol and will precipitate when mixed with this substance. Because DNA is a long linear molecule, the precipitate is stringy.

6. The next step is a bit tricky and requires a steady hand. The technique is outlined in figure 11.3. Nine milliliters of ice-cold ethanol (stock stored in freezer) must be slowly added from a beaker so that it flows down the side of the tilted test tube. If done correctly, the less-dense ethanol will layer on top of the aqueous solution of DNA.

7. Let the tube sit for two to three minutes without disturbing it and you will see a whitish, stringy substance starting to form at the interface of the two liquids. What do you think this is?

Ethanol is a less polar solvent than a 1.5% saline solution. DNA is a polar molecule. As the ethanol diffuses into one saline at the interface, DNA comes out of solution as a gelatinous precipitate.

8. Take a glass stirring rod and extend it into the tube past the interface of the two layers. Do not stir, but roll the stirring rod between your fingers. The DNA fibers will wrap around the rod. The technique is called spooling.

Figure 11.3   Technique for spooling DNA.

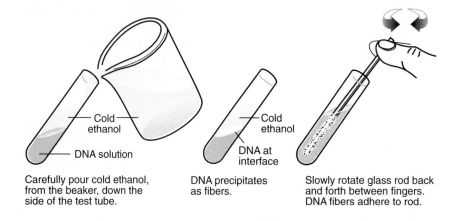

Carefully pour cold ethanol, from the beaker, down the side of the test tube.

DNA precipitates as fibers.

Slowly rotate glass rod back and forth between fingers. DNA fibers adhere to rod.

TABLE 11.1   Mixing table for DNA standard curve (all values in milliliters)

Tube	4% NaCl	Known DNA 500 μg/ml	Unknown DNA from onion	Diphenylamine
1	3.0	0.0	0.0	3.0
2	2.8	0.2	0.0	3.0
3	2.5	0.5	0.0	3.0
4	1.5	1.5	0.0	3.0
5	0.0	3.0	0.0	3.0
6	0.0	0.0	3.0	3.0

---

9. Remove the rod and press it against the side of the tube above the liquids. This will expel excess ethanol from the spooled DNA. The DNA can be redissolved by stirring the rod with the spooled DNA in another tube containing 4% NaCl. This sample can now be tested by the following colorimetric technique that will allow you to quantify the amount of DNA present.

Determining the Amount of DNA in Solution

DNA chemically reacts with diphenylamine to give a blue reaction product in proportion to the amount of DNA present. The amount of blue color can be quantified by reading the absorbance at 600 nm in a spectrophotometer.

To determine the amount of DNA that you have isolated, you will first prepare a standard curve measuring the absorbance of known amounts of DNA. Then you will measure the absorbance of your sample and compare it to the values obtained for the known amounts.

1. Number six test tubes. Add the solutions called for in mixing table 11.1.

2. Cap all six tubes with aluminum foil and place them in a boiling water bath for five to ten minutes. Remove and place the tubes in a beaker of tap water to cool.

TABLE 11.2   Absorbance of DNA solutions at 600 nm

Tube	Amount of DNA (μg/Tube)	Absorbance
1	0	_____
2	_____	_____
3	_____	_____
4	_____	_____
5	_____	_____
6	Unknown	_____

3. Calculate the number of micrograms of DNA added to tubes 2 through 5 (*hint:* quantity = concentration of stock multiplied by the volume used) and record in table 11.2.

4. Review the directions for using a spectrophotometer (fig. 5.5). Set it to 600 nm and use tube 1 to adjust the absorbance to zero, so that any color due to contaminants is blanked. Now read the absorbance of the tubes 2 through 6 and record in table 11.2.

5. Plot absorbance as function of concentration for the known solutions of DNA. Find the absorbance for your unknown on the graph and read the amount of DNA present. This represents the yield from the DNA isolation procedure.

Transformation by Plasmids

In this experiment you will make bacterial cells competent and then expose them to a plasmid that contains a gene conferring resistance to the antibiotic ampicillin. Your assay to tell whether the cells contain the plasmid will be whether they can grow on a nutrient agar containing ampicillin.

Before starting the experimental procedures that follow, you should formulate a testable null hypothesis (H_o) and an alternative (H_a) that relates plasmid treatment to the ability of treated and untreated cells to be grown on nutrient agar containing ampicillin.

H_o

H_a

Note: All procedures are to be performed using sterile technique and sterile solutions and glassware. Disposable pipettes, loops, and tubes should be placed in specially marked bags after they are used so they can be autoclaved before disposal. Why?

Before starting this procedure swab the table where you will work with 10% household bleach. Why?

Making Competent Cells

1. Mark two capped plastic tubes: one + *plasmid* and the other – *plasmid.*

2. Remove the caps one at a time. Use sterile technique and a sterile pipette to add 250 µl of 50 mM $CaCl_2$ to each. After recapping, set the tubes in a beaker containing crushed ice. Discard the pipette.

3. Obtain a petri plate containing Luria agar on which a plasmid-free strain of *E. coli* has been growing for 24 hours at 37°C.

4. Flame a bacteriological loop and let it cool, so that it does not melt the agar when laid on it. If you use sterile disposable plastic loops, do not flame! With the loop, scoop up a visible mass of *E. coli* from the petri plate (fig. 11.4). Be very careful and collect only bacteria and *not any of the agar* on which they are growing. If you collect agar, you will prevent the cells from becoming competent.

5. Transfer the mass to the – *plasmid* tube so the cells are suspended directly in the drop of $CaCl_2$ at the bottom of the tube. Be sure to transfer only cells and not agar.

6. Use a sterile Pasteur pipette to aspirate the suspension until a single-cell suspension is achieved with no cell clumps. If the suspension is not cloudy, add another cell mass and repeat the suspension procedure. Put on the cap and return the – *plasmid* tube to the ice bath.

Repeat procedures 4 through 6 to add cells to the + *plasmid* tube. Discard the pipette and flame the loop.

Transformation Procedure

If the *E. coli* are to take up the plasmid DNA, the temperature and timing in these steps must be followed exactly.

7. Now add plasmid DNA (0.01 µg DNA/µl containing the ampicillin-resistance gene) to the + *plasmid* cell suspension by flaming an inoculating loop, letting it cool, and dipping it in the stock solution of plasmid DNA so that a bubble forms across the loop (about 10 µl).

8. *Carefully* insert the loop into the + *plasmid* tube without touching the sides or you will lose the bubble and the plasmid DNA will not mix with the cells. Swish the loop in the cell suspension. After recapping, tap the side of the tube to mix in the plasmid DNA with the cells (fig. 11.5). Return the tube to the ice bath. Nothing is added to the – *plasmid* tube.

9. After 15 minutes, remove both tubes from the ice bath, and heat shock the cells by immediately suspending both tubes in a 42°C water bath for 90 seconds.

10. After they have been in the water bath for 90 seconds, return the tubes immediately to the ice bath and let them sit for 20 to 30 minutes.

(a) Flame loop until red hot; let cool.

(b) Tilt lid, insert loop, touch to agar surface to cool, then scrape colony off agar.

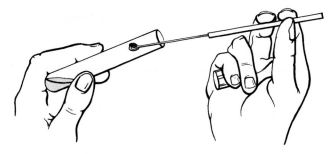

(c) Open tube as shown and swirl loop in medium.

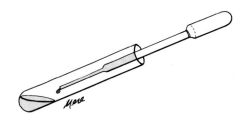

(b) With sterile pipette, aspirate fluid to disperse cells; recap tube.

Selective Growth

11. Use a sterile pipette to add 250 µl of Luria broth to both tubes and cap the tubes. Mix well by tapping the bottom of the tubes (fig. 11.5). Discard the pipette and put the tubes in a beaker.

12. Obtain two petri plates containing *only* Luria agar and label them **norm** for normal growth medium. Obtain two plates containing Luria agar *plus* ampicillin and label them **amp** indicating they contain the antibiotic.

13. Follow the directions below to transfer cells from the – *plasmid* tube to one of each type of plate. Repeat the procedure for cells from the + plasmid tube.

    a. Take one of each type of plate (norm and amp) and label them *No Plasmid* to designate the type of cell suspension to be added. Label the remaining two plates *Plasmid-Treated*.

    b. Use a sterile pipette to add 100 µl of the – *Plasmid* cell suspension to a plate labeled *norm*. Repeat the procedure to add cells to a plate labeled *amp*. Use a sterile glass "hockey stick" to spread the cell suspension uniformly over the surface of the agar (fig. 11.6). When not using the hockey stick, place it in a beaker of 70% ethanol. When using it again, remove it and pass it through a flame to burn the ethanol off. After it has cooled, use it to spread the bacteria. A cool "hockey stick" will not melt agar

Figure 11.5 Technique for mixing solutions added to closed tubes.

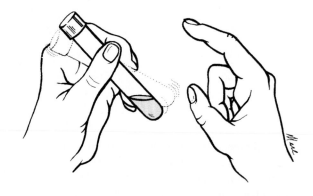

when touched to its surface. A hot one will kill the bacteria.

    c. Use a fresh sterile pipette to repeat step b but now use the +*Plasmid* cell suspension. Discard the pipette. Spread with a glass "hockey stick."

    d. Initial the four petri plates.

    e. Incubate the plates upside down in a 37°C incubator for 12 to 24 hours. Do not overincubate. When the *Amp/Plasmid-Treated* plate has visible individual colonies, analyze all plates or put them in a refrigerator for up to one week before the analysis is performed.

Figure 11.6 Technique for spreading bacteria from plasmid incubation on surface of agar. After placing drop near center of plate (*a*), spin plate while holding hockey stick on surface to spread bacteria evenly on the surface (*b*).

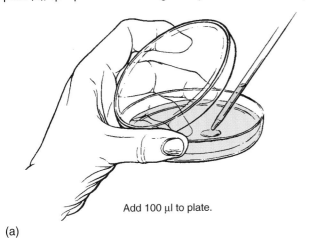

Add 100 μl to plate.

(a)

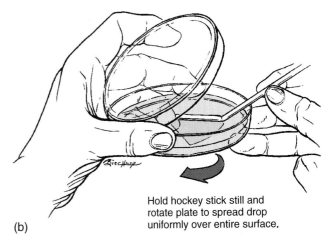

Hold hockey stick still and rotate plate to spread drop uniformly over entire surface.

(b)

Discard all disposable materials according to the directions from your lab instructor. Wash the bench area where you worked with 10% household bleach.

Analysis

After 1 to 2 days when visible colony growth occurs, open the plates and look at the bacterial growth on the surface of the agar. Each colony is the product of one cell. Three results are possible: *no* growth, *selected* colony growth, and bacterial "*lawn*" growth. Colonies are produced when a single cell grows and divides to produce a ring of whitish cells. Lawn growth occurs when there are so many bacteria growing and dividing that they grow into their neighbors, making a continuous "lawn" on the surface of the agar.

Look at your plates and score them according to this classification. On the plate showing selected colony growth, observe the colonies through the bottom of the culture plate and count the number of colonies present, using a marker to put a dot over each colony as it is counted. This technique keeps you from counting the same colony twice. Record the result in table 11.3.

Sit down with your lab partners and systematically discuss the purpose of each plate and cell suspension combination in this experiment.

Which plate demonstrates that the starting *E. coli* are able to grow on the Luria medium? _____

_____

Which plate shows that the transformation procedure did not kill the cells? _____

_____

Which plate shows that the plasmid DNA was not poisonous? _____

_____

Which plate shows that the plasmid DNA was taken up by the cells? _____

_____

TABLE 11.3 in selection experiment	Number of colonies counted	
	Norm Medium	**Amp Medium**
No plasmid	_____	_____
Plasmid-treated	_____	_____

Which plate shows that the antibiotic was active? ____

_____

Which pair of plates shows that transformation is a rare event? _____

_____

Review the H_o and H_a that you made at the beginning of this experiment. Come to a conclusion on whether to accept or reject the null hypothesis. What is your decision? Why?

Genetic engineers often speak of transformation efficiency and express it as the number of transformed colonies per μg of plasmid DNA. The concentration of plasmid DNA solution used in this transformation should be 0.01 μg/μl. To calculate the amount of plasmid DNA used, multiply the number of μl used by the concentration in μg per μl. This is the amount of plasmid DNA in the transformation tube.

To calculate the number of transformed cells in the transformation tube, you must take into account the fact that you used only part of the solution. You pipetted 100 µl onto each plate from a total volume of 500 µl in the tube. Therefore, the tube contained five times more transformed cells than were observed on the plate. What was the transformation efficiency for this experiment?

After you have finished analyzing your plates, flood the surface of the agar with 10% household bleach solution and place the plates in a bucket for disposal. Why?

Learning Biology by Writing

If you did the transformation experiment with the plasmid, you have a well-controlled experiment for a lab report. State the hypothesis that the experiment tests. Indicate the technique used. Give the data obtained and draw a conclusion from the experiment.

Instead of a report, your instructor may ask you to answer the Lab Summary and Critical Thinking Questions.

Lab Summary Questions

1. Explain the difference between genomic and plasmid DNA in bacteria.
2. Why does DNA precipitate when ethanol is added? Why is the precipitate "stringy"?
3. Why were the cells placed in $CaCl_2$ and heat shocked during the transformation experiment?
4. Discuss the experimental design of the transformation experiment, describing the purpose of each petri plate that was inoculated.
5. What evidence do you have that the bacteria were transformed in this experiment?

Critical Thinking Questions

1. Three critical factors required for genetic engineering techniques are described in this exercise. How was each of these factors met in your transformation experiment?

2. Design a procedure that theoretically would allow you to take the gene for insulin and insert it into the cells of a person with diabetes. What would be the major problems encountered in doing such a procedure?
3. What are some possible sources of error in your transformation experiment?
4. What do the "control" plates in the transformation experiment control for?

Internet Sources

The transformation of competent *E. coli* cells is something that is done routinely in research laboratories. For this reason, many researchers are concerned with optimizing this procedure. Use a browser to search the Internet for information on the optimization of transformation.

In your Internet browser, choose a search engine such as **Alta Vista.** Type in efficiency of *E. coli* transformation. You will get a long list back.

Read a few of the articles. What are the authors suggesting be done to improve the efficiency of transformation?

LAB TOPIC 12

Testing Assumptions in Microevolution and Inducing Mutations

Supplies

Preparator's guide available on WWW at
  http://www.mhhe.com/dolphin

Equipment

Incubator at 25°C
Ultraviolet light box (see fig. 12.1)
Desktop computers with Hardy-Weinberg simulation
    program (EVOLVE in BioQuest Collection CD from
    Academic Press recommended)

Materials

White 3" × 5" index cards cut in half with letter **A** on
    half and **a** on half, one for each member of class.
    Six additional cards will be needed, 3 with letter **a**
    and three with α.
Sterile
    Petri dishes with King's agar
    Automatic pipetters and tips
    0.85% saline, sterile packaged 9.9 ml to a
    capped tube
    Glass rod bent into "hockey stick"
Cultures of *Serratia marcescens*
Wax pencils
Alcohol in covered beakers
Alcohol burners
Household bleach
Ultraviolet-shielding safety glasses
Rubber gloves

Prelab Preparation

Before doing this lab, you should read the introduction
and sections of the lab topic that have been scheduled
by the instructor.
  You should use your textbook to review the
definitions of the following terms:

  allele
  dominant
  genotype
  heterozygote
  homozygote
  phenotype
  population
  recessive

You should be able to describe in your own words
the following concepts:

  Allele frequency
  Gene pool
  Genetic drift
  Hardy-Weinberg equilibrium
  Mutation
  Random rating

  As a result of this review, you most likely have
questions about terms, concepts, or how you will do
the experiments included in this lab. Write these
questions in the space below or in the margins of the
pages of this lab topic. The lab experiments should
help you answer these questions, or you can ask your
instructor for help during the lab.

Objectives

1. To simulate the conditions of the Hardy-Weinberg
   equilibrium.
2. To test hypotheses regarding the effects of selection
   and genetic drift on gene frequencies in populations
3. To test a hypothesis that mutations can be induced
   by ultraviolet light
4. To interpret results from bacterial mutagenesis
   experiments

Background

Populations, not individuals, evolve by gradual changes in
the frequency of alleles over time. These changes result
from mutation, selection, migration, or genetic drift. Collectively, these processes comprise **microevolution.** The
mechanisms of microevolution are well understood. In
fact, one of these mechanisms, selection, has been used
for centuries to increase the productivity of crops and
livestock.
  Models of microevolution have been developed since
1858 when Charles Darwin published his version of the theory

of evolution. Darwin's work was monumental in that it established the role of natural selection in evolution. However, Darwin worked 50 years before Mendel's ideas about genetic mechanisms became widely known; thus, Darwin was ignorant of the basic concepts you know about genes, inheritance, and DNA. Researchers in the last 75 years have established the hereditary mechanisms that support Darwin's theory.

According to the microevolution model, a population of organisms can be considered to be a **gene pool,** which is composed of all the copies of every allele in the population at a given moment. In diploid organisms, the genes in the gene pool occur as pairs in each individual, and individuals may be homozygous or heterozygous for a particular trait. Population genetics, the basis of microevolution, deals with the frequency of alleles and genotypes in a population and attempts to quantify the influences of mutation, selection, and other factors causing evolution.

In the early 1900s, when biologists first started to think about the genetics of populations, a common misconception was that a dominant allele would drive a recessive allele out of a population after several generations. People holding this view observed that the recently rediscovered Mendelian genetics indicated that every time two heterozygotes mated, 75% of the offspring expressed the dominant trait. Furthermore, every time a homozygous dominant individual mated with a homozygous recessive individual, all offspring had the dominant trait. They reasoned that the dominant allele over several generations would become the only allele in the population.

In 1908, two mathematicians, G.H. Hardy and G. Weinberg, independently considered this concept and proved it was wrong. They showed that no matter how many generations elapsed, sexual reproduction in itself does not change the frequency of alleles in a gene pool. Changes, if they occur, must be due to the action of other factors—the real agents of evolution.

To understand the insights of Hardy and Weinberg, imagine a hypothetical population of plants that have red flowers and white flowers. Red flowers result from a dominant allele, and white flowers are found only in individuals homozygous for the recessive allele. A field study of 10,000 plants of this species indicated that 84% of the plants had red flowers and 16% had white. Laboratory analysis of the red-flowered plants indicated 36% of the total plants were homozygous for the red allele and 48% of the total plants were heterozygous. Thus, the genotypic frequencies in the population were:

36% *AA*	where *A* = red allele, dominant
48% *Aa*	*a* = white allele, recessive
<u>16% *aa*</u>	
100%	

The frequencies of alleles in the population's gene pool can be obtained by a series of simple multiplications from this information:

Every individual with the genotype *AA* contributes two *A* alleles to the gene pool and every individual with the genotype *Aa* contributes one *A* allele. Given 10,000 individuals, the frequency of the *A* allele in the gene pool equals:

$$2 \times 36\% \times 10,000 + 48\% \times 10,000$$

$$= 12,000 \ A \text{ alleles}$$

In the same population, the frequency of *a* in the gene pool can be calculated by similar reasoning (*aa* individuals contribute two *a*; and *Aa*, one *a*). Therefore,

$$2 \times 16\% \times 10,000 + 48\% \times 10,000$$

$$= 8000 \ a \text{ alleles}$$

On a percentage basis, $60\% \left(\dfrac{12,000}{12,000 + 8000} \right)$ of the alleles

in the gene pool are *A* and $40\% \left(\dfrac{8000}{12,000 + 8000} \right)$ are *a*.

This information describes the gene pool at the time of the study. What happens to the allele frequencies when this population reproduces sexually?

The answer to this question can be determined by creating a theoretical model as did both Hardy and Weinberg. If no selection, mutation, or statistical fluctuations occur and every pollen grain has an equal opportunity to fertilize every egg, we can predict what will be the frequencies of all genotypes in the next generation of our hypothetical population.

To do this, we create a Punnett square for the population to show all the possible combinations of gametes in producing the next generation:

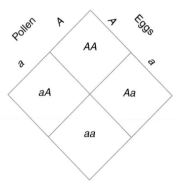

To this calculation, we can add the concept of probability. If 60% of the gene pool is the *A* allele, then 60% of the eggs and sperm will carry the *A* allele and 40% will carry the *a* allele. The *multiplication law of probability* can be used to predict the frequency of each type of fertilization and thus the genotypes of the next generation. This law states that *the relative frequency of two independent events occurring together is equal to the arithmetic product of their individual relative frequencies.*

When this information is added to the Punnett square we get:

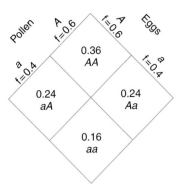

Summarizing the table, we are predicting that the new generation will have 36% AA, 48% Aa, and 16% aa, or 84% red flowers and 16% white. Compared to the parent generation, the genotype and allele frequencies are the same. In fact, the frequencies of alleles and genotypes would remain the same if we repeated these calculations over a hundred generations. The dominant allele does not drive the recessive allele out of the gene pool. The two reside in the gene pool in equilibrium, called **Hardy-Weinberg equilibrium**—as long as no selection, mutation, or other agents of evolution are acting on the population.

The hypothetical situation just described can be generalized. If we say that the frequency of the dominant allele is **p** and that of the recessive allele is **q**, in any population where there are only two alleles at one gene locus:

$$p + q = 1 \text{ and } p = 1 - q \text{ or } q = 1 - p$$

Because of these relationships, we need only measure the frequency of one allele and we can calculate the frequency of the other.

Once the frequency of alleles in a gene pool is known, the frequency of genotypes can be predicted by:

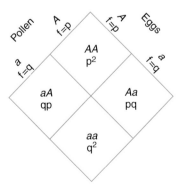

This model indicates that for a population in Hardy-Weinberg equilibrium, the homozygous dominant will be found **p**^{2}% of the time, the heterozygote **2pq**% of the time,

and the homozygous recessive **q**^{2}% of the time. This mathematical representation is called the Hardy-Weinberg law.

This mathematical model is very useful in studies of microevolution. It is used as a null hypothesis—a baseline against which to measure populations. If a population is mating randomly and no other factors change the frequencies of alleles, the frequencies of genotypes in the population should be those predicted by the Hardy-Weinberg equation. If the frequencies are quite different, then this is taken as strong evidence that an agent of change is influencing the population. It is then up to the biologist to seek what is causing the change. This is done by asking such questions as: Is natural selection occurring against a genotype? Is mutation occurring? Is the population small? Is mating not random? Are individuals migrating into or out of the population? By understanding how these factors change gene frequencies in populations, we can understand how evolution occurs on a small scale. Brief discussions of the effects of natural selection, population size, and mutation follow:

Natural selection, as a cause of evolution, acts on the total phenotype of an organism, not directly on its genes or genotype. In essence, it is a test by the environment of the organism's hereditary phenotype. In sexual reproduction, no two offspring are genetically alike, because of biparental inheritance, crossing over, and mutations. This can lead to subtle differences in how well some offspring function. Some function better than others, and produce more offspring. Because offspring tend to resemble their parents more than other members of a population, the next generation in this sequence will be better adapted to its environment. This selection process repeated over generations increases the population's fitness.

Not all changes in allele frequencies are directed by natural selection. Some are random changes, especially in small populations. Biologists call this concept **genetic drift.** This can be explained using statistical sampling theory. Let's consider a classical example, coin tossing. The probability of getting a heads in a coin toss is one half: likewise for tails. Does this mean that when tossing a coin twice, you will get only one heads followed by one tails? It does not. There are equal chances that you will get a tails first followed by a heads, or two heads, or two tails. However, if you were to toss the coin 100 times, there is a better chance that you will get very close to having heads one half of the time and tails one half of the time. The probability of getting the theoretical result increases even more if you were to toss the coin 1,000 times. What statistical sampling theory tells us is that the bigger the sample size, the more likely we are to get the predicted result. When Hardy and Weinberg said that populations must be large in order for their theorem to apply, they were addressing the sample size concern. However, all populations are not large. Many endangered species consist of only a few breeding individuals. Consequently, we would expect to see random changes in gene frequencies in these populations due to statistical sampling error.

Mutation is a never ending process that occurs when random mistakes happen in gene replication before cell division. When mutations occur in cells that will become eggs or sperm, the gametes carry the mutation. If an egg or sperm carrying a mutation participate in a fertilization, then the individual thus created has the mutation as part of its genotype even though the parents did not. Most mutations are harmful but, because they are also recessive, most offspring do not express the mutation and only act as carriers. In later generations if two carriers mate, then the mutation is expressed as a phenotype and natural selection will act on it.

Other factors causing evolutionary change are migrations into or out of the population, and nonrandom mating. Migration can add alleles to a gene pool or it can remove them, depending on the genotypes of those that enter or leave. Nonrandom mating occurs when a species practices mate selection or consists of subpopulations that breed only among themselves.

LAB INSTRUCTIONS

In this lab, you will investigate several aspects of the microevolution model. You will start by playing a game that simulates the conditions of the Hardy-Weinberg equilibrium, demonstrating the effects of random mating, selection, migration, and population size. The long-term effects of natural selection and population size will be then investigated using a computer simulation. The lab ends with an experiment investigating how mutations can be induced in bacteria.

Mating Game

I would like to acknowledge Dave Robinson, a lab instructor at Iowa State University, who suggested this activity.

In this group activity involving all of the students in the lab, you will model the conditions of the Hardy-Weinberg equilibrium. Your instructor has prepared two small cards for each member of the class. Half of the cards have the capital letter **A** on them and half have the lower case **a.** These are to represent the gene pool of a population where the allele frequencies are known. The cards should be distributed two to a student with one-fourth of the students receiving **AA;** one-half **Aa;** and one-fourth **aa.** The two cards represent the genotype of individuals of a hypothetical species, and the population is in Hardy-Weinberg equilibrium. Everyone should note their starting genotype and record it here: _____

We will now set up some conditions for the hypothetical organisms that the students represent. The species, like many insect species, is an annual one that mates once a year and then dies with the new generation reproducing in the following growing season. Mating is random, with every gamete having a chance to combine with any other gamete, as in many sedentary species with external fertilization. The population is at the carrying capacity for its environment, so each season the number of offspring is the same as the number in the parental generation.

▶ Random mating

Now for the fun part. It is time to mate! To simulate random mating everyone should throw their cards into a box, and after all cards have been added, everyone should draw two out to reconstitute a new generation. This is similar to corals on a reef spawning where clouds of eggs and sperm are released, randomly combining to give the next generation. Count the number of individuals in the new generation of each genotype and record below.

	Number	**Ratio**
AA		
Aa		
aa		
Totals		

Has the ratio of genotypes changed from the starting parental generation? Is this population in Hardy-Weinberg equilibrium? Explain.

Count the number of *A* and *a* alleles in the gene pool of the new generation and determine their relative frequencies; *i.e.,* what percent of the gene pool is **A** and what percent is **a?** Have they changed from the first generation? Explain.

What do you think would happen to allele and genotype frequencies if you repeated the mating game? Why? If you cannot imagine what would happen, repeat the game and experimentally determine what would happen.

 What do you think would happen if this scenario repeated over and over again through several generations? Can you see a trend? What is it?

▶ Natural selection

Everyone should exchange cards in order to return to their first genotype. Before mating again, we are going to have a selection event happen. Those with the genotype **aa** cannot compete in life as well as those with a dominant allele in their genotype. Imagine that **aa** can barely get enough food to survive and that they are susceptible to disease. Consequently, when mating time comes around, half of the homozygous recessives cannot mate; they are sick. Decide who among the **aa** genotypes will be sick and have them go to the side. They have died without leaving any offspring. All others get to play the mating game and can throw their cards in the box. When all have been added, those who added cards can draw them back out to form the next generation.

Count the number of individuals in the new generation of each genotype and record below.

▶ Migration and Bottlenecking

Everyone should swap cards to get back to their original genotype. About one-third of the class, based on name or major but not genotype, should separate themselves from the others. They have just migrated to a new locality. Perhaps a hurricane blew them from their homeland to an island. Each population should randomly mate within itself, using the box technique as before.

For the *population remaining* on its homeland, count the number of individuals in the new generation of each genotype and record below.

	Number	Ratio
AA		
Aa		
aa		
Totals		

	Number	Ratio
AA		
Aa		
aa		
Totals		

 Has the ratio of genotypes changed from the starting parental generation? Why?

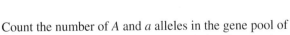 Has the ratio of genotypes changed from the starting parental generation? Why?

Count the number of *A* and *a* alleles in the gene pool of the new generation and determine their relative frequencies. Have they changed from the first generation? Why?

Count the number of *A* and *a* alleles in the gene pool of the new generation and determine their relative frequencies. Have they changed from the first generation? Why?

For the *separated population,* count the number of individuals in the new generation of each genotype and record below.

	Number	Ratio
AA		
Aa		
aa		
Totals		

 Has the ratio of genotypes changed from the starting parental generation? Why?

Count the number of *A* and *a* alleles in the gene pool of the new generation and determine their relative frequencies. Have they changed from the first generation? Why?

▶ *Mutation*

Everyone should return to their original genotype. Simulate a mutation occurring in which the **A** allele mutates to **a** in three cases and to α (alpha) in three cases. When everyone puts their cards in the mating box the instructor will reach in and replace six of the **A** cards with three **a** cards and three α cards, simulating mutations occurring in **A.** Now everyone should draw two cards to simulate the next generation.

Count the number of individuals in the new generation of each genotype and record below.

	Number	Ratio
AA		
Aa		
aa		
Aα		
aα		
αα		
Totals		

 Has the ratio of genotypes changed from the starting parental generation? Why?

Count the number of *A, a* an α alleles in the gene pool of the new generation and determine their relative frequencies. Have they changed from the first generation? Why?

Computer Simulation of Microevolution

While the mating game was instructive for short-term changes from one generation to the next, it did not address the effects of changes over many generations, say 50 or more. Several excellent computer programs have been written to do just this. In this section, you will use a computer program to test hypotheses related to how allele frequencies will change over time as a result of selection against different phenotypes and as a result of varying population size; *i.e.,* genetic drift. Your instructor will describe the program and the computers that you will use. In this section, the experiments that you should conduct will be described.

▶ Before starting, check the computer program that you use to be sure it has the following starting values/conditions set. The simulated population should be:

Large, about 8,000;

No migration should be occurring into or out of the population;

No mutation should be occurring;

Mating should be random;

And initially no selection should occur.

In addition, the program should be set up to:

Have graphic output;

Have the simulation run 50 or more generations;

Plot allele frequencies.

Effect of No Natural Selection in a Large Population

The program is now set to simulate a Hardy-Weinberg equilibrium over 50 generations. Run the simulation and note what happens to gene frequencies during the elapsed time.

Did the allele frequencies change? Describe the change and suggest what were the causes.

State this prediction as null (H_o) and alternative (H_a) hypotheses.

H_o =

H_a =

Run the simulation and study the graphic output. Must you accept or reject your null hypothesis? _____

Your instructor will tell you how to switch the graphic output of your simulation program from plotting the frequency of alleles to plotting the frequency of genotypes under the same selection conditions as before. Run the program. Describe what happens to genotypic frequencies as the allele frequencies change.

Effect of Natural Selection

You will now change the starting parameters for the program by introducing selection into the equation. You should first look at selection against the recessive allele (as expressed through the homozygous recessive phenotype) and then against the dominant allele (as expressed through the homozygous dominant and heterozygous phenotypes). Your instructor will tell you how to introduce selection into the program that you are using. This can be done in two ways. Adults of a certain phenotype may die and thus not be able to breed, or breeding pairs with certain genotypes will simply produce fewer offspring.

Before running the selection program, develop a hypothesis to be tested. If selection is against the recessive allele, what do you think will happen to the frequency of the recessive allele over 50 generations? Will it increase, decrease, or remain the same?

Now remove the selection against the recessive allele. Instead, add selection against the dominant allele. Predict what you think will happen.

Run the program to test your prediction. Describe what happened. Was your prediction supported?

▶ Run the simulation and describe the results. Must you accept or reject your null hypothesis? _____

Run this same simulation again. Did you get the same result? _____

Run it again. Did you get the same result? _____

◆ How do you explain the variation in results that you are seeing?

Genetic Drift

Change the starting conditions of the simulation program back to those you used at the beginning for graphic output and no selection. Now change population size from 8,000 to 40.

▶ Before running the program, predict what you think will happen to allele frequencies over 50 generations in a small population compared to the large one you investigated first.

Genetic Drift and Natural Selection Together

As you did before, introduce a natural selection factor against the homozygous recessive phenotype, using exactly the same value. Do not change the starting population size. Leave it at 40. Before running the simulation, predict what you think will happen to the frequencies of the dominant and recessive alleles.

State these predictions as testable hypotheses, H_o and H_a.

◆ Run the program to test your predictions.

Describe what the results look like. Do they support your prediction? Explain what happened.

Experimental Induction of Mutations

In nature, mutation is the process that creates new alleles. **Mutations** are mistakes that happen during DNA replication before cell division, or abnormalities that develop in chromosome during cell division. Whatever the source, mutations contribute to genetic variability in populations. However, their effect is usually much less than that of genetic recombination events that occur as a result of crossing over and biparental inheritance. Most mutations are detrimental to the organism, but some can be beneficial. Usually, but not always, mutations create recessive alleles. This means that the trait would not be expressed until it appeared in a homozygous individual, a process that could take several generations. Natural selection is the agent in nature that determines whether a mutation is "good" or "bad." A detrimental mutation (allele) would be selected against; *i.e.,* organisms with the mutation would not function as well in the environment and would leave fewer offspring. Over time, the frequency of genotypes carrying a detrimental mutation should decrease in the population. The opposite would be true for a beneficial one.

In this part of the lab topic, you will experimentally induce mutations in a bacterium, *Serratia marcescens.* Bacteria are good organisms to use in mutation studies because they are haploid. If a mutation occurs, it is expressed because there is not a second allele present to mask it. You will study mutations affecting viability and pigment synthesis in *Serratia.* This bacterium is normally red, but when a mutation occurs in the genes producing the enzymes involved in pigment synthesis, no pigment is made and the bacteria are white. Mutations can be induced by several means. Chemicals, called mutagens can change an organism's DNA, causing changes in hereditary information. Ultraviolet radiation has similar effects and will be used in the experiment. Because mutations caused by UV exposures are random, many different mutations will be induced in *Serratia.* Some will affect pigment synthesis. Others will affect viability because the UV exposure damages genes that are essential to life. The assay system you will use will allow you to see both of these effects.

State null and alternative hypotheses to test in your experiment. They should relate the length of exposure to UV light to the amount of mutation expected. State your null and alternative hypotheses.

H_o

H_a

Procedure

▶ About 18 to 24 hours before the lab, *Serratia* was transferred from a slant culture into 100 ml of nutrient broth and cultured at room temperature. You or your instructor should take this culture, and make a sterile, serial dilution. Add 0.1 ml of the culture to 9.9 ml of 0.85% sterile saline. Cap and shake the container 10 times. Take 0.1 ml of this dilution and add to a second 9.9 ml of sterile saline; cap and shake. Again take 0.1 ml of this second dilution and add it to a third 9.9 ml of sterile saline. Shake it well. The cell suspension in the third tube represents a millionfold dilution of the original culture. The first and second dilutions will not be used and should be autoclaved before disposal.

▶ About 30 minutes before you are going to use it, a germicidal UV lamp enclosed in a box should be turned on and allowed to stabilize (fig. 12.1).

<table>
<tr><td align="center"><h2>CAUTION</h2>Never look at a UV lamp, because it can destroy cells in you cornea (surface of the eye).</td></tr>
</table>

The lamp should be in a box, and anyone near the box should wear safety glasses that will filter out ultraviolet light. Anyone reaching into the box should wear a rubber glove and a long-sleeve shirt to protect their skin from the UV light. Why?

▶ Take seven petri plates containing King's agar and number the bottoms 1 through 7.

▶ Shake the third dilution tube well to distribute the cells. Refer to table 12.1 to determine the volume of the *Serratia* dilution to be added to each plate. Note that the plates to be irradiated longer receive a larger volume of

Figure 12.1 Ultraviolet light irradiation box. Insert petri dish into box and remove lid for appropriate time (table 12.1). Replace lid and remove from box. When working at the box protect your eyes by wearing glasses made of ultraviolet-absorbing glass. Protect your skin by wearing a long sleeve shirt and rubber gloves.

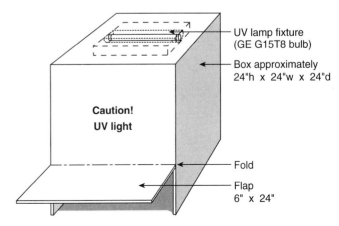

TABLE 12.1 Results from uv irradiation of *Serratia*

Plate	Sample Vol.(ml)	Cumulative UV Exposure (sec)	Total No. Colonies	% Surviving	No. White Colonies	% White Colonies
1	0.025	0	__ × 4*	__	__ × 4*	__
2	0.05	10	__ × 2*	__	__ × 2*	__
3	0.05	20	__ × 2*	__	__ × 2*	__
4	0.10	30	__	__	__	__
5	0.10	40	__	__	__	__
6	0.10	60	__	__	__	__
7	0.20	80	__ × 0.5*	__	__ × 0.5*	__

*Note: Number of colonies should be adjusted to compensate for differences in sample-size plated.

Figure 12.2 Inoculation of petri plates. Label the bottom of the petri dish with a marker. Open the dish from one side and add culture liquid containing bacteria. Use a sterile glass rod bent like a hockey stick to spread bacteria over agar surface.

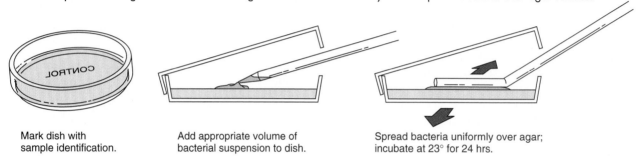

Mark dish with sample identification.

Add appropriate volume of bacterial suspension to dish.

Spread bacteria uniformly over agar; incubate at 23° for 24 hrs.

---

cells to compensate for the lethal effects of UV exposure. Use a sterile pipette to add the required amount to the agar surface. Then use a sterile bent glass rod ("hockey stick") to spread the culture evenly across the surface of the agar (fig. 12.2). The "hockey stick" can be sterilized between the times you spread bacteria on each plate by putting it in a beaker of 70% ethanol. Before using it again pass it quickly through a flame to burn off the ethanol. Let the rod cool for 15 to 20 seconds before spreading the bacteria on a new plate.

Plate 1 is a control and nothing more should be done to it. Plates 2 through 7 will be irradiated with increasing amounts of UV light.

Refer to table 12.1 and determine the exposure time for each plate. Take one plate at a time and slide it into the UV light box. Reach in with a gloved hand and **remove the top** for the time indicated. This is done because glass and many plastics absorb UV light and little UV would reach the bacteria if the lids were left on during the irradiation period. Replace the top and remove the plate. Immediately place the plate in a paper bag or drawer to prevent photoactivation of DNA repair mechanisms. Repeat the procedure for the other plates for the appropriate times.

When all plates have been irradiated they should be placed in a refrigerator until 48 hours before the next lab. When taken out of the refrigerator, they should be incubated for 48 hours at room temperature. They may be incubated longer if growth is slow. (If incubated at temperatures higher than 25°C, no color develops.) When colonies are clearly visible, the plates should be counted.

Analysis

When you examine the plates, count the total number of colonies on each plate and the number of white colonies. Record the results in table 12.1. Note that the plates receiving short UV exposures received lesser volumes of the cell suspension, and the numbers should be corrected by multiplying the colony counts times a volume correction factor. This calculation makes all counts for all plates directly comparable. Calculate the percent surviving by dividing total colony count from plate #1 into colony counts for all other plates and multiplying by 100. Calculate the percent of white colonies by dividing the total colony count into the white colony count for each plate. On one panel of the graph paper at the end of the exercise, plot the percent surviving (from total counts) as a function of irradiation time.

Why do you think there are fewer total colonies formed in those samples that were irradiated longer?

 Now plot the percentage of white colonies as a function of irradiation time on the graph paper at the end of this lab topic. Why does the frequency of white colonies increase? Return to the hypotheses you made and come to a conclusion to accept or reject the null hypothesis.

Learning Biology by Writing

In this lab you did two experiments. One was a computer simulation to test hypotheses related to selection and population size, and the other was a classical "wet lab" experiment demonstrating how mutations can be induced with UV light. Although related, these two experiments are not directly comparable. Consequently, your instructor may ask you to write a lab report on only one of the activities.

For either lab report, follow the directions in appendix D. When biologists sit down to write a scientific paper, they often start by writing an abstract of the work they are reporting. This helps them organize their thoughts. You might try to the same. Start your report with a clear description of the hypotheses that were tested and why these are important ideas. The methods section should be brief and can cite the lab manual as a source for more detail. The results section should contain any tables or graphs of data. Usually one or the other is included for any experiments performed, but not both. The conclusions section should state what was learned from the work and how that relates to what is already known. Future experiments can be suggested to test the ideas that were presented.

As an alternative, your instructor may ask you to turn in answers to the Critical Thinking or Lab Summary Questions that follow.

Internet Sources

Use one of the search engines available through your Internet browser to find information on the topic of Serratia color mutations (or mutants). Google at http://www.google.com is a good search tool. Compare your results to those of others who have done similar experiments. Are your techniques and results similar to those of others? Do any of these sites contain information that would be useful in writing your report? If you use this information, be sure to cite the source. See appendix D.

Lab Summary Questions

1. Define natural selection. Describe what happens to the frequency of a dominant allele in a population when selection is against the homozygous dominant phenotype? Against the heterozygote? Against the homozygous recessive phenotype?
2. How can you determine the frequency of a recessive allele in a population? Of a dominant allele at the same locus, assuming there are only two alleles at the locus in a population?
3. What is genetic drift, and why is the concept important in understanding evolution? Based on your computer simulation work, give comparative examples of what happens to allele frequencies over many generations in small populations compared to large ones.
4. Plot the data from table 12.1 as directed in the **Analysis** section. What is it about these plots that suggests ultraviolet light is mutagenic?
5. Hand in your graphs of the effects of UV irradiation on survival and induction of white colonies in *Serratia marcescens*.

Critical Thinking Questions

1. Huntington's disease is a fatal nervous system disorder causing degeneration of the brain. It is caused by a defective gene on chromosome #4. This is an autosomal dominant trait affecting 1 in 20,000 people. However, it is a delayed action gene that is not expressed until the affected individual is in his or her late 30s or 40s. The homozygous condition is lethal to the fetus. Using the Hardy-Weinberg equation, determine (1) the frequency of carriers in the population, (2) the frequency of fetuses affected by the lethal homozygosity, and (3) the frequency of unaffected individuals.
2. In the Lake Maracaibo region of Venezuela, there is a family of about 3,000 people who are descendants of a German sailor who had Huntington's disease. Would sampling this population reflect the whole population

of Venezuela? What mechanism of microevolution is occurring? Explain.

3. Present prenatal screening (amniocentesis and chorionic villi sampling) allows parents to screen babies for possible genetic defects. As well, reproductive technologies using surrogacy and sperm donation are becoming more common with couples seeking sperm donors or seeking surrogate mothers with "desirable" traits such as high IQ, tallness, and so on. Speculate on the implications these procedures have on the population as a whole with regard to microevolution.

4. Natural selection can be viewed as the sum total effects of the environment on the organism and includes both abiotic (physical) and biotic (other organisms) components. Does natural selection have its effects by acting directly on the genes, genotypes, or phenotypes? Explain your answer.

Testing Assumptions in Microevolution and Inducing Mutations

LAB TOPIC 13

Using Bacteria As Experimental Organisms

Supplies

Preparator's guide available on WWW at
http://www.mhhe.com/dolphin

Equipment

Compound microscopes
Dissecting microscopes
Oil immersion lenses
Incubators or water baths at 37°C and 42°C
pH meter or pH tape

Materials

Alcohol lamps
Diamond pencils
Bacterial loops
Sterile pipettes and suction bulbs
Microscope slides and coverslips
Yogurt (about a 1-to-10 dilution in water)
Whole milk
Living cultures
    Anabaena
    Bacillus megaterium
    Blue green algae from local sources
    Gloeocapsa
    Oscillatoria
    Pseudomonas fluorescens
Soil samples
Immersion oil
Prepared slides for demonstration
    Composite of bacterial types (cocci, bacilli, and
        spirilla)
    Merismopedia (for demonstration)
Autoclave bag for disposal of cultures

Solutions

Gram's stain kit with crystal violet and safranin
    counterstain
25% acetone in isopropyl alcohol
Nutrient agar with 0% and 6% NaCl in petri dishes
Sterile water
Sterile 0.85% saline, packaged 9.9 ml per capped
    tube
India ink
70% ethanol
Test substances, such as antibiotics, metal salts, or
    pesticides; filter disks, 1 cm diameter

Prelab Preparation

Before doing this lab, you should read the introduction
and sections of the lab topic that have been scheduled
by the instructor.
    You should use your textbook to review the
definitions of the following terms:

    bacillus
    bacteria
    coccus
    colony growth
    Cyanobacteria
    Domain
    eukaryote
    Gram's stain
    prokaryote
    spirilla

    You should be able to describe in your own words
the following concepts:

    Agar and growth media
    Bacterial diversity
    Differential growth
    Sterile technique (See figures 11.4 and 11.6)

As a result of this review, you most likely have
questions about terms, concepts, or how you will do
the experiments included in this lab. Write these
questions in the space below or in the margins of the
pages of this lab topic. The lab experiments should
help you answer these questions, or you can ask your
instructor for help during the lab.

Objectives

1. To recognize the diversity of prokaryotic cell types
   among bacteria
2. To investigate where bacteria live
3. To determine the number of bacteria in samples

Background

The diversity of organisms is staggering. In order to organize information about the living world, biologists group organisms taxonomically, creating categories that contain related organisms. In recent years, there has been a complete restructuring of the taxonomy of living organisms. While the system will undoubtedly change as science discovers more information, the present system is organized as follows. The largest of these categories is the **Domain.** Three Domains are recognized. The Domain Archaea contains prokaryotic organisms that were once thought to be bacteria. The Domain Bacteria includes bacteria and blue-green algae. The third, Domain Eukarya, includes all eukaryotic organisms ranging from algae and protozoa to fungi, plants and animals. The Domain Eukarya contains many kingdoms, large taxonomic units containing several phyla. Each phylum contains several classes, and each class several orders. Within the orders are families composed of several genera that are further subdivided into species.

Each type of organism is ultimately identified in this hierarchical classification system by two names, a **binomial** that includes the genus and a species name, both written in Latin and italicized. All binomials are unique; the same name is never given to two different species.

This lab topic covers the Domain Bacteria, true bacteria and blue-green algae. All are prokaryotes; they lack a nucleus and membranous cytoplasmic organelles, such as vacuoles, mitochondria, and chloroplasts. The bacteria and cyanobacteria have cell walls composed of peptidoglycans, and are often surrounded by a gelatinous sheath. Their DNA is arranged in circular molecules and is not complexed with histones to form chromosomes as in the organisms found in Domain Eukarya.

Most bacteria are heterotrophs and depend on other organisms for food. Some are parasitic and cause diseases, such as pneumonia and syphilis, but others are saprophytes and break down organic matter, thus recycling elements, such as carbon, nitrogen, and phosphorus in the environment. A few bacteria and all cyanobacteria are autotrophs and make their own food by photosynthesis. Many bacteria and cyanobacteria can fix nitrogen, converting atmospheric nitrogen gas into nitrate or organic nitrogenous compounds. No eukaryotes can fix nitrogen, so all other living organisms depend on the bacteria to provide nitrogen in a form that can be used in making organic compounds.

LAB INSTRUCTIONS

You will look at representative bacteria and explore some of the techniques used in microbiology.

Bacterial Cell Shapes

Bacteria are small, although not the smallest living organisms. They are very difficult to observe even through the finest light microscopes, let alone those that are normally found in teaching laboratories. Therefore, the identification of bacteria is based on a combination of characteristics. One characteristic used is cell shape. Some bacterial cells are **cocci,** small spheres, while others are rod-shaped **bacilli** or cork-screw-shaped **spirilla.**

▶ Look at the composite slide of bacterial cell types with an oil immersion objective. Objectives such as this allow magnifications of up to ×1000. Your instructor may have this as a demonstration slide. Compare these to the scanning electron micrograph of bacteria in figure 3.1. Sketch the basic shapes of bacteria below.

Gram's Staining

Bacteria are also identified by their staining response to certain dyes. One such staining protocol is known as the **Gram's stain.** One of its dyes is crystal violet. It binds irreversibly to cell wall components of some bacteria. Bacteria that retain Gram's stain when washed with alcohol are said to be **Gram positive;** those that are decolorized by the alcohol wash are said to be **Gram negative.** This is one important tool in bacterial identification. As a child you may have had a throat culture when you were sick. Among other things, the doctor wanted to determine if you were infected with Gram + or − bacteria, because certain antibiotics are effective only against Gram + bacteria.

▶ You will perform the Gram's stain on two 18- to 24-hour cultures of bacteria available in the laboratory. One culture contains *Bacillus megaterium,* and the other *Pseudomonas fluorescens.* Both species are large, but one is Gram positive and the other Gram negative. You will determine which bacterium is Gram + by using the following procedures (fig. 13.1):

1. Wash a microscope slide with soap and water to remove oils. Dip the slide in a beaker of alcohol and let

Figure 13.1   Directions for Gram's staining.

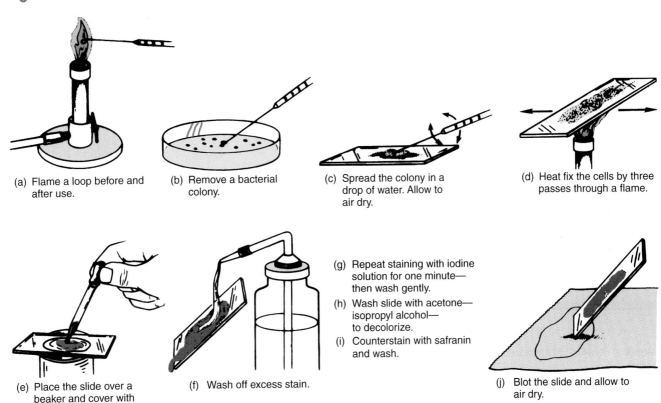

(a) Flame a loop before and after use.

(b) Remove a bacterial colony.

(c) Spread the colony in a drop of water. Allow to air dry.

(d) Heat fix the cells by three passes through a flame.

(e) Place the slide over a beaker and cover with a few drops of crystal violet. Let stand for one minute.

(f) Wash off excess stain.

(g) Repeat staining with iodine solution for one minute— then wash gently.

(h) Wash slide with acetone— isopropyl alcohol— to decolorize.

(i) Counterstain with safranin and wash.

(j) Blot the slide and allow to air dry.

it air dry. Use a diamond pencil to put a "B" on the side of the slide where you will put *Bacillus* and a "P" on the *Pseudomonas* side.

2. If you start with colonies from petri plates, put two tiny drops of distilled water on the slide. If you start from a liquid culture, water is not needed. Flame a bacterial loop and let it cool. Dip the loop into the culture or scoop part of a colony off the agar plate surface. Use the loop to spread one species of the bacteria evenly on one-third of the slide and the other species on another third with a blank area between. Let the slide air dry at room temperature. Gently heat the slide by passing it through a low flame three times. This makes the cells adhere to the surface of the slide, a process known as **heat fixing.**

3. Put the slide on top of a beaker and flood the surface with crystal violet. After one minute, carefully pick up the slide and wash off the stain with a gentle flow of water. Wearing rubber gloves will protect your skin from the stain.

BE CAREFUL!
This stain is very difficult to remove from your skin and clothing.

4. Now the slide should be flooded with Gram's iodine reagent that enhances color development. After one minute, gently wash the slide again with water.

5. Take a squeeze bottle containing 25% acetone in isopropyl alcohol and **gently** squirt it on the surface of the slide for 10 seconds until the solvent running off the slide is colorless. If you squirt too hard, the bacteria will be washed off. Gently wash the slide with water.

BE CAREFUL!
This solvent is flammable and should not be used near an open flame.

6. Flood the surface of the slide with safranin, a **counterstain** that helps you see Gram-negative cells. After 30 seconds, wash off the counterstain with water.

7. Blot the excess at the edge of the slide onto paper toweling and allow the slide to air dry. Examine with a high dry objective, or an oil immersion objective if available.

Which species of bacterium is Gram positive? Which is negative? Sketch a few of the cells below and describe in words how they look.

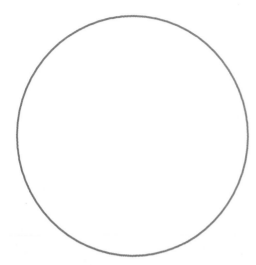

| TABLE 13.1 | Differential growth results for soil bacteria |

TABLE 13.1 Differential growth results for soil bacteria

		Temperature	
		25°C (RT.)	42°C
Salt	0%		
Conc.	6%		

Record growth as 0, +, ++, or +++.

Look at these plates and note the differences between them. Do all growing colonies look the same despite the growth conditions? Is the same number of colonies found on all plates? What other differences can you see? Record your observations below and in table 13.1.

Determinative Microbiology

Identification of bacteria involves looking at a combination of structural, physiological, and ecological characteristics of an unknown microbe. Cell shape, selective staining by one or more dyes, and motility can be used to narrow the possibilities, but then other tests must be employed. Often this involves growth on specialized media or growth under special conditions.

Differential Growth

To illustrate how different species grow under different conditions, a simple experiment can be performed. You or your instructor should take a gram of rich organic soil and add it to 10 ml of sterile 0.85% saline. The suspension should be agitated to extract and suspend the bacteria. About 0.2 ml of the resulting soil extract should then spread onto the surfaces of each of four petri plates containing nutrient agar. Two plates contain 6% sodium chloride, and the other two 0% sodium chloride. One of each pair of plates should be incubated at room temperature, and the other at 42°C for 12 to 24 hours. This is a 2 × 2 experimental design which combines conditions of high and low salt with high and low temperatures to determine if different kinds of bacteria grow under different conditions.

Additional variables may also be tested within this design. You could add to the plates small discs of filter paper that were soaked in antibiotics, such as penicillin, or in salts of heavy metals, such as mercuric chloride, or in organic compounds, such as insecticides. Materials in the discs will diffuse into the agar and affect the growth of bacteria in a circular area around the discs. Bacteria some distance away from the discs should not be affected and can serve as a control for growth under the basic four conditions.

Explain why these differences were found if all plates were inoculated with an extract from the same soil sample. What does this experiment tell you about the general ecological requirements (or tolerances) of different species?

Bacterial Population Counts

Biologists frequently need to estimate the number of individual bacteria in a complex mixture of materials, such as in soil, in food, or in body fluids.

This is done by taking a measured sample of the material and diluting it with sterile water. A small amount of that dilution is spread on the surface of nutrient agar in a petri dish. For example, if 1 g of soil is mixed with 99 ml of sterile water, and 0.1 ml of the mixture is placed on the agar, then 1/1000 of the soil sample has been "plated." Any bacteria in the sample will be spread on the agar surface. They will draw nutrients from the agar medium and each single bacterium will repeatedly divide to form visible colonies. The colonies can be counted easily, and by multiplying by 1000, the number of bacteria in the original gram of soil can be determined.

If the bacteria are plentiful in the soil, a 1/1000 dilution may be insufficient; when a petri plate is covered by thousands of colonies, counting is tedious and subject to error. To avoid this, **serial dilutions** are often made. In this technique, a sample is suspended first in 99 ml of water, and then 1 ml of that suspension is added to 99 ml of water to further dilute it: in this case to 1/10,000. One milliliter of this suspension may then be diluted in another 99 ml to give a 1/1,000,000 dilution and so on. Describe how you would make 1/10,000,000 serial dilution below.

In the lab are samples of ground beef. The problem you must solve is to conduct an experiment that will allow you to estimate the number of bacteria in the food sample.

In order to reduce the number of necessary plates and bottles, each pair of students may be assigned a dilution to use according to table 13.2. In that case, the directions below should be followed with each pair performing one of the serial dilutions.

To reduce the volumes required, all sample dilutions will be based on 0.1 g of sample in 9.9 ml of water, a 1 to 100 dilution.

TABLE 13.2	Number of colonies in 0.1g of ground beef
Dilution	
1/1000	_____
1/1,000,000	_____
1/1,000,000,000	_____

▶ Weigh 0.1 g of ground beef and add it to 9.9 ml of sterile water in a test tube. Mix well.

▶ Using a sterile pipette, take 0.1 ml of the suspension and place it on the surface of nutrient agar in a petri dish to plate 1/1000 of the original sample, the equivalent of a 1/1000 dilution. Spread the inoculum over agar surface using sterile technique illustrated in figures 11.6 and 12.2. Now take 0.1 ml of the first suspension and add it to a second tube containing 9.9 ml of distilled water. Spread 0.1 ml on a separate petri plate for a 1/1,000,000 dilution. Now take 0.1 ml from the second tube and add it to 9.9 ml of sterile water in a third tube. Shake well. Plate 0.1 ml of this third dilution to achieve a 1/1,000,000,000 dilution of the original sample. Label the plates with the dilution, sample, and your initials.

Once the plates are inoculated, allow them to sit covered for several minutes. Then invert the plates (so condensation will not drop onto growing culture) and incubate for one to two days at 37°C. (If growth is too rapid, the plates can be placed in a refrigerator and held until the next lab.) At the end of that time, count the colonies. Best results will be obtained when the number of colonies on a plate is in the range of 30 to 300. Some plates will simply have too many colonies to count and such results should be recorded as "too numerous to count." Record the class results in table 13.2.

 To arrive at the number of bacteria in the original meat sample, what must you now do?

Where do bacteria live

Your instructor will provide you with sterile petri plates of nutrient agar and sterile cotton swabs. Use the swab to wipe a surface that you think is "clean." For example, a desk top, a soda machine, or a pencil you use. After collecting the sample, open the dish and rub the swab on the surface of the nutrient agar. If you use only half the plate, a second sample can be collected with a second swab. Incubate overnight and look at the plates during the next lab.

Applied Microbiology

Fermentation is a term used to describe a collection of metabolic reactions that occur in microorganisms and

Using Bacteria As Experimental Organisms

Figure 13.2 Common cyanobacteria (blue-green algae).

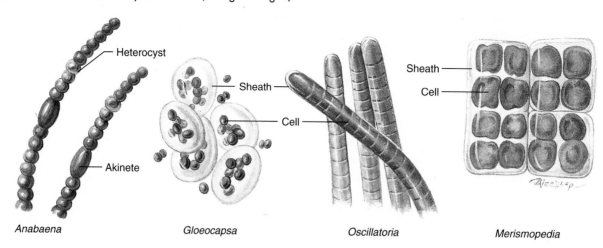

Anabaena   Gloeocapsa   Oscillatoria   Merismopedia

release energy from organic molecules. No oxygen is required and an organic molecule is used as the terminal electron acceptor. The organic molecule that is produced is excreted from the cell and accumulates in the environment.

Fermentation historically has been used as a means of preserving dairy products, resulting in a "sour milk" product. Yogurt is one such product produced when *Streptococcus thermophilus* ferments the sugar lactose in milk to lactic acid (souring) and *Lactobacillus bulgaricus* produces the flavors and aroma of yogurt. As lactic acid accumulates, the pH of the milk drops and the proteins in the milk curdle. Eventually, an acid concentration is reached that prevents the growth of other microorganisms and the yogurt-producing bacteria as well. When this happens, the food is preserved and can be kept at room temperature for extended periods. However, it will not keep indefinitely and the growth of other acid-tolerant microorganisms may eventually cause spoilage.

In the lab is yogurt that has been diluted in water. Place a drop on a microscope slide and add a coverslip. Observe to see the bacteria used in fermenting milk to make this product.

To test for lactic acid, measure the pH of the yogurt and of whole milk. Record the values below.

pH of whole milk _____

pH of yogurt _____

Cyanobacteria

Cyanobacteria (blue-green algae) exist mostly as colonies and filaments and not as single cells (fig. 13.2). Because they contain chlorophyll, they are greenish in color and perform photosynthesis. Like many bacteria, they produce spores that are resistant to drying. Spores allow cyanobacteria to survive unfavorable seasonal conditions. They may also be windblown and distribute the species. Cyanobacteria live in freshwater, marine, and terrestrial environments,

and they flourish when given water, sunlight, carbon dioxide, nitrogen, and certain inorganic salts. Their gelatinous capsules and toxins they produce make them poor food for heterotrophs. They can be a nuisance in water supplies where they cause taste and odor problems.

Make a wet mount from a culture of *Gloeocapsa* (fig. 13.2), a cyanobacterium commonly found on moist rocks and on flower pots in greenhouses. Before adding the coverslip, dip a dissecting needle in India ink and touch it to the water drop on the slide to add just a small amount of ink to the preparation. This technique allows you to see the gelatinous sheaths that surround the cells, because the carbon particles of the ink cannot penetrate the transparent sheaths. Observe under low and then high power. Sketch several *Gloeocapsa* cells below.

Describe the color of the cells in terms of shades of green. Is the green localized in structures in the cytoplasm, or is it spread evenly throughout the cells? Are nuclei visible in the cells? How do the sizes of these cyanobacteria compare to the size of the bacteria that you looked at earlier?

▶ Look at a demonstration slide of *Merismopedia* (fig. 13.2), a cyanobacterium in which the cells are arranged in a regular array within a flat sheet of gelatinous material. Look for a cell that is dividing and describe the division process.

❓ If your microscope were powerful enough, would you expect to see cells undergoing mitosis or meiosis in any blue-green algae? Why?

▶ Make a wet mount of *Anabaena,* a filamentous cyanobacterium (fig. 13.2). Use the India ink method to see the gelatinous sheath. Along the filament, you should be able to see some cells that are differentiated. Elongated cells with thick cell walls are **akinetes** (spores). Enlarged but somewhat spherical cells are **heterocysts,** which have the enzymes for fixing atmospheric nitrogen. They will be most common in cells coming from growth media that are low in nitrate or nitrogen-containing organic compounds. Sketch a filament of *Anabaena* below and include examples of all three types of cells.

❓ Are nuclei visible in the cells? _____

▶ The last cyanobacteria that you will look at is *Oscillatoria* (fig. 13.2). Before making a slide, take a few filaments from a culture and rub them between your thumb and index finger. As you move your fingers, note the slippery ❓ mucuslike feel. What structure found in all cyanobacteria do you think causes this?

▶ Make a slide and examine it with your microscope. Once you have some filaments in view, do not touch your slide but carefully observe the cyanobacteria. Describe any movements that you see. Sketch a filament of *Oscillatoria* below.

Find some short chains of cells that are joined to the filament by a few dead cells. These chains fragment from the filament and continue to grow, producing new filaments, a common form of asexual reproduction in cyanobacteria.

Your instructor may bring some scrapings from flowerpots in the greenhouse to the lab or may have collected some blue-green algae from a lake or stream. Take a few minutes to make a slide and look at these "wild" cyanobacteria.

Using Bacteria As Experimental Organisms

Learning Biology by Writing

Devise a procedure that would allow you to determine the number of bacteria found in potato salad collected from a salad bar in a restaurant. Describe how you would use sterile techniques to avoid false readings.

As an alternative assignment, your instructor may ask you to answer the questions in the Lab Summary and Critical Thinking Questions that follow.

Lab Summary Questions

1. List the characteristics that separate organisms in the Domain Bacteria from those in Domain Eukarya. Which ones were you able to see in this laboratory?
2. If you were given an unknown species of bacterium, how would you go about identifying it in the laboratory?

Internet Sources

Use the **Google** search engine (http://www.google.com) in an Internet browser to locate additional information on cyanobacteria. When connected type in the phrase, cyanobacteria in drinking water. Scan the list that is returned and answer Critical Thinking Question #4.

---

3. What evidence do you have from this laboratory that bacteria are diverse? Cite specific examples as you explain this diversity.
4. Describe how you would determine the number of bacteria present in a sample of coleslaw from a salad bar in a restaurant.

Critical Thinking Questions

1. "To make strawberry jam, mash 4 cups of berries with 7 cups of sugar (sucrose). Bring to a boil and continue boiling for twenty to thirty minutes. Ladle into sterilized jars and cover with melted paraffin wax or sterilized lids."

   Relate jam making to the growth requirements of different bacteria. Would these steps be adequate for preserving (canning) vegetables? Meats? What other methods could be employed to preserve foods?
2. Should a plant nursery that plans to grow pea and bean seedlings use a sterilized potting soil mix?
3. It is well known that many pathogenic bacteria are developing resistance to antibiotics. Explain why this is happening and formulate a recommendation that you would give your physician regarding when antibiotics should be used in the treatment of disease.
4. Why should you be concerned if cyanobacteria are found in your drinking water? Use Internet Sources to locate the information that you need.

LAB TOPIC 14

Diversity Among Protists

Supplies

Preparator's guide available on WWW at
    http://www.mhhe.com/dolphin

Equipment

Compound microscopes
Dissecting microscopes
Desk lamp with 60-watt bulb

Materials

Cultures
    Ceratium
    Euglena
    Chlamydomonas + and − mating types
    Volvox
    Termites with *Trichonympha*
    Vorticella
    Saprolegnia culture kit
    Stentor
    Physarum (slime mold) culture kit
Plastomount of *Ulva* and other marine algae
Prepared slides
    Ulothrix
    Oedogonium
    Spirogyra conjugating
    Entamoeba histolytica, cysts and trophozoites
    Trypanosoma
    Radiolarian strew
Diatomaceous earth
Test tubes, stoppers, and rack
Black paper
Slides and coverslips

Solutions

0.85% saline
Protoslo or methyl cellulose
1% water agar for slime molds in petri dish
Unprocessed (old-fashion) oatmeal

Prelab Preparation

Before doing this lab, you should read the introduction and sections of the lab topic that have been scheduled by the instructor.

You should use your textbook to review the definitions of the following terms:

algae
autotrophic
cilia
flagella
genus
heterogamy
heterotrophic
isogamy
multicellularity
phylum
protist
protozoa
taxonomic kingdom

You should be able to describe in your own words the following concepts:

Differences between bacteria and protists
Diversity among the protists
How ancestral protists might have evolved into
    multicellular organisms

As a result of this review, you most likely have questions about terms, concepts, or how you will do the experiments included in this lab. Write these questions in the space below or in the margins of the pages of this lab topic. The lab experiments should help you answer these questions, or you can ask your instructor for help during the lab.

Objectives

1. To study representatives of several groups of protists
2. To learn the life cycles of biologically important protists
3. To illustrate evolutionary trends among the protists

Background

Protists are eukaryotic organisms. They inhabited the earth a billion years before plants, fungi, and animals. Some fossil evidence indicates that protists might have been present over 2 billion years ago. The ancestors of this group were the first to have a true nucleus, chromosomes, organelles such as chloroplasts, mitochondria, endoplasmic reticulum, cilia, and cell division by mitosis or meiosis.

Most protists are unicellular, although several algae are colonial or truly multicellular. Although protists are often described as being simple organisms, their cellular organization and metabolism is every bit as complex as those found in the so-called higher organisms. In fact, higher organisms are often much simpler at the cellular level because their many cells are specialized to perform particular functions while single protistan cells perform all functions necessary for life as independent organisms.

Protists live everywhere there is water: in the ocean, in freshwater, in puddles, in damp soils, and, as symbionts, in the body fluids and cells of multicellular hosts. Some are autotrophic, making their own food materials through photosynthesis, while others are heterotrophic, absorbing organic molecules or ingesting larger food particles. They are ecologically very important in food chains, especially the algae which are the energy base for most aquatic ecosystems. Most protists can reproduce asexually by mitosis while some reproduce sexually as well, involving meiosis and nuclear exchange. A cyst stage is found in the life cycle of many protists; it allows the species to lie dormant and escape harsh temporary conditions.

Protists that are photosynthetic are commonly called algae. Protists that ingest food are commonly called protozoa. Another group of protists are fungus-like and absorb food materials. They lack a common name. None of the common names have taxonomic status.

The classification of the protists is in a state of flux. The Kingdom Protista, originally proposed by Robert Whitaker in 1969, included eukaryotic organisms that did not fit into Kingdoms Animalia, Planta, or Mycota. It was a category of convenience and not one that represented evolutionary relationships. This means that we are somewhat ignorant about the ancestry of an estimated 60,000 or so species of protists. Modern scientific investigation of cell structure and nucleic acid analysis is starting to show us common characteristics among the 30 or more phyla included in Whitaker's Kingdom Protista. Most biologists now reject Protista as a kingdom. What is not clear is what should replace it. Some authors propose five groupings which have been called "candidate kingdoms." Others refer to seven groupings of what they call "basal eukaryotes." In this lab topic, I will follow the candidate kingdom approach. They are listed below with brief descriptions of the phyla of protists included. Students should recognize, however, that by the time this revision of the lab manual appears in print, new information may be available, resulting in a reorganization of the protists. While this is frustrating, it also demonstrates that science is an ongoing, dynamic process that is not locked into beliefs that betray the facts. Those phyla marked with an asterisk will be studied in lab.

Candidate Kingdom Archaezoa—The several hundred species all lack mitochondria and most, but not all, are parasitic. Includes organisms known as Diplomonads, Trichomonads, and Microsporidians.

Candidate Kingdom Euglenozoa—All have an anterior chamber from which two flagella emerge that are used in locomotion, pulling the organisms through water.

Phylum Euglenophyta* (*Euglena*) flagellated cells with chloroplasts, lacking cell walls but with a proteinaceous pellicle; about 800 species

Phylum Zoomastigophora,* commonly called kinetoplastids (*Trypanosoma* and *Trichonympha*), the thousands of species included have a single mitochondrion and a concentration of extra-nuclear DNA.

Candidate Kingdom Alveolata—All members of this group have small indentations in their cell membranes, called alveoli, beneath the material covering the outer surface.

Phylum Pyrrhophyta,* commonly called dinoflagellates (*Ceratium*), are brown algae having two flagella and outer covering of cellulose plates; about 1,000 species; some authors group these with the Euglenozoa.

Phylum Apicomplexa, commonly called apicomplexans or sporozoans (*Plasmodium*), are parasitic protists with complex life cycles; commonly called sporozoans; about 3,900 parasitic species with a special apical complex used to invade host cells.

Phylum Ciliophora,* commonly called ciliates (*Paramecium*), are protists with cilia and two types of nuclei in cells; about 8,000 species; some authors consider these organisms as a separate group.

Candidate Kingdom Stramenopila (=Chromista of some authors)—All have unusual flagella with fine hairlike projections.

Phylum Chrysophyta,* commonly called diatoms, are photosynthetic freshwater and marine organisms with silica shells; about 11,500 species; some authors consider these organisms as a separate group.

Phylum Phaeophyta, commonly called brown algae and kelps; both unicellular and multicellular marine photosynthetic cells with cellulose cell walls; about 1,500 species; some authors group the red, brown, and green algae together.

Phylum Oomycota,* commonly called water molds, are fungus-like protists; about 500 species with cellulose cell walls and heterotrophic nutrition functioning as parasites and decomposers.

Candidate Kingdom Rhodophyta—commonly called coralline algae because of calcium carbonate in cellulosic cell walls; none of the life cycle stages have flagellated cells; all with cellulose cell walls.

Phylum Rhodophyta, multicellular red algae; sea weeds; about 4,000 species; some authors group the red, brown, and green algae together.

Candidate Kingdom Chlorophyta—All have bright green chloroplasts, and molecular studies show such a close relationship to the plants that some suggest that these should be included in Kingdom Planta and not be included among the protists.

Phylum Chlorophyta,* green algae both unicellular and multicellular; about 7,000 species; some authors group the red, brown, and green algae together.

In addition to those organisms that can be placed into these candidate kingdoms, there is another group of common organisms whose relationships are very difficult to judge at the moment. The one common feature that they share is that they have no permanent locomotor organelles and move by use of pseudopodia, an extension of the cytoplasm caused by cytoplasmic flowing. These are often grouped under the headings:

Sarcodina—Commonly known as amoebas, radialoarians, and forams, the several hundred organisms included here move by pseudopod formation.

Phylum Rhizopoda* (amoebas)

Phylum Actinopoda* (heliozoans and radiolarians)

Phylum Foraminifera* (forams)

Molds—approximately 1,200 species with restricted mobility and cell walls made of carbohydrate material.

Phylum Myxomycota* (plasmodial slime molds)

Phylum Acrasiomycota (cellular slime molds)

Some authors include the water molds here while others place them in the candidate Kingdom Stramenopila as I have done.

LAB INSTRUCTIONS

To develop an appreciation of protist diversity, you will study representatives of the groups listed above that are marked with *.

In preparing some of the materials for this lab, your instructor may mix cultures of two or more organisms so that you will only have to prepare one slide instead of two or three. This will save time and materials.

Candidate Kingdom Euglenozoa

Phylum Euglenophyta

This small group of about 800 species contains flagellated, autotrophic protists that lack cell walls, having instead a flexible outer covering called a **pellicle.** Species are common in waters polluted with organic matter and on the surfaces of wet soils.

Anatomy of Euglena

▶ Make a wet-mount slide of *Euglena* from the stock culture and observe through your compound microscope. If the organisms are swimming too fast to be studied, make a new slide but add methyl cellulose, a thickening agent, or shred a small piece of lens paper or cotton into the drop so as to trap the organisms. Reducing the light intensity will help you see structures in the organism.

Euglena is usually pear-shaped with the blunt end anterior. Does the flagellum push or pull the organism through the water?

Watch the organism closely. What evidence is there that the surrounding pellicle is flexible?

▶ As you study *Euglena,* find the structures indicated in figure 14.1.

What color is the **pigment spot,** also called the stigma, near the base of the flagellum? How many flagella does *Euglena* have?

What evidence do you have before you that *Euglena* is photosynthetic?

Excess sugars produced during photosynthesis are converted into **paramylum,** a unique form of storage starch. Is the chlorophyll of *Euglena* localized in structures, or spread throughout the cell as in cyanobacteria?

Figure 14.1   Anatomy of *Euglena* sp.

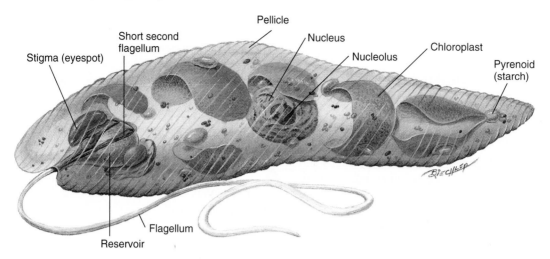

Phototaxis (Demonstration)

▶ Your instructor has set up a simple experiment to demonstrate the phototactic behavior of *Euglena*. Three six-inch test tubes were filled with cultures of *Euglena* and closed with stoppers. One tube is completely wrapped in black paper. Another tube is wrapped with two pieces of black paper, each about 7.5 cm wide, with a gap of about 0.6 cm left between the two pieces of paper to allow light to enter as a band around the center of the tube. The last tube is left unwrapped. Place all three tubes in a test tube rack. Set the rack about 12 inches from a 60-watt light.

❖ If *Euglena* are positively phototactic, predict how they should be distributed in each tube. State these predictions as a pair of hypotheses. What is your null hypothesis? Your alternative hypothesis?

H_o

H_a

After about half an hour, remove the paper from the tubes, and describe the distribution of *Euglena* in each tube. ❖ Must you accept or reject your null hypothesis? Why?

Phylum Zoomastigophora

Members of this phylum use one or more whiplike flagella to propel themselves. All are heterotrophic. Some are free living; others are mutualistic; and others, parasitic. In this section, you will study two members that are from opposite ends of the spectrum in symbiosis.

▶ Species of the genus *Trypanosoma* cause African sleeping sickness. Obtain a blood smear containing this protist and look at it with your compound microscope under high power. Below, sketch what you see.

These organisms are transmitted from one individual to another by the bites of tsetse flies.

At the opposite end of the symbiotic spectrum are species of *Trichonympha,* which live in the guts of termites. While in the gut, they secrete an enzyme that digests the cellulose in the wood fibers ingested by the termite, releasing sugars that are absorbed by the termite and the protistan. This is a form of mutualism in which both members of the association benefit.

▶ Get a live termite from the supply area and place it in a drop of 0.85% saline on a microscope slide. Press on the ter-

Figure 14.2 *Trichonympha* is one of the flagellates found in the guts of termites.

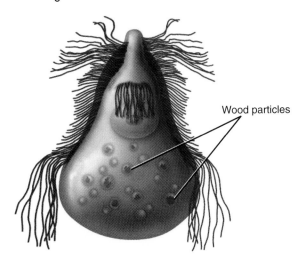

Wood particles

grooves in the wall. *Ceratium* is photosynthetic and contains chlorophyll, but the green color is masked by the presence of orange-brown carotenoid pigments. Make a sketch of *Ceratium* below.

mite's abdomen with a probe while pulling on its head with a pair of forceps. The head should come off, pulling the gut from the animal. Cut the gut free and discard the head and the body. Tease the gut open and mix the contents in the drop of saline. Discard the gut, add a coverslip to the slide, and observe the gut contents with your compound microscope. Look for single motile cells. Compare to figure 14.2.

Candidate Kingdom Alveolata

Phylum Pyrrhophyta

These organisms along with diatoms are the primary components of phytoplankton, forming the basis of marine food chains. About 1,000 species have been described (fig 14.3). Red tide is caused by the explosive growth of certain species of dinoflagellates.

Make a wet-mount slide of *Ceratium* from the stock culture. This dinoflagellate is common in freshwater lakes and streams. Note the tri-spiked appearance of the organisms. These cells have plates of cellulose called **theca,** which surround the cell. Two flagella are present and lie in

Phylum Ciliophora

The members of this large phylum of protists share several characteristics. All move by means of cilia, have a unique multilayered **pellicle,** reproduce sexually, and have two kinds of nuclei, a large **macronucleus** and one or more **micronuclei.**

In lab topic 4, you studied osmoregulation in *Paramecium.* Review your notes. In this lab topic, you will look at *Vorticella,* a sessile-stalked ciliate, and *Stentor,* a funnel-shaped sessile ciliate (fig. 14.4).

To observe either of these species, make a hanging drop slide. Place a drop of culture fluid containing the ciliate on a coverslip. Quickly invert the coverslip and place it on a depression slide so that the drop is suspended above the depression. Attach the coverslip to the slide by putting a drop of water on one corner. Look at the slide with scanning and low power of your compound microscope.

Figure 14.3 Representative dinoflagellates showing the characteristic outer cellulose plates and flagella originating from grooves between plates.

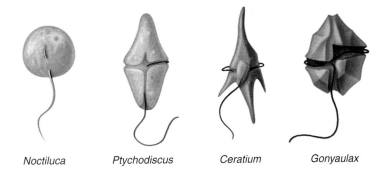

Noctiluca   *Ptychodiscus*   *Ceratium*   *Gonyaulax*

14-5   Diversity Among Protists   **151**

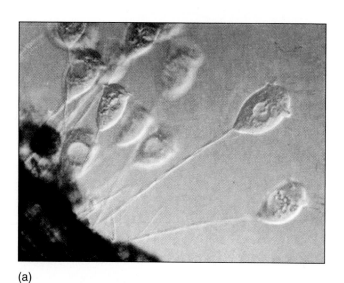

(a)

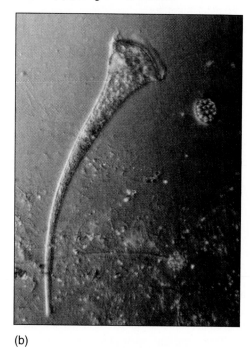

(b)

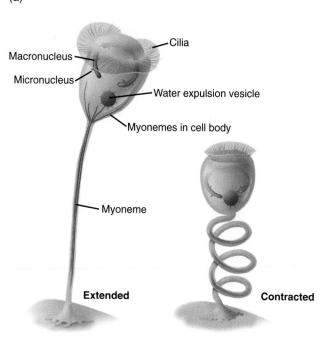

Cilia

Macronucleus

Micronucleus

Water expulsion vesicle

Myonemes in cell body

Myoneme

Extended

Contracted

(c)

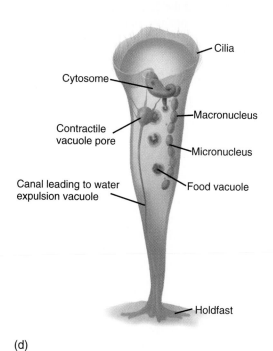

Cilia

Cytosome

Contractile
vacuole pore

Canal leading to water
expulsion vacuole

Macronucleus

Micronucleus

Food vacuole

Holdfast

(d)

As you watch, the ciliate should gradually extend. When extended, tap on the slide as you watch the organism. Describe the differences between *Vorticella, Stentor,* and *Paramecium.* What are the similarities?

Candidate Kingdom Stramenopila

Phylum Chrysophyta

The 11,000 species of diatoms share a common characteristic: a cell wall consisting of two valves made of silica. They are often golden-yellow in color because of an excess of carotenoid and xanthophyll pigments, which mask the green of the chlorophylls that are also present. When the diatom cell dies, the siliceous valves do not disintegrate but accumulate as sediments (fig. 14.5). In California, some deposits of diatoms are 300 feet deep. These are mined to produce diatomaceous earth, a fine powdery

Figure 14.5 Diatoms exist in an exquisite variety of geometrical shapes. Here an artist has arranged several species into a beautiful composition on a microscope slide using a single hair to drag each one into position.

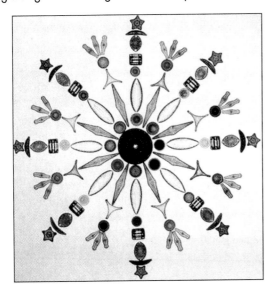

material used as a filtering material in swimming pools and as a fine abrasive in silver polishes and toothpastes.

Make a wet-mount slide from the diatomaceous earth available in the laboratory. Place a drop of water on the slide and then add a very small amount of diatomaceous earth to the drop. Stir well before adding a coverslip and viewing.

Note the exceedingly delicate patterns of the diatom valves. These are the skeletal remains of cells that lived thousands of years ago. In a top down view, some valves will be round, others triangular, ovoid, and irregular. When viewed from the side, these same valves appear rectangular or ovoid.

The ornamentation of the valves is often used as a test of the resolution of microscopes. In a poor microscope, only the outline of the valve will be visible, while in very good microscopes the fine indentations and perforations will be apparent. Sketch a few different types of diatom valves below. Indicate the location of the **girdle,** the region of overlap between the two valves.

Phylum Oomycota: Water Molds

At one time these organisms were considered to be fungi but are now considered protists in the candidate Kingdom Stramenopila. The characteristics that set water molds apart from fungi include: (1) asexual reproduction by bi-flagellated zoospores; (2) cell walls that contain cellulose; (3) diploid "body" cells with meiosis occuring in gametangia; and (4) the production of heterogametes.

Water molds live in freshwater, marine, and moist terrestrial environments. They are important decomposer organisms. Several species are economically important. One species causes potato blight which led to famines in Ireland during the nineteenth century and another causes grape downy mildew which nearly wiped out the French wine industry during the same time period.

The "body" form of water molds superficially resembles that of fungi. Long filaments without cross ways are produced and are called **hyphae,** a term from when they were included with the fungi. The tips of the hyphae may differentiate into zoosporangia which asexually produce motile diploid zoospores. The zoospores swim short distances or are carried by currents and colonize new locations. Under harsh conditions some hyphae differentiate into **gametangia** in which meiosis occurs to produce haploid eggs and sperm. The diploid condition is restored at fertilization. The zygote may become encased in a hard covering and dispersed to produce a new individual.

Saprolegnia is an easily cultured member of this group in which asexual reproduction can be studied (fig. 14.6). Cultures of this mold may be ordered from biological supply houses or grown in the laboratory as follows. Two weeks before class, a few dead insects were placed in petri dishes containing pond water. By lab time, *Saprolegnia* hyphae should be apparent. (Keep culture aerobic, or it will putrefy.) Carefully cut the ends of some of the hyphae from the culture and use a pipette to transfer them into a drop of water on a slide. Add a coverslip and observe with a compound microscope. Sketch what you see.

Diversity Among Protists   **153**

Figure 14.6 Life cycle of the water mold *Saprolegnia* sp.

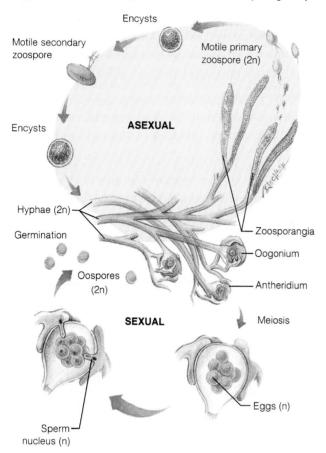

The nuclei in the hyphae of water molds are diploid. Note the absence of cross walls in the hyphae except at the edges of the **zoosporangia.** They should contain a number of motile, asexual spores. If you are lucky, you may find a zoosporangium in which the flagellated spores are starting to escape. The spores will disperse and, if a suitable substrate is found, will germinate to produce new hyphae. Under adverse conditions, spores will encyst and remain quiescent until favorable conditions develop. The asexual spores are called **zoospores** because of their motility.

Sexual reproduction involves production of nonmotile gametes by meiosis inside reproductive structures. Eggs are produced in spherical **oogonia.** Tips of hyphae of the same individual function as male gametangia, or **antheridia.** These grow into the oogonium, the sperm nucleus fuses with the egg nucleus, which then becomes a **zygote.** The zygote develops a thick wall and becomes an **oospore.** The oospore produces a new mycelium on germination.

Candidate Kingdom Chlorophyta

Phylum Chlorophyta

The 7,000 or so species of green algae can be grouped to create a natural progression from single cells to multicellularity. Three lines of evolution are apparent: (1) the forma-

tion of colonies; (2) the formation of multicellular filaments; and (3) the formation of definite multicellular organisms.

Colonial Series

You will examine species from three genera collectively called the volvocine series: *Chlamydomonas*, *Pandorina*, and *Volvox*.

On the supply table, two cultures of *Chlamydomonas* will be found, one labeled + and the other –. Under favorable conditions of light intensity, temperature, and nitrogen starvation, *Chlamydomonas* will undergo sexual reproduction. Place one drop of each culture side by side on a clean microscope slide, but do not mix. While looking at the slide through your dissecting microscope, mix the drops and observe what happens. Add a coverslip and look at the cells under high power of your compound microscope. Sketch the cells below.

Compare your drawing to the life cycle in figure 14.7. *Chlamydomonas* is capable of asexual and sexual reproduction. In asexual reproduction, the cells divide by mitosis. In the initial stages of sexual reproduction, such as you just observed, cells of different mating types come together and their flagella intertwine. The cells are gametes and fuse to produce a zygote. Because the two mating types are morphologically identical, *Chlamydomonas* is described as being **isogamous** (having identical male and female gametes). The zygote will eventually develop a thick wall and go into a period of dormancy, usually over winter. When dormancy ends, the zygote will divide by meiosis to produce four cells (zygospores), which will become typical adult cells. Is an adult cell of *Chlamydomonas* haploid or diploid? What about the zygote? Sketch a *Chlamydomonas* in the circle on page 155.

Figure 14.7   Life cycle of the green alga *Chlamydomonas*.

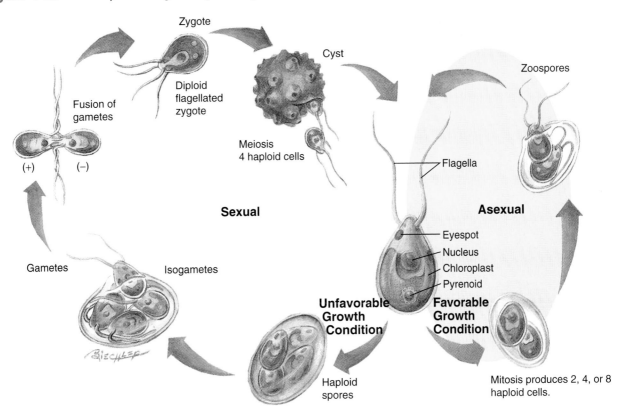

Now make a slide of *Volvox,* but do not add a coverslip. View the slide through your dissecting microscope with reflected light coming from one side. How does *Volvox* differ from *Chlamydomonas?* How is it similar?

A colony of *Volvox* consists of 500 to 50,000 cells. The coordinated stroking of two flagella in each cell allows the organism to move in a direction while spinning on its axis. The cells in the colony are held together by a gelatinous matrix and are connected to neighboring cells by thin strands of protoplasm. Young colonies produced by either asexual or sexual reproduction may be contained inside the parent colony (fig. 14.8). In sexual reproduction, several cells in the colony differentiate into motile sperm and a few others become nonmotile eggs. Sperm swim to the eggs, and fuse with them to form zygotes. Zygotes develop into daughter colonies inside the parent colony and are released when it dies. Because the male and female gametes can be distinguished from one another, *Volvox* is described as being **heterogamous** (having different male and female gametes). Oogamous organisms such as *Volvox* have large, nonmotile eggs. The cells that produce sperm are called **antheridia,** and those that produce eggs are called **oogonia.** Because of this differentiation of cell types and division of labor, the colony has some of the properties of a truly multicellular organism.

Describe how the three species you just observed could illustrate a hypothetical evolutionary series from unicellularity to multicellularity.

Diversity Among Protists   **155**

Figure 14.8   *Volvox: (a) adult colony with smaller daughter colonies inside; (b) differentiation of cells in colony for sexual reproduction.*

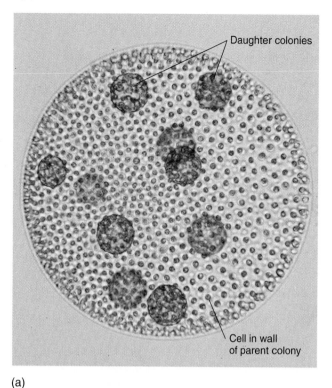

Daughter colonies

Cell in wall
of parent colony

(a)

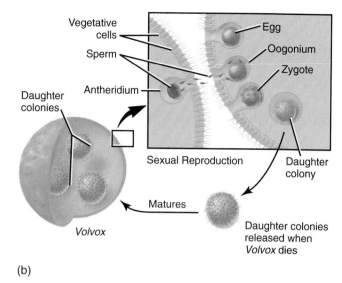

Vegetative cells

Sperm

Antheridium

Daughter colonies

Egg

Oogonium

Zygote

Sexual Reproduction

Daughter colony

Volvox

Matures

Daughter colonies released when *Volvox* dies

(b)

Filamentous Series

In this series you will study two species in two genera: *Ulothrix,* which is isogamous, and *Oedogonium,* which is heterogamous. What differences do you expect?

Obtain a prepared slide of *Ulothrix* and look at it with your compound microscope. This is a common filamentous green algae in streams and lakes. Each unbranched filament is composed of cylindrical cells joined end to end (fig. 14.9). Some filaments have a basal cell that serves as a **holdfast.** Each cell has a single collar-shaped chloroplast that surrounds the cytoplasm and the nucleus.

Scan along the filament and note that you can identify different types of cells. Cells with no specializations are called **vegetative** cells. Others are reproductive cells. Find a cell in which the contents have divided into 2, 4, 8, or 16 cells inside the cell walls. These cells are **zoospores** and are asexual reproductive cells. When released, they swim away by means of four flagella. Each cell can divide and produce a new filament.

You should be able to find some cells, called **gametangia,** in which the contents have divided into 16, 32,

or 64 cells, each with two flagella. These are **isogametes** and when released will fuse with other isogametes to form a **zygote.**

The zygote usually secretes a heavy cyst wall and does not divide until the following spring. The zygote is diploid and will undergo meiosis to produce four zoospores, which each can give rise to new vegetative filaments. The zygote is the only diploid cell in the life cycle. What important process is happening during fusion and meiosis that is an advantage to the species?

Obtain a slide of *Oedogonium* and look at it through your compound microscope. Scan along the filament and find a vegetative cell, which should be examined under high power (fig. 14.10). It contains a net-shaped chloroplast that surrounds the cytoplasm and nucleus.

Some of the cells in the filament will be dark colored and swollen. These are **oogonia** and contain a single egg. In fact, the genus name *Oedogonium* means enlarged egg cell. Find other cells in the filament that are short and disk-shaped. These are **antheridia** and each produces motile sperm. Mature sperm have a crown of flagella on the anterior end. When released, sperm swim to and enter the oogonium

Figure 14.9   Life cycle of *Ulothrix*.

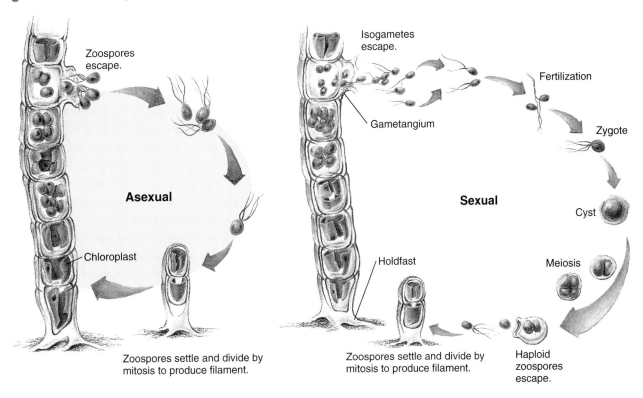

Figure 14.10   Life cycle of *Oedogonium*.

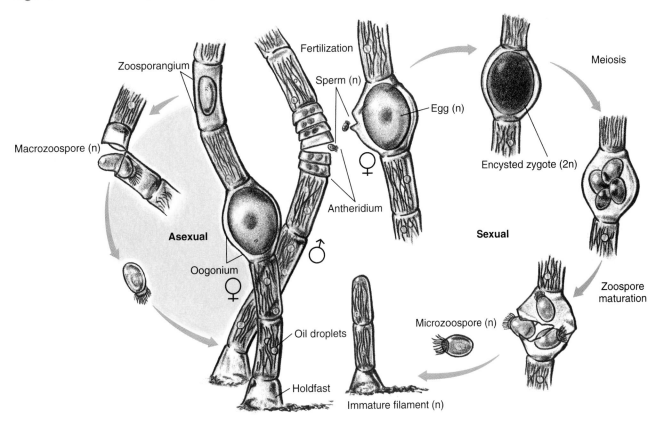

Diversity Among Protists

where fertilization occurs. Zygotes are identifiable by a thick cell wall that surrounds them inside of the oogonia. Zygotes eventually divide by meiosis to produce four **micro-zoospores.** When they escape from the oogonium, they form new filaments. Asexual reproduction occurs when vegetative cells differentiate into large **macrozoospores,** which leave the filament and give rise to new filaments. Is *Oedogonium* isogamous or heterogamous? _____

Before leaving the filamentous green algae, you should examine one other species. Obtain a prepared slide of *Spirogyra.*

Also known as green silk, this species is usually found in cool, clear, running water. Vegetative cells are cylindrical and contain a single, spiral-shaped chloroplast (fig. 14.11). *Spirogyra* is isogamous and does not produce motile gametes. Gametes are transferred by the process of **conjugation.**

Scan your slide until you find two adjacent filaments where tubes are growing outward toward one another (fig. 14.11). One filament is of mating type + and the other −. When the tubes meet they will fuse, forming a connection between the two cells. The contents of one cell will move through this conjugation tube and fuse with the other cell to form a zygote. The zygote overwinters and then will give rise to four haploid nuclei. Three will degenerate and the fourth will grow to produce a new filament.

Leafy Series

Examine either a living or plastic mounted specimen of *Ulva,* the sea lettuce (fig. 14.12). It typifies another line of evolution among the green algae. The leaflike body of this algae, the **thallus,** is multicellular and is only two cells

Figure 14.11 Conjugation between mating types in *Spirogyra:* (*a*) vegetative cells grow as filaments and have a spiral-shaped chloroplast; (*b*) sexual reproduction (conjugation) starts when tubes grow outward from adjacent filaments; (*c*) condensed protoplast leaves male filament and enters female filament; (*d*) zygotes are formed when protoplasts fuse.

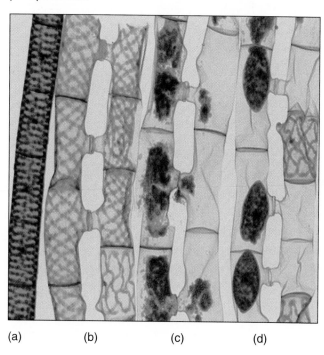

(a)   (b)   (c)   (d)

Figure 14.12   *Ulva,* the sea lettuce: (*a*) mature individuals in the ocean; (*b*) life cycle of *Ulva.*

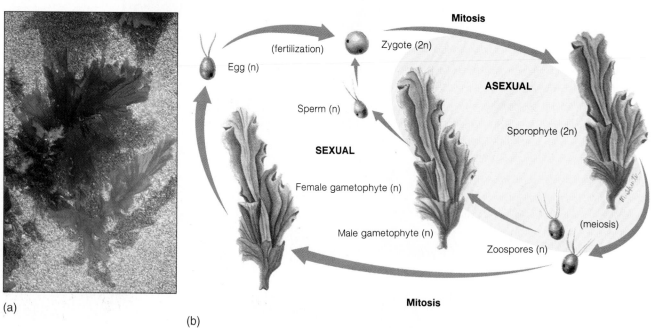

(a)

(b)

thick. Some of the cells at the base are modified into a holdfast, which anchors the thallus in place against tidal currents.

If the number of chromosomes in mature individual *Ulva* are counted, some individuals are diploid and others haploid. The diploid individuals result from the formation of a zygote by the fusion of isogametes. The isogametes are formed by the haploid individuals. Haploid individuals grow from zoospores, motile cells produced by meiosis of cells in the diploid individuals. Thus, *Ulva* shows a clear-cut pattern of **alternation of generations.** The diploid individuals represent the **sporophyte** generation and the haploid individuals are called **gametophytes.** This life cycle pattern is found in all plants and will be studied in detail in lab topics 15 and 16.

Amoeba Group

Phylum Rhizopoda

Commonly known as amoebas, all members of this phylum are unicellular and move by means of pseudopodia. Meiosis and sexual reproduction do not occur in this phylum. All reproduction is asexual. Amoebas are common in soils, freshwater, and marine habitats where they feed on detritus or nonmotile bacteria and algae.

▶ Obtain a prepared slide of *Entamoeba histolytica,* the causative agent of amoebic dysentery, and look at it with your compound microscope. The amoeboid stages on the slide are called **trophozoites.** The other stages on the slide are **cysts.** How do cysts differ from trophozoites? Find examples of both and draw them in the following circle.

Cysts are formed when amoebas secrete a proteinaceous cell wall around themselves and enter dormancy. In the cyst stage, amoebas can survive prolonged periods of dryness and are often dispersed by wind. If a cyst is ingested by a host, the trophozoite is activated and crawls out of the cyst. It burrows into the intestinal wall and causes

cramps and diarrhea, which can lead to death of the host by dehydration. In heavy infections, this organism will sometimes migrate through the tissues and invade the liver, reducing liver function.

Phylum Actinopoda

Meaning ray-feet, the name of this phylum describes the delicate pseudopodia that radiate from the bodies of these organisms (fig. 14.13). Bacteria and detritus stick to the pseudopodia and are engulfed. Radiolarians are marine members of this phylum. They have a delicate shell made of silicates.

▶ Obtain a slide containing a mixture of radiolarian shells. Look at it with your compound microscope. Compare to figure 14.13(b).

Figure 14.13 Radiolarians. (*a*) diagram of cell; (*b*) skeletons.

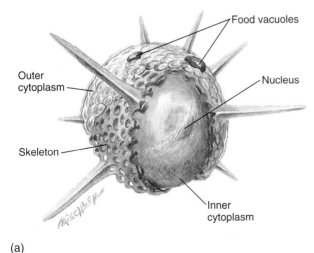

(a)

(b)

Diversity Among Protists　**159**

Mold Group

Phylum Myxomycota

Plasmodial slime molds are common in moist leaf litter and decaying wood. There are about 500 known species. During the growth phase of its life cycle, it has a body form called a **plasmodium,** a nonwalled mass of cytoplasm containing hundreds of nuclei. Plasmodia may be bright orange or yellow in color (fig 14.14). The cytoplasm in a plasmodium streams back and forth and allows the organism to locomote, moving several meters until it "finds" food. They are able to pass through a mesh cloth. If you can imagine a large cytoplasm several inches long "crawling" about, you recognize where the common name slime mold originates. They are not closely related to the cellular slime molds found in another phylum in this grouping. If food or moisture are reduced, plasmodial slime molds differentiate to form **sporangia** which produce spores that are haploid. The

spores are resistant to harsh conditions but will germinate in a moist environment with food. Cells emerging from spores may fuse to start a new plasmodium.

Your instructor will have *Physarum* growing in the lab on 1% water agar with a few flakes of oatmeal for food. Several hours ago a piece of the slime mold was transferred to a fresh petri dish and was set up under a demonstration microscope so that you can see the cytoplasmic streaming. Watch for a few minutes and note how it changes direction. What purpose might this serve in a large organism that feeds by secreting digestive enzymes and absorbing the products of digestion?

Learning Biology by Writing

Your instructor may ask you to answer the questions in lab summary 14 and turn in the answers at the next lab period.

Lab Summary Questions

1. How do protists differ from bacteria and cyanobacteria?
2. Make a list of the genera (plural of genus) observed in this lab. Indicate the candidate kingdom and phylum into which each is classified.
3. Describe what is meant by isogamy and heterogamy. Which is considered more advanced? Why?
4. Why are the protist phyla Rhizopoda, Zoomastigophora, and Ciliophora considered animal-like protists? How do these three phyla differ from one another in terms of locomotion?
5. Describe how the volvocine series of algae (*Chlamydomonas, Pandorina,* and *Volvox*) can be interpreted as representing the evolutionary developmental stages leading to multicellularity.
6. Describe how the filamentous algae series (*Ulothrix* and *Oedogonium*) and the leafy alga *Ulva* may be interpreted as representing the evolutionary developmental stages leading to green plants.
7. Why are protists considered to be important members of ecosystems?

Internet Sources

On the World Wide Web there are several databases that list information about diseases caused by protozoa. The U.S. Center for Disease Control (http://www.cdc.gov) is one and the World Health Organization (http://www.who.int/tdr) is another. The WHO lists ten major tropical diseases. Of these, four are caused by protists that are transmitted by biting insects. The diseases are Chagas disease, leishmania, malaria, and African sleeping sickness. Connect to one of these sites and read about how the diseases are transmitted, what the symptoms are, how they are treated, and where the diseases are common.

Critical Thinking Questions

1. Describe the different adaptations that protists exhibit for withstanding a harsh environment. Compare these adaptations to those of bacteria and cyanobacteria.
2. Diatomaceous earth is classified as an "insecticide" by gardeners and is used to control such pests as ants, snails, slugs, and caterpillars. Explain its insecticidal properties.
3. Why aren't protist cells larger? Relate the shape of protist cells to their function and their environment.
4. Speculate on how colonial aggregates of cells may have led to multicellularity.

LAB TOPIC 15

Investigating Plant Phylogeny: Seedless Plants

Supplies

Preparator's guide available on WWW at
http://www.mhhe.com/dolphin

Equipment

Compound microscopes
Dissecting microscopes

Materials

Living plants
  Coleochaete culture
  Marchantia with gemmae
  Moss gametophytes with sporophytes (*Polytrichum*)
  Moss spores germinated on agar
  Mature ferns
  Where possible, examples of lower seedless plants
  Fern spores
  C-fern sexual differentiation kit from Carolina
     Biological or Fern prothalli growing on agar
Prepared slides
  Coleochaete w.m.
  Marchantia thallus, cross section
  Marchantia antheridia, longitudinal section
  Marchantia archegonia, longitudinal section
  Marchantia sporophyte, longitudinal section
  Moss sperm
  Mature moss capsule, longitudinal section
  Mature moss antheridia and archegonia,
     longitudinal sections
  Fern antheridial and archegonial prothalli
  Fern sporophyte growing from archegonium
  Fern rhizoid, cross section
Slides, coverslips, razor blades
Pasteur pipettes

Solutions

Phloroglucinol stain

Prelab Preparation

Before doing this lab, you should read the introduction
and sections of the lab topic that have been scheduled
by the instructor.

You should use your textbook to review the
definitions of the following terms:

antheridium
archegonium
Bryophyta
gametangium
gamete
gametophyte
lignin
Lycophyta
phloem
protonema
Pterophyta
rhizoid
Sphenophyta
sporangium
spore
sporophyte
thallus
xylem

You should be able to describe in your own words
the following concepts:

Alternation of generations
Vascular tissue (xylem and phloem)
Reproduction in seedless plants
Adaptations to land

As a result of this review, you most likely have
questions about terms, concepts, or how you will do
the experiments included in this lab. Write these
questions in the space below or in the margins of
pages of this lab topic. The lab experiments should
help you answer these questions, or you can ask your
instructor for help during the lab.

TABLE 15.1 Taxonomic divisions of plants

Division	Common Name	Number of Living Species
Nonvascular Plants		
Seedless plants		
Bryophyta	Mosses	10,000
Hepatophyta	Liverworts	6,500
Anthocerophyta	Hornworts	100
Vascular plants (Tracheophyta)		
Seedless plants		
Lycophyta	Club mosses	1,000
Sphenophyta	Horsetails	15
Pterophyta	Ferns	12,000
Seed plants		
Gymnosperms		
Coniferophyta	Conifers	550
Cycadophyta	Cycads	140
Ginkgophyta	Ginkgo	1
Gnetophyta	Gnetae	70
Angiosperms		
Anthophyta	Flowering plants	235,000

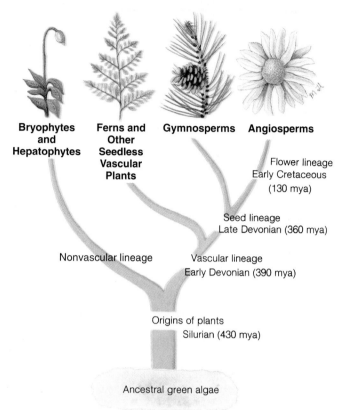

Figure 15.1 A summary of plant evolution. Note the early origin of both Bryophytes and Tracheophytes from the green algae (mya = millions of years ago).

Bryophytes and Hepatophytes

Ferns and Other Seedless Vascular Plants

Gymnosperms

Angiosperms

Flower lineage Early Cretaceous (130 mya)

Seed lineage Late Devonian (360 mya)

Nonvascular lineage

Vascular lineage Early Devonian (390 mya)

Origins of plants Silurian (430 mya)

Ancestral green algae

Objectives

1. To observe the body plans of liverworts, mosses, and ferns
2. To learn the life cycle stages in the alternation of generations found in mosses, liverworts and ferns
3. To collect evidence that tests the following null hypothesis: There is no evidence that seedless plants can be arranged in a phylogenetic sequence demonstrating increasing adaptation to a terrestrial environment.

Background

Plants are a diverse group of multicellular, photosynthetic organisms ranging from simple mosses to complex flowering plants. All plants, and some green algae, share certain characteristics, which include: (1) chloroplasts with thylakoid membranes stacked as grana and containing chlorophylls a and b, (2) starch as a storage polysaccharide in the chloroplasts, (3) cellulose in cell walls, (4) cytoplasmic division by cell plate (phragmoplast) formation, and (5) complex life cycles involving an alternation of generations in which a diploid sporophyte stage alternates with a haploid gametophyte stage; (6) heterogamy with gametes produced in organs called gametangia.

Botanists prefer to use the term "division" rather than phylum for the major taxonomic categories within the plant kingdom and, in turn, divide the divisions into classes, orders, families, genera, and species. Some botanical taxonomists divide the kingdom Plantae into two divisions, the Bryophyta (mosses and relatives) and the Tracheophyta (vascular plants), while others recognize eleven divisions (table 15.1).

Phylogeny is the study of evolutionary relationships among groups of organisms. The evolutionary story of plants is linked to the colonization of the terrestrial environment, an event that started about 400 million years ago (fig. 15.1). Most botanists agree that plants arose from an ancestral green algae. For the first 3.5 billion years, no life existed on the land, but at the end of the Silurian period, the first land plants appeared and rapidly diversified. The appearance of producer organisms in the terrestrial environment established the basis for terrestrial food chains, which were then rapidly exploited by the evolution of terrestrial fungi and animals from aquatic ancestors.

As do many of their algae ancestors, all plants show an **alternation of generations** in their life cycle. In plants, there are two alternating stages. The diploid **sporophyte** generation produces haploid spores by meiosis. These, in turn, produce plants of the **gametophyte** generation, which

Figure 15.2 Generalized concept of alternation of generations, in which a species has distinct haploid and diploid individuals in its life cycle.

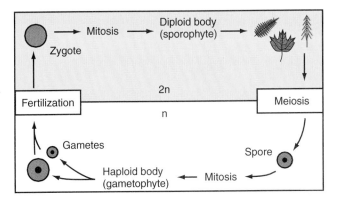

produce haploid gametes by mitosis. When gametes fuse at fertilization, the fertilized egg (zygote) is the first cell of a new sporophyte generation (fig. 15.2).

Plant Adaptations to Land

The critical steps for plants in making the transition from aquatic to terrestrial environments involved the development of adaptations to prevent the loss of water. Land plants are covered by a waxy **cuticle** that slows the evaporation of water from tissues, and the reproductive organs are covered by a protective covering, the **gametangium,** a jacket of sterile (nonreproductive) cells that prevents the drying of gametes and embryos. Furthermore, land plants have modified basal organs (roots) that absorb water and minerals from the substrate to replace what is lost by evaporation. Hereditary variations of this type arising in ancient shoreline algae would have allowed them to enter an environment that, though harsh in its physical characteristics, was free from competitors for resources. Such an environment would have allowed the colonizers to establish themselves and to further diversify.

Once the transition to land occurred, two major lines of evolution developed among the land plants. The direction the nonvascular plants (table 15.1) took saw the gametophyte generation become the dominant photosynthetic stage in the life cycle. Following fertilization, the gametophyte retained the embryo and it developed into a sporophyte that was dependent on the gametophyte. **Spores** that were resistant to desiccation became the primary means of dispersal.

The other evolutionary alternative, the tracheophyte line, gave rise to vascular plants (table 15.1). In these plants, cells located in the central portion of the shoots became elongated and specialized for transport and support, the progenitors of **xylem** and **phloem.** A vascular cambium developed, which allowed secondary growth of the stem and root. These tissues allowed tracheophytes to grow to large sizes, which, in turn, favored development of more efficient anchorage and absorptive systems. At the same time,

reproductive adaptations occurred, giving rise to the seed plants, which you will study in Lab Topic 16.

In seed plants, such as conifers and flowering plants, the gametophyte stage of the life cycle became reduced to a few cells encased in the reproductive organs. Male gametophytes became desiccation-resistant **pollen** which carried sperm that no longer needed water to swim to the eggs. Eggs were retained in a protective organ where, following fertilization, they and surrounding tissues developed into seeds containing an embryo and food reserves. In this group, **seeds,** rather than spores, became the effective dispersal mechanisms for sexually produced offspring. Seeds allowed the plant embryos to be carried overland by wind, water, or animals, allowed a period of dormancy, and provided initial nourishment for germination and growth in the new location.

The efficiency of the adaptations of plants to land over eons of time allowed them to invade many different environmental niches. Relatively simple changes allowed the colonization of swamps and bogs, while more complex adaptations were necessary for life in desert environments. As you study the trends outlined in this lab topic, remember that contemporary plants in different divisions have not descended from one another. They have similarities because they share common ancestors. Present species are related as cousins or more distant relatives, not as you are to your parents or siblings.

LAB INSTRUCTIONS

You will study the terrestrial adaptations of liverworts, mosses, and ferns. You will learn the distinctions between the sporophyte and gametophyte stages in the alternation of generations that occurs in these plants.

Plants May Have Originated from Pre-adapted Green Algae

In Lab Topic 14, you observed colonial, filamentous and leafy green algae that shared many characteristics with plants. Most botanists believe that green algae (Chlorophyta) in the class Charophyceae are the green algae most closely related to plants for a number of reasons: cytokinesis during mitosis is similar in plants and these green algae but differs from that in other green aglae; sperm structure is similar in these algae and mosses; and the fertilized egg is retained and protected by the parent green alga, similar to what happens in plant alternation of generations life cycles. We will start our study of plants by looking at *Coleochaete,* a multicellular, leaf-like green alga that has characteristics representative of the ancestors of kingdom Plantae.

Look at a prepared slide or make a slide from living specimens of *Coleochaete* with the scanning power of your compound microscope. These algae grow as epiphytes on aquatic plants or on rocks in shallow water. They increase

Investigating Plant Phylogeny: Seedless Plants

Figure 15.3 Thallus of the liverwort *Marchantia* in background with stalked sexual reproductive structures.

in size by cell divisions at the margin of the leaf-like disk called a **thallus.** The cells of the thallus are haploid. Sketch a thallus below.

The close relationship with plants is seen during sexual reproduction. *Coleochaete* is oogamous: eggs produced by mitosis near the thallus margin are fertilized by motile sperm that swim to the egg. During maturation of the diploid zygote, non-reproductive (sterile) haploid cells from the surrounding parental thallus grow over and encase the zygote, a situation similar to moss life cycles. Meiosis follows a period of dormancy producing zoospores that can swim off to develop into new haploid thalli. What makes this organism so interesting is the non-reproductive (sterile) jacket of cells that encases the zygote while meiosis occurs. Aquatic algae that protected the zygote would have been pre-adapted to survive occasional conditions of drying as water levels fluctuated. With a few adjustments, these alga/plant-like organisms might have been able to survive and reproduce in moist environments, starting the colonization of land.

Bryophytes

The term "bryophyte" has been used traditionally to describe plants belonging to three divisions: Hepatophyta—the liverworts, Anthocerophyta—the hornworts, and Bryophyta—the mosses. The bryophytes illustrate a clear alternation of generations in which the gametophyte is the predominant stage of the life cycle. The sporophyte stage of the life cycle is reduced and nutritionally dependent on the gametophyte for its survival, a situation similar to that seen in the green alga *Coleochaete*. Reproduction occurs sexually by gametes and asexually by fragmentation or dispersal by spore formation. Although the bryophytes are land plants, they are not completely adapted to a terrestrial way of life. Motile sperm require water films to travel from the male reproductive organs to the eggs held in female reproductive organs in order for the life cycle to be completed. Most bryophytes are small, less than 10 cm tall, due in part to their lack of xylem and phloem.

Division Hepatophyta (Liverworts)

The name of this division is derived from the fanciful resemblance of the shape of some of these plants to the human liver and from the root word *wort* meaning herb. Two body types are found among the species of liverworts: the thallose type and leafy type.

Gametophyte Stage

▶ Examine living specimens of *Marchantia*, a thallose-type liverwort, which is commonly found on the surfaces of damp rocks and soil. Individual plants consist of a body called a **thallus** (fig. 15.3). You are looking at the haploid gametophyte stage of the two stages found in the life cycle. The surface of the thallus has a diamond-shaped pattern, which corresponds to air chambers located beneath the upper epidermis. Note the numerous hair-like **rhizoids** on the lower surface. These are an adaptation for water and mineral absorption. Projecting above the upper surface may be tiny cuplike structures called **gemmae cups.** These produce asexual **gemmae,** small, multicellular discs of green tissue, which develop into new gametophytes when dispersed to suitable locations.

▶ Obtain a slide of a cross section of a thallus and look at it with your compound microscope (fig. 15.4). Sketch the cross section in the circle that follows. Indicate the upper and lower **epidermis.**

? Is there evidence of a **cuticle?** _____ Note the **air pores** on the upper surface, which open into **air chambers.** Beneath the chambers is a layer of **chlorenchyma** tissue where the cells contain chloroplasts. Beneath this layer is the lower **storage tissue.** Are chloroplasts present in these cells? _____ Based on your observations, which of these layers is photosynthetic? _____

? Projecting from the lower epidermis note the **rhizoids,**

Figure 15.4   Stages in the life cycle of the liverwort *Marchantia*.

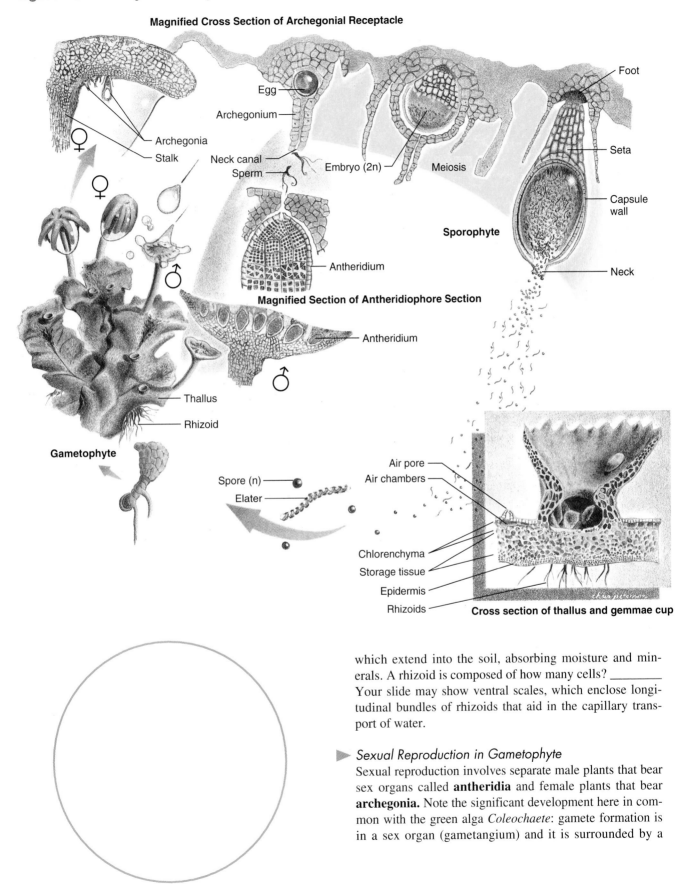

Magnified Cross Section of Archegonial Receptacle

Foot

Egg

Archegonium

Archegonia

Stalk

Neck canal

Sperm

Embryo (2n)

Meiosis

Seta

Capsule wall

Sporophyte

Neck

Antheridium

Magnified Section of Antheridiophore Section

Antheridium

Thallus

Rhizoid

Gametophyte

Spore (n)

Elater

Air pore

Air chambers

Chlorenchyma

Storage tissue

Epidermis

Rhizoids

Cross section of thallus and gemmae cup

which extend into the soil, absorbing moisture and minerals. A rhizoid is composed of how many cells? _____
Your slide may show ventral scales, which enclose longitudinal bundles of rhizoids that aid in the capillary transport of water.

▶ *Sexual Reproduction in Gametophyte*
Sexual reproduction involves separate male plants that bear sex organs called **antheridia** and female plants that bear **archegonia.** Note the significant development here in common with the green alga *Coleochaete*: gamete formation is in a sex organ (gametangium) and it is surrounded by a

protective layer of nonreproductive tissue. What advantage is this development to a land plant?

In the proper season, small stalked **archegoniophores** or **antheridiophores** grow from the notches in the tips of thalli (fig. 15.4) The sex organs develop on the caps of these stalks. Archegoniophores look like fleshy minature palm trees while antheridiophores are somewhat like a large, flat cone, both at the tip of stalks.

Look at a slide of a longitudinal section of an antheridiophore, the male reproductive structure. Note the general shape and the multicellular nature of the organ. Beneath the upper surface are several ovoid **antheridia** surrounded by a jacket of sterile cells. Sperm develop by mitosis inside the sterile jacket and are released through the canal that leads to the upper surface (fig. 15.4). Several hundred sperm are produced in each antheridium.

Now look at a slide of a longitudinal section of an archegoniophore, the female reproductive structure. In rows on the underside are flask-shaped **archegonia** surrounded by a sterile jacket of cells (fig. 15.4). At the base of each archegonium is a swollen area containing the **egg.** Eggs are produced by mitosis, as are sperm. Are eggs and sperm haploid or diploid? _____ Why? Extending downward from the egg is a **neck canal.** Sperm must swim up this canal to fertilize the egg.

Fertilization requires water. Raindrops falling on the antheridia splash sperm onto adjacent archegonia. Using their flagella, the sperm swim to the egg. It develops into the sporophyte stage of the liverwort, embedded in the parent's tissue and surrounded by sterile cells on the un-

derside of the archegoniophore's cap (see yellow area of fig. 15.4).

Sporophyte Stage

Obtain a slide with maturing sporophytes on the under surface of an archegonial receptacle. Following fertilization, the stalks of the archegonial receptacles elongate. The zygote divides repeatedly, forming the multicellular sporophyte stage within the tissue of the gametophyte. Find the sporophyte stage on the slide and study it using low power. Most of the cells in the sporophyte form a **capsule,** which surrounds other cells that become **spore mother cells** and **elaters,** the latter being slender, elongate cells (fig. 15.4). Together these cells make up the **sporangium.** Thus, the sporophyte generation of the thalloid liverwort is a small sporangium attached to the lower surface of the cap of a female receptacle growing from the gametophyte thallus.

Inside the sporangium, each spore mother cell undergoes meiosis, forming four haploid spores. When the capsule dries, it ruptures, exposing a cottony mass of spores and elaters. The elaters, intertwined in this mass, are hygroscopic. Spiral thickenings in their cell walls cause the elaters to twist and coil as they dry or take up moisture, dislodging spores from the mass. The spores are disseminated by the wind and, if they land in a moist area, will germinate to form a new gametophyte generation.

When you finish this section, quiz your lab partner about the life cycle of *Marchantia* with emphasis on the alternation of generations.

Division Bryophyta (Mosses)

Many organisms commonly called "mosses" are not even plants. Reindeer moss is a lichen, Spanish moss is a vascular plant, and Irish moss is an alga. True mosses have the following anatomical and reproductive features.

Gametophyte Stage

Several days before the lab period, moss spores, which are encapsulated haploid cells produced by meiosis, were placed on agar medium in petri plates. By the time of the lab, the spores should have germinated producing **protonemata,** algalike filaments of cells.

Remove one from the culture, make a wet-mount slide of it, and observe through your compound microscope. Each **protonema** is a branching thread of single cells joined end to end. Does this remind you of the general appearance of a filamentous alga? _____ The colorless branches will develop into **rhizoids** (fig. 15.5), which are analogous to the roots of higher plants. Leafy, green **gametophytes** will develop from **buds** on the protonema. As the buds develop, the protonema will continue to grow and form new buds. Protonema can spread to cover an area 40 cm in diameter in a matter of months. Sketch what you see on your slide.

Figure 15.5 Stages in life cycle of a moss.

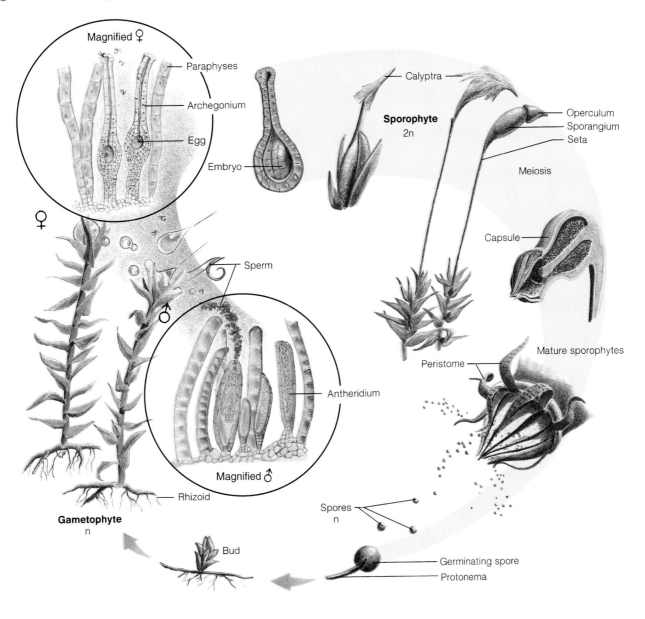

Now obtain a whole-mount slide that shows a young gametophyte developing from a protonema. Because the gametophyte develops by mitotic division and differentiation, all of its tissues are haploid. The leaves are usually a single cell thick and are arranged in whorls of three around the central axis tissue.

As this stage matures, sex organs, or gametangia, differentiate at the tip of the stem. **Archegonia,** the egg-producing organs, will develop on one gametophyte and **antheridia,** the sperm-producing organs, on another, although some gametophytes will bear both. Between the several gametangia at the tips are filaments of nonreproductive cells called **paraphyses.**

Sexual Reproduction in Gametophyte

▶ If mature mosses are available, use a dissecting microscope to observe while dissecting the tips of several stems under a drop of water on a slide. The gametangia will be protected by several leaves. Male gametangia will often be orange in color because of chromoplasts in the sterile jacket cells. Female gametangia often have embryonic sporophytes growing from their tips. To see greater detail look at the dissected tip on the slide, using the 4× objective on your compound microscope. Sketch what you see.

If live mosses are not available for dissection, examine a prepared slide of a longitudinal section of a female gametophyte. Look at it first with your lowest power objective, increasing magnification where necessary to see detail. At the tip look for the bowling-pin-shaped structures, the archegonia between thin filamentous paraphyses. Note that the archegonium is surrounded by a sterile jacket of cells. They protect the egg during development. The single egg is located in the swollen base. The cells of the gametophyte are haploid. Does the egg develop by mitosis or meiosis from the gametophyte cells? _____ The narrow upper region of the archegonium is called the neck and you may be able to see the **neck canal** in some of the archegonia. Sperm must swim down this canal to fertilize the egg.

▶ Now obtain a prepared slide with a longitudinally sectioned tip of a male gametophyte. At the tip, find the oval-shaped antheridia between the thin filamentous paraphyses. The antheridia are surrounded by a thin layer of sterile jacket cells that protect the sperm during development. The tightly packed angular cells inside the jacket are sperm in various stags of development. Do the sperm develop by mitosis or meiosis? _____ If a water film develops on the tip of sexually mature male gametophyte, the sterile jacket of cells splits open releasing sperm into the water film. After a few minutes they begin to actively swim, using two flagella.

Moss sperms must pass from the male gametangia through a film of water to the eggs in the archegonium. ▶ Often raindrops will splash sperm from the antheridium on one gametophyte to the archegonium on another. Obtain a prepared slide of moss sperm. Observe with high power through your compound microscope and sketch a sperm.

Sporophyte Stage

Zygotes develop into embryos surrounded by a protective jacket of sterile archegonial cells, a situation similar to that seen in the green alga *Coleochaete*. The developing diploid embryo marks the start of the next generation, the sporophyte. The growth of the zygote into a multicellular diploid stage by mitosis differs from the situation seen in the green alga *Coleochaete*. It represents an evolutionary progression from the ancestral condition where the diploid zygote becomes a stage in the life cycle. As the embryo grows by mitotic division, differentiation occurs, and a long stalk of sporophyte tissue, the **seta,** grows upward. The end of the seta differentiates into a **capsule** covered by the **calyptra,** derived from surrounding archegonium tissue. The top of the capsule is closed by a lid, or **operculum.** Inside the capsule, meiosis occurs, producing haploid spores. These are the same types of cells that grew and produced the protonemata described at the beginning of this section. When the capsule matures, the operculum drops off, revealing the **peristome.** This ring of toothlike units flexes with changes in humidity releasing the capsule's spores during periods of dryness.

▶ Spores remain viable for several years and are the primary agents of dispersal for the species. If mature capsules are available, observe them under a dissecting microscope, removing the calyptra and operculum to release the spores. Also observe a longitudinal section of a mature capsule on a prepared microscope slide. Sketch the anatomy of the capsule.

ence of lignin relate to the potential of each plant group for reaching larger sizes?

❓ In what ways are mosses adapted to land? In what ways are they still dependent on water?

Nonvascular Nature of Mosses (Optional)

The conduction system of mosses is not nearly as well developed as in the vascular plants. In the vascular plants, the conduction vessels are strengthened by a compound called **lignin,** so that the vessels also serve as support elements. Lignin is not found in the mosses.

▶ Obtain a cross section and, if possible, a longitudinal section of a moss gametophyte stem, such as *Polytrichum.* Note the heavy walled cells in the center of the stem. These cells are specialized for water conduction; most of the larger species of moss have them.

▶ A histochemical test for lignin can be easily performed in the lab. Hold a piece of an older moss stem between the surfaces of a folded sheet of parafilm (see fig. 25.5) and use a fresh razor blade to cut several thin cross sections. Do the same with the stems of some small ferns or woody plants. Place the sections on a microscope slide and add a drop of phloroglucinol stain.

CAUTION

Be careful with this stain since it contains concentrated hydrochloric acid and can damage your skin, eyes, clothes, or microscope.

Wait at least 20 minutes, and then transfer each stem section to a drop of water on a new slide. Add coverslips and observe each stem with a compound microscope. ❓ Which stem sections are red? What does this indicate about the chemical composition of the stems? How does the pres-

Tracheophytes

Division Pterophyta (Ferns)

As a group, the tracheophytes (table 15.1) exhibit a number of adaptations to living in a terrestrial environment. There is a general trend across the group whereby the plant body is differentiated into a subsoil root system for the absorption of water and minerals, and an aboveground stem that can be quite tall in comparison to bryophytes. The leaves are often greatly enlarged and adapted to high rates of photosynthesis. This regional specialization of the plant body required the development of new types of plant tissues, the **vascular tissues** consisting of **xylem** and **phloem,** that form a transport system. The cylindrical cells of the xylem function in carrying absorbed water and minerals to the aboveground parts of the plant, while the elongated phloem cells conduct the products of photosynthesis from the leaves to all other parts of the plant. The cell walls of the xylem are reinforced with lignin, so that the xylem assumes an additional function of being a support tissue. Consequently, most tracheophytes are larger than the nonvascular plants and more diverse.

The tracheophytes include both seedless and seed-bearing plants. The seedless tracheophytes include the club mosses, horsetails, and ferns (see fig. 15.6). Your instructor may have brought living specimens from the greenhouse into the lab. If so, compare them to the photos and note the following features. *Psilotum,* a whisk fern, is unique among the tracheophytes because it lacks roots and leaves. Root function is performed by a stem that grows underground and endomycorrhizal fungi that absorb water and minerals. Leaf function is carried out by the photosynthetic stem. Club mosses, such as *Lycopodium,* are low-growing herbs. They have a belowground stem that gives off roots and aerial branches. Horsetails, such as

Figure 15.6 Sporophyte stages of the fern allies: (*a*) *Psilotum*, a whisk fern, has an aboveground and belowground stem that functions in water and mineral absorption; (*b*) *Lycopodium*, a club moss sometimes called a ground pine or cedar, has a below ground stem that produces roots and an aboveground stem that bears leaves and sporangia; (*c*) *Equisetum*, a horsetail, has jointed aboveground stems that bear whorls of leaves and a belowground true vascular root. Some stems bear a terminal reproductive structure called a strobilus (foreground) which produces spores that give rise to the gametophyte generation.

(a)

(b)

(c)

Equisetum, have underground stems called rhizomes that give rise to aerial branches that have jointed stems bearing whorls of leaves. Various silicon compounds are found in the stem and a common name in colonial times was "scouring rushes" because the abrasive nature of the stems made them useful in cleaning pots.

In this section, you will study the life cycle stages of ferns as the example of seedless, vascular plants. The sporophyte generation is the most conspicuous stage in the life cycle of ferns, but they also have an independent gametophyte stage. In Lab Topic 16 you will study the vascular plants that produce seed.

Gametophyte Stage

▶ Make a wet-mount slide of some fern spores and look at them with a compound microscope. The haploid spores are produced by meiosis and by the differentiation of cells in **sori** (singular *sorus*), organs on the underside of the sporophyte leaf (fig. 15.7b).

▶ If they are available, examine spores that germinated on an agar surface. Make a wet-mount slide of the germinated spores and observe the protonemalike threads. These cells will differentiate into a gametophyte stage called a **prothallus** (fig. 15.7c) that has a heart-shaped leaf and rootlike **rhizoid** areas. This stage is haploid and will grow by mitotic cell division with energy input from photosynthesis. Sketch a germinating spore below.

Sexual Reproduction in Gametophyte

Under appropriate conditions, certain cells in the prothallus stage will differentiate into eggs and multiflagellated sperm.

While many textbooks and figure 15.7c show that a single prothallus will produce both egg and sperm, this may not be the case in nature. In many species, the earliest developing prothalli (plural) produce eggs and the slower developing ones produce sperm. A hormone produced by the faster growing prothalli may influence sex organ develop-

(a)

(b)

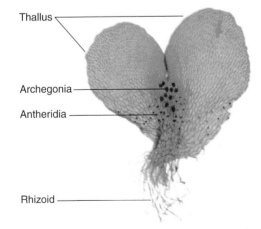

Thallus

Archegonia

Antheridia

Rhizoid

(c)

ment in the slower growing ones. A mating system such as this promotes cross fertilization. What is the evolutionary advantage of cross fertilization?

Obtain two slides showing the different sexual structures of the fern gametophyte (prothallus). One should show the male or sperm-producing gametangia called the **antheridia.** Antheridia are located on the lower surface in the center of the heart-shaped, leaflike thallus near the rhizoids. The other slide will have **archegonia** containing eggs. These are located near the cleft or notch of the thallus. A water film must cover the lower surface of the prothallus for sexual reproduction to occur. The sperm swim through this film to the archegonium, where one sperm penetrates and fertilizes the egg (fig. 15.8).

This fertilized egg is the start of the next generation, the sporophyte, in the alternation of generations. As the diploid sporophyte grows by mitotic division and photosynthesis, the gametophytic prothallus withers and dies.

Gametophyte Stages (Optional)

Stages in the development of the gametophyte may be available in the lab. About two weeks and then every three days before this lab period, your instructor inoculated some spores of the fern *Ceratopteris* sp. (C-fern) on a growth medium. As a result you should have gametophytes in various stages of development, including some that are sexually mature.

Gametophyte stages should be removed from the culture and placed lower surface upward in a small amount of water in a depression slide and observed through your microscope. Sketch a series of developmental stages below.

Figure 15.8   Stages in the life cycle of a fern.

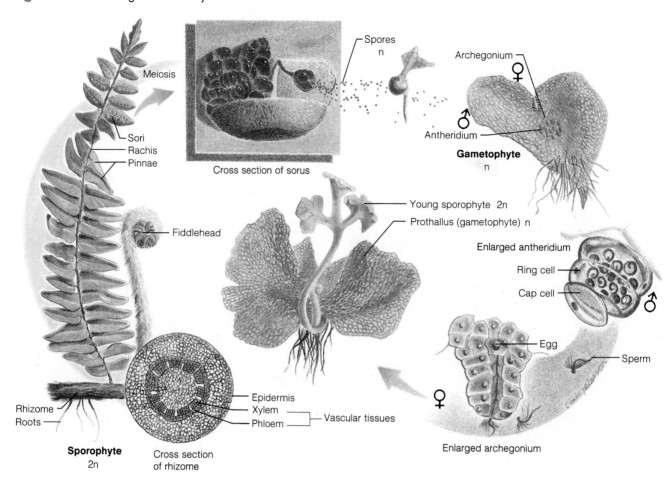

Meiosis

Sori
Rachis
Pinnae

Cross section of sorus

Spores
n

Archegonium ♀

Antheridium ♂

Gametophyte
n

Fiddlehead

Young sporophyte 2n
Prothallus (gametophyte) n

Enlarged antheridium

Ring cell
Cap cell ♂

Egg
Sperm

Epidermis
Xylem
Phloem — Vascular tissues

Rhizome
Roots

Sporophyte
2n

Cross section
of rhizome

♀

Enlarged archegonium

If the oldest gametophytes are sufficiently mature, it may be possible to observe motile sperm and fertilization. As you study the oldest gametophytes, identify the **gametangia** (sex organs), which project from the surface of the leaflike prothallus. When a gametophyte with gametangia is found, study it.

Antheridia may be anywhere on the surface of the gametophyte, and often are most frequent on small, irregular-shaped plants. If it appears to be constructed of three cells—a basal cell, a ring cell, and a cap cell—surrounding a mass of small cells, it is an antheridium (fig. 15.8). You may see motile sperm escaping.

Archegonia are composed of many cells, some embedded in the thallus and others projecting from the thallus as a neck. The neck is composed of four rows of cells with each row consisting of four cells.

Fertilization

▶ To observe fertilization, you must look at the gametangia immediately after placing the gametophytes in water. Select several mature gametophytes of different sizes and place them on a slide. Add a drop of water and immediately observe through your compound microscope.

The water causes the cap cells to pop off the antheridia, releasing sperm. Water also causes the apical neck cells to break open, creating a channel to the egg at the base of the

canal. Sperm swim to the neck and down the channel and fertilization occurs. You will have to scan several gametophytes on your slide to see all of these events. Everyone may not be fortunate enough to see sperm or fertilization.

Sporophyte Stage

Obtain a whole-mount slide of a young sporophyte growing from the archegonium of a prothallus and observe it under low or medium power with the compound microscope. Sketch what you see.

The sporophyte stage will eventually develop into well-defined leaves, stems, and roots with an interconnecting vascular system. The stems of most ferns are horizontal just beneath the soil surface and are called **rhizomes.** However, these structures are quite different from the rhizoids found in the gametophyte fern stage or in the mosses. Rhizoids are single, elongated cells joined end to end, which do not connect with vascular tissue. Rhizomes, on the other hand, are multicellular and contain vascular tissue.

Obtain a slide of a cross section of a rhizome and look at it through your compound microscope. Identify the vascular tissues located near the periphery of the rhizome. Each bundle is surrounded by a layer of endodermal cells and contains large, thick-walled xylem cells and smaller phloem cells. Sketch a vascular bundle below.

The rhizome extends under the surface of the soil and sends up leaves from its buds. These leaves break from the soil in a coiled position called a **fiddlehead** and gradually unroll to produce the fern **frond,** a single compound leaf with **pinnae** projecting from a central **rachis.** Roots with root hairs also develop from the rhizome and provide the sporophyte stage with water, minerals, and anchorage. Since all of the cells making up these structures arise from the zygote by mitosis, they are diploid.

On the undersides of the pinnae, small **sori** develop (fig. 15.8). They are composed of clusters of **sporangia.** Cells in the sporangia divide by meiosis to produce haploid cells that differentiate into spores. These spores are resistant to desiccation and aid in the dispersal of ferns. They are often sold as fern "seed." How do spores differ from seeds?

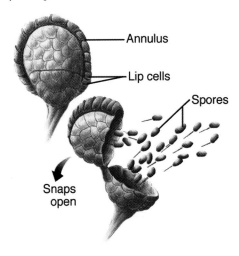

Figure 15.9   Spores are forcibly released by snapping back of reinforced annulus on one side of the sporangium as humidity changes.

As the sporangia mature, they dry and split on the side opposite the **annulus** (fig. 15.9). The annulus with its thickened cell walls snaps back explosively releasing the spores. If ferns with mature sporangia are available, try this experiment. Remove a pinna (leaflet) from a leaf and place it on a piece of white paper. Shine a bright, hot light on it while observing through a dissecting microscope. You may see explosive movements as the sporangia open.

The life cycle of a fern is summarized in figure 15.8. Recall the stages you have observed as you study this illustration. The adaptations of the fern to the terrestrial environment include the desiccation-resistant, dispersible spores; the well-developed vascular system; the root system; and the erect, desiccation-resistant photosynthetic fronds. Most of these adaptations are in the sporophyte stage, which, in contrast to the mosses, is the dominant stage in the life cycle.

When you finish this section, pair off with someone in the lab and quiz one another about the anatomy and alternation of generations in ferns.

Learning Biology by Writing

Write a one or two page essay that describes the life cycle of a moss, liverwort, or fern in detail. Start with the germination of a spore. Describe gamete formation in detail and how fertilization occurs. Describe how the zygote develops into sporophyte and how it produces spores. Use the names of all structures that you studied in lab and indicate the types of cell divisions that occur in development and ploidy levels of various stages.

Lab Summary Questions

1. How is reproduction in *Coleochaete* similar to that in a moss? How is it different?
2. How is reproduction (life cycle) in a moss similar to that in a fern? How are they different?
3. How do spores differ from seeds?
4. Write a short letter to a child describing how to tell the difference between a moss, liverwort, and fern. You could include some sketches.
5. Write a letter to a high school friend describing what is meant by alternation of generations.
6. What adaptations do mosses and ferns have to a terrestrial way of life?

Internet Sources

Read more about an exciting new experimental system for investigating the biology of the life cycle of ferns at *http://www.bio.utk.edu/cfern/index.html* You might want to use a search engine to look for other research work that is being done using ferns as experimental organisms. What are the advantages of ferns as experimental organisms in the study of developmental genetics?

7. List five reasons why the plants studied in this lab are generally not well adapted to hot, dry environments.
8. As you studied the plants in today's lab, what captured your imagination? Why?
9. What event starts the sporophyte stage of the life cycle? When and where does it occur in liverworts? Mosses? Ferns?

Critical Thinking Questions

1. Few, if any, mosses and ferns are found growing in dry biomes. Based on your studies, why do you think this happens?
2. Were plants or animals first to invade terrestrial environments? Why do you say so?
3. Two roommates with little botanical knowledge thought they could make a quick profit by growing ferns for the commercial florist market. They thought it would be very easy. All they had to do was collect the thousands of spores produced by a fern on their desktop and plant these like tiny seeds in individual pots. If they placed the pots in the sun, fertilized and watered the spores, and waited about a year, they would have beautiful plants that they could sell for several dollars apiece. Why are they doomed to failure?
4. Would you expect to find separate male and female sporophyte-stage mosses growing in the wild? Why, or why not?
5. A person interested in study crossing-over during meiosis in ferns prepared microscope slides of an archegonium and antheridium at different stages of development. Why will this study fail?
6. A species of fern has a a diploid number of chromosomes equal to 18. How many chromosomes would you expect to find in a spore? In a zygote? In a cell from a leaf?
7. Explain how vascular tissues are an advantage in terrestrial plants.

LAB TOPIC 16

Investigating Plant Phylogeny: Seed Plants

Supplies

Preparator's guide available on WWW at
   http://www.mhhe.com/dolphin

Equipment

Compound microscopes
Dissecting microscopes

Materials

Living plants
   Gladiolus flowers
   Dicots in flower
   Where possible, examples of gymnosperms other
      than pines
   Various grocery store fruits and vegetables—
      to include apples, peppers, beans, and peas;
      specimens cut in longitudinal and cross
      sections
Plant specimens
   Needles of various conifers
   Staminate and ovulate cones, various species
   Hardwood trees herbarium set
   Gymnosperm herbarium set (or locally collected
      tree specimens)
Prepared slides
   Pine needle, cross section
   Pine stem, cross section
   Pine pollen cone, longitudinal section
   Pine ovulate cone, longitudinal section
   Lily ovary, cross section
   Germinating pollen, whole mount
   Lilium anther, cross section
Slides, coverslips, razor blades
Printed taxonomic keys such as *How to Know the
   Trees,* or *How to Know the Western Trees*

Solutions

10% raw sugar solution (do not use white or brown
   sugar)
Phloroglucinol stain

Prelab Preparation

Before doing this lab, you should read the introduction
and sections of the lab topic that have been scheduled
by the instructor.
   You should use your textbook to review the
definitions of the following terms:

   angiosperms
   anther
   Anthophyta
   carpel
   dicot
   endosperm
   gametophyte
   gymnosperms
   integument
   locule
   monocot
   ovary
   ovule
   pistil
   pollen
   seed
   sporophyte
   zygote

   You should be able to describe in your own words
the following concepts:

   Alternation of generations
   Double fertilization
   How seeds are formed
   How seeds differ from spores
   Plant adaptations to land

   As a result of this review, you most likely have
questions about terms, concepts, or how you will do
the experiments included in this lab. Write these
questions in the space below or in the margins of the
pages of this lab topic. The lab experiments should
help you answer these questions, or you can ask your
instructor for help during the lab.

Objectives

1. To learn the life cycle stages in the alternation of generations found in gymnosperms and angiosperms
2. To appreciate how pollen, ovarian development, seeds, and fruits are reproductive adaptations to the terrestrial environment
3. To collect evidence that tests the following null hypothesis: There is no evidence that plants can be arranged in a phylogenetic sequence that demonstrates increasing adaptation to a terrestrial environment.
4. To use a dichotomous key to identify plant specimens

Background

In this lab topic you will look at the life cycles of the plants that have come to dominate the terrestrial environment on this planet. They are so common that we name major biological regions of the earth, biomes, after them. Grasslands, coniferous forests, and deciduous forests are a few examples to remind us of their prominence. It stands to reason that their success is related to their adaptations to terrestrial living.

The terrestrial adaptations of the seed plants are numerous. They include the ability to grow tall and thus capture sunlight, outcompeting mosses and ferns that are limited in size. What allows this change in growth habit are vascular tissues. Cells of these tissues can carry water to leaves 300 feet above the ground, as in a eucalyptus tree, and can carry photosynthetic products from leaves to non-photosynthetic cells underground in the root systems. Their root systems also anchor their tall bodies so they do not topple in winds and their stems have the ability to grow in girth through a process of secondary growth, providing support for the photosynthetic canopy. However, none of these are truly unique adaptations and can also be found in ferns.

What sets these plants apart from the others is their life cycle strategy. The gametophyte generation of the seed plants has become much reduced in size and is found almost as a parasite living in the moist environment of the sporophyte stage, which also supplies nutrients to the non-photosynthetic gametophyte. Coupled with this trend was the development of a new way of transferring sperm to eggs using pollen as the transfer agent. This development freed the seed plants from depending on water splashes and films to do the job, and allowed them to colonize new habitats where moisture was scarce.

The seed, from which the common name is derived, was also a major evolutionary innovation. It represented a small capsule in which was packaged an embryo and stored food. Its hard outer casing provided protection from desiccation, physical abrasion, and chemical breakdown. This meant that the sporophyte embryo could survive relatively

long periods of harsh conditions and could also be transported great distances, allowing the seed plants to colonize new environments in which growth conditions were intermittent as in seasonal environments.

The appearance of seeds is associated with evolutionary events that affected the ancestral sporangia, the structures where the spore stage of the life cycle are produced. Seed plants are **heterosporous,** producing two kinds of spores from different sporangia. **Microspores,** produced in microsporangia, develop into male gametophytes; **megaspores,** produced in megasporangia, develop into female gametophytes. Evolutionary events affecting both types of sporangia are involved in the adaptation to land.

In seed plants the megasporangium develops not into a chamber, but into a solid mass of cells called the **nucellus.** Within this mass of cells, a single functional megaspore is produced and held. It is not released to develop into a separate individual gametophyte as in the lower plants. The retained megaspore divides to produce the cells of the female gametophyte generation, and that generation remains embedded in the tissues of the sporophyte. One of the cells in the female gametophyte becomes a functional egg. If it is fertilized, it develops into the new sporophyte of the next generation. All of these events occur in a structure called the **ovule** that was new on the evolutionary scene and unique to the seed plants.

A developing ovule initially consists of the megaspore, nucellus, and surrounding protective coats called **integuments,** but it later contains the female gametophyte and, at full maturity, becomes the seed. When an egg inside an ovule is fertilized, the event triggers further development of the integuments of the ovule. They harden to become the outer casing of the seed. Inside the integuments, the fertilized egg develops into an embryo, the next sporophyte generation, while the nucellus breaks down and releases nutrients that are used by the growing embryo. The evolutionary development of seed is directly related to the evolutionary development of the ovule.

Seed plants could also be accurately called the pollen-producing plants because equally important evolutionary events also affected the microsporangium and male gametophyte development. Microspores produced in hollow microsporangia mature to become pollen grains that are the male gametophyte stage of the life cycle. Maturation involves the synthesis of a resistant cell wall impregnated with the waterproofing agent **sporopollenin** and cell division of the initial cell inside.

Each pollen grain starts as a single microspore, and mitotic cell divisions inside eventually produce three or four cells. These cells represent the body of the male gametophyte generation of the seed plant. Because they are inside tiny pollen grains, these gametophyte stages are mobile and can travel great distances independent of moisture conditions.

In the lower plants that were studied last week, water was necessary to transfer sperm from the male gametangium to the egg in the female gametangium. As a result, such plants were limited in the ecological niches that they could inhabit. In seed plants, sperm are no longer naked cells. Contained in desiccation resistant pollen grains, they can blow through the air or be transferred by animals bringing the male to the female gametophyte. When pollen lands on a suitable surface, the male gametophyte becomes active and a long tube grows out through one or more small apertures in the pollen wall. Sperm travel down this tube to fertilize an egg. This reproductive adaptation to a terrestrial way of living has the added benefit of promoting genetic outcrossing between distant individuals.

Needless to say, the sporophyte stage is the dominant, or most easily recognized, stage in the alternation of generations life cycle found in seed plants. What selective pressures might have lead to the decrease in the prominence of the gametophyte and to the dominance of the sporophyte? Sporophytes are diploid. Diploid organisms have two copies of every gene. If one fails, the other is backup. In terrestrial environments, the ultraviolet component of sunlight can cause mutations. Because most mutations are harmful, those plants having an exposed stage that was resistant to the effects of mutations and a "hidden" stage that was susceptible might have had an advantage over others.

If this is true, then we could also ask why has the gametophyte stage been retained in the life cycle of these plants? Some authors think that this might be related to two things. First, haploid cells, as are found in gametophytes, are a good allele-screening mechanism. If an essential gene becomes nonfunctional as a result of mutation, the haploid stage will not be able to perform the essential function and die, removing the genotype from the population. The second line of reasoning addresses the mobility of male gametophytes encased in pollen. For immobile organisms such as plants, the ability to now combine gametes with distant sexual partners would greatly increase genetic variability in populations, providing new gene combinations for natural selection to act upon. It could be that these two advantages were sufficient to allow the gametophyte stage to continue in the life cycle, although in a much reduced prominence.

LAB INSTRUCTIONS

You will study the life cycles of pines and flowers in this lab, learning to apply to these common plants the concept of alternation of generations with distinct sporophyte stages but less conspicuous gametophyte stages. You will also learn how seed plant structures represent adaptations to a terrestrial environment.

Gymnosperms

The gymnosperms appear before the angiosperms in the fossil record. Present-day gymnosperms are classified in four divisions: conifers (which will be studied), cycads,

ginkgo, and gnetophytes (fig. 16.1). All bear naked ovules which develop into seeds at the base of the scales found in the characteristic cones or similar structures of these plants. **Cycads** are tropical plants. They look similar to palms and have a stem that may be 18 m high bearing leaves in a cluster at the top, hence the common name "sago palms." Only one species of **ginkgo** survives today, and all plantings are derived from a few trees found in the Imperial Gardens of China. It has fan-shaped leaves with parallel venation and drops its leaves in the fall as do many dicots. The ovules are born on stalks which give rise to fleshy coated seeds. The **gnetophytes** are a diverse group found in moist tropical areas and arid areas, including North America. Some species are treelike; others, shrubs; and yet others, vines. The **conifers** are by far the largest and most diverse group and include the familiar pines, as well as many others. In this section you will look at pines as examples of the gymnosperms. The sporophyte stage is the conspicuous stage in the life cycle and the gametophyte stage is much reduced to a few cells that give rise to functional eggs and sperm.

Division Coniferophyta (Conifers)

The conifers are an economically important group of about 500 species of plants. They include pine, fir, spruce, hemlock, redwood, sequoia, cypress, and cedar. Most have needlelike leaves, tree or shrublike body plans, well-developed vascular tissues, and secondary growth. Cones are common reproductive structures, but may be modified berry-like structures as in yews and junipers.

Sporophyte Stage of Pine

You will study the anatomy and reproductive cycle of a pine as a representative gymnosperm in the division Coniferophyta.

▶ **Leaves** Look at the demonstration specimens of conifer needles (pines, firs, spruces, yews, etc.). Needles are produced in bunches, collectively called a **fascicle,** of 1 to 8, with the number being characteristic for the species. Note the shiny appearance to the needles. This is caused by the presence of a thick, waxy **cuticle,** which prevents evaporation of water, adapting the plants to dry environments. The reduced surface area of the needles and the pyramidal shape of most conifers are adaptations that enhance shedding of snow in cold climates and allow these plants to hold their leaves year round. Needles are usually retained for 2 to 4 years, although the bristlecone pine may retain its for 45.

▶ Obtain a slide of a cross section of pine needle (fig. 16.2) and look at it with your compound microscope. The cuticle may not be visible because it was extracted by solvents used in the slide preparation process. Note the **epidermis,** which secretes the cuticle, and the underlying **hypodermis** consisting of compactly arranged thick-walled cells. Note the **stomata** that penetrate the surface of the

Investigating Plant Phylogeny: Seed Plants   **179**

Figure 16.1 Examples of gymnosperms. (*a*) Cycad. (*b*) Ginko. (*c*) Gnetae.

(a)

(b)

(c)

Figure 16.2 (*a*) Cross section of a single pine needle. (*b*) From two to eight needles may be found in a fascicle, depending on the species.

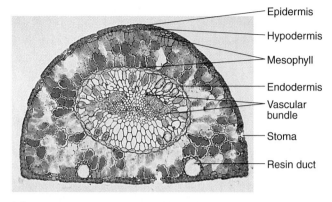

— Epidermis

— Hypodermis

— Mesophyll

— Endodermis

— Vascular bundle

— Stoma

— Resin duct

(a)

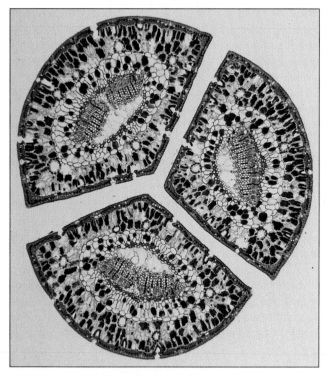

(b) Three needles in fascicle

leaf, forming a passage from spaces inside the leaf to the surrounding air? What is their function?

Beneath the hypodermis is the **mesophyll,** consisting of cells that seem to fit together like pieces of a jigsaw puzzle. These are the photosynthetic cells and contain large, darkly staining chloroplasts. In the center of the mesophyll, note the **endodermis** surrounding one or two **vascular**

Figure 16.3   Cross section of a pine stem.

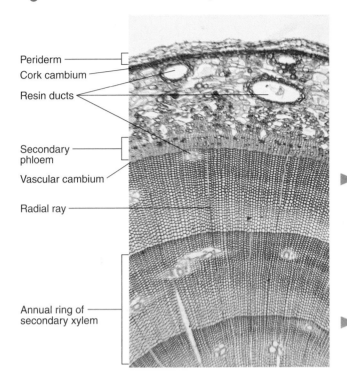

Periderm
Cork cambium
Resin ducts

Secondary phloem
Vascular cambium

Radial ray

Annual ring of secondary xylem

bundles. Each bundle contains xylem with thick-walled tracheids and thinner-walled sieve cells in the phloem cells. Near the flat surface of the needle, two or more **resin ducts** should be visible.

Stem   Vascular tissues are highly developed in the gymnosperms, allowing these plants to grow to great heights. The stems of pines and other conifers are woody, and annually increase in girth due to secondary growth. This leads to the formation of substantial amounts of secondary xylem toward the center of the stem and secondary phloem around the circumference (fig. 16.3). The wood of a pine trunk is composed of secondary xylem.

Obtain a cross section of a pine stem and look at it with your compound microscope. The very center of the stem contains **pith.** It is surrounded by elongated **tracheids** of the **secondary xylem** positioned so that their long axes are parallel to the long axis of the stem. New tracheids are laid down in **annual growth rings.** At the outer edge of the secondary xylem is the **vascular cambium** layer. Its cells divide to form new **secondary xylem** inwardly and **secondary phloem** outwardly. Both tissues are traversed by **rays** that conduct materials across the radius of the stem. Resin ducts should be visible in the layers of xylem. The stem is surrounded by a thick bark or **periderm,** which originates from a **cork cambium** layer outside of the phloem (fig. 16.3). It protects the trunk from invasion by microorganisms, from desiccation, and from damage by fire, which is common in coniferous forests. The roots of gymnosperm sporophytes are also highly developed but will not be studied.

Male Gametophyte Stages

The familiar pine, fir, and spruce trees are all examples of the sporophyte generation of the gymnosperms. To locate the gametophytes, the reproductive organs must be examined. For pines, the reproductive organs are the young cones that develop in the spring of the year. Pine trees develop two types of cones: small **staminate** or pollen cones, which are on the tree for only a few weeks, and the larger, more familiar **ovulate** seed cones ("pine cones"), which remain on the tree for two or more years.

Male Gametophyte

The small male staminate cones appear singly or in clusters near the bases of terminal branch buds in the spring of each year (fig. 16.4a). Pollen cones are usually borne on the lower branches and ovulate cones on the upper. Since pollen falls when it is released, before it is carried by the wind, this promotes genetic outcrossing. Examine the specimens of pollen cones available in the lab. Each cone is composed of whorls of scalelike **microsporophylls,** which contain two **microsporangia.**

Examine a slide of a longitudinal section of an immature pollen cone and locate the swollen microsporangia on the undersurface of the microsporophylls. Many **microspore mother cells** in each microsporangia divide by meiosis, each producing four microspores. Each of these is surrounded by a thick wall, the outer layer of which expands to produce two air-filled bladders (fig. 16.4b).

Can you speculate on what might be the function of these two bladders so typical of pine pollen?

If mature staminate cones are available in the lab, remove a microsporophyll and place it in a drop of water on a slide. Tease it a part, add a coverslip, and observe through your compound microscope. Alternatively look at a prepared slide of pine pollen. Draw a few pollen grains below.

Figure 16.4 Pine reproductive structures: (*a*) staminate (male) cones; (*b*) characteristic pollen of pine.

(a)

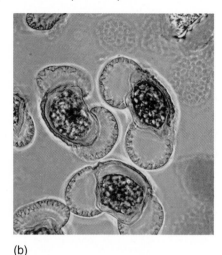

(b)

While in the microsporangium, the microspore develops into a mature pollen grain containing four cells: two **prothallial cells,** a **tube cell,** and a **generative cell.** The prothallial cells will degenerate. These four cells in the pollen grain represent the much-reduced **male gametophyte** stage in the pine life cycle. The sequence of events in the origin of the male gametophyte is summarized in figure 16.5.

When the pollen are mature, the microsporangia, which are now called pollen sacs, rupture, releasing enormous amounts of pollen, some of which are wind-borne to female cones. After releasing pollen, the staminate cones dry and fall off the tree.

The development of pollen was a major adaptation in land plants. The male gametes became highly resistant to desiccation. No longer was a water film required for the sperm to reach the egg. Furthermore, the possibilities of genetic outcrossing beyond the individuals in the immediate vicinity was greatly increased because the wind could carry pollen long distances where it could fertilize eggs borne by individuals outside the neighborhood of the pollen-producing tree.

Female Gametophyte

▶ The familiar pinecone is the female ovulate cone. In it you will find the ovule containing the female gametophyte with its eggs, and the ovule will become a seed following fertilization. Obtain a young ovulate cone from the supply area.

Ovulate cones are borne on the tips of young branches. The cone at first grows with the pointed end upward, but as it matures it inverts. Each cone is made up of whorls of heavy **scales.** Two ovules are borne on the upper surface of each scale. An **ovule** is a small, complex reproductive organ produced by the sporophyte, which surrounds and protects the much-reduced female gametophyte stage as it develops. The outer layer of the ovule is the **integument:** it will become the seed coat if fertilization occurs (fig. 16.6).

Figure 16.5 Schematic showing spore formation and development of male gametophyte generation in gymnosperms.

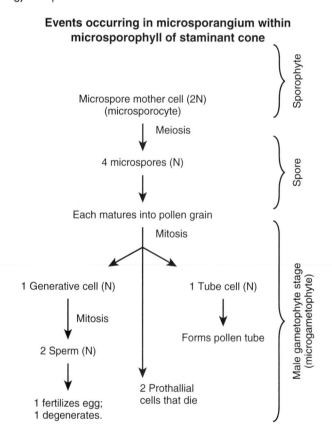

The integuments surround the **nucellus** which is the megasporangium. Here a spore that is produced will develop into a female gametophyte.

Ovules can be observed by cutting away the tip and about one-third of a young cone. While observing through a dissecting microscope, remove scales from the large

Figure 16.6 Life cycle of a pine. If male and female cones appear in the spring of the first year, the mature seeds would not be shed until the fall of the third year, and the seeds might not germinate until the fourth year or later.

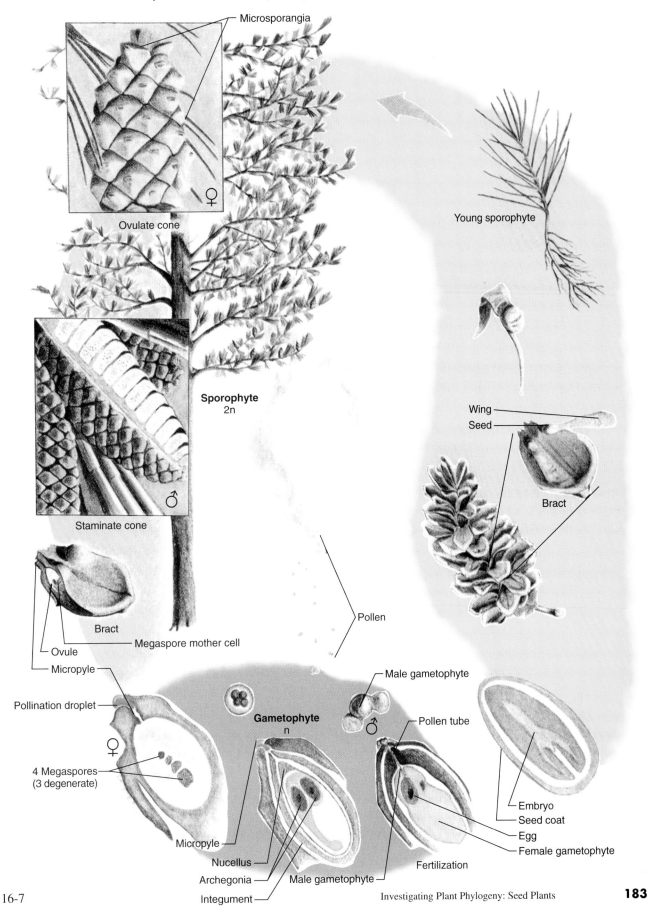

Microsporangia

Ovulate cone

♀

Young sporophyte

Staminate cone

♂

Sporophyte
2n

Wing
Seed

Bract

Bract

Ovule

Megaspore mother cell

Micropyle

Pollination droplet

Pollen

Male gametophyte

Pollen tube

Gametophyte
n

♀

♂

4 Megaspores
(3 degenerate)

Embryo
Seed coat
Egg
Female gametophyte

Micropyle

Nucellus

Archegonia

Male gametophyte

Integument

Fertilization

Figure 16.7   (a) Longitudinal section through an ovulate (female) pinecone. (b) Magnified section through ovule showing micropylar opening and megaspore mother cell.

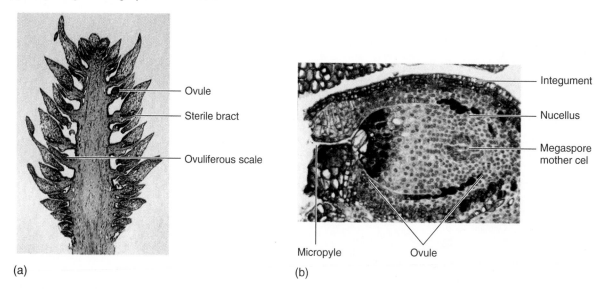

(a)

(b)

remaining piece. Look for two oval structures on the surface of the scale that faces the smaller end (upper surface when the young cone is on tree).

To study the microscopic structure of an ovule, obtain a prepared slide of a longitudinal section of an ovulate cone and look at it through your compound microscope. After determining which end of the section is the tip of the cone and what structures are scales, find the swollen ovules at the base of the scales (fig. 16.7). Identify the integument of the ovule surrounding the female gametophyte. It may be in any stage of development, depending on when the section was made (see fig. 16.7). Because this is sectioned material and all slices do not contain the same structures, you may have to look at several ovules to see the following. If you find one of these structures, share the view with others.

The female gametophyte develops in the following way. At first there is a single diploid **megaspore mother cell.** It divides by meiosis to produce four haploid **megaspores,** three of which degenerate. The remaining functional megaspore undergoes repeated mitotic nuclear divisions, resulting in a multinucleate cell. Cell walls form around the nuclei, yielding the **multicellular female gametophyte.** As the gametophyte continues to develop, it draws nutrients from the surrounding nucellus tissue. At a later stage of development, two **archegonia** form at the end of the ovule that faces the central axis of the cone. Each will contain a functional egg. The sequence of events in the development of the female gametophyte is summarized in figure 16.8.

Look again at the slide of the cone and examine each ovule carefully. When the plane of the section is correct, you should be able to see an opening in the integument on the end of the ovule facing the central axis of the cone (fig. 16.6). Called the **micropyle,** this opening provides sperm with access to the egg. Sketch an ovule with a micropyle in the circle.

Fertilization and New Sporophyte Generation
Fertilization in pines involves a series of events that may take over a year to complete, whereas in spruces, firs, and hemlocks, it may be completed in a few weeks. At the time of pollination, the ovulate cones are small, project upward from the branch tips, and have open spaces between the scales. Windborne pollen enters the cone openings and is trapped in a droplet of fluid secreted at the micropyle (fig. 16.7b). As this droplet dries, pollen is drawn toward the micropyle. The scales of the cone now press tightly together (close) and over the next year fertilization and seed development occur.

As the pollen reaches the micropyle, a tube grows out and penetrates the nucellus to the egg. Two sperm arising from division of the generative cell, pass down the tube and enter the egg cytoplasm. The nucleus of one fuses with the nucleus of the egg, signaling the start of a new

Figure 16.8 Schematic showing spore formation and development of female gametophyte generation in gymnosperms.

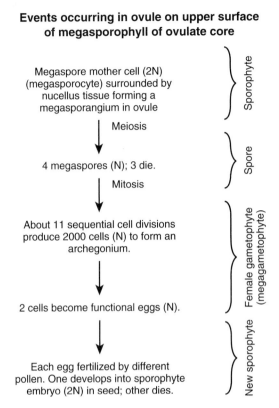

Events occurring in ovule on upper surface of megasporophyll of ovulate core

Megaspore mother cell (2N) (megasporocyte) surrounded by nucellus tissue forming a megasporangium in ovule } Sporophyte

↓ Meiosis

4 megaspores (N); 3 die. } Spore

↓ Mitosis

About 11 sequential cell divisions produce 2000 cells (N) to form an archegonium. } Female gametophyte (megagametophyte)

↓

2 cells become functional eggs (N).

↓

Each egg fertilized by different pollen. One develops into sporophyte embryo (2N) in seed; other dies. } New sporophyte

diploid sporophyte generation. The second sperm nucleus degenerates. The second egg in the ovule also may be fertilized by another pollen, but only one of the zygotes will develop. The zygote that divides to form the embryo pushes into the surrounding nucellus tissue, drawing nutrients to sustain growth.

As the embryo forms, the integument of the ovule develops into the seed coat. It may take up to an additional year for the seed to develop. These seeds are truly remarkable structures containing cells from three generations in the life cycle of the pine: (1) the parent sporophyte contributes the seed coat; (2) the embryo is the sporophyte of the new generation; and (3) the stored food materials in the seed are haploid female gametophyte tissues.

Examine dry, mature pinecones for evidence of seed. Within the seed is the embryonic next sporophyte generation. When air- or water-borne, or transported by an animal to a new location, the seed can germinate, allowing the spread of the species. Thus in gymnosperms, seeds have replaced spores as the means of dispersal when compared to mosses and ferns.

As you finish this section, pair off with someone in the laboratory and quiz one another about gamete production, fertilization, and where you have seen the sporophyte and gametophyte generations. Use figure 16.8 as a guide.

Angiosperms

Angiosperms are seed plants with flowers. They are what we think of when we create a mental image of a plant. They are the most diverse of the plant groups, living in mountain, desert, freshwater, and seawater habitats. They range in size from the tiny duck weeds of about 1 mm tall to the 100 m tall eucalyptus trees in Australia. All angiosperms are classified in Phylum **Anthophyta.** There are two taxonomic classes of anthophytes: **monocots** include the grasses, lilies, bamboos, palms, bromeliads, and orchids; and **dicots** include most nonwoody (herbaceous) plants, broad leaf shrubs, and trees, vines, and cacti.

The sporophyte stage dominates the angiosperm life cycle as in gymnosperms, but unlike gymnosperms the ovules are contained in a tissue-encased ovary that matures into a fruit with seeds after fertilization. The name angiosperm comes from the Greek words angeion (= container) and sperma (= seed) and together emphasize the role of the fruit as a seed container. A colorful floral structure in many species attracts animal pollinators that promotes genetic diversity through cross fertilization. Fertilization in angiosperms involves a **double fertilization** where one sperm fuses with the egg to produce the next generation and another sperm fuses with a second type of cell to produce a nutrient-laden tissue called the **endosperm.** The endosperm fuels the early development of the embryo and seedling during germination of the seed.

Division Anthophyta (Flowering Plants)

The life cycles and structure of flowering plants have many of the same adaptations to terrestrial environment as do gymnosperms. Additional changes include the development of flowers and fruits, and it is these structures that we will focus on in our study.

Gladiolus is recommended as a readily available flower to dissect, but other species will work.

Sporophyte Floral Structure

The flower is a sexual reproductive structure in the sprorophyte stage of anthophytes. Within the male and female floral organs, the gametophyte stages develop. Some species have flowers that contain both male and female structures while others have flowers with only male or female parts. Those species producing separate male and female flowers may have individual plants with flowers of one sex or may have flowers of different sexes on the same plant. *Gladiolus* has both male and female parts in the same flower. Flowers may be single or in clusters as in *Gladiolus*. The flower aggregation is called the **inflorescence.**

Break a single *Gladiolus* flower from an inflorescence in the supply area and compare it to figure 16.9. The stalk of the inflorescence is called the **peduncle** and that of an individual flower is the **pedicel** which you broke to get your specimen. Note that the floral parts arise from a green **receptacle** at the end of the pedicel. The brightly colored **petals** are obvious. These modified leaves function to

Investigating Plant Phylogeny: Seed Plants

Figure 16.9 The sporophyte generation of an angiosperm bears flowers that contain a few cells that act as the male and female gametophyte generation. Male gametophytes are the pollen produced by the anthers of the stamens. Female gametophytes are the embryo sacs contained in the ovules located in the ovaries. Nuclei of the cells in the gametophyte generation are haploid. The diploid number of chromosomes is restored with formation of the new sporophyte generation at fertilization.

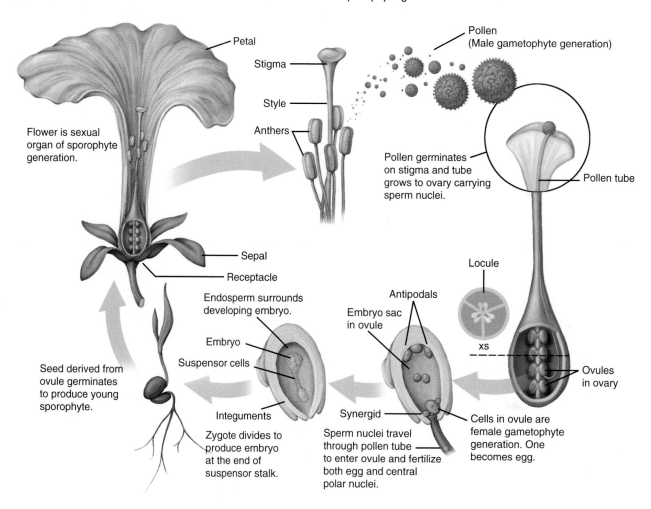

attract pollinators. Many flowers are pollinated by insects, birds, bats, or other small mammals. Animals are attracted to specific flower species as a result of co-evolution of visual and olfactory attractants in the plants and corresponding receptors and behaviors in the animals. Consequently, animals and plants are dependent on each other. In flowering plants that are wind pollinated, such as the grasses, the flowers lack these attractants and are inconspicuous. Beneath the petals are the green **sepals.**

If you pick off the petals, you will see the sex organs inside the flower. The central stalk-like structure, known as the **pistil,** is the female portion of the flower. It may consist of one or more **carpels.** Carpels are thought to have evolved by the folding and fusion of specialized ancestral leaves called **megasporophylls** that bore female sporangia (=megasporangia) along their margins. The base of the pistil contains the ovary where the female gametophyte(s) is (are) located. Egg development, fertilization, and seed development will occur in ovules found here. Above the ovary is a stalk, the **style,** which ends in an expanded tip of glandular tissue, the **stigma.** Alongside the pistil are long filaments with expanded tips. The structures at the tips are **anthers,** the male portion of the flower which will produce microspores that develop into the male gametophyte stage (=pollen). The anthers are thought to have evolved by the fusion of specialized leaves called **microsporophylls** which bore pollen-producing structures.

Coordinate this part of dissection with another person at your lab table so that one of you cuts a cross section of the ovary and of the style on one flower and the other cuts a longitudinal section through all parts of the pistil from another flower. Look at the sections. You may want to use a dissecting microscope to see more detail.

You should be able to see three open chambers in the ovary that are called **locules** (fig. 16.9). Moncots such as *Gladiolus* typically have three locules, and dicots will have four or five. Projecting into the locules from the central shaft are pairs of little bead-like structures, the **ovules.** As in gymnosperms, each ovule contains a megasporangium (=nucellus) that produces a single megaspore. It will de-

Figure 16.10 Hypothetical model for evolutionary development of ovary from carpels, ancestral leaves that bore ovules along edge.

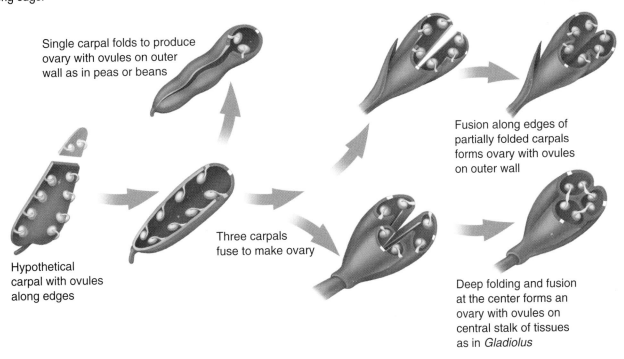

Single carpal folds to produce ovary with ovules on outer wall as in peas or beans

Fusion along edges of partially folded carpals forms ovary with ovules on outer wall

Hypothetical carpal with ovules along edges

Three carpals fuse to make ovary

Deep folding and fusion at the center forms an ovary with ovules on central stalk of tissues as in *Gladiolus*

velop into the female gametophyte that will produce a single egg. Fertilization occurs in the ovule, and each ovule develops into a seed. Now look at the longitudinal section of the pistil. Note that there are several ovules in vertical arrays in the locules. Try to count the number of ovules in one locule. How many did you find? _____ Estimate how many ovules are in the ovary of your plant. _____ How many seeds could this flower have produced? _____

To understand how the carpels might have evolved from megasporophylls, visualize three long narrow leaves that bore ovules on their margins (fig. 16.10). Fold the margins inward in a curling manner so that the ovules are now inside a tube formed by the megasporophyll. Push three of these structures together with the folded edges innermost. You now have a structure that resembles the cross section of the ovary. Over time, the upper parts of the leaf differentiated into the style which had no ovules or locules and into the stigma with its characteristic moist, glandular surface. Examine the surface of the cross section of the style and the surface of the stigma for evidence of an ancestral fusion of three megasporophylls.

The surface of the stigma is moist with sugary secretions from underlying glandular tissues. When mature pollen land on this surface, they germinate, producing a pollen tube. Pollen tubes grow down through the surface of the stigma and follow a pathway toward the ovules, guided by transmitting tissue in the style. What is the distance from the stigma to the lowermost ovule in the *Gladiolus* that you dissected? _____ For the egg in that ovule to be fertilized, a pollen tube must grow this distance, from a microscopic pollen grain caught on the surface of the stigma.

Male Gametophyte

▶ Take a *Gladiolus* anther and place it in a drop of water on a slide. Use a dissecting needle to tease it apart, add a coverslip, and observe with your compound microscope. Sketch what you see.

Investigating Plant Phylogeny: Seed Plants

Figure 16.11 Schematic showing spore formation and development of male gametophyte generation in anthophytes.

Events occurring in microsporangia within anther

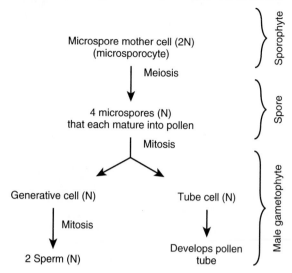

It is possible to artificially induce pollen germination, *i.e.,* to activate the male gametophyte. Your instructor may set this up as a demonstration or you may do the procedure. Harvest some mature pollen or split open an anther from a flower that is shedding pollen. Transfer the pollen to a drop of germinating solution which contains 10% raw sugar. Place some pieces of coverslip in the drop to prevent crushing and then add a coverslip. Observe this preparation for the next hour or so at approximately fifteen-minute intervals for evidence of pollen tube growth. Record your observations below as drawings with marginal time notations. Alternatively, you may look at a prepared slide of a germinating pollen tube.

Examine a prepared slide of a cross section of an anther. Four **pollen sacs** should be readily visible. They serve as **microsporangia** in which thousands, if not more, **microspore mother cells** divide by meiois to produce haploid **microspores.** Depending on the age of the anther used to make your slide, you may find cells in meiosis I or II or differentiating into mature pollen. In mature pollen, three cells are found: a **tube cell** and two **sperm.** Together they are the **male gametophyte** stage of an anthophyte. Because they are contained in the pollen grain, they are mobile and can be carried to distant locations. Often the male gametophyte stage does not fully develop until pollen have been transferred to the stigma. The steps in the formation of the male gametophyte are summarized in figure 16.11. Sketch a cross section of an anther below, drawing detail for only one of the pollen sacs.

Female Gametophyte

Obtain a prepared microscope slide of a cross section of a lily ovary and ovules. Look at it with low power through your compound microscope. Orient yourself by finding the three locules that you saw in your dissection. Each should contain two ovules, although some ovules may be missing because the plane of the section did not cut through where it was located. Compare what you see to fig 16.9 and fig. 26.6.

The anthophyte ovule is similar to those you observed in gymnosperms. Identify the central **nucellus (=megasporangium)** surrounded by **integuments.** At one end of the ovule, the integuments have an opening, the **micropyle,** that allows a pollen tube containing the sperm to reach the egg. The egg develops in the following way.

Within the nucellus, a **megaspore mother cell** divides by meiosis to produce four haploid **megaspore** cells. Three disintegrate, leaving one functional megaspore. It develops into the female gametophyte stage within the ovule. Its nucleus divides three times by mitosis to produce eight nuclei contained in the original megaspore's cytoplasm. What is the ploidy level of each of these nuclei? _____

Three of the nuclei migrate to the end of the megaspore away from the micropyle, two migrate to the center, and three migrate to the end nearest the micropyle where one enlarges to become the egg nucleus. Cell membranes form

Figure 16.12 Schematic showing spore formation and development of the female gametophyte in anthophytes.

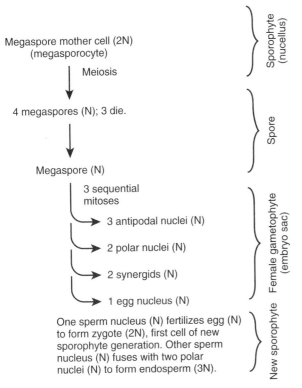

Events occurring in megasporangium within ovules in ovary

Megaspore mother cell (2N)
(megasporocyte)

↓ Meiosis

4 megaspores (N); 3 die.

↓

Megaspore (N)

3 sequential mitoses

→ 3 antipodal nuclei (N)

→ 2 polar nuclei (N)

→ 2 synergids (N)

→ 1 egg nucleus (N)

One sperm nucleus (N) fertilizes egg (N) to form zygote (2N), first cell of new sporophyte generation. Other sperm nucleus (N) fuses with two polar nuclei (N) to form endosperm (3N).

Sporophyte (nucellus)

Spore

Female gametophyte (embryo sac)

New sporophyte

---

around each nucleus separately except for the pair in the center which are surrounded by a single membrane and are contained in a single cell. These seven cells with eight nuclei comprise the **female gametophyte** stage of the flowering plant. Collectively they are referred to as the **embryo sac.** The events in the development of the female gametophyte are summarized in figure 16.12.

Fertilization and the New Sporophyte Generation

Germinating pollen grains are the active male gametophytes of flowering plants. They consist of only three cells. One forms the pollen tube that penetrates the tissues of the pistil navigating to the micropyle of the ovule. The other two are sperm. **Fertilization** occurs when the growing pollen tube enters the micropyle of an ovule and penetrates the embryo sac. Two sperm cells leave the pollen tube. One fuses with the egg to form a diploid zygote, the next sporophyte generation of the plant. The other combines with the two polar nuclei in the single cell at the center of the embryo sac to form a special triploid tissue, the **endosperm.** It nourishes the developing plant embryo. Anthophytes are said to have **double fertilization** because two sperm are involved, but only one diploid embryo is formed.

Fruits

Following fertilization, the embryo develops inside the ovule as the ovule matures into a seed. The integuments surrounding the seed become the hard seed coat. As the seed develops, the ovarian wall enlarges to become the fleshy part of a fruit. Fruits are another example of how plants and animals have co-evolved. Many fruits are colorful, sweet, and fragrant. They attract animals that eat them. The seeds encased in their protective coats pass through an animal's digestive system unharmed. They are often deposited along with a little fertilizer at some distance from the parent plant where they germinate and continue the life cycle.

In the lab are several fruits that have been sectioned either across or longitudinally. Study the fruits and make some quick sketches below in which you show the arrangements of the locules and the seeds in the sections. Count the number of seeds in some representative fruits, such as a pepper, bean, or pea (count the number in one locule and multiply by the number of locules). Record these numbers next to your drawings. Using your seed data, estimate the number of ovules that would be found in a flower of the same species.

In later labs you will look at the tissues, vascular anatomy, and leaf structure of the sporophyte stage of flowering plants. You and your lab partner should now quiz one another about the anatomy of a flower and the life cycle of a flowering plant. Make a list of how gymnosperms and anthophytes are similar as well as different in the details of their life cycles.

Based on your work in this lab topic, fill in table 16.1 by writing *yes* or *no* in each row when the plants in the column have that characteristic.

Plant Key

People wishing to identify an unknown plant often rely first on field guides—books with descriptions, photographs, and drawings of known species. By comparing the description of the plant in the guide with the unknown specimen, a positive identification may be made. However, in many instances, variables such as the age of the plant, the season of the year, the soil type where the plant is growing, local weather patterns, and other factors make it very difficult to correctly identify the species. To assist with these difficult identifications, a "dichotomous" key is often used.

A dichotomous key is a taxonomic classification device that presents the user with two statements (occasionally three or more). Careful observation of the unknown specimen indicates which of the two statements best describes it and is, therefore, the best choice. That choice, in turn, leads to another set of two statements. Again, a proper selection must be made, which leads to other choices. And so on, until the specimen is identified, or "keyed."

Displayed in the lab are various specimens of unknown tree species. They are either fresh specimens collected locally, or preserved specimens mounted on herbarium cards. Using the dichotomous key provided (either *How to Know the Trees* or *How to Know the Western Trees* is recommended), identify—that is, "key"—the trees on display. When using the key remember to always read both choices even if the first statement appears logical. The second choice may be even more appropriate. Be sure of the correct meaning of all terms. A glossary is often provided in the key. When measurements are provided, measure the relevant structures; don't guess. A hand lens or a dissecting microscope is often valuable for observing minute detail. Also, remember that there may be variability within a species. If a choice is unclear, it may be necessary to examine both routes to determine which seems most plausible for your specimen.

Your instructor may also send you out into the field to identify trees "in situ," or ask you to bring local species into the lab to be keyed.

Learning Biology by Writing

A major theme in this lab topic and in the one before was the phylogenetic trend in the plant kingdom that allowed plants to inhabit terrestrial environments. In a short, two-page essay give all of the observations that you made in both labs that support or contradict the following null hypothesis: there is no evidence that plants can be arranged in a phylogenetic sequence that corresponds to a sequence of greater adaptations to a terrestrial environment. Indicate whether you accept or reject this hypothesis based on your observations.

Your instructor may ask you to turn in lab summary 16 at the end of this lab topic.

Internet Sources

The Wollemi pine (*Wollemia nobilis*) is a newly discovered (1994) pine. Forty individuals were found growing in canyon lands to the west of Sydney, Australia. Use a search engine such as Google to locate information on this newest discovery of an ancient group. Why is it considered to be a significant discovery?

Lab Summary Questions

1. What similarities do Gymnosperms and Anthophytes have in their mechanisms of gamete formation, fertilization, and dispersal? What differences?
2. Explain where you would look for the gametophyte generation of a conifer and what it would look like.
3. Explain where you would look for the gametophyte generation of an Anthophyte and what it would look like.
4. Explain how a seed is an important adaptation to the terrestrial environment.
5. Explain how pollen is an important adaptation to the terrestrial environment.
6. What did you find most interesting about this lab and why?
7. Summarize your observation from the previous lab (#15) and this one by filling in table 16.1.

TABLE 16.1 Characteristics of studied plant divisions. Summarize your observations by adding *yes* or *no* to indicate if a feature is found in a group.

Characteristcs	Bryophyta (Mosses)	Pterophyta (Ferns)	Coniferophyta (Conifers)	Anthophyta (Flowers)
Cuticle				
Rhizoids				
Vascular tissue				
Rhizomes				
Flagellated sperm				
Separate male and female gametophyte				
Parasitic or nonphotosynthetic sporophyte				
Photosynthetic gametophye and sporophyte				
Parasitic or nonphotosynthetic gametophyte				
Ovules				
Airborne spores				
Pollen				
Embryo protected by seed coat				
Seed within fruit				

Critical Thinking Questions

1. How does a seed differ from a spore?
2. If a species of flowering plant has a diploid number of chromosomes equal to 30, how many chromosomes will be found in the embryo in a seed? In a pollen nucleus? In a megasporocyte? In nucellus tissue? In endosperm tissue?
3. If a species of pine has a diploid number of chromosomes equal to 40, how many chromosomes will be found in an embryo in the seed? In a pollen nucleus? In a megasporocyte? In nucellus tissue?
4. If all flowering plants depended on the wind to carry pollen from one individual to another, do you think there would be any colorful flowers and fruits? Why?
5. Describe co-evolution and how the concept applies to flowering plants and animals.
6. Why do grasses and conifers lack any showy, sweet, fragrant reproductive structures?
7. How does pollination differ from fertilization?

LAB TOPIC 17

Observing Fungal Diversity and Symbiotic Relationships

Supplies

Preparator's guide available on WWW at
    http://www.mhhe.com/dolphin

Equipment

Compound microscopes
Dissecting microscopes

Materials

Living Specimens
    Mushrooms (from grocery store)
    Rhizopus zygospore culture kit
    Chytridium confervae culture on wet Emerson
        YpSs/4 agar
    Lilac leaves infected with powdery mildew (If out of
        season, try freezing leaves in sealed
        containers for future labs.)
    Miscellaneous fungi samples: puffballs, bracket
        fungi, molds, others as locally available
    Lichen set (crustose, foliose, and fruticose)
Microscope slides
    Peziza section
    Coprinus mushroom, section of gills
    Mycorrhiza and root sections
    Lichen thallus section
    Powdery mildew of lilac
Field guide to mushrooms
Slides and coverslips

Prelab Preparation

Before doing this lab, you should read the introduction
and sections of the lab topic that have been scheduled
by the instructor.
    You should use your textbook to review the
definitions of the following terms:

Ascomycota
ascus
Basidiomycota
basidium
Chytridiomycota
coenocytic
dikaryon
fruiting body
gametangia
hyphae

lichen
mycelium
mycorrhizae
septa
sporangia
spores
Zygomycota

You should be able to describe in your own words
the following concepts:

The life cycle of a fungus from spore stage to spore
    stage.
The general body form of a fungus when it is
    growing and at the time of sexual
    reproduction.
In addition, do some field work and bring
    examples to the lab of fungi you can find on
    campus or where you live.

As a result of this review, you most likely have
questions about terms, concepts, or how you will do
the experiments included in this lab. Write these
questions in the space below or in the margins of the
pages of this lab topic. The lab experiments should
help you answer these questions, or you can ask your
instructor for help during the lab.

Objectives

1. To learn the anatomy and life cycles of several
   representative fungi
2. To observe the symbiotic relationships that fungi
   have developed with various plants
3. To observe fungal diversity

Background

Commonly known as yeasts, mildews, rusts, blights,
mushrooms, and puffballs, the fungi are common organisms in all ecosystems. At one time, fungi were considered

Figure 17.1 Interconnected fungal filaments called hyphae form a mycelium.

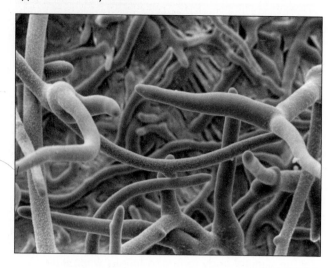

meaning many nuclei are found in a single large cytoplasm. During the life cycle of most fungi, each nucleus is haploid, except at the zygote stage when the nucleus is diploid.

Sexual reproduction in fungi occurs when haploid hyphae of two genetically different mating types fuse. In many fungi the two genetically different nuclei coexist in the cytoplasm of the hypha resulting from the fusion. Such a hypha is said to be **dikaryotic** and **heterokaryotic,** meaning it has two different nuclei. Each nucleus may replicate as the hypha grows. At some point the different nuclei fuse, producing a diploid nucleus which will divide by meiosis to produce haploid spores. When the spores find favorable conditions, they germinate, producing haploid hyphae.

Fungi are heterotrophic; they are completely dependent on preformed carbon compounds from other sources. Consequently, fungi live as saprobes, parasites, and symbionts of dead and living plants, animals, and protists.

Because of their chitinous cell walls, the fungi cannot engulf food materials. Instead, the hyphae secrete digestive enzymes that break down polymeric organic matter into small organic molecules that the hyphae absorb. Somatic hyphae are never more than several micrometers thick; thus, they have a great surface-area-to-volume ratio for efficient absorption. Cytoplasmic streaming, in addition to diffusion, provides efficient transport to nonabsorbing areas. Some parasitic fungi produce specialized hyphae called **haustoria,** which penetrate a host's cells and absorb food materials produced by the host.

Reproduction in fungi can be asexual or sexual. Mitosis and meiosis occur in fungi but are unusual in that the nuclear envelope does not fragment and reform. Instead, the spindle forms inside the nucleus and the chromosomes move to opposite ends of the nucleus. The nucleus then divides in two.

Asexual reproduction may occur by fragmentation of the mycelium with the separated hyphae growing into new mycelia. Under certain conditions, hyphae will differentiate into **sporangia** that produce asexual spores. Spores are single cells in a quiescent state surrounded by a tough, desiccation-resistant wall. The microscopic spores of terrestrial fungi can be borne great distances by the wind. When they settle in moist environments with a suitable food supply, growth ensues.

Sexual reproduction in fungi occurs by: (1) fusion of gametes, and (2) fusion of specialized reproductive structures called **gametangia.** Hyphae are haploid, and meiosis occurs in the zygote to produce haploid spores. In many fungi, spores are formed in or on fruiting bodies, which are the familiar mushrooms, brackets, and puffballs.

Fungi are important components of the geochemical cycles in the biosphere. Their decomposing action releases inorganic compounds and carbon dioxide that would otherwise be tied up in dead organic matter and unavailable to other organisms. Fungi also enter into symbiotic mycorrhizal relationships with higher plant roots, increasing the absorptive capacity of the host plants. Lichens (fungi in association with algae) are important colonizing organisms

degenerate plants that had lost their photosynthetic capability and adapted a saprophytic mode of existence. In recent years, this hypothesis has been rejected in favor of the theory that the fungi are a separate kingdom, the **Fungi (Mycota)** containing four phyla. Recent molecular evidence shows that they may be more related to animals than they are to plants. The oldest fossils are from 400 million years ago.

Fungi are found everywhere. They are extremely important in the recycling of minerals in ecosystems as they break down dead plant and animal matter. It is estimated that an acre of forest soil ten inches deep can contain over two tons of fungi. About 5,000 species are pathogens of crops and 150 species cause animal and human disease. A 1984 study showed that 40% of the deaths from infections acquired while in hospitals were from fungal infections, not bacteria.

The typical body of a fungus is in the shape of long filaments called **hyphae.** The hyphae branch to form a dendritic pattern of interconnecting filaments collectively called a **mycelium** (fig. 17.1). No cell in a fungus is far from its surrounding environment because the mycelium is not a solid mass but more like a tangled mass of filaments. No tissues are found in a mycelium: all cells are generalized to perform the functions of the entire organism. The hyphae of most fungi are surrounded by cell walls composed of the polymer **chitin,** the same material found in the exoskeletons of arthropods. This polymer is not found in the plant, moneran, or protistan kingdoms.

In many fungi, the hyphae are divided by cross walls called **septa,** but the septa can be perforated, allowing cytoplasm to flow from one compartment to the next. This allows the bulk transport of material from one part of the organism to another without crossing cell membranes. New hyphal growth is achieved as new compartments form at the tips. Most fungi, except for the yeasts, are **coenocytic;**

on rock faces, creating an environment for other organisms and building soil.

Parasitic fungi destroy many crops in the field or in storage. A few species are important in manufacturing; yeasts are used in baking and brewing and others in manufacturing antibiotics, such as penicillin.

Some fungi that grow on foodstuffs produce extremely powerful toxins. Aflatoxins produced by fungi that grow on stored grain are carcinogenic at concentrations of a few parts per billion. Ergot is an LSD-like hallucinogen produced by some fungi that grow on stored grains. Some historians think that the Salem witch hunts of the 1600s were induced by ergot contamination of grain supplies, causing mass hallucinations.

Remarkably, the fungi also are able to colonize synthetic environments. For example, some fungi live in jet fuel, on photographic plates, and on other manufactured hydrocarbon materials.

Nearly 100,000 species of fungi have been described. It is estimated that over ten times that number await discovery. Mycologists group the fungi into five phyla. These are:

Phylum Chytridiomycota—chytrids: aquatic organisms that link the protists and fungi; saprobes and parasites; about 1,000 species.

Phylum Zygomycota—Zygomycetes: zygote fungi living in soils or on decaying plant and animal matter; about 600 species.

Phylum Ascomycota—Sac fungi living in a variety of habitats and in symbiotic relationships; about 60,000 species.

Phylum Basidiomycota—Club fungi are important in decomposing wood and in forming mutualistic and parasitic relationships with plants; about 25,000 species.

Deuteromycetes—a nontaxonomic grouping; imperfect fungi with asexual reproduction only. This group will not be studied in this exercise; about 17,000 species.

Traditionally, the lichens are studied with the fungi. Lichens, however, are not single organisms. Instead, they are a unique association of two species: one an alga (or a cyanobacterium) and the other a fungus.

LAB INSTRUCTIONS

In this lab topic, you will examine life cycle stages of several representative fungi and observe the anatomical basis for several symbiotic relationships.

Observation of Field Samples

You should start this lab by looking at locally collected samples of fungi. On your way to class, look under bushes on campus or in the back of your refrigerator for fungi. Bring them to the lab.

Look at your samples under a dissecting microscope and tease them apart. Make slides of small pieces and look at them with your compound microscope.

Describe where you got your sample, what it is growing on, and what it looks like. Make some sketches of it. Write any questions you have about how it grows, what it feeds on, or how it reproduces. Use the space below.

Division Chytridiomycota

The members of this division are thought to be a link between the protists and the fungi. Based on recent molecular data from protein and nucleic acid analysis, the chytrids have been moved from being protists and are now considered fungi. They are aquatic fungi important in decomposition and as parasites of plants and invertebrates.

In the lab is a culture tube with *Chytridium confervae* growing on the surface of a special agar. Its life cycle is depicted in figure 17.2. Gently scrape the surface with a probe and mount the scrapings in a drop of water on a microscope slide. Add a coverslip and observe.

Figure 17.2 Life cycle of *Chitridium*.

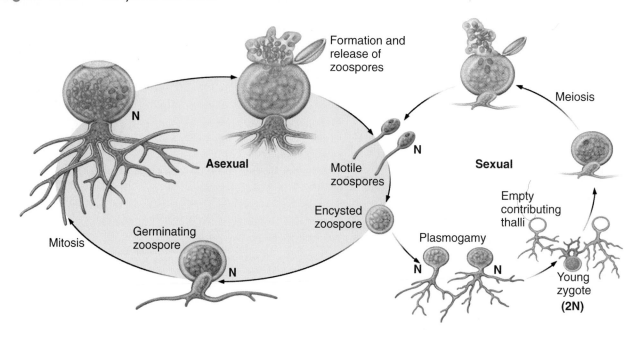

Sketch what you see below.

Division Zygomycota

The 600 members of this division are saprophytic, terrestrial fungi that live in the soil on dead plants or animals. The zygomycetes do not produce flagellated spores nor do they produce a distinct egg or sperm. In sexual reproduction, two mating types of zygomycete gametangia fuse to produce a **zygospore,** a thick-walled sexual spore that develops from the zygote. Asexual spores are also produced from sporangia on the ends of modified erect hyphae.

Rhizopus, the black bread mold, illustrates the life cycle of the zygomycete group (fig. 17.3). These organisms were a problem for bakers before mold inhibitors were used in commercial breads. Now sodium propionate and other materials are added to bread to prevent the growth of this mold and consequent spoilage. This fungus also grows well on fruits and vegetables and causes spoilage during storage.

Before this lab, two mating types of *Rhizopus* (+ and –) were inoculated onto a sterile potato dextrose agar in a petri plate. Observe a culture through your dissecting microscope. Identify the **hyphae** spreading from the points of inoculation to form white **mycelial mats.** All members of this division are coenocytic; many hyphae lack cross walls (aseptate) and the single, large cytoplasm contains many haploid nuclei.

If the culture is old enough, you should see **sporangiophores,** erect hyphae, that bear tiny, black, spherical bodies, **sporangia,** at the tip. Asexual spores are produced by mitosis in sporangia and when released will be wind-carried to new locations where they germinate to produce new mycelia. At the base of each erect sporangiophore, note the rootlike hyphae, the **rhizoids.** Passing laterally from the base of the sporangiophores are horizontal-growing hyphae called **stolons.**

In addition to asexual reproduction, *Rhizopus* can reproduce sexually. *Rhizopus* is **heterothallic,** meaning that there are two mating types. Where hyphae of different types meet, each hypha sends out a bulge of cytoplasm that eventually forms a bridge between the two hyphae (fig. 17.3). The outgrowths of each hypha containing sev-

Figure 17.3   *Rhizopus*, the black bread mold: (*a*) life cycle; (*b*) zygospore photo; (*c*) sporangium photo.

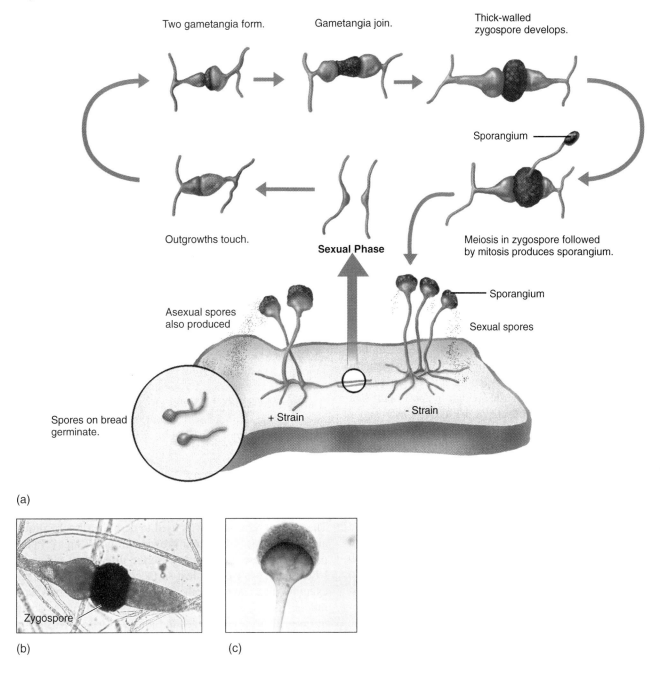

(a)

(b)   (c)

eral haploid nuclei are walled off from the parent hyphae to form **gametangia.** The sexual phase begins when the gametangia fuse, producing a single cell with several haploid nuclei, the **zygospore.** It can become dormant under adverse conditions and is resistant to desiccation for months. Under favorable conditions, the zygospore becomes active: its haploid nuclei pair and fuse to produce diploid nuclei and the diploid nuclei rapidly divide by meiosis. Thus, in a short period of time, genetic recombination occurs in this simple organism. The zygospore then germinates, producing a short **sporangium** that releases haploid spores. These

spores will be air-carried to new locations and germinate to produce new hyphae, thus completing the life cycle.

Zygospore formation should be visible as a black zone at the region where the two mycelial mats overlap. Carefully remove some hyphae from the black line and make a wet-mount slide. If a black zone is not evident, prepared slides of *Rhizopus* zygospores may be substituted. Make a second slide, using hyphae from one edge of the mycelial mat. Look at both slides with the low-power objective on your compound microscope. Can you see septae in the hyphae? Can you see protoplasmic streaming?

Figure 17.7 Life cycle of a basidiomycete mushroom. (a) Haploid hyphae from two mating types fuse to produce a dikaryotic mycelium; each has two nuclei, one derived from each parent. This dikaryotic stage will form a fruiting body, and nuclear fusion occurs in the basidia located on the gills. The zygote thus formed divides by meiosis, producing basidiospores which germinate and form haploid hyphae. (b) Photomicrograph of a basidium, showing basidiospores forming.

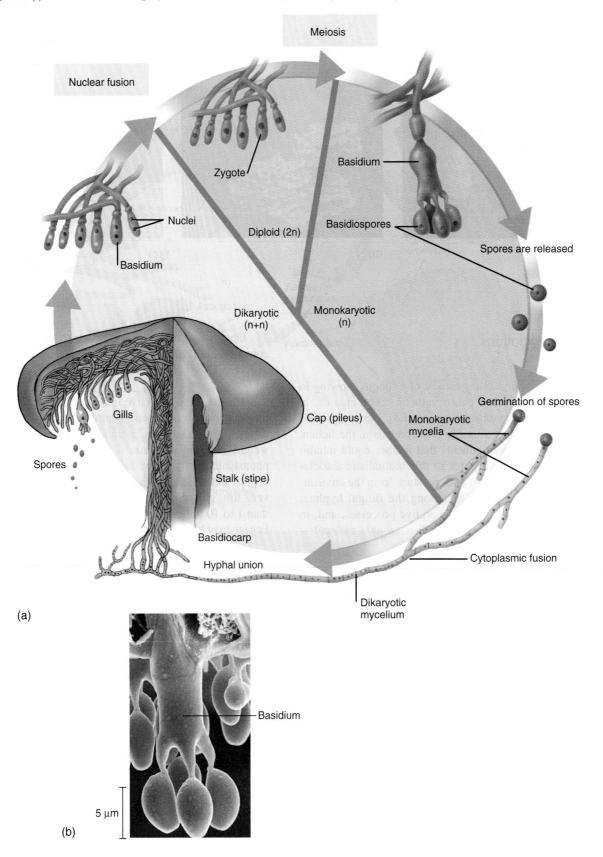

Meiosis

Nuclear fusion

Zygote

Nuclei

Basidium

Diploid (2n)

Basidium

Basidiospores

Spores are released

Dikaryotic (n+n)

Monokaryotic (n)

Germination of spores

Gills

Cap (pileus)

Monokaryotic mycelia

Spores

Stalk (stipe)

Basidiocarp

Hyphal union

Cytoplasmic fusion

Dikaryotic mycelium

(a)

Basidium

5 μm

(b)

Figure 17.8 Lichens: (*a*) a crustose lichen; (*b*) a foliose lichen; (*c*) a fruticose lichen; (*d*) cell structure of a lichen in cross section, showing asexual reproduction by soredia.

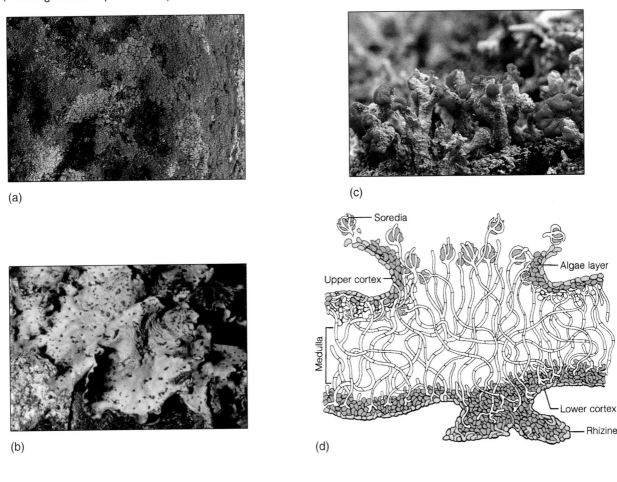

(a)

(b)

(c)

(d)

mineral and water absorption, greatly increasing root efficiency (fig. 17.9). In turn, the fungi obtain carbon compounds from the roots of the plants. Only a few families of vascular plants characteristically lack mycorrhizae. Fossil roots over 400 million years old contain mycorrhizal fungi, attesting to a long and stable coevolution.

There are two types of mycorrhizal relationships found in nature: **endomycorrhizal** and **ectomycorrhizal.** In an endomycorrhizal relationship, the hyphae of a zygomycetous fungus penetrate the root cells, whereas in an ectomycorrhizal relationship, the hyphae of a basidiomycete fungus surround the root cells and extend into the soil. The associations are usually species specific. Endomycorrhizae have been found in fossil roots 400 million years old, indicating that fungi have played a major role in evolution of plants.

Study the demonstration slides of mycorrhizae in association with roots. Sketch the association in the circle.

Observing Fungal Diversity and Symbiotic Relationships

Figure 17.9 Ectomycorrhizal hyphae form a mantle surrounding the root and penetrate between, but do not enter, cortical root cells.

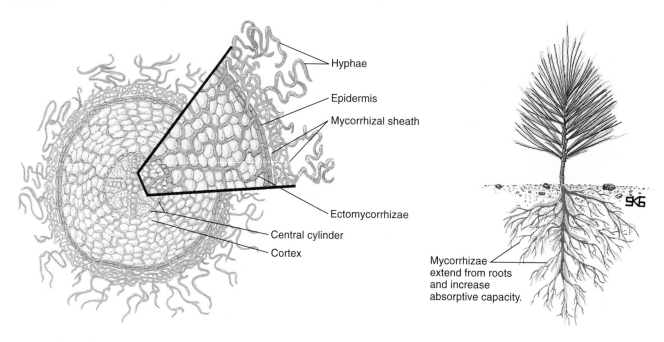

Hyphae

Epidermis

Mycorrhizal sheath

Ectomycorrhizae

Central cylinder

Cortex

Mycorrhizae extend from roots and increase absorptive capacity.

Learning Biology by Writing

Write a short, 200-word, essay describing the importance of fungi in ecosystems as decomposers, parasites, and mutualistic organisms.

Your instructor may ask you to turn in answers to the Lab Summary or Critical Thinking Questions that follow.

Lab Summary Questions

1. What are the distinguishing characteristics of a fungus?
2. List the distinguishing characteristics of the four divisions of fungi studied in this lab.
3. If you were given a sample to identify, what characteristics would you use to determine if it was a lichen or a pure fungus?
4. In general terms, describe a sexual life cycle in fungi. What stages are haploid and which ones are diploid? When is nuclear division by mitosis and when by meiosis?

Internet Sources

Considerable information is available on the World Wide Web on the topic of lichens. A starting place for locating information is the Internet Resource Page for Bryologists and Lichenologists at http://www.unomaha.edu/~abls/resources.html Check out the Images and Information section.

5. Compare the life cycles of fungi from the phylum Ascomycota with those from the phylum Basidiomycota. How are they similar? How do they differ?
6. Give examples of three different symbiotic relationships that fungi have with plants or protists. Indicate which are parasitic and which are mutualistic. Explain why.

Critical Thinking Questions

1. The *Suillus lakei* mushroom is commonly found growing around the base of conifers, especially the Douglas fir. Offer a possible explanation of this association of a fungus and plant.
2. DNA testing revealed that an *Armillaria bulbosa* fungus growing in the Upper Peninsula of Michigan covered 15.9 hectares (38 acres) (it may well be the largest organism in the world). Explain how this basidiomycete could get to be so large. Would you expect to find larger examples?
3. Many fungi produce and secrete antibiotics that inhibit bacterial growth. What is the advantage to the fungus in doing this?
4. Crustose lichens growing on rock surfaces may increase in diameter only a millimeter or so each year. If a lichen started growing on the day you were born, how many millimeters in diameter would it be now? How big is that in inches?

LAB TOPIC 18

Investigating Early Events in Animal Development

Supplies

Preparator's guide available on WWW at
http://www.mhhe.com/dolphin

Equipment

Incubators at 20°C and 37°C
Dissecting microscopes
Compound microscopes
Refrigerator

Materials

Prepared slides
    Sea star development from egg to larval stage
    Early cleavage stages of *Cerebratulus* sp.
    16 hr chick blastoderm (wm)
    18 hr chick (cs)
    39 hr chick (wm)
    72 hr chick (wm)
Live sea urchins in reproductive condition
Photographs, models, or plastic whole mounts of 24-
    and 48-hour stages in the development of a chick
Depression slides
Slides and coverslips
Thermometers
Dropper bottles
Pasteur pipettes
Glass rods
Syringes
Petroleum jelly
Miscellaneous beakers

Solutions

Seawater
0.5 M KCl
3.5% NaCl
50% ethanol as fixative

Prelab Preparation

Before doing this lab, you should read the introduction and sections of the lab topic that have been scheduled by the instructor.

You should use your textbook to review the definitions of the following terms:

    archenteron
    blastodisc
    blastomere
    blastula
    cleavage
    ectoderm
    endoderm
    gastrula
    mesoderm
    primitive streak
    radial cleavage
    spiral cleavage
    totipotent
    zygote

You should be able to describe in your own words the following concepts:

    Cleavage of the zygote to produce a blastula
    Significance of gastrula formation
    Gastrula formation in bird's egg

As a result of this review, you most likely have questions about terms, concepts, or how you will do the experiments included in this lab. Write these questions in the space below or in the margins of the pages of this lab topic. The lab experiments should help you answer these questions, or you can ask your instructor for help during the lab.

Objectives

1. To identify the stages in the early development of sea stars
2. To compare radial to spiral cleavage

3. To observe fertilization and early cleavage in sea urchins
4. To observe the early stages of chick development

Background

Development is one of the truly amazing processes in both plants and animals. Cells from different parents fuse to produce a deceptively simple-looking zygote that, in turn, undergoes a series of cell divisions to produce a multicellular adult. The adult is made up of hundreds of different kinds of cells that perform myriad highly integrated and coordinated functions. For most organisms, although the stages of development have been outlined, the underlying mechanisms that control development remain an enigma. It is known that the developmental program is encoded in the genetic material, but no one yet understands exactly how certain genes are activated and others are repressed at just the right times during development.

An animal's embryonic development can be divided into four major stages: (1) **fertilization,** when the sperm pertrates the egg and is soon followed by syngamy, the fusion of the egg and sperm nuclei; (2) **cleavage,** the mitotic divisions that partition the large cytoplasm of the fertilized egg into smaller cells; (3) **gastrulation,** a morphogenetic cellular movement that produces an embryo with three layers of cells; and (4) **organogenesis,** the process whereby specific organs develop from the primary germ layers.

The specifics of development are different for different animals, but there are many similarities between species in general development. As development progresses, all embryos become more complex as previously nonexistent tissues and organs appear and the embryo becomes capable of performing new functions. This increase in complexity is called **differentiation.**

However, the developing organism is not simply a random assortment of new cell types and organs; these new features always have specific spatial relationships with existing cells and with those that will form later. This development of form is called **morphogenesis** and comes about as a result of cell movement and growth as the embryo uses yolk materials for a source of energy and for chemical building units. During morphogenesis, the cells must "communicate" to inform each other of their location during a migration movement and to trigger the differential use of genetic information.

The physical characteristics of eggs are related to the environment in which an animal lives, the place where the embryo develops, and the stages in the life cycle of the species. Eggs of animals that live in aquatic environments generally have gelatinous coats, which protect the egg from physical and bacterial injury, and moderate amounts of yolk, which supply the embryo with sufficient energy to achieve the developmental stage of self-feeding.

Land animals that release their eggs for development outside of the mother have eggs covered by hard shells that provide protection against physical injury and desiccation. However, such a protective device is not without its problems. The embryo inside must exchange O_2 and CO_2 with the environment, must have sufficient energy and raw materials to develop to an advanced stage, and must have a means of disposing of nitrogenous wastes. Inside such eggs, four extraembryonic membranes grow out from the embryo, surround it, and function in gas exchange, waste storage, and nutrient procurement.

In mammals, since the developing embryo is carried inside the mother's uterus, no outer shell or jelly coats are necessary. When the egg implants in the uterus, extraembryonic membranes surround the embryo and also form the **placenta,** a highly vascularized organ that brings extraembryonic capillaries close to the mother's capillaries where nutrients and wastes are exchanged.

In animals, the way in which an egg undergoes early development is strongly influenced by the amount and distribution of yolk in the egg. Eggs of some species have little yolk and it is uniformly distributed in the egg's cytoplasm, as in the starfish and sea urchin you will study in this lab. Others have a large amount of yolk that displaces the active cytoplasm to one pole of the spherical egg. Such eggs are characteristic of fish, amphibians, reptiles, and birds.

LAB INSTRUCTIONS

You will observe developmental patterns in an echinoderm, a marine worm, and a bird, which have very different kinds of eggs. You will study the early developmental patterns of each organism to see the effects of egg type on development. Because these comparisons are detailed and time consuming, the lab topic may take two laboratory periods if all parts are done.

Sea Star Development

The sea star is used to illustrate early development because the events of cleavage and gastrulation are especially easy to see. More complex development patterns in other organisms may be compared to this simple pattern.

Obtain a microscope slide containing developmental stages of the sea star from the unfertilized egg through the gastrula stages. Scan the slide under low power with the compound microscope and locate the stages described in this section and shown in figure 18.1. These slides are usually *thicker* than most slides you have studied. *Do not use high power,* or you may push the objective through the coverslip while focusing.

The **unfertilized egg** contains a single egg nucleus and a large amount of yolk surrounded by a cell membrane. When the sperm penetrates the egg, a **fertilization membrane** forms as materials stored beneath the cell membrane

Figure 18.1 Stages in echinoderm (sea star) development: (*a*) unfertilized egg; (*b*) zygote surrounded by fertilization membrane; (*c*) two-cell stage (blastomeres); (*d*) four-cell stage; (*e*) morula; (*f*) blastula; (*g*) early gastrula; (*h*) late gastrula.

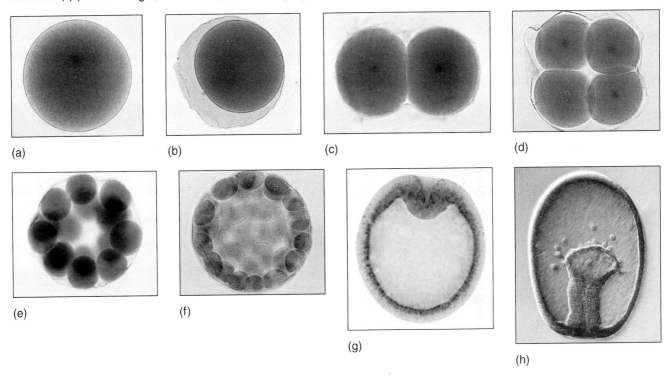

(a)   (b)   (c)   (d)

(e)   (f)   (g)   (h)

are released. This new membrane may be visible as a "halo" above the surface of fertilized eggs on your slide. This membrane prevents the penetration of additional sperm. Once inside the egg, the haploid sperm nucleus migrates through the cytoplasm. It fuses with the haploid egg nucleus, in an event called **syngamy** that forms the diploid nucleus of the **zygote.**

For several hours following fertilization, the zygote undergoes a series of rapid mitotic cell divisions without any intervening periods of growth. This **cleavage** process involves the duplication of chromosomes followed immediately by mitosis and cytokinesis, followed again by chromosome duplication, followed by mitosis, and so on. The result is that the large uninucleate fertilized egg is divided into smaller and smaller cells, each with a single nucleus. If the type of cell division is mitosis, do the nuclei of the developing embryo contain the same or different information than the nucleus formed by syngamy?

The cleavage divisions are usually synchronized in all cells, so that the embryo goes through a series of stages in which it contains 1, 2, 4, 8, 16, 32, and 64 cells. Around the 32-cell stage, the number of cells becomes difficult to

count, and the developing embryo is referred to as a **morula,** meaning mulberrylike.

Find examples of 2-, 4-, 8-, and 16-cell stages on your slide. Note that the size of the developing embryo does not increase as cell number increases. The existing material of the embryo is merely partitioned by cleavage into smaller cells.

As cleavage continues, the cells eventually form a hollow ball of cells that is ciliated on its outer surface. The beating of the cilia allows the embryo to swim out of the enveloping fertilization membrane. This stage is called a **blastula,** the cells are **blastomeres,** and the central cavity is the **blastocoel.** Find a blastula on your slide and identify these structures.

Several hours after blastula formation, a second major morphogenic event occurs. Cells at one end of the blastula undergo rapid growth and move into the blastocoel. This infolding is called **invagination** or **gastrulation** and results in a two-layered embryonic stage called the **gastrula.** The gastrula then elongates, forming a tube within a cylindrical body.

The two layers of cells in the gastrula are called **primary germ layers.** The inner layer is the **endoderm** and will form the lining of the digestive system and digestive glands. The outer layer is the **ectoderm** and will form the skin and the nervous system of the adult. The midgastrula and late gastrula stages can be identified by the elongation of the gastrula and changes in the shape of the endoderm tube. The elongated tube is called the **archenteron,** or primitive gut, and its opening to the outside is called the **blastopore.** It will become the anus of the sea star. The inner end of the archenteron eventually develops two lateral pouches that will grow outward and pinch off, forming a third germ layer called the **mesoderm.** The start of this process is shown in figure 18.1 (*h*). Muscles, connective tissues, and gonads will develop from the cells in this layer. Find a late gastrula stage on your slide and sketch it below. Label all the structures indicated by boldface terms in this paragraph.

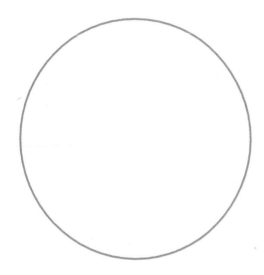

Beyond the gastrula stage, differentiation continues. In the larva stage, the embryo is free swimming or drifting with a mouth and anus. At this stage, it is capable of feeding itself and continues to grow, using external energy sources. At a later stage, a metamorphosis occurs, and the larva changes to a miniature version of the adult sea star.

Cleavage Patterns in Invertebrate Animals

Echinoderms, such as the sea star you just studied, have what is called **holoblastic** cleavage. Holoblastic refers to the fact that when the cells (blastomeres) divide, the daughter cells are usually of equal size and completely separate from one another. Such division processes are characteristic of eggs in which the yolk is evenly distributed. In contrast to holoblastic cleavage, animals that have eggs with large amounts of yolk often have **meroblastic** cleavage. In this case, the daughter cells resulting from a division are often unequal in size with the larger cells containing a greater amount of yolk. You will see an example of meroblastic cleavage when you study avian development.

Echinoderms also have **radial** cleavage (fig. 18.2). Each successive mitotic spindle and cleavage plane forms at right angles to the previous one so that the cells are arranged on radii extending from the center of the cell mass. In contrast to radial cleavage, many invertebrates have **spiral** cleavage. In fact, this basic difference correlates with the two groups of higher animals, the Protostomes (spiral cleavage) and Deuterostomes (radial cleav-

age). (See lab topics 20–22.) In spiral cleavage, after the four-cell stage, the mitotic spindles form at oblique angles to the previous one so that the cleavage planes are also displaced to oblique angles.

To observe evidence of spiral cleavage, look at a slide of the early developmental stages of *Cerebratulus* sp., a marine ribbon worm. Find an eight-cell stage where you are looking down on the organism from the top, a polar view. Note the positions of the top four blastomeres relative to the positions of the lower four. The top blastomeres lie over the grooves between the lower blastomeres. Sketch the stage.

Change slides and look again at the slide of sea star developmental stages. Find an eight-cell stage in polar view and note the relative positions of the upper and lower four blastomeres. They lie directly on top of one another with the cells and grooves in register. Sketch the stage below.

Look again at the eight-cell stages of both the sea star and *Cerebratulus.* Are each of the cells in each embryo the same size? Which of the organisms has holoblastic cleavage?

Figure 18.2 **Comparison of radial and spiral cleavage.** (*a*) In radial cleavage in deuterostomes, the spindles form either parallel or at right angles to the axis running through the poles of the cells. Consequently, cells sit directly on top of cells underneath. (*b*) In spiral cleavage in protostomes after the four-cell stage, the mitotic spindles form at oblique angles to the polar axis of the cells. Because cytokinesis always occurs across the center of the spindle, the oblique spindles result in displacement of the newly forming cells. Arrows indicate the direction of displacement.

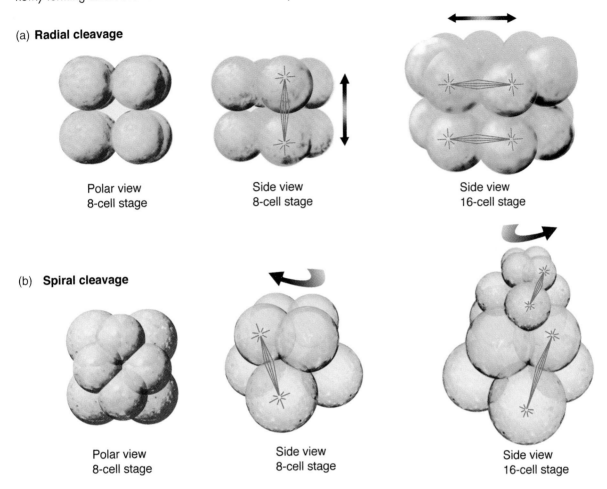

(a) **Radial cleavage**

Polar view
8-cell stage

Side view
8-cell stage

Side view
16-cell stage

(b) **Spiral cleavage**

Polar view
8-cell stage

Side view
8-cell stage

Side view
16-cell stage

Experimental Embryology with Sea Urchins

Sea urchins, as well as sea stars, are members of the phylum Echinodermata and show the same early developmental patterns. Sea urchins are found in both the Atlantic and Pacific oceans. The breeding cycles are such that the West Coast species are fertile in the fall and winter, whereas the East Coast species are fertile from April to September. This makes sea urchins ideal for teaching laboratories because animals in breeding condition are available throughout the year from biological suppliers.

Sea urchins in breeding condition can be induced to release their gametes by a variety of methods. Mild electrical shock, temperature shock, injection of potassium chloride, or surgical removal of the gonads have all been used successfully to obtain eggs and sperm. The separately collected eggs and sperm can be mixed on a microscope slide so that fertilization and development can be observed.

Obtaining Gametes

Because it is difficult to tell the difference between the sexes in sea urchins, several sea urchins may have to be used to obtain both eggs and sperm. One male and one female, however, will provide enough gametes for a whole lab section.

To induce shedding, inject three to four animals, one at a time, through the soft tissue surrounding the mouth with 2 ml of 0.5 M KCl, as shown in figure 18.3. Place the animals in a dry bowl with the **oral** (mouth) surface down and watch the **aboral** surface now on top for the release of gametes through the genital pores. After a few minutes, gametes will be emitted. Males release a white suspension of sperm, and the females a yellowish-brown suspension of eggs (one species has red/purple eggs).

Investigating Early Events in Animal Development

Figure 18.3 Technique for collecting sea urchin gametes. (*a*) Inject 0.5 M KCl into soft tissues near mouth to stimulate gamete release. (*b*) Place animals oral side down in dish and watch for gamete release from aboral surface. (*c*) Place female over beaker of cold seawater with aboral surface down in water. Eggs will sink to bottom. (*d*) Sperm should be removed from male's aboral surface and stored undiluted in a dropper bottle.

(a)

(b)

(c)

(d)

Collect the eggs and sperm as follows (see also fig. 18.3):

Females should be placed over a small beaker filled with cold seawater. Put the aboral surface down and immersed in the water. When shed, the eggs will sink to the bottom of the beaker. To facilitate union with sperm later, these eggs should be washed three times with seawater. Swirl the water in the beaker, allow the eggs to settle and decant the supernatant. The eggs will remain viable for two to three days if stored in the refrigerator in seawater.

Sperm should be collected by holding a male over a sterile petri plate with the oral surface up so that the sperm drip into the plate. They can be transferred to a cold, clean dropper bottle for storage up to 24 hours. No seawater should be added until the sperm are to be used. At that time, add three to four drops of sperm to 25 ml of cold seawater to create a suspension that will remain active for 20 to 30 minutes.

Observing Fertilization

Fertilization events are temperature sensitive and rarely occur above 23°C. If your lab room is hot, cool all slides and suspensions in a refrigerator before starting the experiment.

To observe fertilization, take a depression slide and add one to two drops of seawater containing 10 to 30 eggs. Examine these eggs under the 10× objective on the compound microscope and try to locate the egg nucleus. The yolk is concentrated in one hemisphere of the egg called the **vegetal hemisphere.** It is usually on the bottom as the egg floats. The other hemisphere is called the **animal hemisphere.**

While watching through your microscope, add a drop of the diluted sperm suspension to the slide. Carefully observe the gametes. You will probably not see the single sperm that fertilizes the egg. When a single sperm fuses with the egg, it will release a **fertilization membrane,** which you should be able to observe.

If petroleum jelly is placed around the edge of the coverslip, the slide can be kept and later developmental stages observed during the next 24 hours. (Be sure the temperature does not exceed 23°C.) Refer to the following schedule of developmental events to see the approximate timing for different stages.

Sperm contacts egg	0 minutes
Fertilization membrane forms	2 minutes
1st cleavage (2-cell stage)	1 hour
2nd cleavage (4-cell stage)	1 hour, 30 minutes
4th cleavage (16-cell stage)	2 hours, 30 minutes
Blastula (about 1000 cells)	7 to 8 hours
Gastrula	12 to 15 hours
Pluteus larva	2 days

▶ While waiting for the first cleavage to occur, obtain a regular microscope slide and add a drop of diluted sperm suspension. Add a coverslip and observe the slide under high magnification (or oil immersion) with your compound microscope. Sketch what you see.

Separation of Blastomeres (Optional)

It is possible to separate the blastomeres of a two- or four-cell embryo from an isolecithal egg and to have each cell develop into a normal adult (though the new embryos will be smaller than normal because each cell starts with less yolk material than a normal zygote). The first two divisions in an isolecithal egg pass through both the animal and vegetal poles of the egg and produce four equivalent cells. In other egg types, the cells produced are not equivalent, and separation of the cells does not result in viable embryos. Cells capable of developing into complete embryos are said to be **totipotent.** Since humans have isolecithal eggs and the blastomeres remain totipotent for the first few cleavages, this explains how identical twins, triplets, or quadruplets may be born.

Animals whose blastomeres are totipotent during the early stages of development are said to have **indeterminate** cleavage. Some animals have determinant cleavage: their blastomeres are not totipotent and the fate of the cells is determined or committed early in the cleavage process. The blastomeres cannot develop into whole organisms because their genetic program has committed them to developing into a specific part of the organism.

Hypothesize what will happen if you separated totipotent blastomeres which had indeterminate cleavage.

▶ To separate the blastomeres, the fertilization membrane must first be removed from the cleaving egg. Fertilize about 100 eggs in a 20-ml beaker or test tube. Wait for approximately 30 minutes while they sit at 23°C or less. Take the eggs and draw them rapidly into a Pasteur pipette; then rapidly expel them. Repeat this aspiration process 15 to 20 times. Now examine a few eggs under your microscope to be sure the fertilization membrane has been removed.

▶ Remove the seawater over the eggs and replace it with calcium-free 3.5% saline. (In the absence of Ca^{++}, the blastomeres will not stick together tightly.) Add and withdraw fresh, calcium-free saline several times to ensure that all the old calcium ions are washed away.

▶ After 30 minutes, aspirate the two- or four-cell embryos back and forth in a Pasteur pipette. Observe the cells under the microscope. If the blastomeres have separated, dilute the suspension of blastomeres with seawater. Add about ten blastomeres to a depression slide. Add a coverslip, ring with petroleum jelly, and observe the slide over the next 48 hours. Alternatively, the stages can be incubated in a test tube and samples can be periodically withdrawn for observation. Your instructor can add alcohol to fix them 24 and 48 hours from now.

Does the data that you collected lead you to support or reject the hypothesis made earlier?

Compare the rate of development of the separated blastomere embryos to that of the normal embryos in the first part of this exercise. Compare the sizes of the blastula and gastrula stages in these two experiments. Describe the differences.

Chick Development

Because of time limitations, it will not be possible to observe all of the developmental stages of the chick. Three stages have been chosen to illustrate chick development: 16-, 18- and 24-hour embryos. You will look at prepared slides.

Investigating Early Events in Animal Development   **211**

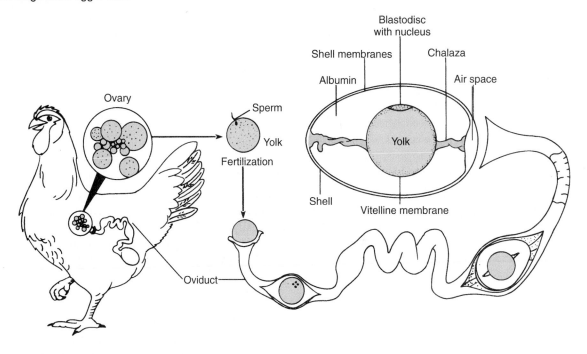

Chick Cleavage and Gastulation

Since the eggs of birds (also fish and reptiles) contain large amounts of yolk, the nucleus and cytoplasm are confined to a small disc-shaped area called the **blastodisc,** which sits on top of the yolk. Fertilization occurs at the blastodisc, and cleavage divisions are confined to this area. The large yolk does not divide, as in the starfish; the developing chick embryo "rides" atop the yolk during development (fig. 18.4). When slides are prepared, the yolk is removed and only the blastodisc area is on the slide.

Chick Cleavage and Gastulation

Obtain a whole mount slide of a 16-hour old chick embryo. You will be looking at the entire blastodisc after it has undergone several cell divisions. As the cleavage divisions occur in the blastodisc, a flat layer of cells called the **blastoderm** is produced. As the cleavage continues, the blastoderm becomes several cell layers thick with a fluid-filled space, the **subgerminal space,** developing between it and the yolk. The blastula is formed when the cell layers of the blastoderm separate into an upper **epiblast** layer and a lower **hypoblast** layer (fig. 18.5). The space between them is the avian equivalent of the **blastocoel.**

As development continues, cells in the center of the long axis of the pear-shaped blastoderm migrate down from the surface and away from the center area in a process that is equivalent to gastrulation. These movements result in the recognizable **primitive streak** stage (fig. 18.6). Find the primitive streak on your slide.

Now get a slide of a cross section through the blastoderm for an 18-hour chick. Compare what you see to figure 18.7. Identify the three primary germ layers: ectoderm

Figure 18.5   Side view of developing chick embryo. (*a*) Blastodisc confined to top of yolk. (*b*) Several cell layers accumulate. (*c*) A cavity appears in cell mass when blastula forms.

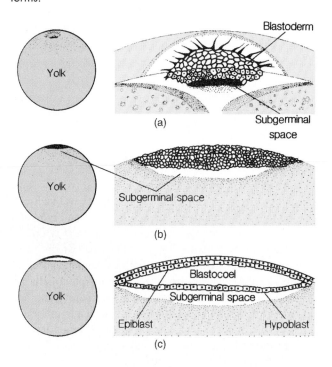

Figure 18.6 Gastrulation in bird egg: (*a*) blastodisc on yolk; (*b*) ectoderm and endoderm form by splitting of blastodisc into two layers; (*c*) mesoderm originates from cells migrating into cavity through primitive streak.

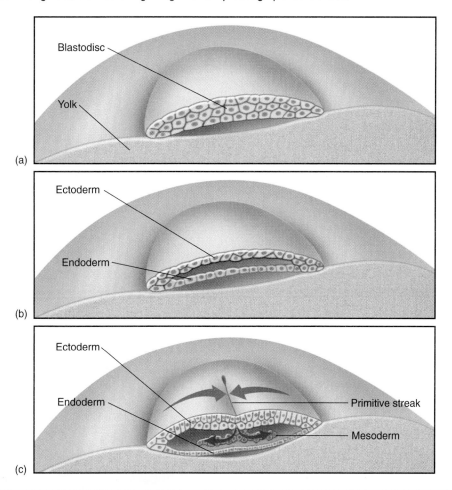

Blastodisc

Yolk

(a)

Ectoderm

Endoderm

(b)

Ectoderm

Endoderm

Primitive streak

Mesoderm

(c)

Figure 18.7 Germ layers formed during gastrulation fold inward raising the developing embryo off the surface of the yolk as the archenteron, notochord, neural tube and coelom differentiate.

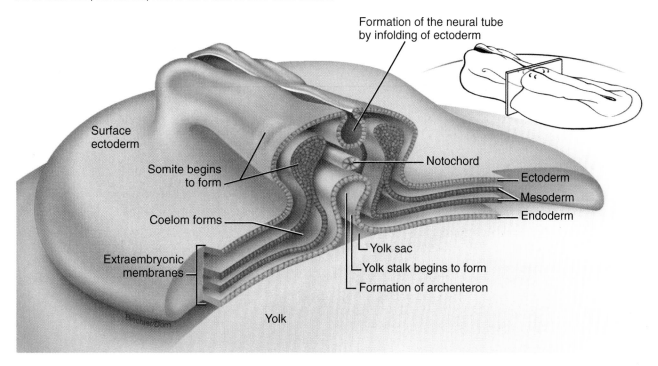

Formation of the neural tube by infolding of ectoderm

Surface ectoderm

Somite begins to form

Coelom forms

Extraembryonic membranes

Notochord

Ectoderm

Mesoderm

Endoderm

Yolk sac

Yolk stalk begins to form

Formation of archenteron

Yolk

on the outside; mesodermal layers; endoderm surrounding the yolk and starting to form a digestive tube. Are the **neural tube** and notochord visible?

Later Developmental Stages

Study photographs or whole mount microscope slides of 33-hour and 72-hour chick embryoes. Note the developing nervous system and the anterior-posterior axis.

By 33 hours, the relatively unorganized structure of the blastoderm is gone, replaced by an obviously animal-like embryo (fig 18.8). The **neural tube** runs the length of the embryo and its anterior portion is beginning to differentiate into regions of the brain. A crude **heart** has formed. **Somites,** segments of body muscle, are beginning to appear.

By 72 hours, note that the anterior half of the chick has rotated and lies on its left side, while the posterior half remains dorsal side up (fig. 18.9). Considerable differences may also be noted in the nervous and circulatory systems. The neural tube has differentiated into a five-part brain and a nerve cord. **Optic cups** are developing from the optic vesicle, and the optic lens is starting to form. An auditory vesicle should be visible. The heart has changed from a muscular tube to a two-chambered organ, and aortic arches with associated gill slits have developed. Note the vitelline arteries and veins. What are their functions?

Figure 18.8   Dorsal view of 33-hour chick.

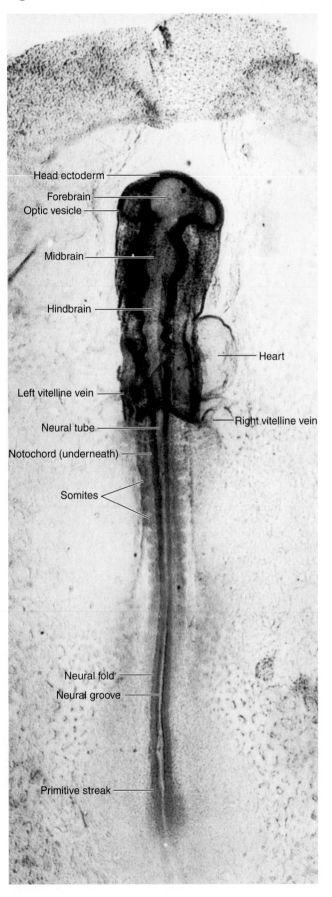

Figure 18.9 Whole mount of a 72-hour chick embryo.

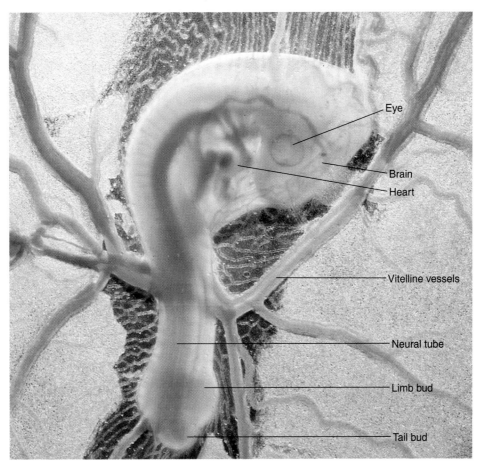

Eye

Brain

Heart

Vitelline vessels

Neural tube

Limb bud

Tail bud

Learning Biology by Writing

In this lab topic, you observed the early developmental stages in echinoderms and birds and determined if totipotency is retained by cells after cleavage begins. Write a report in which you compare the general developmental patterns of these two organisms and correlate these patterns with the egg type of each organism (isolecithal or telolecithal). Discuss the concept of totipotency, including in your discussion experimental results obtained from the lab topic.

As an alternative assignment, your instructor may ask you to answer the Lab Summary and Critical Thinking Questions that follow.

Lab Summary Questions

1. Describe how a fertilized sea star egg develops into a blastula.
2. Why is the gastrula considered an important stage in development? How does the gastrula stage reflect the general adult body plan of most animals?
3. Describe the events that occur when a sea urchin's egg is fertilized.
4. Compare and contrast radial and spiral cleavage.
5. If a four-cell stage is separated into individual blastomeres that each develop into a complete embryo, what does it prove?
6. In your own words, describe how an embryo develops from the "yolk" of a bird's egg.

Critical Thinking Questions

1. Describe the different ways that frog, chicken, and human embryos have for procuring nutrients, exchanging gases, and treating wastes.
2. Twins may be identical or fraternal. What is the difference? Are the cells in early human embryonic stages totipotent? Support your answer with reasons.
3. Reflect on the general life cycles of a bird and an invertebrate such as a starfish. Recognize that many adult birds are smaller than starfish. Why are bird eggs always larger than invertebrate eggs? Consider in your answer the energy budgets for development and the role of larval free-living stages.

4. Many invertebrates produce several thousand eggs during their lifetimes. Why is the world not over run by invertebrates?

Internet Sources

Check the WWW for current research on early developmental stages in animals. Use the search engine Google at www.google.com and enter the phrase research on blastula or research on gastrula. Scan the sites returned and write a summary that is of most interest to you.

LAB TOPIC 19

Animal Phylogeny: Investigating Evolution of Body Plan

Supplies

Preparator's guide available on WWW at
http://www.mhhe.com/dolphin

Equipment

Compound microscopes
Dissecting microscopes
Saltwater aquarium with living demonstration
  specimens of sponges and anemones

Materials

Preserved specimens
  Leucosolenia
  Gonionemus
  Taenia (tapeworm—demonstration)
  Ascaris
Prepared slides
  Leucosolenia, longitudinal section
  Hydra, longitudinal section
  Obelia, whole mount
  Cnidocyte, demonstration slide
  Dugesia, cross section and whole mount
  Clonorchis sinensis or other fluke, whole mount
  Tapeworm, scolex and proglottids (demonstration)
Living *Dugesia*
Living *Hydra*
Petri dishes
Raw meat
Dissecting pans and instruments
Watch glasses
Spring water
Baerman funnel apparatus
Sieves, 25- and 200-mesh
Soil samples, 100-g packets

Prelab Preparation

Before doing this lab, you should read the introduction and sections of the lab topic that have been scheduled by the instructor.

You should use your textbook to review the definitions of the following terms:

acoelomate
bilateral symmetry
body plan
Cnidaria
coelom
Nematoda
Platyhelminthes
Porifera
pseudocoelomate
radial symmetry

You should be able to describe in your own words the following concepts:

How body plans become more complex as you
  move from sponges to nematodes
The advantages of bilateral symmetry and a body
  cavity
The advantages of cellular specialization to form
  tissues and organs

As a result of this review, you most likely have questions about terms, concepts, or how you will do the experiments included in this lab. Write these questions in the space below or in the margins of the pages of this lab topic. The lab experiments should help you answer these questions, or you can ask your instructor for help during the lab.

Objectives

1. To study the functional anatomy of representatives from four simple animal phyla
2. To trace the phylogenetic development of general body plan, levels of organization, and coelomic cavity in the animal kingdom
3. To illustrate the life cycles of simple invertebrates
4. To collect evidence that tests the following null hypothesis: There is no evidence that the animals in the four phyla (studied during this lab) can be arranged in an evolutionary sequence from simple to complex.

The next four lab topics are a coordinated unit on the animal kingdom. By looking at a few selected phyla, you

Figure 19.1   An interpretation of the phylogenetic relationships among the major animal phyla.

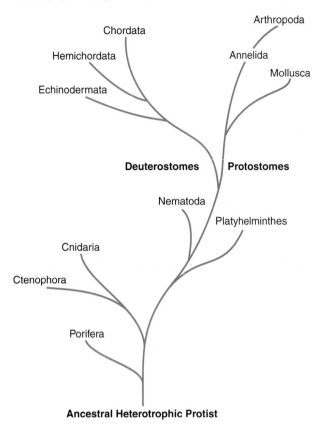

will be able to identify and trace several major evolutionary changes through the animal kingdom. In this lab, you will examine eight representatives from four phyla that illustrate the phylogenetic development of general body organization.

Background

As you look at specimens and learn new anatomical terms, resist the temptation to simply memorize the names of the structures; keep your viewpoint broad and look for similarities and differences. Ask yourself which characteristics probably evolved first and which evolved more recently. Remember that no living species is ancestral to another nor is any phylum more primitive than another. Rather, because some living species have changed less over time than others, they are more similar to ancestoral forms. Consequently, it is possible to envision hypothetical ancestral forms by studying living species. The heterotrophic protists alive today are not ancestors of jellyfish or reptiles, but they are more similar to the one-celled organisms that once lived on earth than are today's jellyfish or reptiles. Figure 19.1 shows one interpretation of the relationships among phyla in the animal kingdom.

The most general trait that is used to classify animals is the **level of organization** in the adult. All multicellular animals (Metazoa) start life as single cells (fertilized eggs), but

further organizational levels appear in sequence during development. In the Parazoa branch of the animal kingdom, which is represented by sponges, development results in cell specializations, but the cells are not organized into tissues or organs. The adults in the Eumetazoa branch, which includes all other phyla, have tissues, organs, and organ systems (fig. 19.2).

A second general trait that is used to categorize animal phyla is **body symmetry.** During embryonic life, all eumetazoic animals are solid or hollow balls of cells with radial symmetry. The **Radiata** (jellyfish, hydras, and anenomes) retain this radial symmetry into adult life, whereas all other eumetazoans become bilaterally symmetrical as they develop from embryo to adult. Bilateral symmetry is most often accompanied by the development of a nerve cord and increased cephalization. These organisms are usually more motile than radially symmetrical organisms.

A third trait used in studying phylogenetic relationships is the **pattern of alimentary structures.** Sponges have a **channel network** through which water from the external environment is pumped and food particles are removed by phagocytosis. This is a very simple mechanism. Radiates and flatworms have a **one-hole sac plan** (incomplete) in which one opening serves both as mouth and anus. Though this type of digestive system is not complex, it does establish the existence of a **diploblastic body** composed of at least two layers: the **endoderm** lining the gut and the **ectodermal** covering. The alimentary tract in all other phyla is a **two-hole tube** (complete) with a mouth and an anus. Food travels one way in a two-hole alimentary canal and the tube becomes specialized, with digestion occuring in one region and absorption in another.

If embryonic development is studied in the phyla with tubular systems, two developmental patterns may be seen. In the **protostomes,** the blastopore of the gastrula stage becomes the mouth, and the anus develops as a secondary opening. In the **deuterostomes,** the blastopore becomes the anus, whereas the secondary opening is the mouth.

After the appearance of a definite ectoderm and endoderm, a third layer, the **mesoderm,** appears in the bilaterally symmetrical phyla. Organisms with three embryonic layers are called **triploblastic.** The way the mesoderm develops and distributes itself in the embryo further characterizes each animal phylum (fig. 19.3). In the **acoelomate** phyla, the mesoderm fills the space between the endoderm and ectoderm; there is no body cavity.

In all other phyla, mesoderm only partially fills the body cavity, and the rest of the cavity is filled with fluid. This arrangement allows the animals to attain a larger size because the fluid in the body cavity can serve as a hydraulic skeleton and as a transport medium for nutrients and wastes. In the **pseudocoelomate** groups, the body cavity is bound by the body wall on the outside (the ectoderm) and the digestive tract on the inside (the endoderm). Mesoderm is found in the body wall and as cords in the cavity but it is not found in the wall of the digestive system.

Figure 19.2 A classification system based on tissue organization, symmetry, and body cavities for phyla of the animal kingdom.

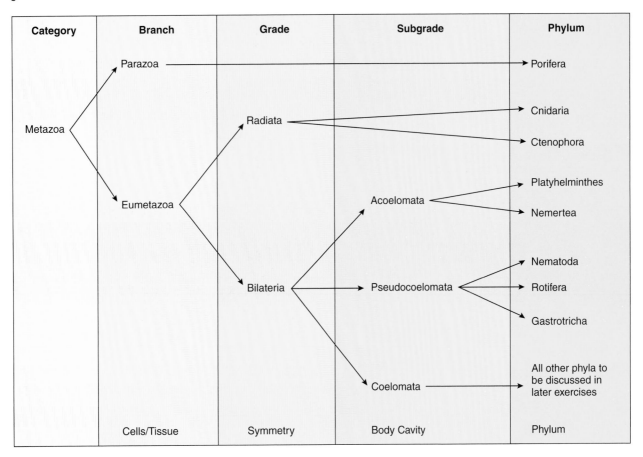

Figure 19.3 An artist's conception of true body cavities in the animal kingdom. From P. Weisz, *Science of Biology*, 4th ed. Copyright © McGraw-Hill Book Company, New York, NY.

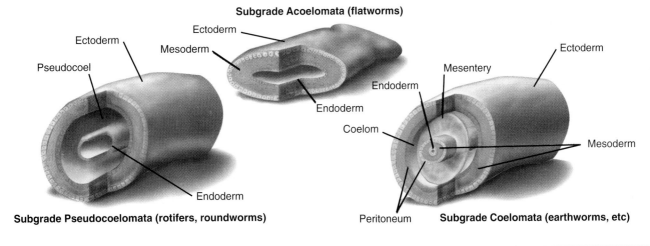

In the **coelomates,** a layer of mesoderm called the **peritoneum** lines the body cavity and surrounds the digestive system. This mesoderm layer often forms sheets of **mesenteries** that support the internal organs. The coelomate phyla are discussed in lab topics 20, 21, and 22.

LAB INSTRUCTIONS

In this lab topic, you will study representatives from four phyla: **Porifera, Cnidaria, Platyhelminthes,** and **Nematoda.**

Figure 19.4 Longitudinal sections through different types of sponges showing the relationship of spongocoel to choanocytes and canals: (*a*) ascon body type; (*b*) sycon body type; (*c*) leucon body type.

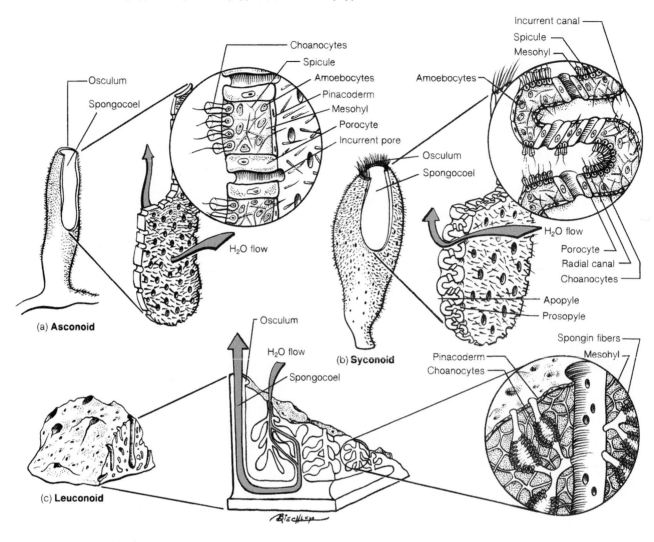

Phylum Porifera

The 9,000 or so species of sponges in the phylum Porifera are among the simplest multicellular animals because they lack distinct organs. Phylogenetically, the sponges are an offshoot of the main evolutionary patterns seen in the animal kingdom. They consist of cells surrounding a hollow central area called the **spongocoel.** The spongocoel is not a coelom nor a digestive system; it is part of the water movement system of the sponge. The body walls of sponges are perforated by numerous openings that connect to a canal system. Unique cells with flagella, called **choanocytes** (fig. 19.4), are found only in this phylum. They move water in through the pores on the body surface and out through a large body opening, the **osculum** (see fig. 19.4). Individual cells engulf and digest food particles entering the canal system with the water stream. Furthermore, the water stream supplies the respiratory needs of the cells and removes wastes.

The body walls of sponges are supported by a skeleton consisting of either (1) calcium carbonate crystals, (2) silica

Before starting this lab, I will propose a null hypothesis for you to test, to falsify if you can. The hypothesis is: *"There is no evidence that animals in the four phyla studied here can be arranged in a sequence from simple to complex."* If you find evidence that you can, then you have falsified the null hypothesis and must accept an alternative. Formulate an alternative hypothesis and state it below.

crystals, or (3) fibers of a protein called **spongin.** These three types of skeletal elements form the basis for dividing sponges into three taxonomic classes, the **Calcarea,** the **Hexactinellids,** and the **Demospongia.** Approximately 95% of the species are in the class Demospongia.

▶ Look at a preserved specimen of one of the simplest sponges, *Leucosolenia.* Note the base, the pores on the surface, and the **osculum.** The skeleton made of calcium carbonate should be visible as **spicules.** You may want to use the stereoscopic microscope to observe the specimen.

▶ Now obtain a slide of a longitudinal section of *Leucosolenia* and look at it first with the dissecting microscope. Compare the section to figure 19.4. What type of body plan does *Leucosolenia* have? _____

Using the compound microscope, identify the **choanocytes,** which line the **spongocoel.** The coordinated beating of their flagella drives water out the osculum and draws water containing food and oxygen in through the surface pores.

Sponges feed by removing small particles of organic matter from the water. The movement of the flagellum on each choanocyte causes a vortex to form in the collar region of the cell. Particles settle into the base of the collar where they are engulfed by extensions of the choanocyte's membrane to form food vacuoles. Do sponges have intracellular or extracellular digestion? _____

▶ Other cell types should be visible on the slide. The external surface of the sponge is composed of **pinacoderm** cells. The pinacoderm layer is interrupted by **porocytes,** cells that form hollow cylinders through which water enters the spongocoel. Between the pinacoderm and choanocyte layers is a gelatinous matrix, the **mesohyl,** in which spicules are embedded and often extend through the pinacoderm. Amoebocytes in the mesohyl manufacture the spicules. These cells also can transform into gametes during the reproductive season. What level of organization do sponges show? _____

Parts (*b*) and (*c*) of figure 19.4, show the basic body organization of more complex sponges. The **sycon** body type can be thought of as an accordionlike folding of the ascon body type. The **leucon** type is a more complex folding where individual chambers are created. Zoologists think that the location of choanocytes in distinct chambers makes them more efficient in capturing food particles. Examine the specimens of bath sponges in the lab. What type of body plan do they have? _____

In your study of sponges did you find any evidence that they have organs or organ systems? Describe the reasons for your answer.

How do you think sponges perform the important physiological functions of respiratory gas exchange, circulation, and excretion?

Phylum Cnidaria

The Cnidaria (previously known as the Coelenterata) are animals with rudimentary organ development and radial symmetry. The phylum includes about 10,000 species commonly known as jellyfish, sea anemones, and corals. Most species are marine. The name *Cnidaria* comes from the term **cnidocytes,** which are unique stinging cells found in these animals. Cnidarians have bodies consisting of two well-defined layers, the outer epidermis and an inner gastrodermis that lines the digestive system. Between these cell layers is a layer of gelatinous material called **mesoglea.** The rudiments of a neuromuscular system are also seen for the first time in this group.

The cnidarians have three basic body forms: a planula larval stage and two adult forms—a sedentary **polyp** stage and a free-swimming **medusa** (jellyfish) stage. The life cycles of many species involve one or more of these stages. The phylum Cnidaria contains three taxonomic classes: **Hydrozoa,** which includes *Hydra, Obelia,* and the Portuguese man-of-war; **Scyphozoa,** which consists mainly of marine jellyfish; and **Anthozoa,** which includes sea anemones and corals.

A Polyp: Hydra

Hydra has only a polyp stage in its life cycle; it does not have a medusa stage. *Hydra* is more complex than the sponges just studied because it has well-defined tissues and rudimentary organs (gastrovascular cavity, nervous system, and gonads) and is capable of involved behaviors.

▶ Get a living *Hydra* from the supply area and put it in a small watch glass with a small amount of water. Study the animal with your dissecting microscope. Depending on its condition, it may be asexually budding another *Hydra* or it may have swellings of the body wall which is a developing gonad. Gently poke it with a probe and watch its response.

Animal Phylogeny: Investigating Evolution of Body Plan

Figure 19.5 Anatomy of *Hydra*.

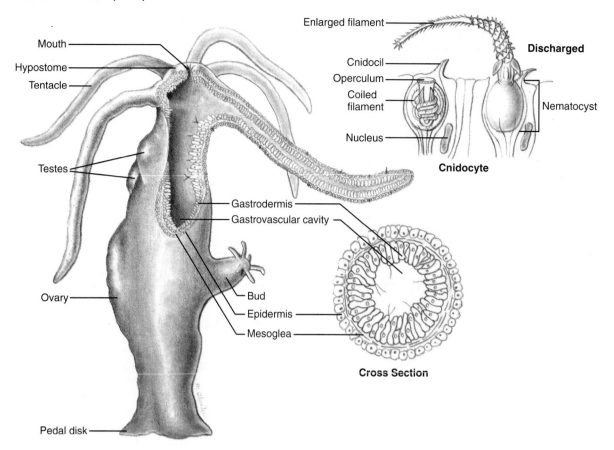

Identify the mouth, tentacles, and body column (fig. 19.5). Describe the behavior you observed. From this behavior would you conclude that *Hydra* has a nervous system? Muscles? Return the *Hydra* to the aquarium.

Look at a slide of a longitudinal section of *Hydra* (fig. 19.5). Identify the following structures: **gastrovascular cavity, mesoglea, epidermis, mouth,** and **tentacles.** Examine the tentacles closely. You may be able to find a cnidocyte containing a **nematocyst,** an apical organelle containing an extensible thread that can entangle or penetrate small prey. Sketch a cnidocyte in the circle.

How do you think that *Hydra* performs the important physiological functions of gas exchange, circulation, and excretion?

A Medusa: Gonionemus

Members of this genus are small marine jellyfish found throughout the world. Obtain a preserved specimen of *Gonionemus* and put it in a small dish filled with water. Handle it carefully—jellyfish are fragile.

Examine *Gonionemus* with your dissecting microscope. Note the radial symmetry. Compare the specimen to the cutaway view shown in figure 19.6. The jellylike consistency of the medusa is due to a thick layer of **mesoglea.**

Figure 19.6   The anatomy of the jellyfish *Gonionemus* sp.

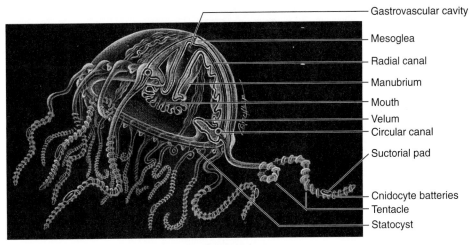

- Gastrovascular cavity
- Mesoglea
- Radial canal
- Manubrium
- Mouth
- Velum
- Circular canal
- Suctorial pad
- Cnidocyte batteries
- Tentacle
- Statocyst

Subumbrellar Surface

The carbohydrates and proteins of this gel-like material are highly hydrated, and water contributes most of the bulk of the jellyfish body.

Find the **mouth** at the end of the **manubrium** on the subumbrellar surface. Trace the digestive system into the **gastrovascular cavity** to the four **radial canals** and to the **circular canal** with extensions into each of the tentacles.

The medusa swims by slowly contracting neuromuscular cells in the body wall, which forces water out of the subumbrellar cavity. The **velum** is a skirtlike membrane that narrows the area through which water escapes when the muscles contract, thus increasing propulsive velocity.

The tentacles of *Gonionemus* contain batteries of cnidocytes and adhesive pads that sting and immobilize prey.

The sexes are separate in *Gonionemus,* but are hard to distinguish. Find the gonads attached to the subumbrellar surface of the radial canals. *Gonionemus* gametes are released into the sea, are fertilized externally, and develop into ciliated planula larvae, which attach to submerged objects. These larvae then grow into microscopic polyps. During the polyp stage, *Gonionemus* may reproduce asexually by budding to form additional polyps or may differentiate to produce miniature medusae that detach and grow.

How do you think jellyfish such as *Gonionemus* carry out the important physiological functions of respiratory gas exchange, circulation, and excretion?

A Colonial Form: Obelia

Many cnidarians are colonial and have life cycles in which the polyp and medusa stages alternate. Observe a slide of *Obelia* under low power with the compound microscope. You should see the branching pattern of a colony and two types of polyps: one with tentacles and one without. Polyps with tentacles are feeding polyps called **hydranths;** those without are reproductive polyps called **gonangia** (fig. 19.7). Because the gonangia lack any means of collecting food, they depend on the hydranths. Food enters the hydranth's gastrovascular cavity, which is continuous with a canal system throughout the colony. The similarity of a single hydranth to *Hydra* should be apparent.

Obelia is surrounded by a chitinous exoskeleton, the **perisarc.** The living tissue inside is the **coenosarc.** Where the perisarc surrounds the hydranth it is called a **hydrotheca,** and where it surrounds a gonagium it is called a **gonotheca.**

The life cycle of *Obelia* illustrates the alternation of generations commonly found among the Cnidaria: an asexually reproducing **polyp** stage is followed by a sexual **medusa** stage. The gonangia of the polyp stage produce tiny, short-lived medusae by budding from a central stalk, the **blastostyle.** The medusae, in turn, produce sperm or eggs. Fertilized eggs develop into ciliated planula larvae that swim about and settle to produce a new polyp stage. As the polyp develops, asexual buds grow but do not detach, thus forming a colony. What are the advantages of a reproductive strategy in which there is both sexual and asexual reproduction?

Figure 19.7 Anatomy and life cycle of *Obelia*, showing vegetative and reproductive polyps and medusa stages.

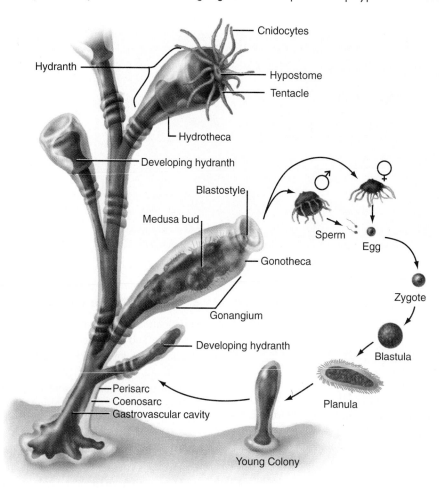

Phylum Platyhelminthes

The approximately 20,000 species in the phylum Platyhelminthes, the flatworms, exhibit several structural advances over the cnidarians. The nonparasitic forms contain well-developed digestive, excretory, reproductive, nervous, and muscular systems. In the parasitic species, many of these systems have been reduced through adaptation to a parasitic way of life. For example, the neuromuscular system is reduced in parasitic flatworms allowing only limited movement. In tapeworms, the digestive system is absent since nutrients are obtained from the host's intestine by absorption. The enlarged reproductive system of parasitic species produces enormous numbers of eggs, insuring that the species will infect other hosts and continue to thrive.

The organisms in this phylum are bilaterally symmetrical and have the three distinct tissue layers (endoderm, mesoderm, and ectoderm) characteristic of all higher phyla. They have no body cavity and are said to be **acoelomate.** There are three taxonomic classes: **Turbellaria,** including planaria, which are free-living flatworms; **Trematoda,** or parasitic flukes; and **Cestoda,** or parasitic tapeworms.

Turbellaria

The common freshwater planarians in the United States are members of the genus *Dugesia.* They live under rocks and sticks in streams and lakes.

Obtain a living planarian, place it in a petri dish in some spring water, and observe its locomotion. Is there a definite "head" end? What type of symmetry do you see?

Observe what happens when you gently touch the planarian's head with a probe or when you turn the organism over. Normal locomotion occurs by cilia located on cells covering the ventral surface. Muscles in the body wall allow the twisting, shortening, and extensions seen when the animal is disturbed.

Place a piece of raw meat in the dish and observe. What does the planarian's ability to move and behavior tell you about its muscle nervous systems?

Figure 19.8   Internal anatomy of planaria, a flatworm. Nervous and excretory systems are not shown and will not be visible on your slide. Diagram shows digestive system on left and reproductive system on right. Both systems are found on both sides.

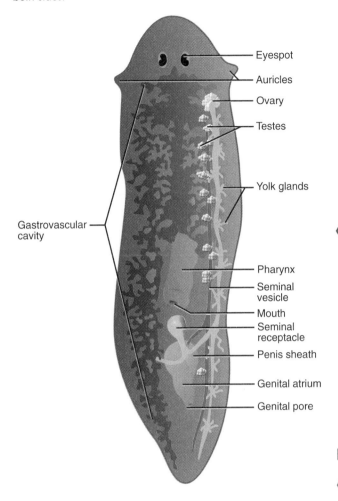

- Eyespot
- Auricles
- Ovary
- Testes
- Yolk glands
- Gastrovascular cavity
- Pharynx
- Seminal vesicle
- Mouth
- Seminal receptacle
- Penis sheath
- Genital atrium
- Genital pore

Figure 19.9   Cross section through anterior region of the planarian. Note absence of body cavity.

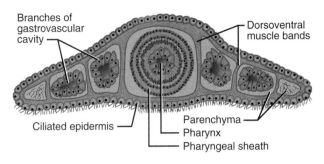

- Branches of gastrovascular cavity
- Dorsoventral muscle bands
- Ciliated epidermis
- Parenchyma
- Pharynx
- Pharyngeal sheath

The excretory and nervous systems will not be apparent in this specimen. They require special staining techniques to be clearly seen. The reproductive system will not be studied in detail in this organism. No specialized respiratory gas exchange organs are present nor is there a distinct circulatory system. How do you think planarians carry out these important functions?

The planarian's **mouth** is found at the end of a **protrusible pharynx** located about midway along the animal's body on the ventral surface. Two light-sensitive **eyespots** are found on the anterior, dorsal surface.

Obtain a stained whole mount of *Dugesia* and compare it to figure 19.8. The branched digestive system should be clearly visible. How many lobes extend anteriorward? _____ Posteriorward? _____

Flatworms lack an anus. Do they have a complete or an incomplete digestive system? _____

Each of the lobes of the saclike gastrovascular cavity is highly branched. What is the advantage of this branching in an animal that lacks a circulatory system?

Now obtain a microscope slide of a cross section of a *Dugesia* and note the obvious organ and tissue organization. Can you see well-defined layers of cells? Are organs present? Does the planarian have a body cavity distinct from the gastrovascular cavity? Are the tissues and organs seen in this cross section more or less complex than those seen in sponges? Cnidarians? Compare the slide to figure 19.9.

Figure 19.10 Life cycle of the human liver fluke *Clonorchis*. Infected human passes eggs in feces. A miracidium larva, hatching from an egg, infects a snail. The miracidium asexually reproduces in snail, giving rise to cercaria larvae. These escape from snail and penetrate fish. In fish, cercaria encyst in muscle. Humans are infected when they eat raw fish containing encysted larvae.

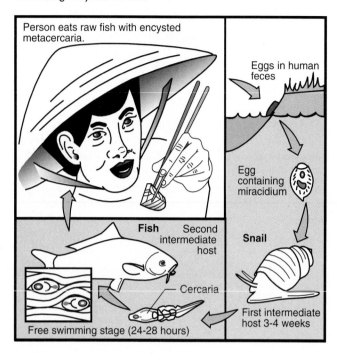

Person eats raw fish with encysted metacercaria.

Eggs in human feces

Egg containing miracidium

Fish Second intermediate host

Snail

Cercaria

Free swimming stage (24-28 hours)

First intermediate host 3-4 weeks

Figure 19.11 Anatomy of an adult human liver fluke.

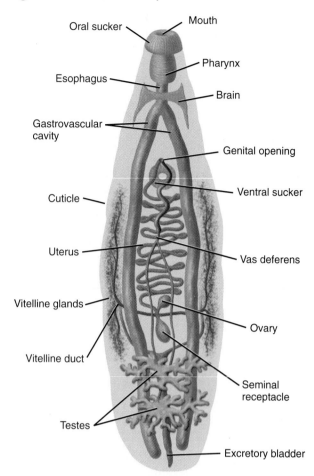

Oral sucker
Mouth
Esophagus
Pharynx
Brain
Gastrovascular cavity
Genital opening
Cuticle
Ventral sucker
Uterus
Vas deferens
Vitelline glands
Ovary
Vitelline duct
Seminal receptacle
Testes
Excretory bladder

The ventral surface of the worm can be identified by finding the **ciliated cells** on the surface: these allow the worms to move in a gliding fashion. **Gland cells** on the ventral surface secrete mucus, which aids in locomotion. Some cells on the dorsal surface contain darkly staining **rhabdites.** When provoked, planarians release the sticky contents of the rhabdites, producing a repellent slime.

Beneath the surface find the ends of the **longitudinal muscles.** What happens to the animal's shape when they contract? Also find the **circular muscles,** which pass around the animal's body. What happens when they contract? **Dorsoventral muscles** should also be visible passing from the dorsal to the ventral surfaces. When they contract, what happens to the shape of the worm?

they go through complex life cycles involving larval stages that infect intermediate hosts, usually snails.

Clonorchis is a fluke that lives in human bile ducts, releasing eggs into the intestine where they are voided with the feces. Figure 19.10 shows the life cycle of this organism. Eggs develop into intermediate larval stages that first inhabit snails and then fish as intermediate hosts. Eating raw or improperly prepared fish containing *Clonorchis* larvae leads to infection in humans. The adults mature in the human body, producing eggs that start the cycle over again.

Obtain a slide of *Clonorchis sinensis.* Look at the slide under the low-power objective and find the structures shown in figure 19.11.

The **mouth** is located in the center of the **oral sucker.** A short **esophagus** leads to two branches of the **gastrovascular cavity.** No anus is found, so the animal has a saclike digestive system. How are nutrients distributed throughout the organism?

Trematoda

All adult members of class Trematoda are called **flukes** and are parasitic in vertebrate hosts. After the eggs are shed,

A second sucker is located on the ventral surface. Can you find a body cavity?

Locate the paired **testes** in the posterior third of the worm. Small ducts lead from the testes to the **genital** opening located near the ventral sucker. These animals are monecious, that is, individual worms have both male and female reproductive organs, though self-fertilization is not normal.

The **ovary** is anterior to the testes. It produces eggs that pass into the **uterus** where they are fertilized by sperm, from previous copulations, held in the **seminal receptacle.** Yolk from the **vitelline glands** combines with the fertilized egg and a shell is formed before the eggs are released through the genital opening.

As is the case with most parasitic organisms, the digestive, nervous, and muscular systems of *Clonorchis* are reduced, and the reproductive system is enlarged. This ensures that large numbers of eggs are produced to compensate for the great odds against successfully completing a life cycle. Each egg ingested by a snail will produce several larvae, amplifying the reproductive potential even more. Further development depends on the larvae infecting a fish and ultimately being eaten by a human. Many larvae simply die, but the more larvae there are, the greater the chances of survival for the species.

Cestoda

Like trematodes, cestodes—tapeworms—are highly adapted to a parasitic way of life. The adults live in the intestines of vertebrates and have features that prevent them from being swept out with fecal material (fig. 19.12). Figure 19.13 shows the life cycle of the dog (or cat) tapeworm, which has a flea as its intermediate host.

Look at the preserved demonstration specimens of tapeworms in the lab. The **scolex** at the anterior of the worm anchors the worm in the intestine. The "body" consists of **proglottids,** or segments, that are youngest at the anterior end and oldest at the posterior end. A digestive system is completely lacking. The worm grows by producing new proglottids in the **neck** region. Proglottids essentially contain only reproductive structures and produce massive amounts of eggs. Why doesn't a tapeworm require a digestive system? How does it obtain nutrients? How is its body structure adapted to the environment in the intestine of its host?

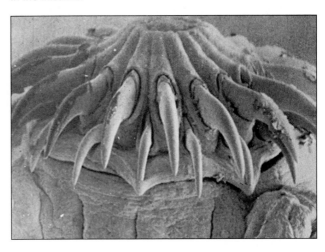

Figure 19.12 Scanning electron micrograph of the anterior end of the cat tapeworm, *Taenia taeniaeformis,* showing an adaptive feature in the form of hooks that anchor it in the intestine.

Phylum Nematoda

All 90,000 species in this phylum show two characteristics not found in the previously studied three phyla: (1) a tube-within-a-tube body plan in which the digestive system has a separate mouth and anus, and (2) a body cavity in which the internal organs reside in a free space between the endoderm and ectoderm. Because this cavity is not lined with the mesoderm, it is called a **pseudocoel.**

Ascaris will be used to demonstrate the functional anatomy of nematodes. It is a parasite living in the intestines of swine and humans. It is similar in appearance to most free-living nematodes, except it is much larger.

Obtain an *Ascaris* and identify the **mouth** and the **anus.** Males will have a hooked posterior end. What is the sex of your specimen? _____ The body is covered by a noncellular **cuticle,** which is secreted by underlying cells.

Place the specimen in a dissecting pan and pin the anterior and posterior ends. Cut the animal longitudinally along the middorsal line and pin the body wall down to expose the internal organs. Flood your dissecting tray with water, so that the organs float. This will make structures easier to see. Identify the structures shown in figure 19.14.

Note the open body cavity where the organs are free-floating inside the space surrounded by the body wall. The tube-within-a-tube body plan should also be obvious with the digestive tube running from the anterior mouth to the posterior anus and, in normal life, surrounded by the body wall tube. Be sure to trace the digestive system from mouth to anus. Is this system more or less efficient than the digestive system of *Clonorchis?* _____ Why?

Figure 19.13 Dog or cat tapeworm: (*a*) life cycle; (*b*) photograph of a mature proglottid; (*c*) gravid proglottids have an enlarged uterus filled with eggs.

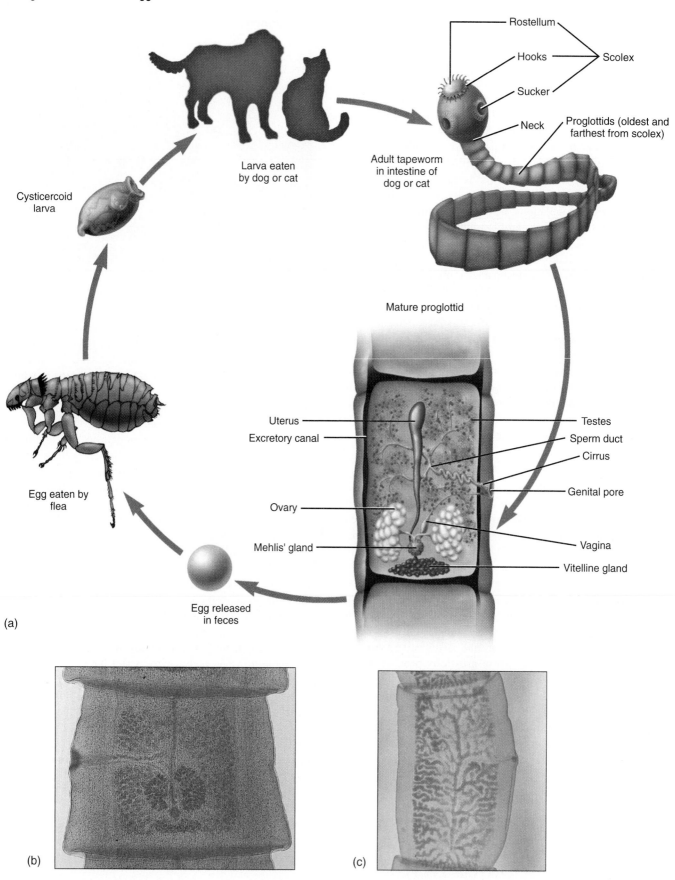

(a)

(b)

(c)

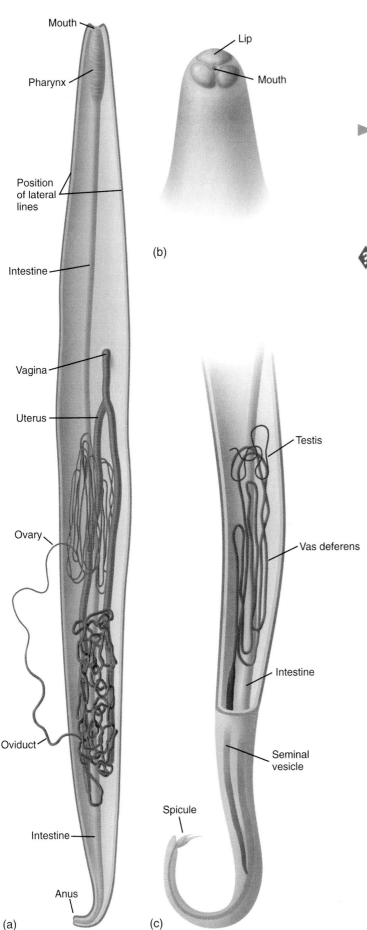

Mouth

Pharynx

Position
of lateral
lines

Intestine

Vagina

Uterus

Ovary

Oviduct

Intestine

Anus

(a)

Lip

Mouth

(b)

Testis

Vas deferens

Intestine

Seminal
vesicle

Spicule

(c)

Figure 19.14 Anatomy of *Ascaris*: (*a*) internal anatomy of female; (*b*) oral view of female; (*c*) posterior end of male.

▶ Figure 19.15 shows the worm in cross section. Note the open body cavity with the free-floating digestive and reproductive systems unattached to the body wall and surrounding tissues. The body wall consists of a noncellular **cuticle** covering the animal, a cellular **epidermis** that secretes the cuticle, and a **longitudinal muscle** layer. No circular muscles are found in the body wall. What types of movements would you predict that *Ascaris* can make? ◆ What are the structural and protective advantages of having a semirigid cuticle?

Two **lateral canals** run the length of the worm and carry excretory products to an excretory pore located near the mouth. Most of the body cavity is filled with tubules of the reproductive system that have been cut in cross section.

Nematodes lack specialized organs for gas exchange and circulation. How do you think they perform these important physiological functions?

Animal Phylogeny: Investigating Evolution of Body Plan

Figure 19.15 Cross sections of *Ascaris:* (a) male and (b) female.

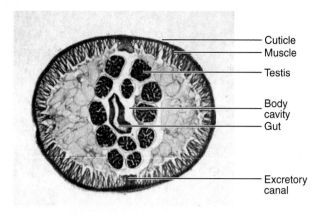

(a)

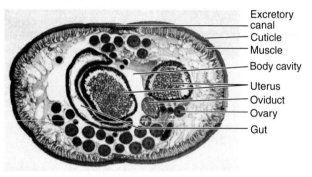

(b)

In your own words, describe the major differences in body plan between a nematode and planarian when viewed in cross section.

Collecting Nematodes from Soil Samples

Free-living nematodes are common in the upper few inches of a number of soils where many are detritus feeders and others are plant parasites. The population densities of nematodes in samples from various habitats can be determined by fairly simple techniques which you will use in this portion of the lab.

Your instructor will have collected soil samples from different habitats. These samples will have been processed by adding about 100 grams of pulverized soil to 400 ml of water in a large beaker. The suspended soil sample is poured through two fine sieves: first through a 25-mesh sieve and then the filtrate is passed through a 200-mesh sieve. The larger mesh sieve retains detritus and small stones but allows the nematodes to pass. The small mesh screen retains the nematodes but allows fine clay and mud particles to pass. The 200-mesh screen is inverted over a beaker and washed with a water stream from a squeeze bottle so that the material on the top surface of the sieve is transferred into the beaker.

The contents of the beaker are then poured into a Baerman apparatus which consists of a 6-inch funnel fitted with a layer of very thin bed sheeting or tissue paper and rubber tubing with a pinchcock on the stem (as in fig. 19.16). The funnel is filled with water so that the sheeting is covered and allowed to stand overnight. Any nematodes in the sample will burrow into the sheeting, pass through, and fall to the bottom where they will collect in the rubber tubing.

To observe the nematodes, take a small plastic petri dish that has had its bottom scored into a 10 mm by 10 mm-grid on the outside. Drain a few milliliters of water from the rubber tubing into the dish and observe the dish through your dissecting microscope.

Assuming that the nematodes are randomly distributed in the dish, you can count the nematodes in several of the squares etched in the dish. Record ten such counts below and calculate an average.

 What is the area in square mm for one square?

_____

Measure the diameter of the dish in mm and calculate the total bottom area in square mm for the dish. Remember the area of a circle is equal to π times the radius squared. _____ How many squares are found on the bottom of the dish (divide the second answer by the first above).

_____

If you multiply the number of squares per dish by the average number of nematodes counted per square, you will know the total number of nematodes in the sample. If you know the sample size in grams, you can calculate the number of nematodes per gram of soil.

How many nematodes do you estimate per gram of soil? _____

Figure 19.16 Technique for collecting nematodes from a soil sample.

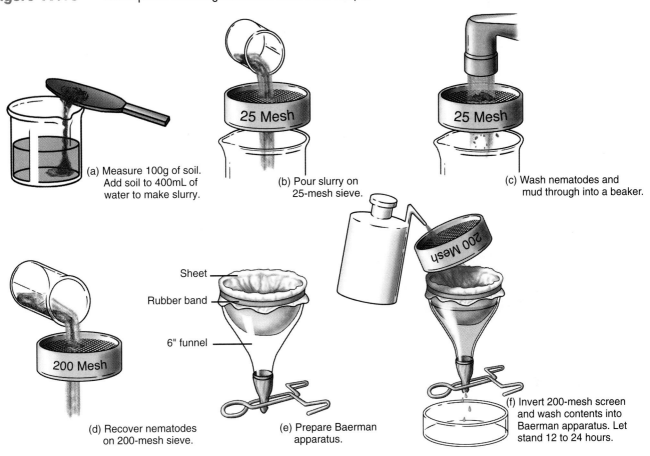

(a) Measure 100g of soil. Add soil to 400mL of water to make slurry.

(b) Pour slurry on 25-mesh sieve.

(c) Wash nematodes and mud through into a beaker.

(d) Recover nematodes on 200-mesh sieve.

Sheet
Rubber band
6" funnel

(e) Prepare Baerman apparatus.

(f) Invert 200-mesh screen and wash contents into Baerman apparatus. Let stand 12 to 24 hours.

Does this number differ for soils from different habitats? Cite examples below.

Hypothesize what factors are involved in determining the population density of nematodes in soils.

Learning Biology by Writing

Write an essay summarizing your observations that support or fail to support the null hypothesis that these four phyla can be arranged in a phylogenetic sequence from simple to complex. Be sure to recognize those characteristics related to parasitism as secondary developments and not as major phylogenetic trends.

Your laboratory instructor may ask you to answer the Lab Summary and Critical Thinking Questions that follow.

Lab Summary Questions

1. Fill in table 19.1 to summarize your observations on the evolution of body organization.
2. Describe the changes in the digestive system as you move from the sponges through the cnidarians and flatworms to the nematodes.
3. What are the advantages of a tube-within-a-tube body plan compared to a body plan in which the gut is saclike?

TABLE 19.1 Summary of animal characteristics

Internet Sources

Many of the species in phylum Platyhelminthes and phylum Nematoda are parasites. Scientists who study these animals belong to professional societies that promote research and dissemination of information. Use your browser to connect to one of the following: American Society of Parasitologists at http://www-museum.unl.edu/aspl Society of Nematologists at http://www.ianr.unl.edu/son/ What did you find most interesting at these sites?

4. Contrast radial and bilateral symmetry. Why is nervous system development coordinated with bilateral symmetry?

5. What is a body cavity, and why is it beneficial to an animal?

6. None of the animals studied in this lab had organs specialized for respiratory gas exchange or circulation. How do these animals perform these important physiological functions?

7. At the beginning of this lab, the following null hypothesis was made: There is no evidence that animals in the four phyla studied here can be arranged in an evolutionary sequence from simple to complex. What observations have you made today that falsify this hypothesis?

Critical Thinking Questions

1. Radially symmetrical animals lack a brain and nerve cord. Why?

2. Which of the animals do you consider simple? Which complex? Why?

3. Sponges are sessile and often green in color. Why aren't they considered plants?

LAB TOPIC 20

Protostomes I: Investigating Evolutionary Development of Complexity

Supplies

Preparator's guide available on WWW at
http://www.mhhe.com/dolphin

Equipment

Compound microscopes
Dissecting microscopes

Materials

Preserved specimens for demonstrations
    Polychaetes
    Chiton
    Snail
    Squid, whole and dissected
    Tusk shells
Preserved clams for class
Live earthworms (night crawlers) for class
Live leeches for demonstration
Prepared slides Live *Lumbriculus variegatus*
    culture kit
    Earthworm, cross section
    Clam, cross section
Dissecting trays and instruments
Capillary tubes

Solutions

10% ethanol

Prelab Preparation

Before doing this lab, you should read the introduction and sections of the lab topic that have been scheduled by the instructor.

You should use your textbook to review the definitions of the following terms:

    Annelida
    blastopore
    coelom
    ectoderm
    endoderm
    enterocoelom
    gastrula
    mesoderm
    Mollusca
    protostome
    schizocoelom

You should be able to describe in your own words the following concepts:

    What is a coelom and why it is important
    The body plan of an annelid
    The body plan of a mollusc

As a result of this review, you most likely have questions about terms, concepts, or how you will do the experiments included in this lab. Write these questions in the space below or in the margins of the pages of this lab topic. The lab activities should help you answer these questions or you can ask your instructor for help during the lab.

Objectives

1. To illustrate the development of a true coelom
2. To dissect an earthworm and a clam
3. To compare the animals studied here to those studied in lab topic 19
4. To collect data to test the null hypothesis that annelids and molluscs are more complex than the animal phyla studied previously

Background

Within the **coelomates,** those animals with a true coelom, two major groups of phyla can be discerned. These groupings are based on embryological evidence of how the mesoderm lining of the coelom develops and whether the blastopore of the gastrula stage develops into the oral or the anal opening of the adult animal.

In the **protostomes,** which include the phyla Annelida, Mollusca, and Arthropoda, the mesoderm originates as a solid mass of cells that grows inward from the endoderm near the blastopore (fig. 20.1). This group of cells then splits to form a sac that grows out to line the body cavity. The term **schizocoelomate** describes how the body cavity develops in protostome animals by splitting of mesoderm

Figure 20.1 Two patterns of coelom formation. In the schizocoelomates (protostomes), mesoderm cells originate near the blastopore and split to form a pouch, while in the enterocoelomates (deuterostomes), mesoderm cells originate as pouches from the gut. The end result in both cases is the same: mesoderm completely lines the body cavity (coelom) of the tube-within-a-tube organism.

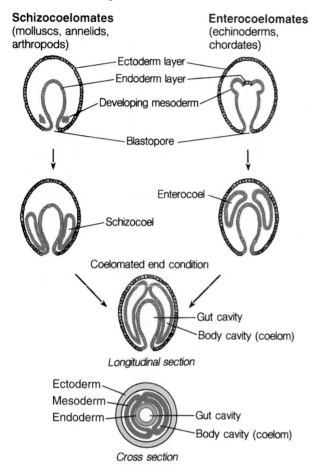

Schizocoelomates (molluscs, annelids, arthropods)

Enterocoelomates (echinoderms, chordates)

- Ectoderm layer
- Endoderm layer
- Developing mesoderm
- Blastopore

Enterocoel

Schizocoel

Coelomated end condition

- Gut cavity
- Body cavity (coelom)

Longitudinal section

- Ectoderm
- Mesoderm
- Endoderm
- Gut cavity
- Body cavity (coelom)

Cross section

buds to form sacs. The term *protostome* ("first mouth") describes the fact that the blastopore of the gastrula becomes the mouth, whereas the anus develops secondarily.

In the **deuterostomes,** which includes the echinoderms and chordates (lab topic 22), the mesoderm arises as an outpocketing of the endoderm to form sacs at the end away from the blastopore (fig. 20.1). These sacs grow out to line the coelom. The term **enterocoelomate** describes the origin of the coelom from the digestive lining in these animals. The term *deuterostome* ("second mouth") describes the fact that the mouth develops as the second opening of the digestive system. The anus is derived from the first opening—the blastopore of the gastrula. See lab topic 12 for a full discussion of embryology.

LAB INSTRUCTIONS

You will study three representatives of the schizocoelomates: an earthworm, a clam, and a squid.

Phylum Annelida

The phylum Annelida includes about 15,000 species of segmented worms, which exhibit several structural advances over the simpler animal phyla. These advances include a well-developed, fluid-filled coelom with the organs held in suspension by membranes called mesenteries. Muscles in the body wall are well developed and they have a tubular digestive system. In addition, the annelids have a separate, closed circulatory system, an efficient excretory system, and a highly developed central nervous system with a concentration of ganglia (nerve centers) at the anterior end. There are three taxonomic classes in the phylum: the **Polychaeta** (marine worms), the **Oligochaeta** (earthworms), and the **Hirudinea** (leeches) (fig. 20.2).

Earthworm Dissection

Earthworms can be obtained inexpensively at bait stores. Live worms are preferable for dissection. They can be anesthetized by submersion in 10% ethanol for a few minutes prior to dissection.

External Anatomy

▶ Note the obvious segmentation of the earthworm. External structures in the earthworm are usually located by reference to the segments numbered from the anterior end. You can identify the anterior end by locating the **clitellum,** a swollen band covering several segments in the anterior third of the worm (fig. 20.3). The clitellum secretes a cocoon around the eggs when they are released. Earthworms are hermaphroditic, having both male and female organs, so every worm has a clitellum.

The mouth is located in the first segment, and overhanging the mouth is a fleshy protuberance called the **prostomium.** The anus is in the last segment. There are no obvious external sense organs at the anterior end. The dorsal surface is identifiable by the **dorsal blood vessel,** which appears as a dark reddish line.

▶ The surface of the worm has an iridescent sheen because light is refracted by the **cuticle,** a thin noncellular layer of collagen fibers. They are secreted by the cells of the underlying epidermis. Mucus secreting cells in the epidermis lubricate the body surface and keep it moist.

▶ Run your fingers back and forth along the sides of the worm and feel the projecting chitinous bristles called **setae.** Each segment has two pairs of setae on the ventral surface and two pairs on the side. How would the setae help the earthworm in locomotion?

Figure 20.2 Representatives of phylum Annelida: (a) polychaete marine annelid; (b) a freshwater leech.

(a)

(b)

Find the **excretory pores** located ventrolaterally in each segment, except for the first few and the last one. In the region of segments 9 to 15, the reproductive system openings are found on the ventral surface. Depending on the species, the **male pore** is usually in segment 15 surrounded by fleshy lips, and the **female pore** is in segment 14 just anterior to the male pore. In the grooves between the ninth and tenth and eleventh segments are the small openings of the seminal receptacles, part of the female reproductive system where sperm are stored following copulation. You may need to study this area with your dissecting microscope to see the openings indicated.

Internal Anatomy

To examine the internal anatomy of the worm, lay it in a dissecting pan dorsal side up, pin it through the prostomium, stretch it slightly, and pin it through the last segment. Open the worm as in figure 20.3.

Cut to one side of the dorsal blood vessel but avoid cutting internal structures. In the area of segments 1 through 5 be very careful, because if you cut too deep you will destroy the "brain" and pharyngeal area. After the incision is made, pin the body open by putting pins at a 45° angle through every fifth segment of the body wall. This will provide a quick reference as you try to find various structures. When the specimen is opened and secure, add enough water to the pan to cover the open worm. This prevents the tissues from drying out and floats organs and membranes for easier viewing.

Look at the internal organization of the worm and note the concentration of structures in the anterior portion (fig. 20.4). Also observe how the body cavity is divided into compartments by cross walls called **septae.**

Figure 20.3 Procedure for dissecting an earthworm.

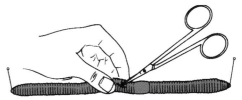

(1) Pinch skin with fingers. Cut through skin (off center) to the anus. Do not damage internal organs by jabbing points downward.

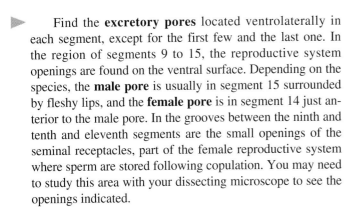

(2) With scalpel, cut through septa on both sides. Pin body wall to tray.

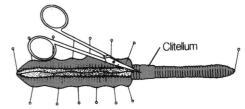

(3) Cut through clitellum toward anterior end. Cut septa and pin to end.

Figure 20.4   Dorsal view of the internal anatomy of an earthworm.

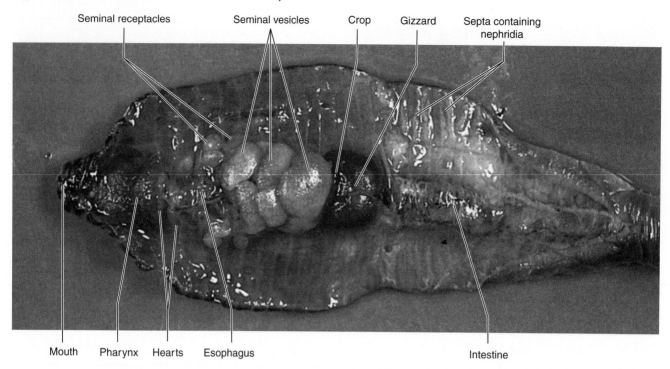

Seminal receptacles   Seminal vesicles   Crop   Gizzard   Septa containing nephridia

Mouth   Pharynx   Hearts   Esophagus   Intestine

Each of the compartments is normally filled with fluid, which acts as a hydrostatic skeleton. Well-developed **circular** and **longitudinal muscles** in the body wall exert pressure on this fluid and allow a variety of movements.

If only the circular muscles contract, how will the shape of the worm change? How does the shape change when the longitudinal muscles contract?

Find the tubular digestive tract running from the mouth to the anus (fig. 20.5). Behind the mouth is the muscular **pharynx.** It appears to have a "fuzzy" external surface because several dilator muscles extend to the body wall. Contraction of these muscles expands the pharynx and sucks in particles of soil and detritus. Peristaltic waves of muscle contraction sweep ingested material down the **esophagus** into a thin-walled storage area, the **crop.** The material passes into the muscular **gizzard,** which grinds it into a fine pulp. Digestion and absorption of organic material take place in the **intestine** and undigestable soil particles are voided through the **anus.** What is the longest part of the digestive system? How does its length relate to its function?

Does the earthworm have a complete or incomplete digestive system?

Earthworms have no special respiratory structures. Respiratory gases are exchanged by diffusion across the body surface. Earthworms have **closed circulatory** systems and their blood is red because it contains the oxygen-carrying protein hemoglobin similar to the protein found in human red blood cells. Oxygen diffuses into the capillaries near the surface of the body wall and is carried throughout the body.

Find the **dorsal blood vessel** overlying the digestive tract. It collects blood from the capillaries in each segment. Blood flows anteriorward in this vessel. Surrounding the esophagus are five pairs of pulsatile arteries, the **"hearts."** They pump blood from the dorsal vessel into the **ventral blood vessel.** Blood in the ventral vessel flows posteriorward, passing into capillaries in the body wall in each segment and into subneural vessels under the nerve cord.

If you are dissecting a live earthworm, you should be able to see the waves of contraction in the hearts. Time the contractions per minute in one of the hearts and record.

Your instructor may have live California blackworm or mudworms *(Lumbriculus variegatus)* in the lab. These small oligochaetes are found in organic sediments in shallow water near the shores of lakes, ponds, and marshes throughout North America. They are easily raised for classroom use.

Figure 20.5   Lateral view of the internal anatomy of an earthworm.

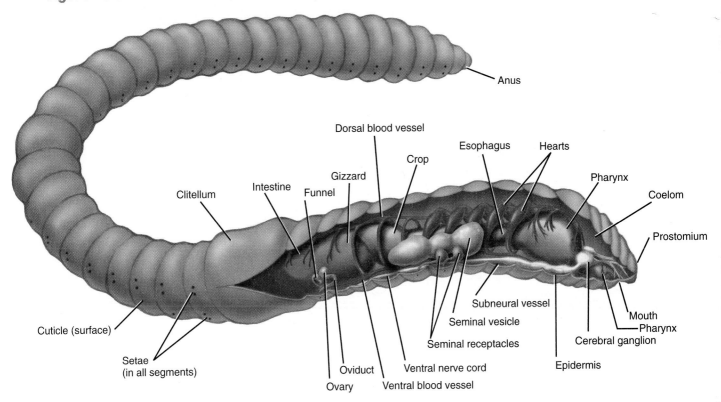

Use a disposable pipet to pick one up and transfer it to a petridish in a drop of water. Obtain a capillary pipet and coax the animal to enter it. This will require some trial and error.

When the animal is in the capillary tube, place the tube on your compound microscope stage and observe it with the scanning objective. You should be able to see blood flowing in its vessels and to locate the "hearts" at the anterior end. Study the pattern of blood flow. Below, make a quick sketch of the worm and indicate the pattern of circulation from the hearts until the blood returns to the hearts.

Each segment except for the first few and the last contain a pair of **nephridia,** or excretory structures, that filter coelomic fluid and remove waste products. (See fig. 20.7 and lab topic 29.)

Refer to figure 20.5 and note the position of the nerve cord and ganglia. Carefully remove the alimentary canal from the posterior third of the worm and look for the ventral nerve cord with its string of segmented ganglia. Remove a ganglion, mount it in a drop of water on a slide, and observe it with your compound microscope.

How many lateral nerves come from each ganglion? _____ Does the arrangement of the ganglia correspond to the segmentation? _____

Return to your worm and trace the nerve cord up into segments 1 through 5. Locate the **subpharyngeal ganglion.** Trace it forward to the **circumpharyngeal connectives** that connect to a pair of **cerebral ganglia** lying above the pharynx. This complex is considered the worm's "brain."

Because earthworms are hermaphroditic, their reproductive systems are complex (fig. 20.6). Self-fertilization does not occur. The male system consists of three pairs of **seminal vesicles,** which span body segments 9 to 13. They surround the two pairs of tiny **testes** located on the posterior surfaces of the septae separating segments 9 and 10, and 10 and 11. Sperm are released into the chambers of the seminal vesicles where they mature before being released via very small ducts, the **vas deferens.** Sperm exit through

Figure 20.6   Lateral view of reproductive organs in an earthworm.

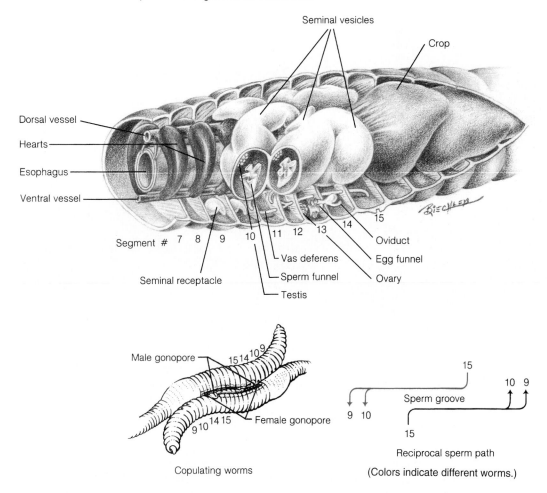

the **male gonopores** located in grooves on the ventral surface of segment 15.

The female system consists of a small pair of **ovaries** located on the septa between segments 12 and 13. Eggs released from the ovaries enter the coelom before passing into the **oviducts,** which carry them to the **female gonopores** located in grooves on the ventral surface of segment 14. In segments 9 and 10 are two pairs of the **seminal receptacles,** which receive sperm from the partner during copulation.

When two worms copulate, there is a reciprocal exchange of sperm (fig. 20.6), which are stored by the receiving partner. Observe the longitudinal **sperm grooves** on the external ventral surface, which facilitate this exchange.

Fertilization occurs externally sometime after copulation. Eggs are released through the female gonopore into a collar of mucus produced by the clitellum. The collar containing the eggs travels forward over the worm and sperm are released into it as it passes over the openings of the seminal receptacles, fertilizing the eggs. The collar slips off the anterior of the worm and becomes a cocoon in which the embryo develops.

Histology

▶ Obtain a slide with stained cross section of the earthworm. Using the scanning objective of your microscope, compare the section to figure 20.7.

First note the obvious. The worm has open spaces surrounding the organs. This is the **coelom** or body cavity. The organs are free to move independent of body wall movements. Correlate the structures in the slide with those you saw in the dissection. Note the composition of the body wall, including the **cuticle,** the **epidermis,** the **circular muscles,** the **longitudinal muscle** layer, and the **peritoneum,** which lines the coelom. You may see some **setae** and their associated musculature in the body wall. Observe the **typhlosole,** the dorsal infolding of the intestine, which greatly increases the area of absorption. Surrounding the intestine, find the **chlorogogue layer,** which functions like the vertebrate liver, storing glycogen and fat and metabolizing amino acids. Locate the three major longitudinal blood vessels: the **dorsal, ventral,** and **subneural vessels.** Finally, identify the **ventral nerve cord** and locate the three giant fibers plus the numerous smaller associated fibers. The giant fibers convey impulses quickly from one end of the organism to the other.

Figure 20.7 Cross section of an earthworm.

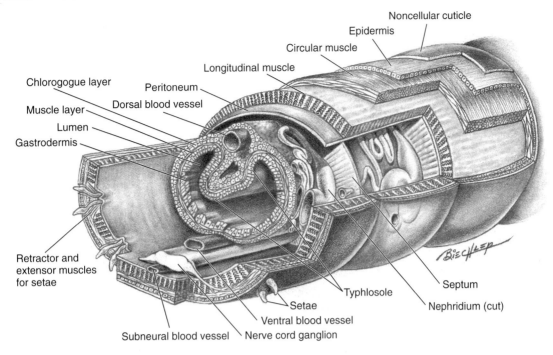

Noncellular cuticle
Epidermis
Circular muscle
Longitudinal muscle
Peritoneum
Dorsal blood vessel
Chlorogogue layer
Muscle layer
Lumen
Gastrodermis
Retractor and extensor muscles for setae
Subneural blood vessel
Nerve cord ganglion
Ventral blood vessel
Setae
Typhlosole
Septum
Nephridium (cut)

Annelid Comparisons

▶ Your instructor will have demonstration specimens of leeches and polychaetes in the laboratory. Look at the specimens and note the obvious similarity to the earthworm. ◆ What characteristics do all annelids share?

◆ Based on external anatomy, how do leeches differ from earthworms? How are they similar?

◆ Based on external anatomy, how do polychaetes differ from earthworms? From leeches? How are they similar?

You are finished with your dissection of the earthworm. Dispose of the animal according to the directions of your lab instructor.

Phylum Mollusca

There are about 150,000 species of molluscs. It is a diverse phylum containing chitons, oysters, snails, slugs, octopuses, and other less well-known animals. Molluscs are bilaterally symmetrical and have a body that can be divided into four anatomical regions: head, visceral mass, foot, and mantle. In many species, the mantle secretes a calcareous shell that surrounds the animal. Molluscs have a three-chambered heart and an open circulatory system (except for the squid and octopus). Hemocyanin is the oxygen-carrying pigment in the circulating fluid, hemolymph.

Molluscan Diversity

▶ Several specimens of different kinds of molluscs are on demonstration in the laboratory (fig. 20.8). They will be arranged by taxonomic class, and you should write brief descriptions of each under the appropriate following headings. As you look at these specimens, be sure to identify these six regions of the body: **head, mouth, visceral mass, foot, mantle,** and **shell.** Also note the type of symmetry exhibited.

1. Class *Polyplacophora.* These marine animals are commonly known as chitons. They live in intertidal areas. They are subjected to wave action, and alternate drying and submerging caused by tidal movements.

Figure 20.8 Representatives of phylum Mollusca:
(a) a chiton; (b) a tusk shell; (c) a conch or whelk.

(a)

(b)

(c)

How does their anatomy equip them for such a harsh environment? How many plates are on the dorsal surface? _____ Where are the gills? _____ Write your description of the body regions below.

2. Class *Scaphopoda.* These marine animals are commonly known as tusk shells for obvious reasons. Look at the examples of shells on display in the lab.

3. Class *Gastropoda.* Commonly known as snails, slugs, limpets, conchs, and whelks, the gastropods are the largest class in the phylum. All gastropods undergo torsion and spiraling during embryological development. Torsion involves the 180° rotation of the visceral mass so that the digestive and nervous systems have roughly a U-shape. Spiraling involves the coiling of the visceral mass inside a shell. Write your descriptions of body regions below. What is the adaptive advantage of the shell?

4. Class *Cephalopoda.* Commonly known as nautiluses, squids, and octopuses, these animals are active marine predators. Nautiluses have an external shell, but in others the shell is internal or absent. Write your descriptions of body regions below. A demonstration dissection of a squid is available in the lab, and you will look at it later.

5. Class *Bivalvia.* This group includes clams, oysters, mussels, and scallops. They are sedentary, filter-feeders characterized by a laterally compressed body surrounded by a two-piece shell. The shell is lined by the mantle. The head is indistinct. A large muscular foot can be extended and is used in burrowing. You will dissect a clam in the lab and do not need to write a description.

Figure 20.9 Anatomy of a partially dissected freshwater mussel.

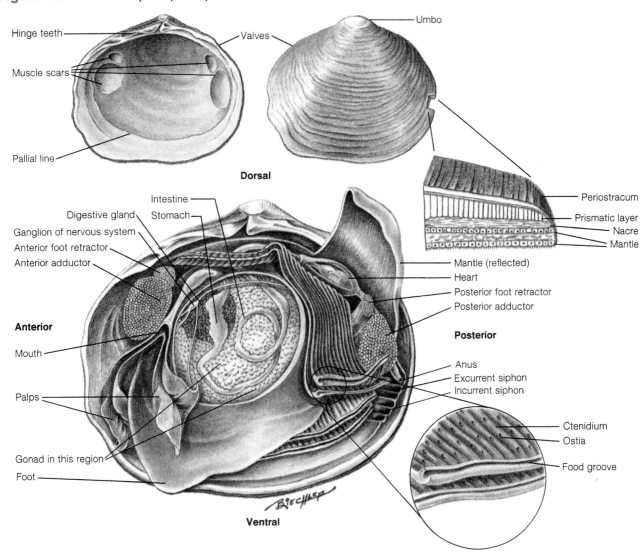

Clam Dissection

External Anatomy

▶ Obtain a specimen of a freshwater clam. Note the two **valves** that make up the halves of the shell. The shells are hinged on the **dorsal** surface of the animal, near the swollen **umbo** area (fig. 20.9). The ventral surface is opposite the hinge, along the opening between the two valves.

In some preserved specimens, the muscular **foot** will be extending from the ventral gap between the valves. It also extends anteriorward. On the posterior surface, toward the dorsal surface, darkly pigmented **siphons** may also be visible in the gap between the shells. Holding the animal dorsal up and anterior away from you, identify the right and left valves.

▶ Now study the external surface of one of the valves. Use a scalpel to scrape the surface. The outer layer that you are removing is the **periostracum.** Composed of the insol-

uble protein conchiolin, it protects the calcium carbonate portion of the shell from chemical and physical erosion. Is $CaCO_3$ eroded by acids or bases? _____ The lines on the shell represent periods of growth, but are not annual growth lines. The shell and its periostracum are secreted by cells at the edges of the mantle along the gap between the valves. The umbo region is oldest and the regions at the edge have just been added to the shell.

Internal Anatomy

▶ To open the clam for study, you must reach in through the gap between the valves and cut the anterior and posterior **adductor muscles,** which pass from shell to shell. See figure 20.9 for their locations. Insert a scalpel through the gap, pass its point along the inside of the left valve to the muscle region, and slice.

┌───┐
│   **C A U T I O N**   │
│ Be careful: Sloppy technique may result in your dam- │
│ aging internal organs or, worse yet, cutting yourself. │
└───┘

After opening, free the soft mantle adhering to the left valve and cut the elastic ligament that forms the hinge. You should now be able to remove the left valve leaving the soft body parts in the right valve. Note how the soft tissue of the **mantle** forms an envelope around the visceral mass. The mantle performs three important functions: (1) it secretes the shell; (2) it has cilia on its internal surface, which help draw in water; (3) it is a major site of respiratory gas exchange along with the gills. Examine the left valve to see the **pallial line** about 1 cm from the edge. This is where the mantle attaches to the valve.

Examine the dorsal-posterior edge of the mantle and find the thickened, pigmented area. If you match up the left and right mantles, they form two tubes, a ventral **incurrent siphon** and a dorsal **excurrent siphon.** Cilia on the surfaces of the siphons, mantle, and gills (ctenidia) draw water into the mantle cavity over the gills and out the excurrent siphon. This water brings a stream of oxygen and suspended, microscopic food to the animal and carries away wastes. In some marine clams, the siphons may be a third of a meter long, allowing the clam to live buried in sand or mud.

Lay back the mantle to expose the **visceral mass** and muscular foot. When circular muscles in the foot contract, it extends in much the same way that a water-filled balloon elongates when squeezed at one end. Once extended into mud or sand, the tip of the foot expands to anchor itself and retractor muscles pull the body toward the anchored foot, providing a reasonable form of locomotion. Some clams can burrow a foot or more in wet sand in as short a time as one minute.

Most of the visceral mass is covered by two flaps of tissue, the **ctenidia,** or gills. Although they also function in respiratory gas exchange, the primary function of these structures is in gathering food. Clams are filter-feeders, obtaining food by filtering algae, detritus, and bacteria from water entering the mantle cavity through the siphons.

The structure of the ctenidia is best seen in slides of cross sections of bivalves viewed through a microscope (fig. 20.10). Get your microscope and a prepared slide. The shell was removed before the animal was sectioned to make the slide. Find the visceral mass hanging like a pendant with curtains of tissue surrounding it on both sides. The inner two curtains are the ctenidia and the outermost is the mantle. Sometimes the ctenidia will have brood pouches, called marsupia, in which the eggs develop.

Look carefully at one ctenidium on the microscope slide. Note that it is hollow with a **water tube** passing up the center. Close examination will reveal that the surface of the ctenidium has small porelike openings, the **ostia.** Water

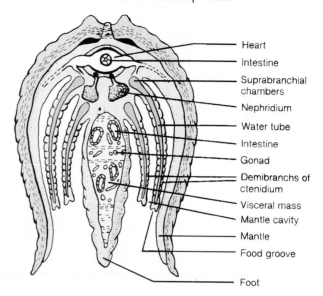

Figure 20.10   Cross section of a freshwater mussel. Shell will not be included on microscope slide.

- Heart
- Intestine
- Suprabranchial chambers
- Nephridium
- Water tube
- Intestine
- Gonad
- Demibranchs of ctenidium
- Visceral mass
- Mantle cavity
- Mantle
- Food groove
- Foot

entering the mantle cavity passes through the ostia into water tubes and then upward to **suprabranchial chambers** (fig. 20.10). From there the water flows to the excurrent siphon.

Any particles contained in the water stream are filtered out at the ostia where they are trapped in mucus that coats the surface of the ctenidia. Cilia on the surface of the ctenidia move the mucus down to the food groove that passes along the tip of a ctenidium. Find the **food groove** on your slide.

Now return to your dissection specimen. Look at the tip of the ctenidia where the food groove is located. Follow the food groove anteriorward to the region of the cut anterior adductor muscle. In this region between two flaps of tissue called **palps,** you should find the **mouth** of the clam. Food particles trapped in mucus enter the digestive system here and pass to the **stomach.**

To see the stomach, remove the ctenidia covering the visceral mass. Scrape and pick away the tissue of the visceral mass to expose the digestive tract. Tissue of a large, greenish-brown **digestive gland** surrounds the stomach and a lighter **gonad** surrounds the **intestine.** The intestine winds its way dorsally and then posteriorly to the anus located near the excurrent siphon in the region of the posterior adductor muscle. Does the clam have a complete or incomplete digestive system? _____

Clams have open circulatory systems. The circulating fluid is **hemolymph.** It enters the heart directly from the surrounding tissue spaces, the **hemocoel,** and is pumped through arteries to the tissues. At the tissues, the hemolymph does not enter capillaries. Instead, it is released directly into the tissue spaces from where it eventually percolates back to the heart.

Figure 20.11   External anatomy of a squid.

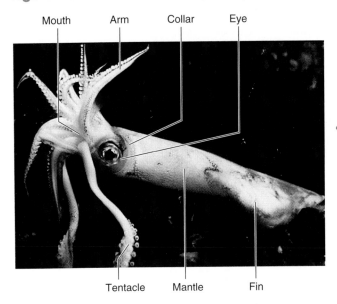

Mouth   Arm   Collar   Eye

Tentacle   Mantle   Fin

The body is covered by a **mantle** that forms a sheath around the **visceral mass.** At the junction of the mantle and the head, on the **posterior** surface, find the **siphon.** Water is drawn into the mantle cavity through the **collar** at the ventral edge of the mantle and is forcibly expelled through the siphon by contraction of muscles in the mantle. This allows the animal to swim rapidly by jet propulsion. Steering is accomplished by varying the direction of discharge from the funnel, by movement of the arms, and by movement of the fins. Give two reasons why this speed and direction of locomotion would be advantageous.

The squid's body is stiffened by a longitudinal skeleton, the **pen.** It consists of chitin, a complex polysaccharide. Feel the pen embedded in the **anterior** mantle (side opposite that bearing the siphon).

Now look at a partially dissected specimen of a squid in which the posterior surface of the mantle has been slit lengthwise to reveal the visceral mass (fig. 20.12). Can you identify a body cavity? Examine the anterior edge of the mantle and imagine how water enters here and then is expelled through the siphon. Two **gills** should be visible in the mantle cavity. Water passing through the cavity is not only used in jet propulsion, but also brings oxygen to these organs and carries away carbon dioxide. At the bases of the gills, the **branchial hearts** should be visible. They pump blood through the gills. Squid have closed circulatory systems, unlike other molluscs that have open systems.

Just posterior to the siphon, find the dark **ink sac.** When disturbed, a pigment is released from the sac into the water in the mantle cavity. The cloudy suspension leaves the siphon, forming a confusing cloud that allows the squid to escape predators. Near the ink sac, find the **anus.** What advantage is there to having the anus in this location?

To see the clam's heart, look at the dorsal surface of the visceral mass just anterior to the posterior adductor muscle. Remove the thin membrane in this area to reveal the **heart** sitting in the **pericardial cavity.**

The heart is three-chambered: one **ventricle** and two **atria.** As the intestine travels from the visceral mass to the anus, it passes through the ventricle. Two **aortas** leave the ventricle, one passing anteriorward and the other posterior.

Beneath the heart, find the paired dark brown **nephridia** (kidneys). They remove nitrogenous wastes from the hemolymph. Urine passes from these to a pore in the suprabranchial chamber above the ctenidia. Water flowing through this chamber carries wastes out the excurrent siphon.

You will not study the nervous or reproductive systems of the clam. Dispose of your clam as instructed.

General Anatomy of a Squid

Squids are highly active marine predators. They can swim rapidly by jet propulsion and have well-developed eyes to spot prey.

Look at a whole specimen of a squid and note the **head, tentacles,** and two large **eyes** (fig. 20.11). The orientation of the squid is unusual. The head and tentacles represent the **ventral** surface and the opposite end bearing two fins is the **dorsal** surface. Examine the appendages and note these are two types: four pairs of **arms** and two elongated **tentacles.** Note the suckers on the surfaces of the appendages. Look in the center of the base of the tentacles and find the **mouth.** A chitinous **beak** should be visible. It is used to grasp prey and to tear flesh into pieces, which are swallowed.

Most of the digestive system is not visible as it passes from the mouth to the anus in an elongated U-shape. At the base of the U, in the end of the mantle cavity opposite the collar, the **intestinal caecum** should be visible. This large, thin-walled sac stores ingested food and is the site of much digestion and absorption. The paired **nephridia** (kidneys) are located on the midline, dorsal to the ink sac and anus.

Figure 20.12 Anatomy of the visceral mass in a partially dissected squid.

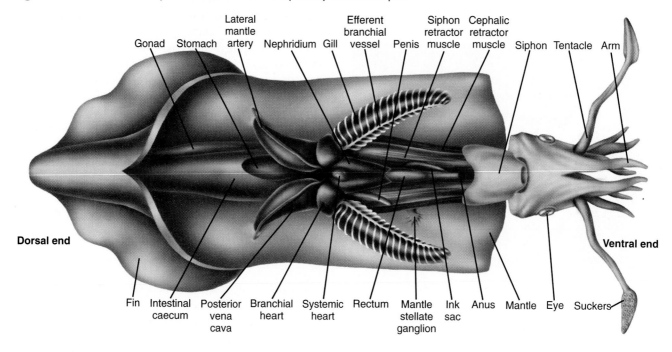

Gonad　Stomach　Lateral mantle artery　Nephridium　Gill　Efferent branchial vessel　Penis　Siphon retractor muscle　Cephalic retractor muscle　Siphon　Tentacle　Arm

Dorsal end

Ventral end

Fin　Intestinal caecum　Posterior vena cava　Branchial heart　Systemic heart　Rectum　Mantle stellate ganglion　Ink sac　Anus　Mantle　Eye　Suckers

Learning Biology by Writing

You have had the opportunity to study the general anatomy of an earthworm, clam, and squid as well as animals in lab topic 19. As you consider the representatives of the six animal phyla that you have studied, what evidence do you have to accept or refute the null hypothesis that annelids and molluscs are more complex than the sponges, cnidarians, flatworms, and nematodes? Summarize your arguments in a one-page essay.

Your instructor may ask you to turn in answers to the Lab Summary and Critical Thinking Questions that follow.

Internet Sources

Zoological Record is an indexing service that has a long history of publishing summaries of zoological research. They have applied their many talents on organizing information on the WWW. Use your browser to connect to their site at:

http://www.biosis.org/zrdocs/zooinfo/ZOOLINFO.HTM

When connected, click on <u>invertebrates</u>. Explore the information available for Annelida and Mollusca.

Lab Summary Questions

1. What are the characteristics of a protostome animal? Which phyla are considered protostomes?
2. From your own observations, offer reasons why an earthworm is considered more evolutionarily advanced than a nematode.
3. What are the advantages of body segmentation?
4. What are the diagnostic features of the phylum Mollusca?
5. Explain how a clam is a filter feeder. Draw a diagram that shows the pathway of food particles from water into the mouth of a clam.
6. What evidence can you present from your observations to indicate that annelids and molluscs are more complex than cnidarians or nematodes?

Critical Thinking Questions

1. Based on your knowledge of an earthworm's anatomy and physiological systems, explain why large earthworms come to the surface after a heavy rain.
2. In general terms, explain how annelids are similar to molluscs but different from cnidarians and nematodes.
3. Do all molluscs have a calcareous shell? Is any animal that has a calcareous shell a mollusc? Give examples of any exceptions.
4. Describe a hydrostatic (hydraulic) skeleton. Explain how an animal like an earthworm can locomote using its hydrostatic skeleton.

LAB TOPIC 21

Protostomes II: Investigating A Body Plan Allowing Great Diversity

Supplies

Preparator's guide available on WWW at
   http://www.mhhe.com/dolphin

Equipment

Compound microscopes
Dissecting microscopes

Materials

Horseshoe crab
Spiders
Acorn and gooseneck barnacles
Crayfish
Lubber grasshopper
Miscellaneous insects for identification to order
Living *Daphnia*
Dissecting pans and instruments

Solutions

Petroleum jelly
Congo red, 0.3 g per 100 ml H_2O
Yeast
Test tubes

Prelab Preparation

Before doing this lab, you should read the introduction and sections of the lab topic that have been scheduled by the instructor.
   You should use your textbook to review the definitions of the following terms:

   Arachnida
   Chelicerata
   coelom
   Crustacea
   cuticle
   exoskeleton
   Insecta
   open circulatory system
   Uniramia

You should be able to describe in your own words the following concepts:

   The general body plan of an arthropod
   The diversity of the Phylum Arthropoda
   Taxonomic key

As a result of this review, you most likely have questions about terms, concepts, or how you will do the experiments included in this lab. Write these questions in the space below or in the margins of the pages of this lab topic. The lab experiments should help you answer these questions, or you can ask your instructor for help during the lab.

Objectives

1. To illustrate the diversity of the arthropods by studying the external anatomy of a horseshoe crab, a spider, *Daphnia*, and a barnacle
2. To study the internal anatomy of a crayfish as a representative arthropod
3. To note the evolutionary significance of the arthropod anatomy
4. To use a taxonomic key to identify insects to the level of order
5. To collect evidence to accept or refute the hypothesis that all Arthropods have an exoskeleton, segmented body, and jointed appendages

Background

The phylum Arthropoda is the largest among the proto-stome phyla. In fact, approximately 75% of all living animals are arthropods. Nearly a million species have been described and the arthropod population of the world is estimated at a billion billion (10^{18}) individuals. Crustaceans, insects, millipedes, spiders, and ticks are but a few examples of this diverse phylum that has representatives in virtually every environmental habitat from the ocean depths to mountain tops, including aerial environments.

   As in the annelids, the bodies of arthropods are segmented, but the segments are often fused to make distinct body regions, such as the **head, thorax,** and **abdomen.** Some arthropods have a **cephalothorax** in which the head and thorax regions are fused.

Figure 21.1 Horseshoe crab:(a) dorsal view; (b) ventral view.

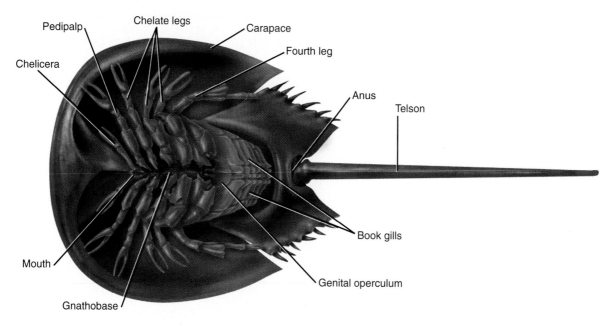

Arthropods are encased in a **cuticle** composed of chitin, a complex polysaccharide stiffened by calcium salts. Acting as an exoskeleton, the cuticle protects and is jointed to allow movement. For terrestrial species, the cuticle is an effective barrier preventing desiccation and may explain why the arthropods have been so successful on land. Muscles are in distinct bundles rather than being part of the body wall and allow a variety of movements; especially in combination with the jointed appendages. In order for individual arthropods to grow, they molt their exoskeletons, and most species have several developmental stages in their life cycles.

The digestive system of arthropods is more or less a straight tube leading from the mouth to the anus. The nervous system consists of two ventral cords with ganglia serving as integrating centers. Arthropods have an open circulatory system with a dorsal artery and heart that pumps hemolymph anteriorward. Because the exoskeleton limits diffusion, gas exchange is through special structures, such as gills, book lungs, or small tubules called tracheae. The nervous system and sensory organs are well developed allowing complex behaviors and locomotion. Arthropods have distinct excretory organs that remove nitrogenous wastes from body fluids and which function in electrolyte balance. The sexes are usually separate.

The phylum is divided into four subphyla: the **Trilobitomorpha** (extinct trilobites); **Chelicerata,** whose first appendages are modified for feeding; **Crustacea** or crustaceans; and **Uniramia,** which have unbranched appendages. Both the Crustacea and Uniramia are mandibulate or jawed arthropods. Several classes occur within each subphylum. Insects are the largest class among the arthropods with 800,000 described species.

The five classes with the greatest number of species are:

Subphylum Chelicerata
 Class Arachnida—spiders, scorpions, mites, and ticks

Subphylum Crustacea
 Class Crustacea—crustaceans

Subphylum Uniramia
 Class Chilopoda—centipedes
 Class Diplopoda—millipedes
 Class Insecta—insects

LAB INSTRUCTIONS

You will study horseshoe crabs and spiders as examples of chelicerates and various crustaceans and insects as examples of mandibulates.

Chelicerate Arthropods

Horseshoe Crab

Horseshoe crabs are marine scavengers that consume molluscs, worms, and plant material as they slowly move over the sandy bottoms of shallow bays and oceans. Despite its name, it is not a true crab but belongs to subphylum Chelicerata, class Merostomata (fig. 21.1).

Look at the horseshoe crab on demonstration in the laboratory. From a dorsal view, the exoskeleton is fused into a **carapace** that covers the 7 segments of the cephalothorax and the 12 segments of the abdomen. A

Figure 21.2   Spider: (a) dorsal view; (b) ventral view; (c) sagittal section showing internal organs.

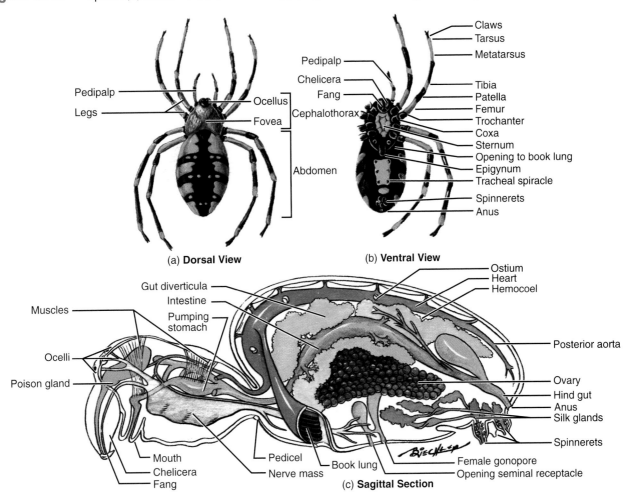

(a) **Dorsal View**

(b) **Ventral View**

(c) **Sagittal Section**

hinge joint marks the dividing line between the body regions and allows the animal to bend. A tail-like **telson** extends from the tip of the abdomen. A pair of **compound eyes** are prominent on the anterior of the cephalothorax and two simple eyes (**ocelli**) are found on either side of a small spine located on the midline anterior to the compound eyes.

Turn the animal over and look at its ventral surface (fig. 21.1). Six pairs of jointed appendages surround the medially located mouth. The small first pair are **chelicerae** used in manipulating food. All but the last pair end in **chelae,** pincerlike structures used to grasp food. The base of the appendages bear spiny **gnathobases,** which macerate food and force it into the mouth as the appendages are moved in a kneading fashion.

The ventral surface of the abdomen is covered by six pairs of flat plates. The first of these is the **genital operculum.** Lift the plate to see the two **genital pores,** the openings of the reproductive system. Under the remaining five plates are **book gills,** so-called because in life the gills fan back and forth as you would fan the pages of a book.

Spider

▶ Obtain a preserved garden spider in a dish of water and examine it with your dissecting microscope. Most specimens will be female because males are smaller and not routinely sold by biological supply houses.

▶ Examine the external anatomy with your dissecting microscope. Note the two body regions, the **cephalothorax** (also known as the prosoma) and the **abdomen** (also known as the opisthosoma), which are connected by a slender **pedicel,** a modified first abdominal segment (fig. 21.2). Locate the eight **ocelli,** simple eyes, on the anterior dorsal surface of the cephalothorax.

Turn the spider over and examine the ventral surface. Six pairs of appendages should be visible. The first are the **chelicerae,** which terminate in hollow, hardened **fangs** that inject a paralytic poison into prey. The second pair of appendages are the **pedipalps,** which serve a sensory function in feeding. Behind the pedipalps are four pairs of **walking legs,** each composed of several segments.

Figure 21.3   Barnacle: Internal anatomy of an acorn barnacle.

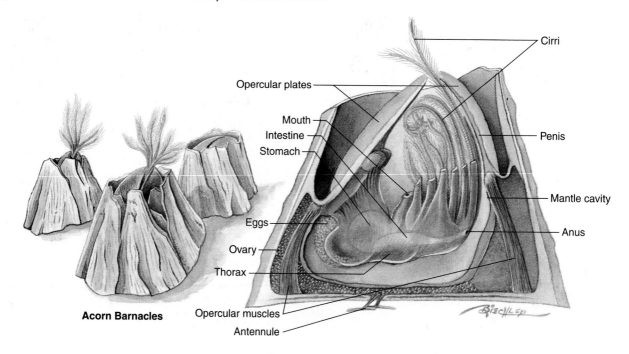

Acorn Barnacles

Cirri

Opercular plates

Mouth
Intestine
Stomach

Penis

Mantle cavity

Anus

Eggs

Ovary

Thorax

Opercular muscles

Antennule

On the ventral abdominal surface, find the paired slits, **spiracles,** that open into the **book lungs.** Insert the tips of fine forceps into the opening and fold back the exoskeleton to reveal the plates of the book lungs. Hemolymph that contains the respiratory pigment hemocyanin circulates in the plates and exchanges respiratory gases with the air in the book lung compartment.

On the median line of the abdomen near the spiracle, find the platelike **epigynum,** which covers the female genital opening. The openings of three pairs of **spinnerets** should be visible on the ventral surface near the tip of the abdomen. Silk glands beneath the openings exude a viscous solution of silk protein that hardens when exposed to air.

Find the small paired **spiracles** on either side of the midline of the ventral surface of the abdomen. These open into a system of tubules, the **tracheae,** that supplement the gas exchange capabilities of the book lungs. How does this system of gas exchange differ from that of the earthworm?

The internal anatomy of the spider will not be studied, but it is shown in outline in figure 21.2. Look it over and quiz your lab partner about how hemolymph circulates in the spider.

Mandibulate Arthropods

Crustaceans

The crustaceans are a diverse group of arthropods. You will study the anatomy of a barnacle, a microcrustacean, and a crayfish to develop an appreciation of this diversity. As you study each animal, keep in mind the evolutionary trends discussed earlier.

Barnacles

Sessile members of the class Crustacea, these animals live attached to rocks, ships, whales, and seaweeds. Some barnacles have a long stalk, the gooseneck barnacles; and others have a compact body, the acorn barnacles (fig. 21.3). Examples of both are on demonstration in the lab.

Obtain a specimen of a preserved acorn barnacle in a dish of water. Although it superficially resembles a mollusc because of its calcified exoskeleton, dissection will show that it has jointed appendages, which are diagnostic of the phylum Arthropoda.

The barnacle is surrounded by a **carapace** composed of calcified lateral and upper plates. The four upper plates comprise the **operculum,** which is open when the animal is feeding. While looking through your dissecting microscope, carefully pry open the opercular plates to expose the animal.

Figure 21.4 *Daphnia*: Internal anatomy.

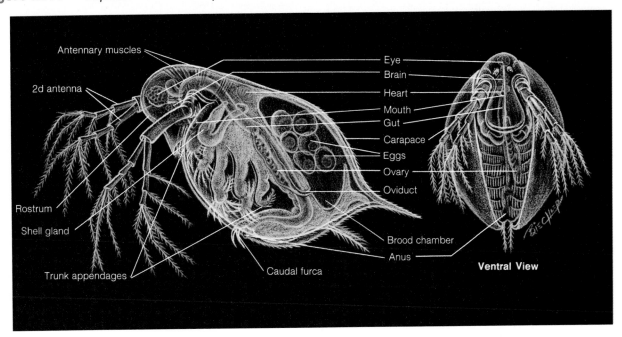

The anatomy of the barnacle can be confusing. Picture an arthropod lying on its back with its legs projecting upward inside the surrounding plates. Each of the thoracic segments bears branched appendages called **cirri.** To feed, the animal opens the opercular plates, extends its thorax, and rapidly sweeps the cirri through the water, netting small particles, which are conveyed to the mouth. Find the mouth, and opposite it, the anus. A single, long penis should also be visible. Barnacles are hermaphroditic, and possess ovaries as well as testes. Cross fertilization is the rule, but being sessile, barnacles can only mate by extending the penis to nearby barnacles.

Daphnia
Commonly known as the water flea, this microcrustacean inhabits ponds and lakes in North America and Europe.

Make a slide of living *Daphnia* by catching one in a pipette from the lab culture and transferring it with some water to a slide. Use your dissecting microscope to watch it swim. Add a coverslip and observe it through your compound microscope (fig. 21.4). Ranging in size from 0.5 to 5 mm, *Daphnia* are covered by a transparent **carapace,** which wraps around the body in a bivalve fashion. The head bears a conspicuous, single compound **eye,** and two large **antennae.**

Water fleas swim in a jerky fashion by stroking the antennae back and forth. These animals are filter-feeders. Legs inside the carapace terminate in comblike **setae,** which sieve bacteria, algae, and small particles from the water. Food balls are transferred to the **mouth** and take about one-half to three hours to pass through the digestive system.

Water fleas live about 100 days under optimum conditions. Because the exoskeleton constrains growth, it is periodically molted: once every 24 hours in juvenile forms and once in two or three days in adults. Identify the structures labeled in figure 21.4.

Reproduction is usually parthenogenic (unfertilized eggs develop) and results in female offspring. Only under stressful conditions are males produced and sexual reproduction follows their appearance.

Feeding behavior can easily be observed. Prepare a suspension of stained yeast by adding 3 g of wet yeast and 30 mg of Congo red stain to a test tube containing 10 ml of water. Boil gently. A *Daphnia* can be immobilized by using a dissecting needle to place a very small dab of petroleum jelly on a microscope slide. Cover it with a drop of water. Add a *Daphnia* and carefully begin to remove the water. The animal should fall on its side and become stuck to the jelly. Add back enough water to cover the animal.

Dip a dissecting needle into the suspension and remove a small drop. Transfer this to the *Daphnia* slide while watching through your dissecting microscope. You should be able to see the sieving action of the leg combs and red yeast should be taken into the digestive tract. Congo red is is a pH indicator: red above pH 5, but blue at pH 3. Hypothesize what color the yeast will be as they pass through the digestive system. Test your hypothesis by carefully observing the yeast in different parts of the digestive system. You may have to greatly reduce the light intensity to see the colors. Record your observations below.

cause they bear **chelae** (pincers), which are used in grasping and tearing food as well as in defense.

Several appendages surround the mouth. The heavy **mandibles** are the jaws that grind food. There are also two pairs of large **maxillipeds** and two pairs of **maxilla** that handle food and carry sensory organs for taste and touch. Probe the mouthparts to find the second maxilla, which bears a large elongated plate, the **gill bailer.** The sculling action of the bailer draws water over the gills that are located under the sides of the carapace. The **antennae** and smaller **antennules** are the last appendages on the head. They have sensory functions.

The gills are contained in branchial chambers, located laterally in the cephalothorax. The lateral carapace covering the gills is called the **branchiostegite.** What are the advantages and disadvantages of having the gills enclosed by the exoskeleton?

Do your observations support or disprove your hypothesis?

Crayfish

You will use a crayfish for your demonstration dissection of a crustacean. Obtain a crayfish, dissecting instruments, and a dissecting pan filled with wax from the supply area.

External Anatomy   Examine the external anatomy of the crayfish (fig. 21.5). The body is divided into two major regions, the posterior **abdomen** and the anterior **cephalothorax** covered dorsally by the **carapace.** The abdomen is composed of six segments and ends in a flaplike projection, the **telson.** Thirteen fused segments make up the cephalothorax. The **cervical groove,** a depression in the carapace, marks the separation of the head segments from the thorax. The head ends anteriorly in the pointed **rostrum.**

Lay the animal on its back and examine the appendages. On the last abdominal segment are two flattened lateral appendages, the **uropods.** Contraction of the ventral abdominal muscles draws the telson and the extended uropods under the body, allowing the crayfish to swim rapidly backwards. On the other abdominal segments are small **swimmerets.** If your specimen is a male, the swimmerets on the first abdominal appendages will be pressed closely together and pointed forward. They serve as sperm transfer organs (gonopods), receiving sperm from openings at the base of the last pair of walking legs. If your specimen is a female, these swimmerets are small. Be sure you observe both males and females in the lab.

The large appendages of the cephalothorax are the **walking legs.** The first pair of legs are called **chelipeds** be-

Cut away the branchiostegite on one side of the crayfish. Be careful not to damage the underlying gills. Gradually clear away the material to reveal the feathery gills. Water is drawn in through an opening at the posterior end of the branchial chamber, passes over the gills, and exits anteriorly as a result of the action of the gill bailer. Cut off a gill and mount it in a drop of water. Add a coverslip and look at it with your compound microscope. Sketch what you see below.

Figure 21.5   External anatomy of male crayfish:(a) dorsal view; (b) ventral view.

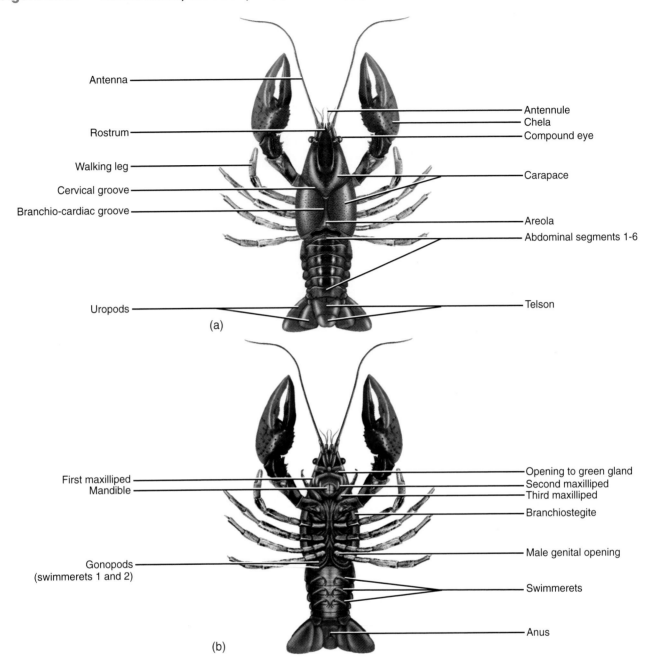

(a)

(b)

Hemolymph, crustacean blood, circulates through the gills and comes in close proximity to water passing over the gills, where carbon dioxide and oxygen exchanges occur. How is the respiratory system similar to that of the molluscs? How is it different?

▶ *Internal Anatomy*   You will now study the internal organs of the crayfish. With scissors, cut very carefully to one side of the dorsal midline of the carapace from where it joins the abdomen forward to a region just behind the eyes. Use a scalpel or a probe to carefully separate the carapace from the underlying **hypodermis,** the tissue that secretes the carapace. Completely remove the carapace. Now cut through the exoskeleton covering the abdomen from the first to last segments. Carefully remove the exoskeleton to expose the underlying structures. Remove the gills and the membrane separating them from the internal organs. Flood the tray with water so that the organs float.

21-7   Protostomes II: Investigating A Body Plan Allowing Great Diversity   **251**

Figure 21.6   Dorsal view showing internal organs in a partially dissected crayfish.

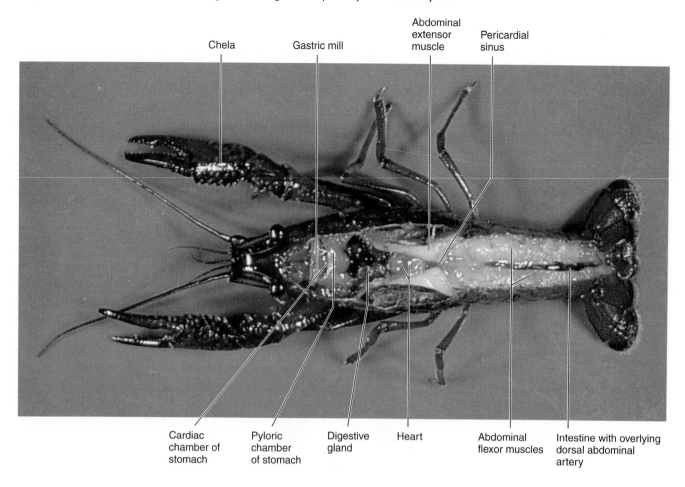

Chela   Gastric mill   Abdominal extensor muscle   Pericardial sinus

Cardiac chamber of stomach   Pyloric chamber of stomach   Digestive gland   Heart   Abdominal flexor muscles   Intestine with overlying dorsal abdominal artery

Compare the opened crayfish to figures 21.6 and 21.7. Starting in the abdomen, note the two longitudinal bands of **abdominal extensor muscles** that extend forward into the thorax. Contraction of these muscles straightens the abdomen from a curled position. Beneath the extensors, find the segmented **abdominal flexor muscles.** Contraction of these muscles bends the abdomen ventrally and allows the crayfish to swim rapidly backward in escape reflexes. Passing along the dorsal midline of these muscles is the **intestine** with the overlying **dorsal abdominal artery.** If you are dissecting an injected specimen, this artery should contain colored latex. Carefully remove the long bands of the extensor muscles. How has segmentation led to increased mobility?

Trace the dorsal abdominal artery anteriorward to where it joins the delicate membranous **heart** located in the posterior third of the cephalothorax. The heart sits in the **pericardial sinus.** Hemolymph in the sinus enters the heart through small openings, the **ostia,** which should be visible. When the heart contracts, small flaps of tissue seal the ostia from inside the heart, and hemolymph in the heart is forced out into the arteries. These arteries are often difficult to locate, but should be visible after careful study. As figures 21.6 and 21.7 indicate, these include the single **ophthalmic artery,** the **paired antennary arteries,** paired **hepatic arteries,** and a single **sternal artery.**

Arthropods have open circulatory systems. When hemolymph is pumped away from the heart in arteries, it leaves the arteries and enters the spaces in tissues at the periphery of the animal. Hemolymph in the tissues is displaced by new hemolymph being pumped in and percolates back to the pericardial sinus where it enters the heart and is pumped to the periphery again. Hemolymph that contains the respiratory pigment hemocyanin is pumped through the sternal artery and enters the **sternal sinus** from which it flows into the gills before returning to the pericardial sinus. This circuit allows respiratory gas exchange to occur.

Figure 21.7 Sagittal section of crayfish showing positions of internal organs.

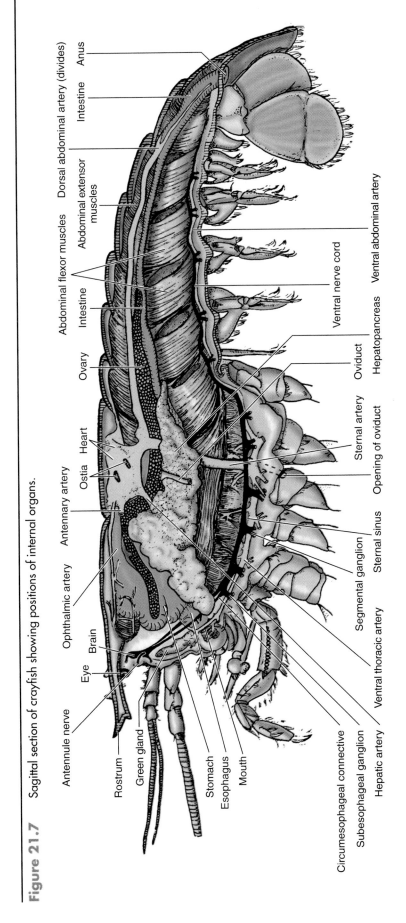

From your examination of your specimen's external anatomy, you should know whether you have a male or female. In females, the paired ovaries lie ventral and lateral to the pericardial sinus. If collected in breeding season, the ovaries may be enlarged and filled with eggs. In males, the paired testes should be visible on either side of the midline beneath the pericardial cavity.

Trace the digestive system from the mouth through the **esophagus** to the **stomach** and into the **intestine.** Find the anus ventrally at the base of the telson. Locate the large **hepatopancreas,** a large gland that secretes digestive enzymes into the stomach. Cut the esophagus and intestine to remove the stomach. Open the stomach and observe the hardened areas of the wall, the **gastric mill** that grinds food. At the posterior end of the stomach, fine bristles prevent large pieces of ingested food from entering the intestine.

Remove the organs from the cephalothorax. Find the **antennary glands** at the base of the antennae. These are the crayfish's excretory organs and also are known as green glands, though they are not green in color.

Look on the floor of the body cavity and find the ventral nerve cord. **Segmental ganglia** are collections of neurons that function as integration centers. Lateral nerves leave the ganglia and innervate the appendages. If you follow the nerve cord anteriorly, you will find ganglia beneath and above the stub of the esophagus. These function as the crayfish's brain.

Many arthropods have **compound eyes.** Such eyes are made up of hundreds of individual light-sensitive units called **ommatidia.**

Cut the surface off one of the crayfish's eyes and mount it in a drop of water on a slide. Look at the slide with your compound microscope. Sketch what you see in the circle to the right. Compound eyes give an animal mosaic vision. What does that mean?

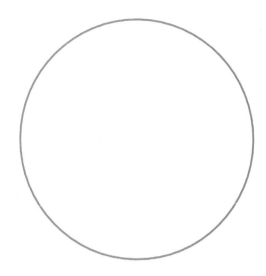

Uniramian Arthropods

Insects

External Anatomy

Obtain a grasshopper and study its external anatomy, comparing it to figure 21.8. Note the chitinous exoskeleton consisting of hardened plates or **sclerites** separated by thin membranous areas called **sutures.** The sutures allow movement of the body segments and appendages. Describe several ways that the exoskeleton adapts the grasshopper to life in the terrestrial environment.

The grasshopper's thorax consists of three fused segments: the large anterior **prothorax,** the **mesothorax,** and the **metathorax.** Each segment bears a set of legs, and both mesothorax and metathorax have a pair of wings.

The jointed legs consist of five main parts. The short **coxa** attaches to the ventral body wall and to the small **trochanter.** The **femur,** the largest leg segment, is attached distally to the spiny **tibia,** which in turn attaches to a **tarsus** with terminal claws. The metathoracic legs are specialized for jumping. In males, legs are also used for stridulation—rubbing the teeth on the femur against the edge of the front wing to produce their familiar call. Observe the heavy **forewings** and the membranous **hindwings** with their char-

Dispose of your crayfish as directed by your instructor Pair off with another student in the lab and explain to each other how a crayfish gets oxygen to its tissues and the pathways of circulation of hemolymph.

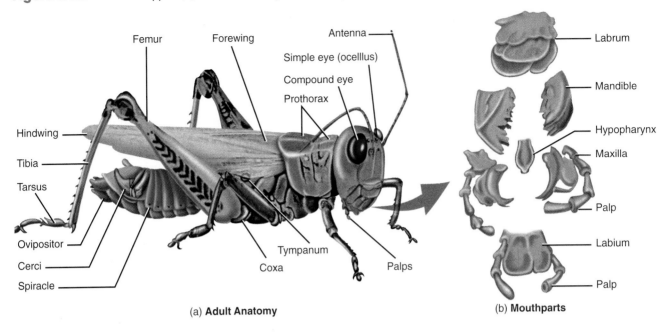

(a) **Adult Anatomy**   (b) **Mouthparts**

acteristic venation pattern.How do the appendages of the grasshopper differ from those of the crayfish?

Many receptors are found on the head, including the compound eyes; the simple eyes, or ocelli; touch, taste, and smell receptors on the antennae; and chemical receptors on various mouthparts. To observe the mouthparts, look at your specimen under a dissecting microscope. These mouthparts are those of a chewing insect and are quite different from those of a piercing and sucking insect, such as a mosquito. Compare the mouthparts of the horseshoe crab, the crayfish, and the grasshopper. How is each adapted to its particular feeding habits?

Along the sides of the eleven abdominal segments, locate the ten respiratory **spiracles,** openings into the **tracheal respiratory system** (see lab topic 27). The last three segments are modified in the two sexes either for copulation or for egg laying. A pair of sensory **cerci** occur laterally on the eleventh segment.

Insect Orders

There are over 20 orders in the class Insecta. For example, butterflies are in the order Lepidoptera, and beetles in the order Coleoptera. A dichotomous taxonomic key follows. This key will allow you to identify the correct order for insect specimens that your instructor will provide.

Keys are used by biologists to identify unknown organisms. They have been made for the phyla and for taxonomic groups within the phyla down to the species level. With the right key and some knowledge of the anatomy of the group, it is possible, with practice, to identify any specimen to the species level. Keys are based on diagnostic characteristics that are shared by certain members of a group, but not by others.

To use the key, obtain a specimen. While observing it read the first couplet of statements. Decide which statement describes your specimen and go to the couplet indicated by the number that follows the correct description. Eventually, you will arrive at a statement that indicates the taxonomic order in which your specimen is classified, *i.e.* you have identified it. For some specimens, a dozen or more descriptions must be read and decisions made. Never randomly skip around when using a key. Follow the order specified. A dissecting microscope may help you view smaller specimens.

Key to Orders of Insects

Source: Muesebeck, C.F.W. 1952. What Kind of Insect Is It? *Insects: The Yearbook of Agriculture, 1952. A. Stefferud, ed. Washington, U.S. Government Printing Office.*

1. Wings present, the front wings often in the form of hard, leathery, or horny, wing covers 2
   Wings absent or represented only by minute pads 19

2. With only one pair of wings, these always membranous 3
   With two pairs of wings, the front pair often with hard wing covers, beneath which the hind wings are concealed 5

3. End of abdomen with two or three slender, but conspicuous, backwardly projecting filaments 4
   End of abdomen without such filaments . . . .
   Order **Diptera** (mosquitoes, midges, flies) . . . .

4. Wings with a network of veins, including many cross veins . . . .
   Order **Ephemeroptera** (mayflies)
   Wings with few longitudinal veins and no cross veins . . . .
   Order **Hemiptera,** in part (males of scale insects)

5. Front wings horny, rigid, opaque, without veins, meeting in a line over middle of body and concealing membranous hind wings 6
   Front wings usually membranous although often covered with scales or hairs; if leathery, with the veins distinct, and not meeting along a line over middle of body 7

6. Tip of abdomen with a pair of prominent forceps-like appendages; front wings (wing covers) very short . . . .
   Order **Dermaptera** (earwigs)
   Tip of abdomen without such appendages; front wings (wing covers) usually covering most of abdomen . . . . .
   Order **Coleoptera** (beetles)

7. Front wings more or less leathery 8
   Wings membranous 9

8. Mouth parts in the form of a sharp piercing and sucking beak . . . .
   Order **Hemiptera,** in part
   Chewing mouthparts without a sharp projection . . . . . .
   Order **Orthoptera**

9. Wings covered with minute overlapping scales, often in beautiful color patterns . . . .
   Order **Lepidoptera** (moths and butterflies)
   Wings not covered with scales 10

10. Wings narrow, bladelike, and fringed with long bristles; tarsus ending in a large bladderlike structure . . . .
    Order **Thysanoptera** (thrips)
    Wings not bladelike; tarsus without a bladderlike structure 11

11. Mouthparts in the form of a beak fitted for piercing and sucking . . . .
    Order **Hemiptera,** in part
    Chewing mouthparts 12

12. Wings with numerous longitudinal veins and many cross veins forming a network 13
    Wings with few cross veins and usually with few longitudinal veins, not net-veined 17

13. Antennae very short and inconspicuous, composed of few segments 14
    Antennae conspicuous, composed of many segments 15

14. Hind wings small; tip of abdomen with two or three long filaments extending backward . . . .
    Order **Ephemeroptera,** in part (mayflies)
    Front and hind wings of about equal size; abdomen without terminal filaments . . . .
    Order **Odonata**

15. Wing veins mostly membranous and faint; front and hind wings of same size and shape; tarsi four-segmented . . . .
    Order **Isoptera** (termites, winged form)
    Wing veins strongly developed 16

16. Tarsi two- or three-segmented . . . .
    Order **Plecoptera** (stoneflies)
    Tarsi five-segmented . . . .
    Order **Neuroptera** (lacewings, dobsonflies)

17. Tarsi five-segmented 18
    Tarsi two- or three-segmented . . . .
    Order **Corrodentia** (psocids)

18. Wings covered with fine, long hair and held rooflike over abdomen . . . .
    Order **Trichoptera** (caddis flies)
    Wings transparent, not covered with long hairs, not held rooflike over abdomen . . . .
    Order **Hymenoptera**

19. Tip of abdomen with two or three long appendages directed backward 20
    Tip of abdomen without such appendages 21

20. Abdominal appendages thick, rigid, in the form of forceps . . . .
    Order **Dermaptera** (earwigs)
    Abdominal appendages delicate, flexible, antenna-like . . . .
    Order **Thysanura** (silverfish)

21. Tarsus composed of only one to three segments 22
    Tarsus composed of four or five segments 26

22. Antennae conspicuous, projecting in front of head . .23
    Antennae very short, inconspicuous, not projecting in front of head 25

23. Antennae composed of three to six segments 24
   Antennae with more than six segments; very tiny insects that sometimes occur by the thousands in damp houses . . . .
   Order **Corrondentia** (psocids)

24. Mouthparts in the form of a distinct beak; body greatly flattened . . . .
   Order **Hemiptera** (bedbugs)
   Mouthparts not in the form of a beak; body not flattened . . . .
   Order **Collembola** (springtails)

25. With biting mouthparts . . . .
   Order **Mallophaga** (biting lice)
   With piercing and sucking mouthparts . . . .
   Order **Anoplura** (sucking lice)

26. Antennae prominent 27

Antennae inconspicuous, not projecting 28

27. Body noticeably constricted at base of abdomen, antennae elbowed, the basal segment very long; tarsus five-segmented . . . .
   Order **Hymenoptera** (ants)
   Body not constricted at base of abdomen; antennae not elbowed, basal segment short; tarsus four-segmented
   Order **Isoptera** (termites)

28. Body strongly compressed from the sides; abdomen distinctly segmented; coxae very large and strongly flattened; legs fitted for jumping . . . .
   Order **Siphonaptera** (fleas)
   Body not compressed; abdomen not distinctly segmented; legs not fitted for jumping . . . .
   Order **Diptera,** in part; wingless forms (sheep-tick and its relatives)

Learning Biology by Writing

Write an essay that summarizes the evidence gathered in this lab that supports or refutes the hypothesis that all Arthropods have an exoskeleton, segmented body, and jointed appendages. Indicate how the diverse species studied are similar and how they differ from each other.

   As an alternative, your instructor may ask you to turn in lab summary 21 at the end of this lab topic.

Internet Sources

Check the WWW for information about arthropods as vectors of diseases. Use your Internet browser to connect to Google at www.google.com. Enter the word disease and enter the name of an insect order such as Diptera, Hemiptera, or Siphonaptera, or the name of the family of ticks, Ixodidae. You will get hundreds of citations back. Scan the citations for examples of insects that are carriers of human diseases. Connect to one of the sites and read about the disease and the vector. Alternatively, enter the name of an arthropod-borne disease such as malaria, Chagas, sleeping sickness, or Lyme disease. Write a short paragraph describing what you found. Include the URL in case you wish to return to the site.

Lab Summary Questions

1. What are the distinguishing characteristics of the phylum Arthropoda?
2. What are the general differences between the chelicerate and the mandibulate arthropods?
3. Describe the circulatory and respiratory systems of a crayfish.
4. What are the similarities between annelids and arthropods? What are the differences?
5. What is a taxonomic key and how is it used?
6. What evidence do you have from your observations to refute the hypothesis that all arthropods have an exoskeleton, segmented bodies, and jointed appendages?

Critical Thinking Questions

1. Why are there no really large arthropods? What could be engineering-type factors limiting body size?
2. How does the segmentation of an arthropod compare with the segmentation of an annelid?
3. What correlation may be made between body symmetry and locomotion?
4. Arthropods are the most diverse (greatest number of species) and the most common (greatest number of individuals) of all the animals. List several reasons why you think they have been so successful.

LAB TOPIC 22

Deuterostomes: Investigating the Origins of the Vertebrates

Supplies

Preparator's guide available on WWW at
http://www.mhhe.com/dolphin

Equipment

Compound microscopes
Dissecting microscopes

Materials

Preserved specimens
    Demonstration specimens of the five classes of
        echinoderms
    Sea star
    Branchiostoma (amphioxus)
    Adult sea squirt (*Ciona* or *Molgula*)
    Perca flavescens (perch)
Prepared slides
    Sea star arm, cross section
    Ascidian tadpole larva
    Branchiostoma, whole mount
    Branchiostoma, cross section
Dissecting pans and instruments
Photographs of various echinoderms

Prelab Preparation

Before doing this lab, you should read the introduction
and sections of the lab topic that have been scheduled
by the instructor.

    You should use your textbook to review the
definitions of the following terms:

    bilateral symmetry
    cephalization
    Cephalochordata
    Chordata
    deuterostome
    Echinodermata
    radial symmetry
    Urochordata
    Vertebrata

You should be able to describe in your own words the
following concepts:

    The general body plan of an echinoderm
    The general body plan of a chordate
    The organ systems you would expect to find in a
        complex animal

As a result of this review, you most likely have
questions about terms, concepts, or how you will do
the experiments included in this lab. Write these
questions in the space below or in the margins of the
pages of this lab topic. The lab experiments should
help you answer these questions, or you can ask your
instructor for help during the lab.

Objectives

1. To illustrate the anatomy of echinoderms
2. To study the anatomy of the primitive chordates
3. To study the anatomy of a bony fish
4. To relate the anatomy of echinoderms, primitive
   chordates, and bony fish to evolutionary trends
5. To collect evidence to test the hypothesis that all
   chordates have a notochord, post anal tail, and
   gill slits, at some stage in their life cycle and a
   dorsal hollow nerve cord

Background

The deuterostome phyla contain only about 4% of the
species in the animal kingdom. Included in this group are
the echinoderms, or spiny-skinned animals, such as sea
stars, sea urchins, sand dollars, sea cucumbers, and sea
lilies. Also included are the chordates, which encompass
relatively simple marine animals, such as the sea squirts,
through the various classes of vertebrates, animals with
backbones. These animals are the familiar fish, amphibians,
reptiles, birds, and mammals.

    The name *deuterostome* describes a characteristic
shared by the approximately 59,000 species in this group:
the mouth forms as the second opening (after the anus) of
the digestive system during embryonic development (See
figure 20.1). Other embryological characteristics also unite
the group: all have radial cleavage, bilaterally symmetrical
larvae or adult stages, and mesoderm tissue forming as an
outpocketing of the primitive gut (enterocoelous).

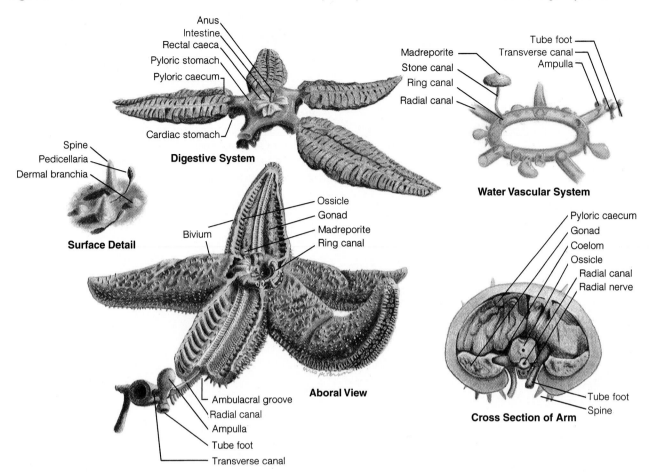

Digestive System

Water Vascular System

Surface Detail

Aboral View

Cross Section of Arm

LAB INSTRUCTIONS

You will observe the anatomy of four deuterostome animals: an echinoderm (a sea star) and three chordates (a tunicate, an amphioxus, and a perch).

Phylum Echinodermata

The approximately 7,000 species of echinoderms are spiny-skinned marine animals that are slow-moving but voracious feeders. Adults have a five-part (pentamerous) radial symmetry that develops secondarily from a bilaterally symmetrical larval stage. Echinoderms have true coeloms lined by ciliated peritoneal cells. Review your notes on lab topic 18, which described the development of echinoderms in detail.

Echinoderms have a tubular digestive system. Respiratory gas exchange is across the body surface. They lack a circulatory system and an excretory system. The nervous and sensory systems are not well developed although they have reasonable behaviors. The sexes are separate individuals.

Peculiar features of the animals in this phylum are the protruding spines derived from the mesoderm, pentamerous

symmetry, and a unique **water vascular system.** Living species are divided into six classes: the sea stars (**Asteroidea**), the brittle stars (**Ophiuroidea),** the sea urchins and sand dollars (**Echinoidea),** the sea cucumbers (**Holothuroidea),** the sea lilies (**Crinoidea),** and the sea daisies (**Concentricycloidea),**

Sea Star

Since sea stars, also commonly called starfish, are readily available, they are the echinoderms most often used for dissection.

External Anatomy

▶ Obtain a sea star and observe its pentamerous radial symmetry. The ventral surface, the underside, is called the **oral** surface, and the dorsal is called the **aboral** surface. Examine the central disc aboral surface and find the **madreporite,** a light-colored calcareous disc, which acts as a pressure-equalizing valve between the water vascular system and seawater (fig. 22.1). The anus is located more or less in the center of the aboral surface of the central disc.

Observe the many spines protruding from the surface of the organism. These spines give the echinoderm ("spiny-skinned") phylum its name. If a sea star is submerged in water and its surface examined with a dissecting microscope, many small, pincerlike **pedicellariae** may be seen projecting from the surface of the animal. The pedicellariae remove encrusting organisms from the skin, keeping it extremely clean in comparison to many other invertebrates.

Scrape away some of the tissue at the surface and mount the material in a drop of water on a microscope slide. After adding a coverslip, observe the slide under the compound microscope and sketch some pedicellariae in the circle. Some of the scrapings may contain **dermal branchiae,** which are hollow, bubblelike extensions of the body wall that aid in gas exchange.

The body wall contains the skeleton of the sea star. It is made of a large number of interconnecting **ossicles.** These can best be observed in a dried specimen.

Turn the animal over and observe its oral side. The mouth is surrounded by a soft membrane called the **peristome** and is guarded by oral spines. Five **ambulacral grooves** radiate from the area of the mouth down the center of each arm. Each groove is filled with **tube feet,** extensions of the water vascular system. Though they are difficult to see, a sensory tentacle and a light-sensitive eyespot are at the end of each arm.

Internal Anatomy

Cut 1 to 2 cm off the tip of an arm and then cut along each side of the arm to where it joins the central disc. Join the cuts by cutting across the top of the arm at the edge of the disc, and remove the body wall to expose the coelom within the arm (fig. 22.1). Note how the organs are free in the coelom. Ciliated peritoneal cells line the coelom and move fluid through the cavity. Coelomic fluid carries oxygen, food, and waste products to and from the organs. There is no separate circulatory system.

Now remove the aboral body wall covering the central disc without removing the madreporite. Do so by cutting carefully around the periphery of the disc and then cutting to and around the madreporite, freeing it from the body

wall. Lift and remove the body wall carefully, noting where the intestine joins the anus in the piece you are removing.

In each arm are pairs of **pyloric caecae,** or digestive glands, that empty into the pyloric portion of the stomach. The **pyloric stomach** connects with the **intestine** on the aboral side; on the oral side, it connects with the **cardiac stomach** and **esophagus.** When the sea star feeds, the cardiac stomach and the esophagus are everted through the mouth and surround the material to be digested. Enzymes are secreted by the cardiac stomach so that partial digestion occurs outside the body. Food particles then are ingested and digestion is completed inside the body. Carefully remove the digestive glands to expose the next layer of organs in the arm. Is this a complete or incomplete digestive system? _____

Under the digestive glands in each arm lie two **gonads.** Depending on the stage of the breeding cycle at the time the sea star was caught, the gonads may either fill the arm or be quite small. **Gonoducts** lead from each gonad to very small **genital pores** located at the periphery of the central disc on the aboral surface between the arms. Though the sexes are separate, they are difficult to distinguish except by microscopic observation of the gonad contents.

Make a slide by mounting a small piece of the gonad in a drop of water. Tease the tissue fragment apart, add a coverslip, and observe with the compound microscope. Do you see eggs or flagellated sperm? Sketch what you see below. Compare your sketch to those of classmates until you find a sea star of the opposite sex. Do the gonads look different between the two sexes?

The water vascular system of the echinoderms is a unique internal hydraulic system associated with the functioning of the tube feet. Water enters the madreporite and travels via the **stone canal** to a **ring canal** that encircles the mouth (fig 22.1). From the ring canal, the water passes out into **radial canals** in each arm, and from each radial canal, several **transverse canals** lead to each pair of **tube feet.** Remove the gonad from the dissected arm to reveal the radial canal and the ampullae of the tube feet.

Figure 22.1 shows a cross section of a sea star and the relation of the radial canals to the tube feet and to the **ampullae** on the inside of the animal. Contraction of the ampullae extends the tube feet, whereas relaxation retracts them, providing a means of slow locomotion and a means of adhering to substrates.

▶ Obtain a prepared microscope slide of a cross section of a sea star arm and compare what you see to figure 22.1.

The nervous system of the sea star lacks cephalization, as is characteristic of radially symmetrical animals. Though you will not be able to see them, a **nerve ring** surrounds the mouth with **radial nerves** passing into each arm. At the junction of the ring and radial nerves are ganglia. The only differentiated sense organ is the eyespot at the tip of each arm, though there are other sensory cells located throughout the epidermis.

Note that in your almost complete dissection of this animal that you did not see a heart or blood vessels, nor were there any excretory organs. How do you think the animal performs the functions of these organs?

Echinoderm Diversity

▶ Examples of other echinoderms will be available in the laboratory as specimens or photographs (fig. 22.2). After looking at these materials, you should be able to identify unknown echinoderms to the level of taxonomic class. The characteristics of the classes are listed below.

Class Holothuroidea (sea cucumbers)
  —no arms, although oral tentacles are present
  —body soft and elongate

Class Crinoidea (sea lilies and feather stars)
  —branched arms radiate from body as multiples of five
  —madreporite absent
  —sea lilies have body at end of jointed stalk
  —feather stars are free swimming

Class Echinoidea (sea urchins and sand dollars)
  —no arms
  —body encased in a rigid calcareous test
  —test may be globular (sea urchins) or disc-shaped (sand dollars)

Class Ophiuroidea (bracket and brittle stars)
  —five thin flexible arms
  —no digestive glands in arms

  —lack anus
  —no ambulacral grooves on undersides of arms

Class Concentricycloidea (sea daisies)
  —A new class of echinoderms discovered in 1986 in 1200 m of water off the coast of New Zealand. The class is represented by one species *Xyloplax medusiformis.*

Phylum Chordata

The phylum Chordata shows a remarkable diversity, including both simple marine animals and highly developed birds and mammals. However, all members of the phylum share some common, distinctive characteristics. These include (1) a **dorsal hollow nerve cord;** (2) a flexible, dorsally located longitudinal supporting rod called a **notochord;** (3) the occurrence of paired **pharyngeal gill slits** either during embryological development or in the adult organism; and (4) a **postanal tail.**

The phylum Chordata is divided into three subphyla: the tunicates (**Urochordata**), lancelets (**Cephalochordata**), and vertebrates (**Vertebrata**).

Urochordata

Most tunicates are sessile marine animals that live attached to rocks, pilings, and ships. They are filter-feeders, sieving out particles from a constant stream of water entering their basketlike pharynxes.

The four chordate characteristics are found only in the larval stages of the approximately 1,300 species in the urochordate group. In the adult, the notochord disappears, so that the adult organism might be mistaken for a member of a lower phylum. In this lab, you will study the tadpole larva of a tunicate. The adult form will be available as a demonstration.

Larval Anatomy

▶ Obtain a slide of a tunicate tadpole larva and look at it under low power with your compound microscope (fig. 22.3). At the anterior end of the body are adhesive glands that attach the larva to a substrate prior to its metamorphosis into a sessile adult. Near the dorsal surface is a sensory vesicle containing a light-sensitive **ocellus** and **otoliths** that aid in balance. The mouth (actually the opening of the incurrent siphon) is located dorsally a short distance from the anterior end. It opens into a pharynx with a few gill slits in its walls. The mouth is closed for most of larval life. When it opens, water is pumped through the pharynx by the action of beating cilia. It exits through the slits into a developing chamber, the atrium. Water leaves the atrium through an opening, the **excurrent siphon.**

Observe the tail region and identify the **dorsal nerve cord,** the **notochord** beneath it, and bands of muscle cells. The nerve cord, notochord, and muscles break down, and new organs develop when the tadpole stage metamorphoses into the adult. It is the notochord and position of the nerve

Figure 22.2 Diversity of echinoderms: (*a*) sea cucumber; (*b*) sea lily; (*c*) feather star; (*d*) sea urchin; (*e*) brittle star; (*f*) sea daisy.

(a)

(b)

(c)

(d)

(e)

(f)

Figure 22.3 Stages of the life cycle of a tunicate: (*a*) adult sea squirt; (*b*) internal anatomy of an adult; (*c*) anatomy of a tadpole larva.

(a)

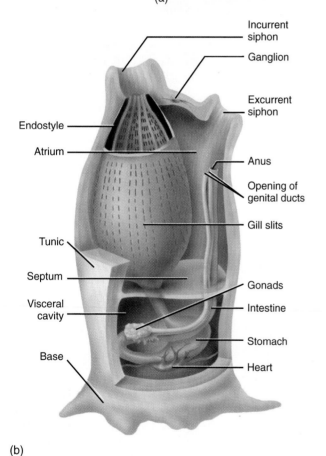

(b)

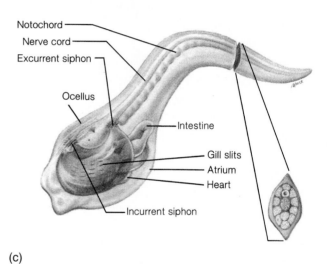

(c)

cord that lead zoologists to place these animals in the phylum chordata.

Adult Anatomy (Demonstration)

▶ Look at the adult tunicate (*Ciona* sp. or a similar species) available as a demonstration specimen. Commonly known as sea squirts, these animals are covered by an external man-

tle composed of an outer noncellular **tunic,** underlying epidermis, and muscle. The name *sea squirt* describes the fact that water beneath the mantle can be forcibly expelled from the siphons by muscular contraction when the animals are disturbed. The name *tunicates* describes the outer covering.

Identify the two siphons opposite the base. Small tentacles surround the upper **incurrent siphon** where water enters the organism. The **excurrent siphon** is located laterally.

▶ If the mantle is cut open from the excurrent siphon to the base, the internal organs will be exposed. Note the large **pharyngeal basket** sitting in a large open space, the **atrium.** On the side of the pharyngeal basket opposite the excurrent siphon is a deep groove called the **endostyle,** which is the principal site for producing mucus used in feeding. Water drawn in through the incurrent siphon by the beating of cilia on the wall of the basket passes through the gill slits into the atrium. Suspended organic matter is filtered out of the water and is caught in mucus, which passes to the bottom of the basket and enters the **esophagus.** The **stomach** and **intestine** are in the **visceral cavity,** which is separated from the atrium by the **septum.** The intestine passes through the septum and empties through the anus into the atrium near the excurrent siphon. How does this digestive system compare with that of the sea star? What is the advantage of having the anus located near the external siphon?

There is no specialized respiratory system. Gas exchange occurs by diffusion across the body surfaces that are exposed to the water stream passing through the animal. Tunicates have an open circulatory system. A tubular heart is located below the stomach.

Tunicates are hermaphroditic, each possessing an ovary and a testis lying in very close proximity to each other and located near the loop of the digestive system. These closely placed gonads are shown as one structure in figure 22.3*b.* Gonadal ducts pass from the gonads to the region of the excurrent siphon where they open near the anus. When eggs and sperm are released, they are carried by the water stream into the surrounding water. Fertilization is external. Development is rapid and tadpole larvae settle within a few days to metamorphose into the adult body form. What is the advantage of having the gonoduct open near the excurrent siphon?

Cephalochordata

The subphylum cephalochordata contains a few small fishlike species commonly known as lancelets. Lancelets live in shallow marine environments throughout the world. In some places, lancelets occur in sufficient numbers to be used as a human food source. The common American species is in the genus *Branchiostoma,* though it is often called amphioxus, the former generic name for the group.

External Anatomy

▶ Immature lancelets are small enough to be mounted whole on microscope slides. Adults can be chemically treated so that the tissues are rendered transparent. They are then mounted in plastic. Obtain a slide or plastic mount and look at it through a dissecting microscope.

Note its general form and the absence of paired appendages (fig. 22.4). The **rostrum** is the anteriormost part of the lancelet. Beneath it is the **oral opening** surrounded by a fringe of **oral cirri.** Behind the opening is the **vestibule,** which is lined by bands of ciliated cells collectively called the "wheel organ." The beating of the cilia draws a water stream containing oxygen and food particles through the oral opening into the vestibule. Mucus secreted into the vestibule traps the food particles. The mucus passes into the mouth at the rear of the vestibule. Dorsal and ventral fins occur caudally. On the ventral surface, there are two pores: the **anus** is the most posterior opening, and the **atriopore** is anterior to the anus.

Internal Anatomy

▶ Study a whole mount of a young lancelet with the dissecting microscope on high power with transmitted light and then with the compound microscope on low power (fig. 22.4*a*). Do not use high power.

Trace the digestive tract from the mouth into the **pharynx,** a filterlike basket consisting of **gill bars** and **gill slits.** Water containing suspended food material passes from the vestibule through the mouth into the pharynx and is filtered through the slits and enters the **atrium,** the space surrounding the gill basket. Water then flows from the atrium to the outside through the atriopore. Food particles cannot pass through the gill slits. Food thus collected and concentrated passes as a mucous string posteriorly into the esophagus, stomach, and intestine (fig. 22.4*b*). Respiratory gas exchange occurs as the water passes over the blood vessels of the gill bars. What is the advantage of having so many gill bars?

The body wall of the lancelet is composed of V-shaped **myomeres.** When contracted, these muscles bend the body to allow the animal to swim and burrow. Lancelets commonly burrow into sediments tail first with the oral opening remaining at the surface. The dorsal **notochord,** a cartilaginous rod, runs the length of the organism just dorsal to the pharynx and intestine. It stiffens the animal along its length so that contraction of the myomeres does not shorten the body. Instead, myomere contraction causes side to side movement that allows swimming locomotion. Dorsal to the notochord is the **nerve cord** with a small enlargement at the anterior end, a primitive brain.

The circulatory system is difficult to observe (fig. 22.4*c*). The system is closed but lacks a definite heart. Blood is collected from the digestive system tissues by a **subintestinal vein.** It flows into the **hepatic portal vein,** which drains through the **hepatic vein** into the **ventral aorta**

Deuterostomes: Investigating the Origins of the Vertebrates

Figure 22.4 Lancelet anatomy: (*a*) cross section in region of pharyngeal basket; (*b*) internal organs; (*c*) diagram of ciculatory system; (*d*) cross section in posterior half.

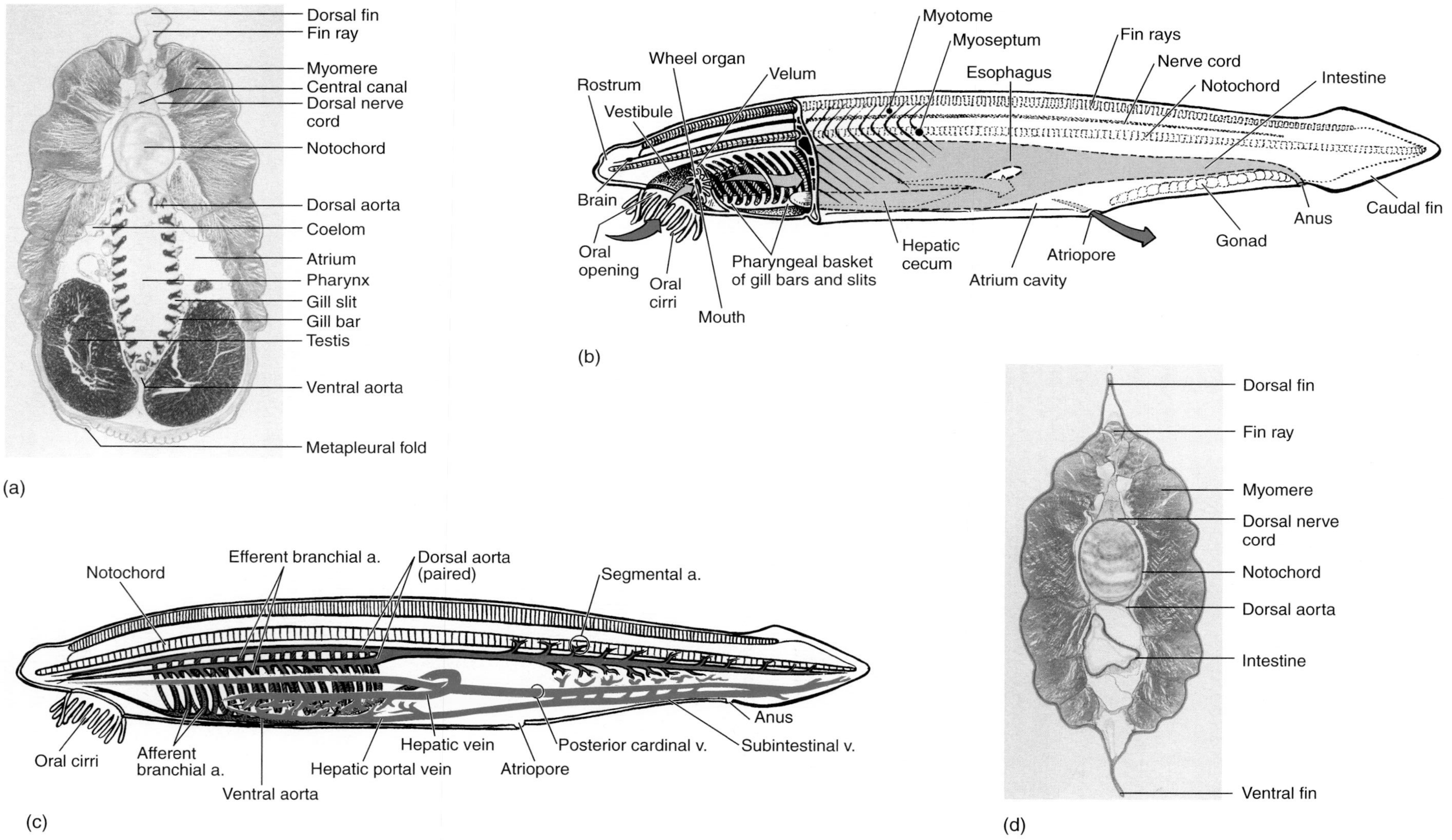

(a)

Dorsal fin
Fin ray
Myomere
Central canal
Dorsal nerve cord
Notochord
Dorsal aorta
Coelom
Atrium
Pharynx
Gill slit
Gill bar
Testis
Ventral aorta
Metapleural fold

(b)

Intestine
Caudal fin
Anus
Notochord
Nerve cord
Fin rays
Gonad
Myotome
Myoseptum
Esophagus
Atriopore
Atrium cavity
Hepatic cecum
Pharyngeal basket of gill bars and slits
Velum
Wheel organ
Rostrum
Vestibule
Brain
Oral opening
Oral cirri
Mouth

(c)

Notochord
Dorsal aorta (paired)
Efferent branchial a.
Segmental a.
Anus
Subintestinal v.
Posterior cardinal v.
Atriopore
Hepatic vein
Hepatic portal vein
Ventral aorta
Afferent branchial a.
Oral cirri

(d)

Dorsal fin
Fin ray
Myomere
Dorsal nerve cord
Notochord
Dorsal aorta
Intestine
Ventral fin

below the pharynx. **Afferent branchial arteries,** each with a contractile bulb at its base, leave the aorta and pass up through the gill bars, where gas exchange takes place. Pumping occurs by contraction of the branchial bulbs and the ventral aorta. Blood from the gills is collected by the **efferent branchial arteries** and flows into paired (right and left) **dorsal aortas** that join to form a median dorsal aorta posterior to the pharynx. This vessel supplies blood to the posterior tissues and to the muscles through the **segmental arteries.**

Locate the gonad located beneath the intestine in the caudal region (fig. 22.4b). The sexes are separate in lancelets and the gonad may be either an ovary or a testis. In sexually mature individuals it enlarges and may fill the atrium. Eggs or sperm are released into the atrium, and pass out through the atriopore with feeding water currents. Fertilization is external.

Look at the cross sections of *Branchiostoma,* in figure 22.4a and d. Locate the notochord (roughly just above the center in the section). Dorsal to it is the nerve cord. Observe the nerve cord closely and note the hollow center, or **central canal,** within the nerve cord (a characteristic of chordates). Beneath the notochord will be the tube of the digestive system. In the section through the gill area, note the arrangement of the gill bars, slits, and atrium.

If you reflect on the general anatomical plan of amphioxus, its similarities to a fish are obvious. Except for the segmented backbone, jaws, and appendages, its general body plan is that of a fish. The similarity to the tunicate tadpole larva also should be apparent. This similarity probably means that both amphioxus and cartilaginous fish are related ancestrally through some organism much like a tadpole larva. Each group has evolved since then to the forms we see today. During the long periods of time (500×10^6 years ago) since the Ordovician, when fish appeared as the first vertebrates, other evolutionary changes have occurred to give rise to the other classes of vertebrates, including amphibians, reptiles, birds, and mammals.

Vertebrata

In the subphylum Vertebrata, all four chordate characteristics are present at some stage of the life cycle. Also in vertebrates, the notochord is replaced by the **vertebral column** (backbone) surrounding the nerve cord. Vertebrates are highly cephalized with well-developed sense organs and a distinct brain located at the anterior end of the dorsal nerve cord. The brain is encased in a skull, which together with the vertebral column, makes up the axial skeleton. Most vertebrates have two pairs of appendages supported by an appendicular skeleton. In contrast to the exoskeleton of many invertebrates, the vertebrate skeleton is a living endoskeleton capable of growth and self-repair. What are the advantages of this type of skeleton for a terrestrial existence? Describe why this type of skeleton places limits on the size of a vertebrate.

The circulatory system of vertebrates is a closed system consisting of a ventral heart and a closed vessel system of arteries, veins, and capillaries, with the respiratory pigment, hemoglobin, contained within red blood cells. The sexes are usually separate and reproduction is sexual.

This subphylum contains the largest number of chordate species. It includes seven classes: (1) Agnatha (jawless fishes), (2) Chondrichthyes (cartilaginous fishes), (3) Osteichthyes (bony fishes), (4) Amphibia, (5) Reptilia, (6) Aves, and (7) Mammalia. You will dissect a perch, a representative of the 30,000 species of bony fish. In later labs, you will study vertebrate anatomy in a mammal, the fetal pig.

External Anatomy of a Perch

Obtain a perch and wash it in tap water to remove the preservative.

> # CAUTION
> **Be careful:** the dorsal fin contains spines that can give you a painful puncture wound.

Note the main regions of the body: head, trunk, and tail. Find the two **dorsal fins** and the ventral **anal fin** (fig. 22.5). Note the paired **pectoral** and **pelvic fins.** The thin membranes of the fins are supported by skeletal elements, the **fin rays.** Cut the tips of the dorsal fin rays off with scissors to protect your hands during the subsequent dissection.

Pry open the mouth and examine the internal edge of the upper surface to see the **buccal valve,** a small flap of tissue that seals the mouth opening when the jaws are closed.

On each side of the head are the **opercula,** large plates used to ventilate and protect the gills. Along the posterior and ventral edges of the opercula, find the **branchiostegal membranes.** Fish pump water over their gills by the following mechanism. When the opercula are raised with the mouth open, the branchiostegal membranes close off the posterior gill openings so that water flows through the mouth into the gill chambers. The mouth is then closed and sealed by the buccal valve. The opercula are then lowered, forcing water out the gill openings. Do you think that the gills of a fish are more efficient in gathering oxygen than the gill bars of the lancelet? Why?

Figure 22.5   Internal organs of a perch.

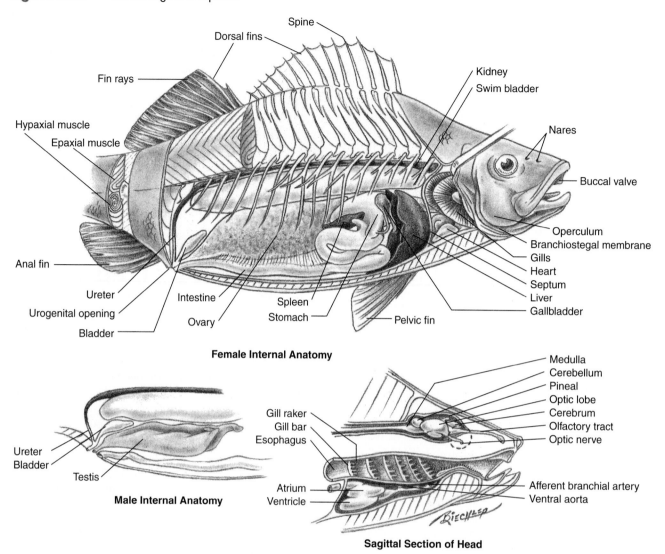

Female Internal Anatomy

Male Internal Anatomy

Sagittal Section of Head

Examine the side of the body to find the **lateral line.** Sense receptors located here are sensitive to water pressure and weak electric currents.

Internal Anatomy of a Perch

To expose the internal organs, lay the perch on its left side and insert the tips of your scissors just anterior to the origin of the anal fin. Cut forward along the midventral line to the pectoral fins. Make a second cut parallel to the first but above the lateral line on the same side of the perch. Now connect the two cuts with a third at the posterior edge. Lay the right body wall back, free it from underlying tissues, and remove it.

Note the various organs suspended in the coelom by **mesenteries,** thin, transparent projections of the shiny **peritoneum,** a tissue that lines the body cavity. Dorsal to the other organs is the long **swim bladder** (fig. 22.5). The air content of the bladder is actively regulated to maintain the buoyancy of the fish, allowing it to remain motionless in the water without settling to the bottom.

Remove the right operculum to expose the gills attached to the **gill arches** of the **pharyngeal wall.** Insert the tips of your scissors into the mouth and cut through the angle of the jaw back through the gills to expose the inside of the pharynx.

Examine the inside of the pharynx and visualize how water entering the mouth flows through the slits in the pharyngeal wall, over the gills on the gill bars, and out through the external gill openings. On the inside surface of the gill bars, note the **gill rakers,** which act as crude sieves in this species and direct ingested food toward the opening of the esophagus. In species that feed on suspended algae and detritus, the gill rakers may be highly developed and resemble a fine-tooth comb, which effectively removes small particles from water passing through the pharynx. In a predatory species, such as the perch, the gill rakers are widely spaced.

Pass a blunt probe down the **esophagus** into the **stomach** located dorsal to the liver. Trace the **intestine** from the stomach to the **anus.** The **spleen** (not part of the digestive system) is the triangular-shaped organ located in the region

Figure 22.6   Circulatory system of perch. Major arteries are in red and veins are in blue.

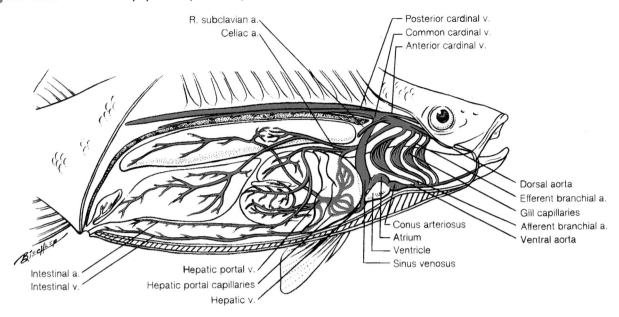

R. subclavian a.
Celiac a.
Posterior cardinal v.
Common cardinal v.
Anterior cardinal v.
Dorsal aorta
Efferent branchial a.
Gill capillaries
Afferent branchial a.
Ventral aorta
Conus arteriosus
Atrium
Ventricle
Sinus venosus
Intestinal a.
Intestinal v.
Hepatic portal v.
Hepatic portal capillaries
Hepatic v.

between the stomach and the small intestine. Lift the lobes of the **liver** to find the globular, dark green **gallbladder.** What is its function?

Cut open the stomach to determine what your perch had eaten. List the contents below.

What adaptations have you observed so far in this dissection that would explain why bony fish, in general, are larger in size than lancelets?

Fish have a closed circulatory system. Find the **two-chambered heart** anterior to the **septum** in front of the liver, and ventral to the pharynx (fig. 22.6). The anterior chamber is the muscular **ventricle,** which pumps blood to the gills via the **ventral aorta** and **afferent branchial arteries.** Blood rich in oxygen and low in carbon dioxide leaves the gills via the **efferent branchial arteries.** Find the **dorsal aorta** dorsal to the gills. Blood collected from the gills flows via this major vessel to the rest of the body.

Blood is collected from the tissues via the **cardinal vein,** which runs just ventral to the vertebrae and from the digestive system via the **hepatic portal vein.** The latter flows into capillaries in the liver and is collected by the **hepatic vein.** The cardinal vein and hepatic vein flow into the **sinus venosus,** which returns blood to the **atrium** of the heart.

The excretory system consists of the paired **kidneys,** which are located dorsal to the swim bladder behind the peritoneum. Remove the swim bladder and find the dark brown organs firmly attached to the dorsal wall of the body cavity and extending almost its entire length. Along the median border of the kidneys are the **urinary ducts.** As they pass posteriorly, they unite into a single duct, which passes ventrally to the urinary opening. A urinary **bladder** may be seen branching from the duct's distal end.

Perch have distinct males and females. Females have a single large **ovary** lying between the intestine and swim bladder. If your fish is a female and was reproductively "ripe" when collected, the ovary will be filled with yellowish eggs. Eggs leave the posterior of the ovary via an oviduct that leads to a urogenital opening posterior to the anus.

TABLE 22.1 Comparison of the characteristics of various phyla

Phylum	Symmetry	Coelom	Digestive System	Circulatory System	Respiratory System	Nervous System	Segmentation
Porifera	None	None	Intracellular	Absent	Body surface	Absent	Absent
Cnidaria	Radial	None	Gastrovascular cavity (sac) (incomplete)	Absent	Body surface	Ladder-like	Absent
Platyhelminthes	Bilateral	Acoelom	Gastrovascular cavity (sac) (incomplete)	Absent	Body surface	Ladder-like	Absent
Nematoda	Bilateral	Pseudocoelom	Complete; (tube) mouth develops from blastopore	Closed	Body surface or gills	Ventral cord	Present
Annelida	Bilateral	Schizocoelom	Complete; (tube) mouth develops from blastopore	Closed	Body surface or gills	Ventral cord	Present
Mollusca	Bilateral	Schizocoelom	Complete; (tube) mouth develops from blastopore	Open	Body surface or gills	Cord	Present
Arthropoda	Bilateral	Schizocoelom	Complete; (tube) mouth develops from blastopore	Open	Body surface or gills or book lungs or trachea	Ventral cord	Present
Echinodermata	Secondarily radial	Enterocoelom	Complete; (tube) anus develops from blastopore	None	Body surface	Not easily seen	Absent
Chordata	Bilateral	Enterocoelom	Complete; (tube) anus develops from blastopore	Closed	Lungs or gills	Dorsal cord with cephalization	Present

If your fish is a male, paired **testes** will be visible caudal to the stomach. A **vas deferens** passes caudally in a longitudinal fold in each testis. These fuse to form a single duct that passes to a genital pore posterior to the anus.

To observe the brain of the perch, shave the skull bones off the front of the head to expose the brain. The major divisions of the brain are: **telencephalon, optic lobes, cerebellum,** and **medulla oblongata** (fig. 22.6).

Table 22.1 summarizes the major evolutionary themes that you have observed in your brief survey of the animal kingdom in lab topics 19, 20, 21, and 22.

Learning Biology by Writing

All general biology textbooks indicate that chordates have a set of common characteristics: (1) a notochord; (2) a dorsal hollow nerve cord; (3) gill slits at some stage of the life cycle; and (4) a post-anal tail. In a short essay, describe the evidence you have from your lab work that supports or refutes this generalization.

Your instructor may also ask you to answer the following Lab Summary and Critical Thinking questions.

Internet Sources

Use the WWW to locate information about current research being done on cephalochordates. Use your browser to connect to www.google.com. Type in the search terms, *Branchiostoma* and research. Scan several of the sites that are listed and write a few paragraphs summarizing what you think are the most interesting sites. List the URL's at the end of your summary.

Lab Summary Questions

1. Why are echinoderms and chordates grouped together?
2. Describe the water vascular system of a sea star.
3. Why are tunicates considered to be chordates?
4. Make a table of the organ systems in a sea star and a perch. When an organ system is absent, explain how its function is carried out in the organism.
5. Why is *Branchiostoma* considered an important evolutionary link among the chordates?
6. Have you observed any evidence that would allow you to refute the hypothesis that all choerdates have a notochord, postanal tail, gill slits, and a dorsal nerve cord? Explain.

Critical Thinking Questions

1. Although echinoderms are considered advanced, they are simple organisms that often lack some traits found in "lower" phyla. What are some of these traits?
2. What are the similarities between a tunicate tadpole larva and a lancelet?
3. Why are animals as different as a sea squirt and an eagle classified into the same phylum?

LAB TOPIC 23

Investigating Plant Tissues and Primary Root Structure

Supplies

Preparator's guide available on WWW at
   http://www.mhhe.com/dolphin

Equipment

Compound microscopes
Dissecting microscopes

Materials

Wisconsin Fast Plants planted two weeks before. Root
   system to be observed in this lab.
Green onions with root sprouts about one week old to
   be used as unknowns
Demonstrations of tap- and fibrous root systems
Corn root system from germinating corn
Fresh onions with 2-inch roots
Fresh apples and pears
Fresh celery
Prepared slides
   Ranunculus root tip, longitudinal section
   Ranunculus root, cross section
   Corn root, cross section
   Alfalfa *(Medicago)* stem cross section
   Spiderwort *(Tradescantia)* leaf trichomes
      (whole round)
   Dicot leaf cross section
   Grape stem *(Vitis)* cross section
   Pumpkin or grape stem macerate
   Pine wood macerate
Razor blades and forceps
Slides and coverslips

Prelab Preparation

Before doing this lab, you should read the introduction
and sections of the lab topic that have been scheduled
by the instructor.
   You should use your textbook to review the
definitions of the following terms:

   cambium
   casparian strip
   collenchyma
   companion cell
   dermal tissue
   endodermis
   epidermis
   ground tissue

   meristem
   parenchyma
   pericyole
   periderm
   phloem
   root cap
   root hair
   sclerenchyma
   sieve tube
   stele
   tracheid
   vascular tissue
   vessel
   xylem

   You should be able to describe in your own words
the following concepts:

   Plant tissue systems
   Primary growth
   Secondary growth
   How water enters root vascular systems

   As a result of this review, you most likely have
questions about terms, concepts, or how you will do
the experiments included in this lab. Write these
questions in the space below or in the margins of the
pages of this lab topic. The lab experiments should
help you answer these questions, or you can ask your
instructor for help during the lab.

Objectives

1. To study the basic tissues and cell types found in
   plants
2. To observe the structural differences in primary
   roots of monocots and dicots
3. To recognize the basis for primary and secondary
   growth in roots
4. To identify unknown plants by their root anatomy

Background

Following their evolution from aquatic algae, land plants have adapted to a wide variety of habitats, producing a great impact on the evolution of life. By dominating most terrestrial ecosystems, they have influenced the evolution of bacteria, fungi, and animals that depend on them in various ways for food and shelter. They have also contributed to major changes in the physical environment, not the least of which is their role in the oxygen cycle and the associated UV-screening ozone layer.

In the next three lab topics, you will study the development of the plant body. As you saw in Lab Topics 15 and 16, the plant body has evolved into two systems, the **root** system which absorbs, anchors, and stores, and the **shoot** system, the stem, which positions the leaves (the photosynthetic organs)—and flowers (the reproductive structures). In this lab topic, you will look at the cellular bases for these systems and study roots in angiosperms.

Although the root and shoot systems vary considerably among angiosperms, all consist of the same cell types and tissues. If you compared a leaf to a root or a monocot to a dicot, you would not find different cell types. Instead, you would find that there are different arrangements and proportions of basic cell types. All plants consist of four tissue systems: the dermal system, the vascular system (only in vascular plants), the ground system, and the meristematic systems.

The **dermal tissue system** is the surface of the root and shoot systems. It protects them from invasion by microorganisms, prevents desiccation, allows respiratory gas exchange, and may secrete noxious compounds that defend against herbivores. The **vascular tissue system** allows for water transport throughout the plant body. Water is important for several reasons. It is the source of hydrogen atoms for carbohydrate synthesis in photosynthesis; it is the medium for most cellular chemical reactions; it is the transport medium for mineral nutrients and organic compounds so that they can move from one plant region to another; and it is the primary agent causing enlargement of plant cells following cell division. The **ground tissue system** consists of most of the cells found in plants. The pith of stems and roots, the flesh of fruits, and the photosynthetic layers of leaves are examples of the ground tissue system. Far more than being a "filler" tissue, the ground system includes most of the metabolically active cells found in plants.

The **meristematic tissue system** is not considered a tissue by some, but its unique characteristics and importance in the plant life cycles warrant its elevation to this status. Meristems contain cells that continue to divide, producing all other cells of the plant body. Division of cells in **apical meristems,** located at the tips of the roots and shoots, cause what is called **primary growth,** elongation of the root or shoot. When an apical meristem cell divides, one daughter cells remains meristematic and pushes forward with the tip. The other daughter differentiates into one of three promeristems: protoderm, procambium, and ground meristem. Division of the promeristems, in turn, produces the tissue systems of the shoot and root, and their growth lengthens the shoot or root. The **protoderm** produces the outer covering of the plant. The **ground meristem** gives rise to the metabolically active cells of the plant. The **procambium** differentiates into two cylinders of tissue in the stem and roots. Called **cambiums,** these cylinders of tissue contain cells that continue to divide. They are responsible for **secondary growth,** growth that increases the diameter of stems and roots. One cambium just beneath the outer surface, the **cork cambium,** produces the bark that replaces the epidermis. The other is the **vascular cambium.** It will divide to produce additional vascular tissues. The additional vascular tissues allow roots to grow out from the plants, gaining access to new sources of minerals and water. The vascular tissues also enlarge root and stem diameters so that these do not collapse under the weight of branches and leaves. Secondary growth has evolved as an adaptation to increased competition in complex communities such as forests. Plants with the capability to grow tall can outcompete other species and gain access to light.

Before studying the differences between various tissues types, one should review the elementary organization of plant cells. Plant cells are eukaryotic cells. Therefore, they contain a nucleus and all of the membranous organelles that you would expect. In addition some plant cells, but certainly not all, contain chloroplasts. In addition, maturing plant cells usually contain a prominent central vacuole. Beside storing water and minerals, vacuoles are part of the growth strategy of plant cells. Surrounded by a cell wall, plant cells can be restricted in their ability to grow. However, by quickly storing water in a vacuole, a plant cell can rapidly enlarge before the wall is fully synthesized, creating room for growth. Later as the cell adds cytoplasm, it need only release an equivalent volume of water to keep from becoming crowded in its "home."

The living portion of a plant cell, the cytoplasm and nucleus, is the **protoplast.** The protoplast and cell wall together make up the plant cell. Cell walls are made when protoplasts deposit cellulose microfibrils outside the cell membrane in a matrix of resin-like materials. Modern engineering studies indicate that such composite materials have great strength for their weight; for example, various fiberglass products are analogous, containing glass fibers embedded in a plastic matrix. The first cell wall that is deposited is called the **primary cell wall** (figure 23.1). The primary cell walls of adjacent cells are held together by pectin, a glue-like polysaccharide, forming a layer called the **middle lamella.** Some cells have only a primary cell wall and it can vary from thick to thin. Other cells produce a **secondary cell wall.** Because it is made after the primary wall, it is located on the inside of the primary wall. It consists of cellulose microfibrils laid down in layers, each with a different orientation than the previous layer. The layers are impregnated with **lignin,** a polymer of smaller phenolic compounds which forms a rigid matrix. After cellulose lignin is the second most common material found in plant

Figure 23.1 Transmission electron micrograph showing primary and secondary plant cell walls. Middle lamella is composed of pectin.

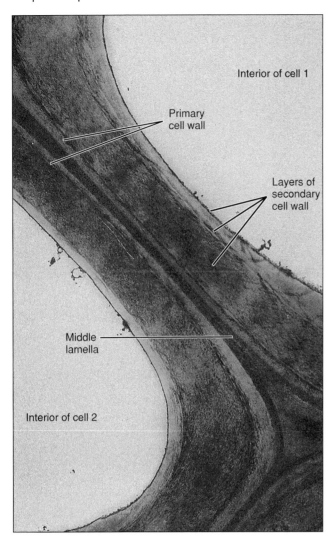

Interior of cell 1

Primary cell wall

Layers of secondary cell wall

Middle lamella

Interior of cell 2

cells. Wood is composed primarily of cellulose and lignin, *i.e.* secondary cell wall material.

Some plant cells retain their protoplasts at maturity. In others, especially xylem, the protoplasts dies and it is the remaining cell wall that performs the cell's intended function. Therefore, when studying plant material, the word *cell* is used in a slightly ambiguous fashion. When looking at mature plant material, some cells may be living protoplast with cell walls while other cells will be only the remaining cell wall after the protoplast has died. Those cells that retain their protoplasts at maturity also exhibit another special anatomical feature. Living cells in plants tend to be interconnected by tiny cytoplasm bridges from one cell to the next (Figure 23.2). These are called **plasmodesmata.** Through plasmodesmata, the cytoplasm of one cell is continuous with the cytoplasm of another. Consequently, materials can move quickly from one cell to another without passing through cell membranes that can be bottlenecks to molecular transport. Most living cells in plants may be interconnected in this way.

You will learn to recognize the basic cell types and tissues that are found in angiosperms. The knowledge gained will be applied to studying root structure.

Plant Cells and Tissues

In this section, you will look at several examples of the basic tissue systems in plants. The characteristics of the cell types are summarized in figure 23.3.

Dermal Tissue System. The cell layer that covers the primary plant body is the **epidermis.** In the primary plant body, it arises from division of cells in the protoderm. The epidermis has a number of functions, including gas exchange, protection against desiccation, absorption of water and minerals, and secretion of wax. Epidermal cells retain their protoplasts at maturity. The outer walls of epidermal cells are covered by a fatty material called cutin that forms a **cuticle.** Often a wax coat is found outside the cuticle. The cuticle is thick in plants that grow in dry environments and thin in those the grow in wet environments. What do you think is its function?_____

_____ The whitish deposit on the surfaces of some house plants is an accumulation of this epicuticular wax. Look at examples in the lab. What happens when you take a tissue and polish the surface of one of these plants?

▶ Prepare a slide of onion epidermis. Onions are modified stems with layers of epidermal cells separating parenchyma cells. Take a slide and place a drop of water on it. Cut an onion half and peel the outer layers to expose a moist inner layer. With forceps, strip off a small piece of the outermost layer from the central core, taking only a cellophane-thin piece. Put it in the water on the slide and add a coverslip. Observe. Below, sketch several of the cells. How does the three-dimensional shape of these cells reflect the function of epidermis?

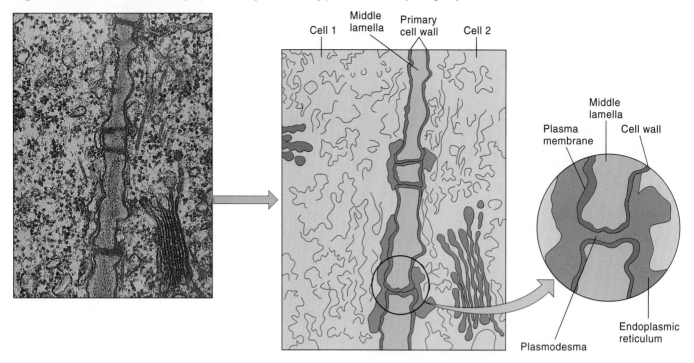

Epidermal cells are often modified to produce unicellular and multicellular projections from the plant surface, called **trichomes.** They have a variety of functions. On leaf surfaces, they reflect light and cool the leaf; they can absorb moisture; and they can be defense mechanisms. Root hairs are trichomes specialized for water and mineral absorption. Get a prepared slide of *Tradescantia* (spiderwort) from the supply area and use your microscope to observe the trichomes on the surface of its leaves. You may have had contact with nettles, or stinging plants. The leaves of these plants have glandular trichomes on their surface, and when you brush against them, the pointed trichomes penetrate your skin, releasing a toxin that can cause a painful (though temporary) swelling.

Obtain a cross section of a leaf and look at the epidermis through your compound microscope. The cuticle may not be visible because it is often dissolved by the solvents used in making a slide. As you scan along the lower leaf surface, you will see the specialized epidermal **guard cells.** What do you think is their function?

The epidermis is a primary tissue system and is short lived in the stems and roots of plants with secondary growth. As the plant stem or root increases in girth from within, due to an increase in the amount of vascular tissue, the epidermis is stretched and splits. However, before this happens, a secondary dermal tissue, the **periderm (= bark),** develops and takes over the functions of the epidermis. You will see this when you study roots and stems.

Ground Tissue System.   Arising from cell divisions in the ground meristem of the primary plant body, ground tissues are the most common types of cells found in plants.

TABLE 23.1 Basic tissue types found in plants

Cell Types	Description
Dermal tissue	Includes living epidermal cells, trichomes, root hairs, and guard cells; protects internal cells and functions in water absorption and transpiration; arises from apical meristem; in secondary growth, epidermis is replaced by periderm arising from cork cambium.
Parenchyma	Component of ground tissue; most common living cells in plant body and found in leaf mesophyll, herbaceous stems, fruits, and rays of woody plants; functions in storage, secretion, and wound healing; arises from apical meristem and retains capability to divide (typically no secondary cell wall) and differentiate into other tissue types.
Collenchyma	Component of ground tissue; living cells forming distinct strands beneath epidermis in stems and petioles that support the primary plant body; arises from apical meristem.
Sclerenchyma	Component of ground tissue; includes living and dead fibers and sclerids; thick secondary cell walls with lignin provide support and mechanical protection; found throughout the plant, often in association with xylem and phloem; arises from apical meristem.
Xylem	Component of vascular tissue; includes tracheids and vessels; functions in transport of water and minerals from roots to shoots; primary xylem arises from apical meristem and secondary xylem from vascular cambium; protoplast produces thick secondary cell wall and then dies; lumen of dead cells provides pathway for water transport.
Phloem	Component of vascular tissue; includes living sieve tube members and companion cells; functions in distributing photosynthetic products and storage compounds throughout plant; primary phloem arises from apical meristem and secondary phloem from vascular cambium.
Meristems	Cells that continue to divide and grow, forming tissues described above; apical meristems at root and stem tips allow growth in length (called primary growth); lateral meristems (also called cambiums—vascular and cork) found in roots and stems allow growth in girth (called secondary growth). (See fig. 23.3.)

Figure 23.3 Locations of meristems in a dicot.

They function in basic metabolism, storage, and support in plants. These functions are carried out by three basic cell types: parenchyma, collenchyma, and sclerenchyma.

Parenchyma cells are the most common in plants. They retain their protoplasts at maturity and can live for long a time. Parenchyma cells in some long-lived plants are estimated to be over 100 years old. Parenchyma cells have only a primary cell wall. They retain the ability to divide when they are old and are active in wound healing. When stimulated by wounding, the parenchyma cells dedifferentiate and divide, and then differentiate into new tissues. For example, when a root develops on a cutting, the new root results from parenchymal cell division. Specialized parenchyma cells called chlorenchyma cells perform photosynthesis. Others called aerenchyma cells are characterized by prominent intercellular spaces which provide pathways for gas diffusion in active regions of the plant, such as the spongy mesophyll of the leaf.

To observe parenchyma cells that are active in storage, take an apple or pear and cut it in half. Mount a very small dab of the pulp in a drop of water on a slide and tease it apart to make a cell suspension. Add a coverslip, press on it to spread the cells, and observe with your microscope. Sketch a few cells below. What is their shape?

Now get a prepared slide of a cross section of an alfalfa *(Medicago)* stem and observe with your microscope. Look at the large cells in the center of the section. This is the pith of the stem and is composed of large, rounded parenchyma. Look for the intercellular spaces between adjacent cell walls. Such spaces are characteristic of this tissue. Are the cells of uniform size? _____

Collenchyma cells are another type of ground tissue. These living cells have unevenly thickened primary cell walls lacking lignin. They are found in growing leaves and their petioles, and elongating stems often have collenchyma cells arranged as cylinders beneath the epidermis. This arrangement produces greater support than a solid mass of cells located at the center of the stem.

Collenchyma cells can be seen in the cross section of the alfalfa stem you just looked at. Look at the slide again and note how the section is angular. Look in the corners of the angle to the outside of the vascular bundles. Just beneath the epidermis are cells surrounded by irregularly thickened cell walls. These are strands of collenchyma. Describe how collenchyma cells differ in appearance from the parenchyma cells in the pith.

Obtain a piece of celery *(Apium).* The part of celery that we eat is the petiole of a compound leaf. Note the ribs on the outside of the "stalk." Use your fingernail to peal away a rib and then strip it out along the length of the petiole. This is a collenchyma strand. Grab opposite ends of the strand in each hand and try to break it. Describe how tough it is.

The third type of ground tissue, **sclerenchyma,** is dead at maturity, lacking protoplasts. Sclerenchyma cells have very thick, rigid, secondary cell walls impregnated with lignin. There are two cell types. **Sclerids** are short and occur singly or in clumps. **Fibers,** the second type, are long, slender cells forming strands. The commercial fibers, hemp, jute, and sisal, used to make ropes, are strands of sclerenchyma fibers.

Sclerids are found in some fruits. The grit that you feel when you eat a pear is due to aggregates of sclerids called stone cells in the fruit pulp. Put some pear pulp on a slide and add a drop of phloroglucinol stain. It stains only cells that have lignin in their cell walls. Tease the cells apart and let the preparation sit for 10 to 15 minutes. Add a coverslip and gently press on it to spread the cells. Observe through your microscope. What is the shape of the stone cells (sclerids)?

Obtain a prepared slide of a cross section of a grape stem *(Vitis).* Look at the fiber bundles through your microscope. Find the small cells with very thick cell walls. These are sclerenchyma fibers. What do you suppose their function is?

Vascular tissues.   Arising from division of cells in the procambium in primary growth and from the vascular cambium in secondary growth, vascular tissue includes both xylem and phloem. These vascular tissues form a continuous conduction system throughout a plant's body. The tubular shape of their cells reflects their function as conduits. Through these tubes, water, nutrients, and minerals move from the roots to other parts of the plant body and photosynthetic products are transported from the leaves to other living, but nonphotosynthetic, cells.

Xylem includes several cell types but two types predominate: tracheids and vessel elements. **Tracheids** are long, narrow spindle-shaped cells, whereas **vessel elements** look like open barrels joined end to end forming continuous tubes called **vessels.** Both types of cells lack a protoplast when mature, and have thick, lignified cell walls. The thick walls, often with spiral ridges, resist collapsing when water is pulled though the xylem by the negative pressure generated in transpiration. The function of the protoplast during xylem development is to make the primary and secondary cell walls and then to die, leaving a conduit for water and dissolved materials movement. Both types of cells have thin portions of the cell wall called **pits,** which allow lateral movement of water between adjacent conduits. Vessel elements have actual **perforations** (holes) in the cell walls.

Obtain a slide of a maceration of pine wood. Gymnosperms generally lack vessel elements in their xylem. It is composed almost entirely of **tracheids,** whose tapered ends overlap to form a conduit. Water moves from tracheid to tracheid through pits. In the left half of the circle that follows, sketch some tracheids. Can you see the pits?

Vessel elements are considered more evolutionarily advanced than tracheids. They are composed of short cell wall sections from different cells that are joined end to end without any cross walls. Vessel elements may allow water movement for distances of up to three meters without passage through a pit. This allow for rapid water movement and may be the reason that angiosperms have become the dominant terrestrial plants.

Compare the pine macerate to a macerate of a pumpkin or grape stem, a dicot. Look for the vessel elements among many other different cell types on the slide. In the right half of the previous circle, add a sketch of some of the pumpkin vessel elements.

Phloem cells transport dissolved organic materials in plants from leaves to other parts of the plant body during photosynthetic periods, and also transport materials from storage areas to other parts of the plants during periods when photosynthesis is not occurring. Phloem consists of two types of cells, sieve tubes and companion cells. **Sieve tubes** are living at maturity, but the protoplasts are somewhat unusual. They lack nuclei, ribosomes, and the Golgi apparatus. Furthermore, sieve tubes are joined end to end with perforated cell walls, called **sieve plates,** between the stacked cells. Extensive plasmodesmata join adjacent cells so that a stack of sieve tube members actually represents one continuous cytoplasm. Dissolved materials are thus able to move long distances without having to cross a cell membrane.

Look again at the slide of a macerate of the pumpkin stem. Look for cells that are joined end to end, but with cross walls. These are **sieve tube elements.** If the angle is right, you may be able to see the holes in the cross walls (= **sieve plates**). Associated with some of the sieve tube cells on the side, you may see small **companion cells.** They contain nuclei and all of the membranous organelles, except chloroplasts, that you would expect to find in plant cells. Extensive plasmodesmata between the companion cell and sieve tube element allow the companion cell to supply many of the metabolic needs of the reduced cytoplasm in the sieve tube element. The lives of the two cells are tightly coupled, and when one dies, the other does as well. In most dicots, the two cells live about a year. If you cannot find these cells on the slide, look at a cross section of *Cueurbit* stem. You should be able to see sieve plates in some of the phloem cells. As you look at the section, note the prominent vascular bundles arranged in a circle around the hollow center. Use high power to observe a single vascular bundle. In the center of a single bundle is heavy walled xylem. On two sides (toward center and outside of stem) are phloem cells. If you look closely, you will see sieve plates in some of the phloem. You should also be able to see smaller companion cells next to a sieve tube element. Sketch some sieve tube elements and companion cells below.

You are now finished with your study of tissues and cell types in vascular plants. Before continuing to the next section, where you will look at the organization of tissues in the root system, clean up your work area. Return all prepared slides to their trays and clean any slides that you made.

Root Systems

A plant requires not only carbon dioxide, oxygen, and light to grow but also sources of water and dissolved minerals. Algae obtain water and minerals by direct absorption across their cell surfaces, whereas terrestrial plants, except for the mosses, have root systems that absorb these materials from the soil.

Some roots serve as storage depots for photosynthetic products, and these reserves fuel growth over winter and during leafing out in the spring. The tapping of maple trees in the spring takes advantage of the maple's capacity to store carbohydrates in its roots one season and mobilize them as sugars in another. Roots anchor and stabilize the shoot in soil, an impressive accomplishment in a mature oak that may withstand wind gusts of over 50 miles per hour.

Compared to stems, we know much less about roots because they are more difficult to study. Their role in ecosystems where they unlock the soil's store of ions such as phosphate, sulfate, and calcium cannot be overstated. Herbivores not only gain energy and essential amino acid and vitamins from eating plants, they also gain ions necessary for their metabolism. Some scientists have observed that because of roots, animals, including humans, do not have to eat soil!

Root systems are often highly developed and account for 80% of the biomass of the plant. For example, a researcher found that a mature winter rye plant had a root system with a total length of 387 miles and a surface area of 2,554 square feet, containing 14 billion root hairs with a combined length estimated to be 6,603 miles. The most amazing fact, however, is that all of this was contained in less than two cubic feet of soil. Obviously, intimate contact occurs between soil and plant.

Mere absorption of water and minerals, however, is not sufficient for large terrestrial plants to survive. These materials must be transported to the actively metabolizing tissues where water is used in photosynthesis and where ions are necessary cofactors in many enzyme-catalyzed reactions. A transport system is also necessary to move photosynthetic products from the leaves to nonphotosynthesizing tissues, such as the roots, to supply their energy and growth requirements.

Two specialized tissues, **xylem** and **phloem,** perform these functions in vascular plants, or tracheophytes. In simple terms, the xylem contains nonliving tubular cell walls (left in place after the protoplast dies) which transport water and dissolved minerals from the roots to the leaves; the phloem contains living tubular sieve tube cells, which actively distribute photosynthetic products throughout the plant. Depending on whether the plant is a monocot or a dicot, these tissues will be found in different arrangements or locations in the root and stem. Among the dicots, there are further differences, depending on whether the plant is herbaceous or woody.

Whole Roots

Your lab instructor will set up demonstrations of the types of root systems found in plants.

Examine the specimen showing a **taproot** system consisting of a simple long conical root that penetrates the soil with many lateral branches from the main axis. Taproots are characteristic of gymnosperms and many dicots. Side branches bear root hairs at their tips. The parenchyma cells of taproots store carbohydrates; some taproots are used by humans for food. Can you name five?

Now look at the **fibrous root** system of a grass. Note how the root system repeatedly branches to form a complex network. Fibrous roots are characteristic of monocots.

In addition to their primary root system, many plants have **adventitious** roots, which arise as lateral extensions from the base of the stem. Examine the root system of a mature corn plant to see these.

If you planted Fast Plants two weeks before this lab, harvest one quad to observe the roots. Soak the quad in water to loosen the growth medium. Use a stream of water to knock off particles as you gently pull on the stem. After freeing the plant from the quad, lay the plant in a dish of water. What kind of a root system does it have? Can you see root hairs? How far back from the tip are they found?

Root Histology

In this section, you will observe both longitudinal and transverse sections of roots from dicots and monocots.

Longitudinal Section of a Young Root

Examine a prepared slide of a longitudinal section of the dicot buttercup (*Ranunculus*) under low-power magnification. Compare your specimen to the diagram in figure 23.4. Note how the cell shapes change as you scan from the root tip toward the base of the root.

Identify the **root cap,** a thimblelike covering of cells that protects the tip of the root as it pushes through the soil.

Figure 23.4   Tissue organization in a young root.

Mature Region

Stele

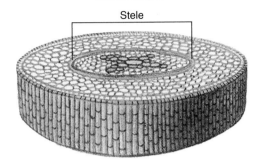

Region of Differentiation

Xylem
Phloem
Endodermis
Pericycle
Cortical parenchyma
Epidermal cells
Root hairs

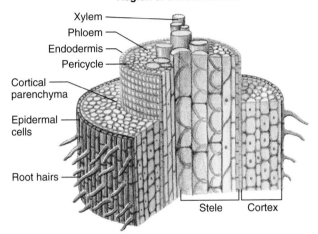

Stele   Cortex

Region of Elongation

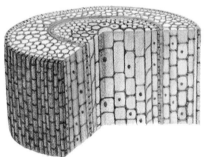

Region of Cell Division

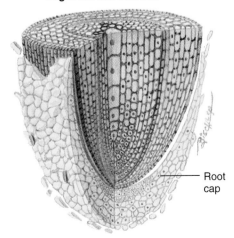

Root cap

The cells of the root cap are constantly renewed. It has been estimated that an elongating corn root can shed 10,000 root cap cells in a day. Once formed, they live four to nine days before they die, secreting a slimy material, called mucigel, that also lubricates the passage of the root through the soil. It is estimated that 20% of the carbohydrate made in photosynthesis gets used to make mucigel. Root growth, obviously, must contribute substantial amounts of carbohydrate to the soil where it supplies energy for the growth of soil fungi and bacteria after serving its lubrication function.

In the root tip behind the root cap is the **apical meristem,** the layer of cells that divides during root growth. You studied mitosis in the cells from this region in onion roots in Lab Topic 8. This tissue and the cells derived from it are responsible for the root's **primary growth,** the elongation of the root. After cells are formed by division, they increase in size in the **region of elongation** behind the meristematic region. Cells in this region can increase in size 150-fold by water uptake. The elongation of these cells pushes the root forward up to 4 cm per day through the soil. Root growth is indeterminate, stopping only when the plant dies.

Further up the root, note how cellular differentiation is apparent. Several types of cells should be visible. The **epidermis** is the outer cell layer of the root. As the tip of the root penetrates new soil and epidermal cells mature, **root hairs** grow out from the epidermis and capture previously untapped water and mineral resources of the soil. Root hairs are epidermal cells (trichomes) that are specialized for absorbing water and dissolved minerals. The roots of pines, birches, willows, and oaks often do not have well-developed root hairs. Instead, they have mycorrhizal roots in which filaments of a symbiotic fungus carry out the functions of root hairs. The fungal hyphae extend into the soil and penetrate between cells of the root cortex. They convey water and minerals into the root from the soil. (See fig. 17.9.) Plants may die when transplanted because the delicate root tips, with their root hairs, are torn off or dry in the moving process. Note the regions of **absorption** marked by the root hairs. How many cells are in a root hair?_____ (Look for cross walls.)

The **cortex** consisting of **parenchyma** cells extends from beneath the epidermis to a central vascular cylinder, the **stele.** The parenchyma cells of the cortex store starch and other materials in many plants. Note the loose arrangement of cells in the cortex. Water and dissolved minerals can travel through these spaces. Are the parenchyma cells normally living (with a protoplast) or dead?_____

The stele is composed primarily of **xylem** and **phloem,** which conduct materials from and to the roots. Parenchymal cells, also found in stele, are living cells with relatively thin walls. They can divide to repair wounds and are physiologically active in storage. In young roots such as this, the xylem and phloem are primary tissues derived from the apical meristem. As the root ages and thickens, secondary xylem and phloem will develop from a vascular cambium.

Figure 23.5 Pathway of water and dissolved mineral absorption in a dicot root. Radial walls of endodermal cells are impregnated with the wax suberin, preventing water from passing between cells as it moves from the cortex to the stele. The only pathway for water is through the waxless tangential surface walls and the protoplasts of the endodermal cells, allowing them to regulate water uptake.

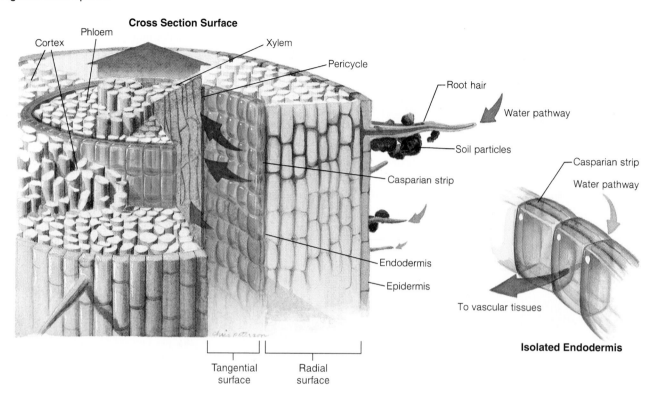

Cross Section Surface

Cortex
Phloem
Xylem
Pericycle
Root hair
Water pathway
Soil particles
Casparian strip
Endodermis
Epidermis

Casparian strip
Water pathway
To vascular tissues

Isolated Endodermis

Tangential surface
Radial surface

When this happens, the epidermis is shed and replaced by the **periderm,** a bark-like covering. Because the root hairs are lost at this time, the older portions of roots with periderm do not absorb water and minerals.

Cross Section of Roots

Dicot Roots

Obtain a prepared slide of a cross section of a buttercup root *(Ranunculus)* and look at it with low power. Identify the epidermis, cortex, and stele as you did in the longitudinal section. Much of the bulk of young roots is contributed by the parenchyma cells of the cortex.

The cylindrical stele should be readily visible at the center of the root (fig. 23.4, top section). Examine it with the high-power objective. At the center of the stele the large, heavy-walled **xylem** tissue may be seen in an x-like configuration, although it will sometimes have a symmetry based on three or five radiating poles. Between the radiating poles of the xylem locate the phloem. In dicots, a **vascular cambium** develops between the primary xylem and phloem. During secondary growth of dicot roots, this cambium produces new **secondary xylem** to the inside and **secondary phloem** to the outside. Eventually, the addition of vascular tissue forms expanding concentric rings of secondly xylem and phloem. The expanding central cylinder

crushes the primary phloem, cortex, and epidermis that were on the outside of the primary root. These layers slough off as a secondary growth continues. The pericycle also produces another cambium, the **cork cambium.** Division of its cells produces a bark layer on the roots.

The outer boundary of the stele is composed of two cell layers, the pericycle and endoderm. The **pericycle** is a narrow zone of parenchymal cells just outside the phloem. These cells maintain their ability to divide and form a meristematic tissue. Just outside of the pericycle is the **endodermis.** It is a physiologically important boundary that regulates water and mineral absorption. With a shape like a flattened box, the rectangular cells of the endodermis have a cell wall on each of their six faces; four sides have a band of waxy substance called **suberin,** but two are not suberinized (fig. 23.5). The waxless walls face the outside of the root on one side and the vascular tissues on the other. A good analogy to remember in understanding the structure of the endodermis is a brick wall: bricks are like cells with six faces and in a wall, four of those faces are covered by mortar which is analogous to suberin. Because the wax-containing walls of adjacent endodermal cells abut one another, they form a barrier called the **Casparian strip.** It is not possible to see the Casparian strip on your slide.

Figure 23.6   Cross section of a water hyacinth root showing formation of a branch root from pericycle.

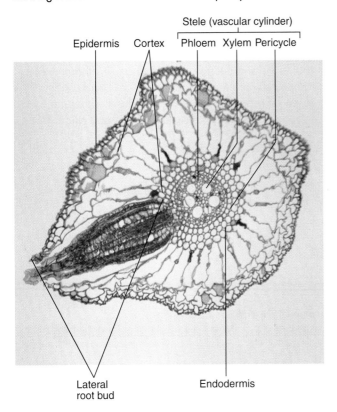

When water and dissolved minerals enter the root, they travel the paths shown in figure 23.5. At the endodermis, suberin prevents water from diffusing around the cells, so water and dissolved minerals must pass through the plasma membranes into the protoplast of the endodermal cells, allowing them to regulate what passes to the xylem in the stele. Thus, these cells control absorption by the plant. Water and minerals that enter the stele and pass to the xylem are transported to other parts of the plant. As a root starts secondary growth in its aging regions and root hairs are lost from the surface, the endodermis also changes. Suberin will be secreted on all faces of the endodermal cells, essentially making it waterproof in the older parts of the root.

Lateral roots arise internally in the pericycle about one inch back from the tip. The growing tissues push their way to the surface through the endodermis, cortex, and epidermis (fig. 23.6). This process differs from lateral branching in stems where the branches arise from buds located at nodes on the stem surface. Nodes and associated buds are absent in roots.

Monocot Roots

Obtain a slide of a cross section of a corn (monocot) root. Identify the outer epidermis and the parenchyma cells of the cortex. Note the intracellular spaces among the cortical parenchyma cells. Note the general similarity to the dicot root but the obviously different organization to the stele. It is surrounded by an endodermis as in the dicot root, but the xylem and phloem cells are found in clumps around the periphery of the stele rather than at its center. The large cells to the inside of the pericycle are xylem cells. Phloem cells are located between the xylem vessels, just beneath the pericycle. No vascular cambium is found in monocot roots between the xylem and phloem. Consequently, they do not undergo secondary growth. The central area of the stele is filled with fairly uniform parenchymal cells and is called the **pith.** Sketch a section of a corn root below.

Investigating Living Roots

If time permits, investigate the microscopic anatomy of roots by preparing freehand sections. Your instructor will have isolated roots from dicots and monocots in the lab, but will not tell you which is which. If you planted the Wisconsin Fast Plants, it can be prepared by this technique and compared to the unknowns. Your task is to examine the microscopic anatomy and to identify the unknowns.

Remove a root tip about 2 inches long from one of the unknowns. Take a two-inch square of Parafilm, a flexible plastic used to seal lab containers, and fold it in half. Place the root on the inside of the fold so that it is sandwiched between plastic. Place the preparation on a microscope slide and press on the Parafilm to hold the root in place. Cut a cross section of the root by slicing across it and the Parafilm. Discard the cutoff material. Add several drops of water to form a puddle at the newly cut edge. Now make a second cross section cut, making the section as thin as possible. Mount this thin section on a new slide in a drop of water and add a coverslip. Observe the cut face of the cross-sectioned root at first with the 10× objective of your compound microscope.

Identify the epidermis, cortex, stele, and xylem in your preparations. Which unknown has a tissue arrangement like a dicot? Which look like a monocot?

Learning Biology by Writing

Because the activities in this lab were more descriptive than experimental, your instructor may want you to write a summary rather than a report. In about 200 words, describe how roots grow. Describe how differentiation of tissues occurs. Distinguish between primary and secondary growth. Describe the basic tissue systems and cell types found in roots.

As an alternative assignment, your instructor may ask you to turn in answers to the Critical Thinking and Lab Summary questions that follow.

Internet Sources

Root hairs are the sites in roots where most absorption of minerals and water occurs. Many researchers study root hairs and the transport mechanisms that function to accumulate minerals from the soil. Locate a research report on the World Wide Web dealing with root hairs. Write an abstract of the report and include the URL.

Lab Summary Questions

1. For each tissue type studied, add the names of the cell types and fill in the blank spaces in the table to summarize your observations.

Tissue	Cell type	Alive or dead	Cell wall (Primary or secondary) with or without lignin	General three-dimensional shape	Functions
Dermal					
Ground					
Vascular					
Meristematic					

2. Distinguish between primary and secondary growth in plant roots.

3. Draw diagrammatic cross sections of primary monocot and dicot roots on a piece of paper. Label the major tissues. Write a few sentences that describe how they differ.

4. Using both cross-section and longitudinal-section diagrams, describe the pathway for water and mineral absorption in plant roots.

Critical Thinking Questions

1. A student is trying to develop a tissue culture of pear cells. He is using sclerenchyma cells to start. Why is he not having any luck?

2. Pines depend on mycorrhizal relationships. A plant nursery was sterilizing soil before planting Christmas tree seedlings and found that their trees took on average two to three years longer to reach sellable size. Why?

3. Many trees will die after their roots are flooded with water or cars are parked under them, compacting the soil. Why?

4. People often spread fertilizer under large trees to make them grow even faster. Sometimes they put the fertilizer close to the trunk, thinking it will be absorbed faster. Why is this erroneous thinking?

LAB TOPIC 24

Investigating Primary Structure, Secondary Growth, and Function of Stems

Supplies

Preparator's guide available on WWW at
   http://www.mhhe.com/dolphin

Equipment

Compound microscopes
Dissecting microscopes

Materials

Slides and coverslips
Prepared slides
   Coleus shoot tip, longitudinal section
   Medicago stem, cross section
   Basswood *(Tilia),* cross section
   Cross and radial sections of oak
   Corn stem, cross section
Twigs of hickory or buckeye with apical bud
Potted, growing *Coleus* plants
Sunflower plants about 12" tall; one well watered and
   other water deprived
Wisconsin Fast Plants planted by students three weeks
   earlier
Fresh celery
100-ml beaker
Razor blades and forceps
Shallow pans

Solutions

0.5% methylene blue
5% NaCl
Phloroglucinol stain

Prelab Preparation

Before doing this lab, you should read the introduction
and sections of the lab topic that have been scheduled
by the instructor.
   You should use your textbook to review the
definitions of the following terms:

   apical meristem
   bark
   cork
   cork cambium

   dicot
   guard cell
   herbaceous
   monocot
   phloem
   stomata
   transpiration
   vascular cambium
   vascular tissue
   woody
   xylem

   You should be able to describe in your own words
the following concepts:

   Primary growth in stems
   Secondary growth in stems
   The transpiration-tension-cohesion theory of water
      movement in xylem
   How guard cells regulate stomatal opening

   As a result of this review, you most likely have
questions about terms, concepts, or how you will do
the experiments included in this lab. Write these
questions in the space below or in the margins of the
pages of this lab topic. The lab experiments should
help you answer these questions, or you can ask your
instructor for help during the lab.

Objectives

1. To observe the differences in stem anatomy
   between monocots and dicots, including examples
   of herbaceous and woody dicots
2. To recognize the basis for primary and secondary
   growth in stems
3. To demonstrate that water rises through the xylem
   tissues because of transpiration
4. To determine experimentally the osmotic response
   of guard cells

Background

Evolutionary theory tells us that life in natural environments is an intense competition for natural resources. The evolution of plants is no exception to this rule. As we saw in the last topic, roots expand outward from plants during primary growth, "seeking" untapped sources of water and minerals used in metabolism, and competition occurs. Above ground, another competition occurs for that limited resource necessary for all plants to grow, light. The goal of the game for those engaged in it, if it can be called a game, is to gain unobstructed access to the maximum amount of light. One way to gain access to light is to grow taller than nearby plants and then to spread branches over a large area to intercept light. The winners in this competition have set some impressive records. Redwood trees in California and eucalyptus trees in Australia have achieved heights in excess of 100 meters. Lastly, and this is impressive, a single banyan tree (stem) in India has a canopy (branches and leaves) that covers an area equal to about five football fields. This is not to imply that all species are driven by natural selection to achieve such impressive sizes. The record holders simply emphasize the concept that tallness can be a way to outcompete other species and be evolutionarily successful. Obviously, many plants have achieved success on a much more modest scale.

Tallness as an evolutionary strategy for maximizing light capture requires evolutionary development of additional systems: a vascular system for the transport of water and minerals to the leaves so that they can carry on photosynthesis and for transport of photosynthetic products to the living cells of the root; secondary growth that strengthens the stem to carry the weight of the leaves and branches; and an enlarged root system to anchor and absorb.

As in roots, stems elongate by adding cells to their tips when the stem apical meristem cells divide. Elongation of the stem in this way is called **primary growth.** Differentiation of the new cells gives rise to numerous other cell types, tissues and organs. Some become the primary dermal system that covers the plant. Others eventually give rise to the various tissues of the leaves, vascular tissues, and ground tissues of the stem.

Two vascular tissues, **xylem** and **phloem,** perform support and transport functions in most vascular plants. In simple terms, xylem contains nonliving, rigid tubular cell walls (left in place after the protoplast dies) which transport water and dissolved minerals from roots to leaves and support the plant; phloem contains living sieve tube cells, which actively distribute photosynthetic products throughout the plant.

In a fescue grass, transport of water and nutrients is perhaps not an impressive task, but in a 300-foot tree, such as *Sequoia,* transport poses special problems. For example, to raise a water column to a height of 300 feet requires a pumping or vacuum force of 300 pounds per square inch. How does the plant accomplish this? Apparently, two mechanisms are involved.

Root pressure, due to the osmotic uptake of water by a plant's root, can account for raising the water column several feet in some species. Since root cells contain dissolved organic solutes, such as sugars and proteins, the *osmotic gradient* in wet soil goes from the soil into the root, literally forcing water into the xylem elements.

The second and more important mechanism for transporting water with dissolved minerals through the xylem to greater heights involves the combined effects of **transpiration** and water cohesion. Leaves have small porelike openings called **stomata** (singular: stoma), usually on their undersurfaces. (See lab topic 25). Stomata are involved in gas exchange during photosynthesis but also allow water vapor to escape by diffusion. Lost water is replaced by vapor diffusion from the xylem into the leaf, resulting in a tension on the water column in the xylem. Water, along with dissolved minerals, is pulled from roots to leaves. This mechanism depends on the cohesiveness of water due to hydrogen bonding between adjacent water molecules. The opening and closing of the stomata are regulated by specialized **guard cells,** which are responsive to water availability.

Transport of organic molecules to other tissues occurs through the sieve tube members of the phloem. Carbohydrates are actively transported from the photosynthetic leaf cells into the phloem. In the roots and regions of growth, these materials are removed from the phloem and used as energy sources or for making other kinds of molecules. Thus a solute concentration difference exists between the two "ends" of the phloem: high concentrations of the solute are present in the leaf, and low concentrations are in the root. Water diffuses from the xylem and enters the phloem by osmosis at the leaf end. This creates a pressure that drives the phloem fluid toward the roots where carbohydrates are removed and metabolized, and water reenters the xylem system. This mass flow results in the **translocation** of photosynthetic products. When plants break their dormancy after winter, flow of materials is reversed. Sugars from starches in the roots are transported to buds and cambiums where it fuels cell growth and division.

Secondary growth is the process that thickens stems and roots after they elongate by primary growth originating in the apical meristem. Secondary growth originates in the **vascular cambium** and the **cork cambium.** The vascular cambium is a ring of tissue that is located between the primary xylem and primary phloem in the primary stem—that is formed at the stem tip. Divisions of the vascular cambium produce secondary xylem and phloem that add to the existing xylem and phloem. This addition increases the girth of the plant cell body. Again, the world records are quite impressive. A chestnut tree in Sicily has a circumference of 58 meters (= a diameter of 18+ meters). Approximately 90% of the mass of this tree would be secondary xylem due to activity of its vascular cambium. The vascular tissues formed from the vascular cambium not only support, they also conduct materials throughout the plant.

If you consider the geometry of secondary growth, an interesting problem emerges. The vascular cambium is a cylin-

Figure 24.1 Anatomy of a woody stem: (a) external features; (b) detail of terminal bud scar; (c) longitudinal section of terminal bud.

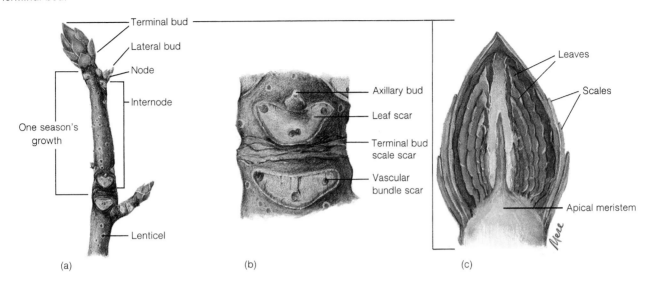

(a)

- Terminal bud
- Lateral bud
- Node
- Internode
- One season's growth
- Lenticel

(b)

- Axillary bud
- Leaf scar
- Terminal bud scale scar
- Vascular bundle scar

(c)

- Leaves
- Scales
- Apical meristem

der of tissue located inside another cylinder, the outer cell layers of the stem consisting of phloem and epidermis. If the inner cylinder enlarges (*i.e.,* divides to produce new cells), the surface of the outer cylinder will crack and split. This is what happens in plants. The cork cambium located to the outside of the vascular cambium and near the surface of the plant body divides to produce new cells that fill the cracks. This covering is the outer layer of the bark and protects the plant body from attack by microorganisms and from desiccation. As the stem cylinder continues to enlarge, cracks do eventually appear at the surface. This is why bark on many trees is rough with vertical striations or may even be shed, but new bark formed beneath the surface protects the tree.

LAB INSTRUCTIONS

In this lab topic, you will look at the anatomy of stems, learning the differences between herbaceous and woody stems found in dicots and between dicots and monocots. In dicots you will consider both primary and secondary growth. At the end you will do some simple experiments demonstrating water movement through the xylem.

Stem Structure

Stems support the flowers, fruits, and photosynthetic leaves in most plants. Conducting tissues in the stem bring water and minerals to the leaves and distribute photosynthetic products to the roots. Some stems are photosynthetic, as in cacti, and others are important in vegetative reproduction, food storage, and water storage. In this section of the lab, you will examine the basic organization of two types of dicot stems and a monocot stem.

External Dicot Stem Structure

Examine a woody twig from a hickory, buckeye, or other tree. At the tip of the stem or its branches, find the **terminal** or **apical bud.** In the bud is the **apical meristem,** the source of cells for primary growth that elongates the stem. In most plants, the apical bud produces a hormone that inhibits development of lateral buds, a phenomenon called **apical dominance.** Gardeners have long recognized this phenomenon. They obtain bushy shrubs and trees by clipping off the shoot tips, which removes the inhibition of the lateral buds, allowing branches to develop. Conversely, tall plants can be "forced" by trimming lateral branches, thus removing the lateral buds and directing the plant's energy reserves to the apex of the shoot.

Note the prominent **leaf scars** where leaves were attached (fig. 24.1). If you look carefully at a leaf scar, you can see the vascular bundle scars where stands of xylem and phloem entered the petiole of the leaf. The leaves grow out from locations on the stems called **nodes;** the stem segments between the nodes are **internodes.** Small pores called **lenticels** should be visible in the bark of the internodes. These loosely organized areas of bark allow metabolically active tissues in the twig to exchange respiratory gases with the atmosphere. The angle formed between a leaf and the internode above it is called the **axil.** Examine the area just above a leaf scar and you should see **lateral** or **axillary buds,** which can form branches of the stem.

Internal Herbaceous Dicot Stem Structure

Now obtain a prepared slide of a longitudinal section of a *Coleus* shoot tip. *Coleus* is a **herbaceous plant,** meaning that the stem shows limited secondary growth and the stem does not thicken very much. Annual plants such as these usually die at the end of the growing season. Using the

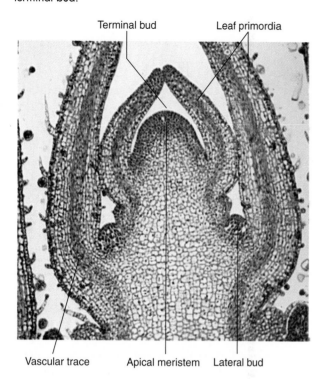

Figure 24.2 Longitudinal section of *Coleus* stem terminal bud.

Terminal bud   Leaf primordia

Vascular trace   Apical meristem   Lateral bud

scanning objective on your microscope, identify the lateral buds and terminal buds as you did for the woody stem. Note several **leaf primordia,** which surround and protect the terminal bud (fig. 24.2). What important group of cells necessary for continued growth is found the terminal bud?

As the *Coleus* matures, flower buds may develop in the terminal bud. Note the **lateral bud** in the angle (axil) between the top surface of the leaf and stem. Lateral buds near the tip of a stem are in a state of dormancy, held in check by hormones produced by the terminal bud. As the stem elongates the distance between the terminal and lateral buds increases and the inhibitory effect is lessened allowing the lateral buds to develop into branches. In pines, firs and spruces this effect produces the pyramidal shape of the tree crown. If the terminal bud is damaged, the lateral buds quickly grow.

Examine the terminal bud under high power and find the **apical meristem** composed of darkly staining, rapidly dividing small cells. As cells are produced at the tip and subsequently elongate as they mature, the stem undergoes **primary growth** or growth in length. Examine the cells below the meristem and note their larger size and evidence of differentiation. Some will become vascular tissues; others will give rise to parenchyma cells, leaf primordia, and lateral buds. Trace the developing vascular bundles consisting of xylem and phloem passing into each of the leaf primordia.

Now obtain a cross section of a differentiated region of *Medicago,* an herbaceous dicot stem and look at it under

your 4× objective (fig. 24.3). Herbaceous stems undergo little secondary growth and are nonwoody. Note the general organization. In young stems resulting from primary growth, the outside of the stem is covered by a single layer of living cells, the **epidermis.** The epidermis protects the underlying tissues from drying, and often contains stomata openings. Epidermal cells on many stems form trichomes. Perhaps you have noticed the "hairy" stems of tomatoes which bear many trichomes. Waxes often are secreted by the epidermal cells to form a **cuticle.** In older stems during secondary growth the epidermis is replaced by the periderm that contains openings called **lenticels.** Just beneath the epidermis is the **cortex,** a complex region. The bulk of the stem is **pith,** a soft matrix of living parenchyma cells. In the stems of some species the pith dies, producing a hollow stem.

The **vascular bundles** are arranged in a circle a short distance inside the cortex. Examine a single vascular bundle under high power. Xylem makes up the inward side of the vascular bundle. Most xylem is dead at maturity. The protoplast breaks down and the walls remain as water conduits. The larger of these are **vessels** and the smaller ones **tracheids** (fig. 24.4). The outer portion of the vascular bundle is phloem. Phloem is composed of larger **sieve tube members** and smaller **companion cells** (fig. 24.4). These cells must remain alive to function at maturity. Separating the xylem and phloem is a layer of **vascular cambium.** (see fig. 24.3). During stem secondary growth when the stem increases in diameter, division of the vascular cambium produces secondary xylem and secondary phloem. The vascular tissues of the stem are continuous with those of the leaves and roots, forming an effective transport system throughout the plant.

Examine one of the "ribs" of the stem and identify the **collenchyma** cells with their unevenly thickened walls. These cells are living at maturity and provide support.

Internal Woody Dicot Stem Structure

Woody dicots, in contrast to herbaceous dicots, have stems that increase in girth year after year by secondary growth. As a plant increases in girth, most of its cells are buried in the added layers and cannot be supplied with oxygen. This physiological limitation has been solved by an interesting evolutionary adaptation. Only the cells near the surface of the trunk retain their protoplasts and are alive. Those near the center die, but their cell walls remain structurally intact and provide mechanical support for the crown of the plant. What functional reason can you suggest that would explain why most cells in the stem are elongated rather than being cuboidal?

Figure 24.3 Tissue organization in a herbaceous dicot stem: (*a*) photomicrograph of a representative stem (alfalfa—*Medicago*); (*b*) artist's interpretation of tissues.

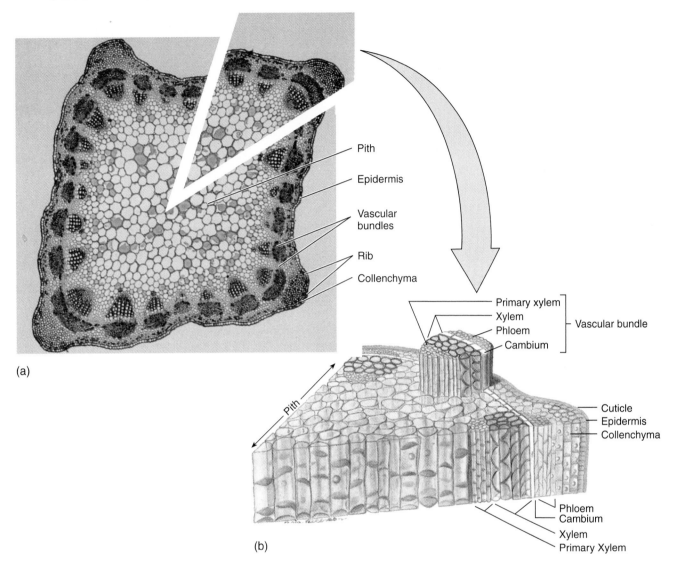

(a)

Pith
Epidermis
Vascular bundles
Rib
Collenchyma

Primary xylem
Xylem
Phloem
Cambium
Vascular bundle

Pith

Cuticle
Epidermis
Collenchyma

Phloem
Cambium
Xylem
Primary Xylem

(b)

A woody stem has three basic regions: bark, vascular cambium, and wood. In young twigs before secondary growth is initiated, the surface of the twig is covered by epidermis. With the onset of secondary growth, bark develops as the epidermis splits and sloughs off. **Bark** is divided into an outer region, the periderm, and an inner region, the secondary phloem. The **periderm** consists of an outer **cork** layer, which prevents evaporation and protects the underlying tissues, and a **cork cambium,** which gives rise to the outer cork cells. **Secondary phloem** consists of sieve tubes, companion cells, sclerenchyma fibers, and storage parenchyma cells. These cells function in food conduction, storage, and support.

Beneath the phloem is the **vascular cambium.** Division of these cells produces secondary phloem to the outside and secondary xylem to the inside. Wood, the innermost region, consists entirely of **secondary xylem.** Its functions are water conduction and support.

Obtain a prepared slide of a cross section of a two-year-old basswood (*Tilia*) twig. Find the pith at the center of the section (fig. 24.5). The life span of the parenchyma cells forming the pith varies by species. As the pith ages, the protoplasts die and the cells accumulate tannins and crystals. Surrounding the pith is a narrow layer of primary xylem; both pith and primary xylem originated from the apical meristem and represent primary growth during the first year of life. To the outside are relatively wide layers of secondary xylem laid down in annual growth rings. Each annual ring consists of larger springwood cells and smaller summerwood cells. At one time the vascular cambium was next to the primary xylem, surrounding it as a cylinder of cells. As the vascular cambium cells divided, the daughter cell to the inside became secondary xylem, and the cell to the outside remained vascular cambium to divide again and again. Thus the cylinder of vascular cambium is always enlarging and is located just outside of the xylem regardless of how

Figure 24.4 Cells of the xylem and phloem: (*a*) scanning electron micrograph of cucumber xylem vessel; (*b*) scanning electron micrograph of phloem sieve plate in pumpkin and broken companion cell (cc).

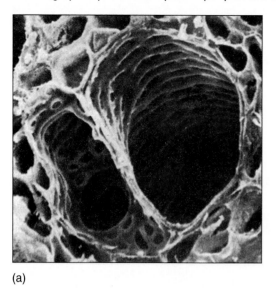

(a)

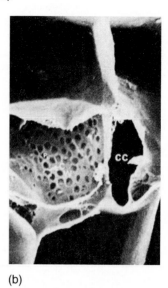

(b)

Figure 24.5 Tissue organization in a cross section of a two-year-old basswood stem, a woody dicot.

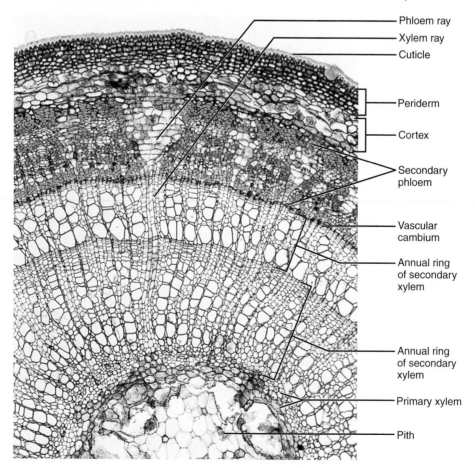

Phloem ray
Xylem ray
Cuticle
Periderm
Cortex
Secondary phloem
Vascular cambium
Annual ring of secondary xylem
Annual ring of secondary xylem
Primary xylem
Pith

Figure 24.6 Tissue organization in a mature oak stem.

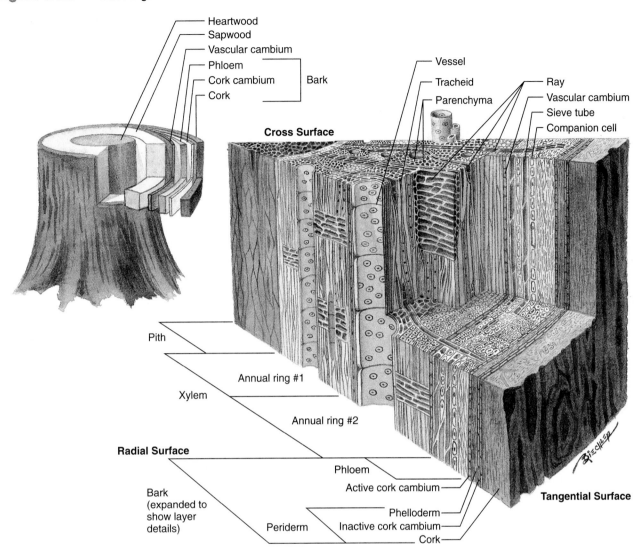

many xylem cells are added in secondary growth. The vascular cambium separates the xylem from the outer phloem.

The phloem is composed of triangular sections, some of which point inward and others outward. Those that point inward, ending in sharp points, are **phloem rays** and are continuous with radial lines of cells, the **xylem rays,** which cross the xylem toward the pith. These cells conduct materials radially in the stem and its branches. In between the phloem rays are other triangular sections, ending in blunt points, that point outward. These consist of heavily stained **phloem fibers** and large, thin-walled **sieve tube members.** These join end to end to form **sieve tubes.** Small **companion cells** are associated with the sieve tube.

In young dicot woody stems, just outside the phloem is the **cortex,** a zone of loosely arranged, large parenchyma cells. The cortex is replaced by the **periderm** as the stem grows in diameter due to cell division in the vascular cambium (fig. 24.6). The long axis of the cells in this zone follows the circumference of the stem rather than its long axis as do the vascular tissues.

The periderm is divided into three layers. Adjacent to the phloem are four to six layers of thick-walled cells, the **phelloderm.** External to the phelloderm are the thin-walled cells of the **cork cambium,** which gives rise to phelloderm and the outermost **cork cells,** which cover the mature stem. Cork cells secrete the waterproofing compound, suberin. It is a lipid. The bark of a woody plant consists of all the layers external to vascular cambium and thus includes the phloem and the components of the periderm.

As a tree grows in girth, what must happen to the existing bark as secondary xylem is added on the inside of the vascular cambium?

Figure 24.7 Monocot stem: (*a*) cross section of a corn, *Zea mays*, shows vascular bundles are scattered throughout parenchyma; (*b*) enlargement showing a vascular bundle.

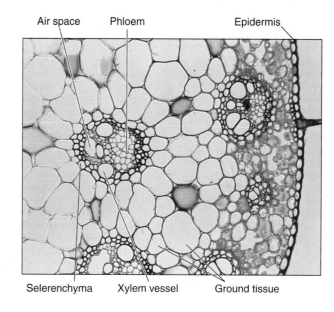

Air space   Phloem   Epidermis

Selerenchyma   Xylem vessel   Ground tissue

(b)

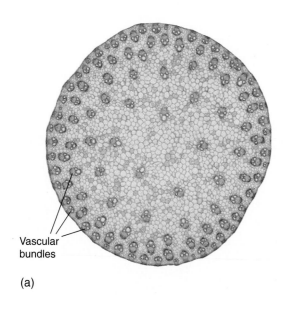

Vascular bundles

(a)

Structure of Wood

Wood is composed exclusively of years of accumulation of secondary xylem to the inside of the vascular cambium. If you look at a cross section of a tree trunk or branch two types of wood are visible (fig. 24.6). In the center the **heartwood** is darker in color due to the accumulation of resins, tannins and other metabolic wastes. It is the older wood. Just outside the heartwood, but not including the bark, is the **sapwood.** It is lighter in color and is younger, not having accumulated the metabolic products. It will gradually darken and become heartwood as new sapwood accumulates from division of the vascular cambium.

Commercial lumber is classified as softwood or hardwood. Softwoods are coniferous woods taken from pine, spruce, fir, or hemlock. Hardwoods are taken from dicotyledonous angiosperms, such as oak, cherry, ash, and several other species.

Both types of wood are prepared by sawing the secondary xylem layers into boards. The xylem of softwood has **tracheids** as the main structural cell type and lacks the **wood fibers** and **vessels** characteristic of hardwood. In this section you will look at the structure of oak, studying the wood in cross and radial sections (fig. 24.6).

Cross Section

Obtain a prepared slide of a cross section of oak wood and look at it first with low power. Bark is probably not included in this section. Compare it to the top of figure 24.6. Vessels and wood fibers make up the bulk of the xylem, although some tracheids are also present. The **pores** in the wood are the ends of the vessels. Identify the **annual**

growth rings composed of larger cells in springwood and smaller in such summerwood. Note the **wood rays,** which pass between rings.

Radial Section

A radial section is taken from a cylindrical stem by cutting a longitudinal section from the outer surface through the center of the cylinder. Such a section cuts the wood fibers and vessels in longitudinal section and shows the rays passing at right angles to the wood fibers and vessels.

Obtain a slide of a radial section of oak wood and look at it with low-power objective and then high-power. Identify the wood fibers, vessels, and rays. Look carefully at the walls of the vessels and note the **bordered pits,** which allow water to pass out of the vessels. Which cells are larger in diameter: vessels or wood fibers?_____

Internal Monocot Stem Structure

Examine a prepared slide of a cross section of a corn (maize) stem. Use the 4× objective. Refer to figure 24.7 and identify the cell types in your specimen.

Note how the monocot stem organization differs from the dicot stems studied earlier. The vascular bundles are not arranged in concentric circles, but are scattered, instead, throughout the ground tissue of the stem. Examine a **vascular bundle** under higher magnification. (Some people think these bundles resemble a monkey's face.) The bundle is surrounded by thick-walled sclerenchyma fibers. The xylem typically consists of four conspicuous vessels. The phloem is found in the area between the two larger vessels and extends from there toward the edge of the bundle. Look

Figure 24.8   Anatomy of a grass plant. Stem is hollow in most species.

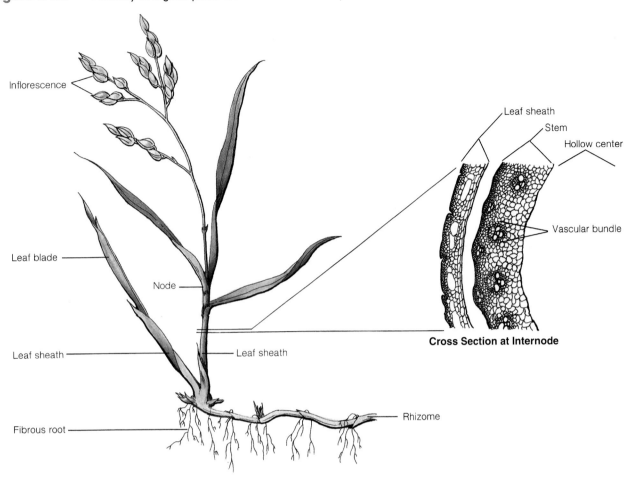

Inflorescence

Leaf blade

Node

Leaf sheath — Leaf sheath

Fibrous root

Rhizome

Leaf sheath
Stem
Hollow center

Vascular bundle

Cross Section at Internode

for sieve plates in some of the sieve tubes. A vascular cambium does not occur in monocots. Consequently, monocots do not produce secondary xylem or secondary phloem; *i.e.*, they have no secondary growth.

The stem of corn is not representative of the stems of many grasses. Grass stems are often hollow, except at the nodes (fig. 24.8). In hollow stems, the vascular bundles are arranged as concentric circles in the walls of the stem, but vascular cambium is not present.

In grasses, a meristem is found at the base of each internode and at the base of the leaf sheath. These are located at the base of the grass plant. When we cut grass, the tip of the leaf and stem is cut off but the meristems close to the soil are not harmed. They produce new cells and the leaf and stem "shoot up" so that the lawn must be cut again, and again, and . . .

Branches arise from the lower portions of grass stems at the soil level. Usually, the branches grow more or less horizontally for a short time, often underground, to form **rhizomes** from which leaf and flower stems grow upward and fibrous roots grow downward. This type of growth is called tillering and allows a single plant to spread out from a single seed, crowding out competing plants.

Transpiration (Demonstration)

In the previous sections, you studied the anatomy of the transport system in plants. Now you will investigate the functions of the transport system.

During periods of active photosynthesis, the stomata of the leaves are open, allowing carbon dioxide to enter. At the same time, water escapes from the leaf and is replaced by conduction up the xylem.

Obtain a fresh piece of celery stalk with leaves. Cut off and discard the bottom two inches of the stalk while holding the stalk under water. Quickly place the cut end of the leafy stalk in a beaker containing a 0.5% solution of methylene blue dye. Place the celery in sunlight.

Let the stem sit in the dye for 30 to 60 minutes. Hypothesize how far the dye will move in an hour. At the end of an hour make cross sections at 1 cm intervals up the stem. How many centimeters did the dye travel in an hour? Does this value match your hypothesis?

Use a razor blade to make a thin section of a region that had a high dye concentration. Mount the section on a slide and look at it with your compound microscope. In what tissue is the dye located?

Testing the Water Tension Hypothesis

Water is thought to be pulled through the xylem system by tension created in the leaves when water is lost through evaporation. If this is the case, then it can be easily demonstrated. If sunflower plants are grown in pots until they are about one foot tall, they can then be subjected to different watering regimens. One pot will be well watered for a week before lab. The other will not be watered. If the theory is correct, the water in the xylem of the unwatered plant should be under greater tension than in the watered one.

To test this hypothesis, fill a tray with a methylene blue dye solution. Take the plants one at a time and bend the stems so that they are under the surface of the dye solution. Cut the stem while it is immersed. The xylem which is under tension will take up the dye.

Students can then take the stems and cut them every 5 mm to see how far the dye has traveled.

Describe the results below.

How do the results of this experiment support or refute the water tension hypothesis for water movement in xylem?

cause of the orientation of cellulose fibers in the cell wall, have a crescent shape, which creates an opening next to adjacent cells (fig. 24.9). When water is not readily available, the guard cells become flaccid, losing their crescent shape and with their edges becoming straight. When both guard cells straighten, they cover the opening of the stoma.

These conditions can be simulated by treating a strip of leaf epidermis containing guard cells with distilled water and a 5% salt solution.

Place a drop of distilled water on a slide. Obtain a *Zebrina* leaf. Fold it in half with the underside inward (fig. 24.10). As it begins to break, pull the upper half away from the break. This should produce a cellophane-thin piece of the lower epidermis. Mount it in the water drop. Add a coverslip and observe. In the left half of the circle below, sketch your observations of the guard cells.

What would you hypothesize would happen if you added several drops of a 5% salt solution next to and touching the coverslip?

Do so. Touch a piece of paper towel to the opposite side of the coverslip. This will wick the salt solution under the coverslip. Observe the guard cells again and sketch a few cells in the right half of the circle above. How are the cells different? What caused this change in the guard cells? Must you accept or reject the hypothesis that you made?

Guard-Cell Response to Osmotic Stress

The opening and closing stomata is related to the osmotic condition of the guard cells. When water is readily available in the leaf, the guard cells are turgid. Turgid guard cells, be-

Figure 24.9 Stoma and guard cells from leaves: (*a*) artist's representation; (*b*) transmission electron micrograph of stoma and guard cells.

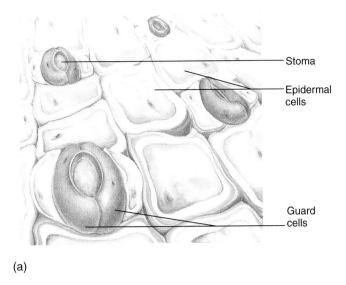

Stoma

Epidermal cells

Guard cells

(a)

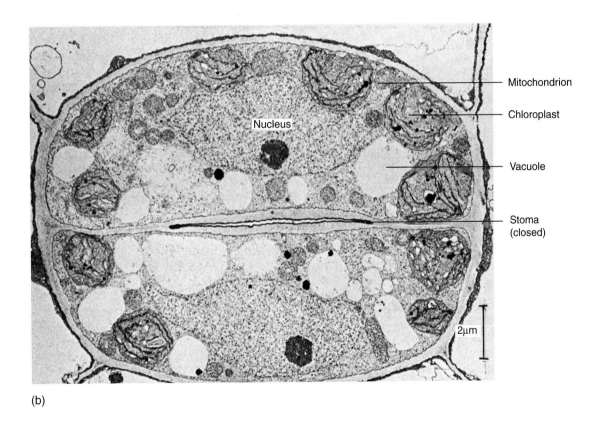

Mitochondrion

Chloroplast

Nucleus

Vacuole

Stoma (closed)

2μm

(b)

Figure 24.10   Technique for removing epidermis from underside of a leaf: Pinch leaf.

It is sometimes fun to work with numbers to estimate the magnitude of an event or process. Some numbers are available regarding transpiration. A researcher counted the number of leaves on a 47-foot silver maple tree. He found that there were 177,000 leaves, with an average leaf area of 26 cm². If the transpiration rate per silver maple leaf is 0.01 ml/hour/cm², how many liters of water would be lost from a mature tree in 12 hours?_____ How many gallons move through the vascular system and out the stomata of a tree this size in a day? (A gallon equals 3.9 liters.)_____ Another researcher working with tobacco estimated there were 12,000 stomata per cm². Assuming the same number for maple leaves, how many stomates did the silver maple have?_____

Learning Biology by Writing

Because the activities in this lab were more descriptive than experimental, your instructor may want you to write a summary rather than a report. In about 200 words, describe how water enters a dicot plant and moves to the leaves. Include anatomical diagrams of the root, stem, and leaf showing the water pathway. Discuss the mechanism of movement and your estimate of the amount of water transpired in a day by a mature tree.

As an alternative assignment, your instructor may ask you to turn in answers to the Critical Thinking and Lab Summary questions.

Lab Summary Questions

1. Distinguish between primary and secondary growth in stems.
2. As a stem undergoes secondary growth, what will happen to cell layers in a woody stem located to the outside of the vascular cambium and phloem?

Internet Sources

Locate a research report on the World Wide Web dealing with how phloem transports organic molecules. Write an abstract of the report and include the URL.

3. Draw diagrammatic cross sections of monocot and herbaceous dicot stems below. Indicate how they differ.
4. Draw cross sections of both a woody dicot stem and a herbaceous stem. Indicate how they are different.
5. Describe the process of transpiration and why it is an important process in plants. Include a brief discussion of how water and minerals move from the roots to the leaves.

Critical Thinking Questions

1. The tallest trees in the world are around 100 m high. Do you think this is the maximum possible height or could trees grow even higher? Explain.
2. How can information about previous climate can be obtained by studying the growth rings of trees? By observing stem structures?
3. If you cut the bark of a tree down to the vascular cambium, in a circle completely around the tree, the tree dies. Why?
4. A tree grows about one foot each year. You carve your initials in the trunk of a tree that is 15 feet tall. Your initials are four feet above the ground today. How far above the ground will they be 10 years from now?
5. What has happened to the epidermis, cortex, and primary phloem of a 100-year-old maple tree?

LAB TOPIC 25

Investigating Leaf Structure and Photosynthesis

Supplies

Preparator's guide available on WWW at
http://www.mhhe.com/dolphin

Equipment

Spectrophotometers
Kitchen blender
Compound microscopes
Desk lamps with outdoor flood lamps

Materials

Several different kinds of fresh leaves
Fresh *Elodea* cuttings
Frozen spinach
Prepared slides
    Dicot leaf, cross section
    C_4 plant, corn, leaf cross section
Flat-sided tanks at least 10 cm across
Ring stand and two clamps
Test tubes with one-hole stoppers (2 cm × 15 to 20 cm)
Syringe, 1 ml
1-ml pipettes with bend at the top end above 0 as in
    figure 25.8
800-ml beakers
Cheesecloth
Graduated cylinders
250-ml separatory funnel
125-ml Erlenmeyer flasks

Solutions

Acetone
Petroleum ether
80% methanol
10% NaCl
Na_2SO_4 powder (anhydrous)
MgO powder
0.1 M $NaHCO_3$

Prelab Preparation

Before doing this lab, you should read the introduction
and sections of the lab topic that have been scheduled
by the instructor.

You should use your textbook to review the
definitions of the following terms:

    chlorophyll
    compound leaf
    dicot
    epidermis
    guard cell
    mesophyll
    monocot
    palmate
    pinnate
    simple leaf
    spectrophotometer
    vascular bundle

You should be able to describe in your own words
the following concepts:

    Light reactions
    Dark reactions (Calvin cycle)
    Absorption spectrum

As a result of this review, you most likely have
questions about terms, concepts, or how you will do
the experiments included in this lab. Write these
questions in the space below or in the margins of the
pages of this lab topic. The lab experiments should
help you answer these questions, or you can ask your
instructor for help during the lab.

Objectives

1. To observe the structure of dicot and monocot leaves

2. To measure the absorption spectrum of chlorophyll

3. To test the hypothesis that the rate of photosynthesis is proportional to light intensity

4. To determine by graphic analysis the light intensity at which the plant's oxygen production equals its oxygen consumption and to confirm this prediction by direct experimentation

Background

Photosynthesis supplies energy to virtually every ecosystem. Not only do plants use this energy to grow and reproduce, but the plants also are eaten by animals or decomposed by bacteria and fungi, thus supplying these organisms' energy needs as well. In photosynthesis, light energy from the sun is captured through complex biochemical reactions and used to make new covalent chemical bonds. Carbon dioxide serves as a source of carbon, and water as a source of hydrogen in these reactions. It is estimated that 150 billion tons of sugar are produced annually by plants on a worldwide basis. An important by-product of this process is molecular oxygen, which enters the atmosphere replacing that which is consumed in respiration and various other oxidations. In addition, oxygen in the upper layers of the atmosphere is converted to ozone, which blocks the transmission of mutagenic ultraviolet light to the earth's surface.

Realize that energy, unlike materials such as carbon dioxide or water, does not cycle in an ecosystem. Instead, energy flows in a direction toward dissipation. Plants are able to convert light energy to chemical bond energy where it is stored in the bonds of sugars. As plants use these sugars or are consumed by animals and these animals are in turn consumed by other animals or die and decompose, energy is lost as heat when the chemical bonds of the plant biomass are broken and re-formed in animal biomass. As a general rule, only 10% of the energy available in ingested food is used to make new biomass; the remaining 90% dissipates as low-grade heat (entropy) throughout the universe. If photosynthesis suddenly ceased on this planet, life would eventually end because all energy stored in chemical bonds of biomass would be released as heat as one organism fed on another in ever shortening food chains.

Plants (along with algae and some bacteria) contain **chlorophyll,** a remarkable green-colored molecule that can absorb light energy and transfer it to its electrons. Chlorophyll is located in the membranes of the **grana** found in those unique plant organelles called **chloroplasts** (fig. 25.1). These electrons can perform cellular work that ultimately results in using carbon dioxide to make organic molecules with energy stored in the chemical bonds. Work is done when the electrons pass through the **photosynthetic electron transport chain,** a collection of compounds located in the thylakoid membranes of the chloroplasts. The energy in the electrons is used to form **ATP** (adenosine triphosphate) by chemiosmosis, and the high-energy electrons may be transferred to the electron carrier **NADP**

Figure 25.1   The structure of a chloroplast: (a) transmission electron micrograph; (b) artist's three-dimensional reconstruction.

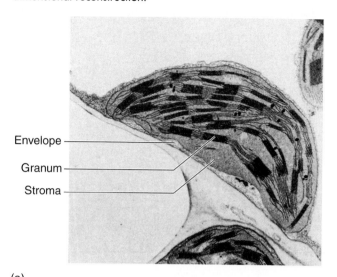

(a)

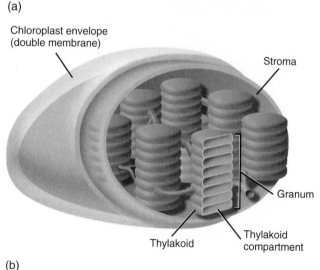

(b)

(nicotinamide adenine dinucleotide phosphate), producing NADPH. These processes, collectively known as the **light reactions,** produce oxygen as a by-product. It escapes from the chloroplast into the atmosphere. Review the diagrams of the light reactions in your textbook and determine what is the source of oxygen released during photosynthesis.

The ATP and NADPH produced during the light reactions are used as energy and hydrogen sources to build complex organic molecules. The set of reactions involved in building such molecules is called the **Calvin cycle,** or **dark reactions.** In these reactions, CO_2 molecules are bonded one at a time to already existing organic molecules, a process known as CO_2 fixation. The result is an increase in the size and number of organic molecules in the cell, using energy and hydrogen coming from the light reactions.

Depending on the type of plant, one of two types of molecules is the product of the dark reactions. In most plants, glyceraldehyde-3-phosphate is the only product,

Figure 25.2 Relationship between photosynthetic CO_2 fixation and the synthesis of other biochemical molecules.

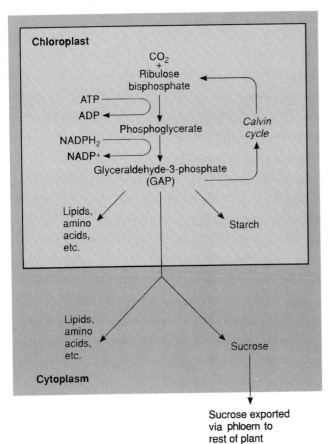

which is then converted into hundreds of other compounds by other enzymes in the chloroplast or cytoplasm (fig. 25.2). Since glyceraldehyde-3-phosphate is a three-carbon compound, this type of plant is said to carry out **C_3 photosynthesis.**

In many grasses, a four-carbon compound, oxaloacetic acid, is the product of CO_2 fixation in the dark reactions in certain cells, while other cells produce glyceraldehyde-3-phosphate. Your text explains how these reactions are related. Such plants have **C_4 photosynthesis.** The differences between C_3 and C_4 photosynthesis are also reflected in the leaf structure and habitat of the plants.

LAB INSTRUCTIONS

In this lab topic, you will observe the anatomy of leaves, determine the absorption spectrum of chlorophyll, and measure how light intensity influences photosynthetic rate.

Types of Leaves

▶ Several types of leaves will be on the demonstration table in the laboratory. Look at them and be sure that you understand the features indicated in boldface in the following paragraphs.

A leaf consists of a flat **blade** and a stalk, or **petiole,** which attaches the leaf to the stem. Xylem and phloem traces pass from the leaf through the petiole and stem to the roots. At the point where a leaf attaches to a stem, except in grasses and their relatives, an axillary bud is found. The bud gives rise to branches of the stem. Leaves may be **simple,** consisting of a single blade and a petiole, or **compound,** consisting of several leaflets joined to a single petiole (fig. 25.3).

Vascular tissues (xylem and phloem) will be visible as **veins** in the leaves. Three venation patterns are found in leaves:

1. **Parallel**—veins pass from the petiole to the tip of the blade in a more or less parallel fashion. Monocots (grasses and their relatives) have leaves with parallel venation. If a vein branches into two equal parts, then it is called dichotomous venation. Ginkgo leaves (fig. 25.3*e*) have dichotomous venation; they are gymnosperms.

2. **Net venation**—veins are not parallel. Dicots (broadleaf plants) have leaves with net venation.

   a. **Pinnate**—a single main vein gives off smaller branch veins that run parallel to each other as in a feather.

   b. **Palmate**—several main veins radiate from where the petiole joins the blade as in fingers radiating from a palm.

Be sure to find examples of all types of venation among the demonstration materials. Record below the names of the species observed and whether they have simple or compound leaves and the type of venation.

Species	Simple or Compound	Venation
1.		
2.		
3.		
4.		
5.		
6.		
7.		
8.		
9.		
10.		

Many leaves display adaptations to specific habits. Leaves from dry areas are often thickened and have hairs that minimize water loss. Aquatic plants with floating leaves will have large air spaces in the leaf. There are often

Investigating Leaf Structure and Photosynthesis   **301**

Figure 25.3 Types of leaves and venation: *(a)* palmately veined maple leaf; *(b)* simple but lobed leaf of tulip tree; *(c)* opposite, simple leaves of dogwood; *(d)* whorled leaves of bedstraw; *(e)* fan-shaped leaf of *Ginkgo,* with dichotomous venation; *(f)* palmately compound buckeye leaf; *(g)* pinnately compound black walnut leaf; *(h)* a grass leaf; *(i)* linear leaves of yew; *(j)* needles of pine.

differences in the leaves of plants that are adapted to full sunlight or shade. Conifers have needles that have reduced surface areas, which reduce moisture loss in cold winters and dry summers.

Internal Leaf Anatomy

Although chloroplasts are found in any green parts of plants, the leaves are the primary solar energy collectors, whether the plants are ferns, pines, birches, or grasses. Leaves are remarkable organs having a high surface area per unit volume. Furthermore, many plants carry and position their leaves to capture sunlight effectively. Water evaporation through leaf surfaces creates a tension which draws water upward, through the xylem—an important transport function.

In this section, you will study the anatomy of leaves in both dicot and monocot flowering plants.

Structure of a Dicot Leaf

Obtain a slide of a cross section of a dicot plant. Look at it with the low-power objective of your compound micro-

scope, switching to high power to see detail (fig. 25.4). How many cells thick is the leaf?_____

In nature, the surface of the leaf is usually covered by a **cuticle,** a layer of wax secreted by the underlying cells, which slows water loss through the large surface area of the leaf. The slide preparation process often extracts the wax from the leaf surface. Is a cuticle visible on your slide?_____

Find the **epidermis,** usually a single layer of cells at the surface of the leaf. These cells lack chloroplasts and have a thick cell wall on their outer surface. Closely examine the epidermal cells on the lower surface of the leaf and find the specialized **guard cells** surrounding an opening called the **stoma** (plural: stomata). (See fig. 24.9.) Are stomata found more frequently on the upper or lower surface of a leaf?_____ Hundreds of stomata may be found in a square millimeter of leaf surface. Carbon dioxide enters a leaf through these openings and water and oxygen escape through them.

The inside of the leaf is composed of **mesophyll** tissue made of thin-walled parenchyma cells. When parenchyma cells contain chloroplasts they are called chlorenchyma.

Figure 25.4   Tissues in a leaf: Artist's three-dimensional drawing of leaf organization.

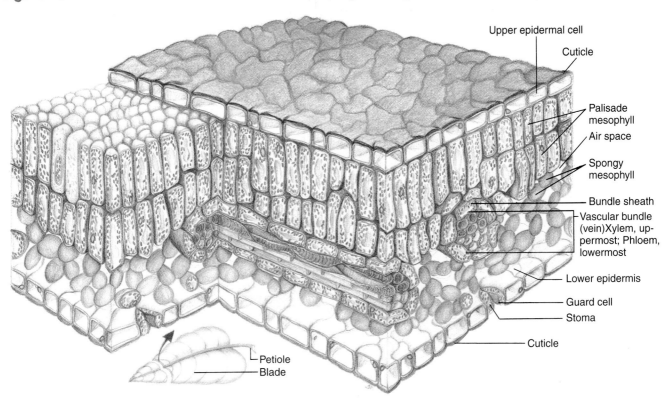

Upper epidermal cell
Cuticle
Palisade mesophyll
Air space
Spongy mesophyll
Bundle sheath
Vascular bundle (vein)Xylem, up-permost; Phloem, lowermost
Lower epidermis
Guard cell
Stoma
Cuticle
Petiole
Blade

Mesophyll chlorenchyma accounts for most of the photosynthetic activity of a plant. Find the **palisade** mesophyll composed of one or more layers of closely packed cells. These are always near the upper surface of the leaf when it is on the plant. Beneath the palisade cells is **spongy** mesophyll composed of irregularly shaped cells separated by a labyrinth of air-filled, intercellular spaces which allow carbon dioxide, oxygen, and water vapor movement. Note how these spaces are continuous with the stomata. Which of the mesophyll layers has more chloroplasts?

You should also be able to see **vascular bundles** composed of xylem and phloem that transport materials to and from the mesophyll and the rest of the plant. What materials do they bring to the mesophyll?

What materials do they translocate from the mesophyll?

Note the tightly packed **bundle sheath** cells that surround the vascular bundles. They control the exchange of materials between the leaf and the vascular system.

Use different colored pencils to add arrows to the diagram in figure 25.4, showing the path of carbon and path of water through a leaf.

To determine if the structure of leaves is similar in a number of different plants, you will now hand section leaves from several different species of plants. To do this, follow the procedure outlines in figure 25.5. By sandwiching a leaf between two offset microscope slides, you can use a razor blade to cut thin cross sections of the leaf. The leaf sections should then be mounted on a third slide in a drop of water and a coverslip added. Look at the edge of the sections first with the scanning objective, switching to the 10× objective of your compound microscope to identify the tissue layers in the leaf.

How can you distinguish between palisade mesophyll and the spongy mesophyll? Is the palisade layer toward the top or the bottom of the leaf? How many cells thick is the dermis in all leaves examined? Are all leaves about the same number of cells thick? Record your observations below.

Figure 25.5 Two-slide method for making cross section of a leaf.

(a) Sandwich a leaf between two microscope slides so that the leaf is exposed at one end. Add several drops of water at the edge.

(b) Take fresh razor blade and cut along edge of top slide. Discard tip of leaf.

(c) Move top slide slightly to the left to expose a very small part of leaf. Cut again to make a very thin cross section. Transfer section to another slide.

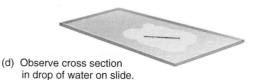

(d) Observe cross section in drop of water on slide.

Figure 25.6 Scanning electron micrograph of a bluestem grass, *Andropogon*.

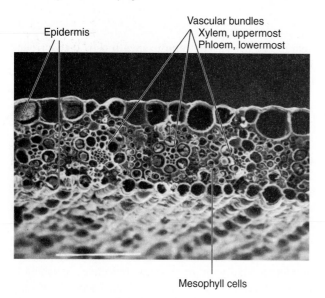

Epidermis

Vascular bundles
Xylem, uppermost
Phloem, lowermost

Mesophyll cells

Structure of a Monocot Leaf

About 65,000 of the 235,000 species of flowering plants are monocots. These include the bamboos, grasses, palms, lilies, tulips, orchids, and several other flowers. The leaves of these plants differ from the broadleaf you just studied (fig. 25.6). Monocot leaves are usually long and narrow with parallel venation rather than broad with a netlike vascular system.

Obtain a cross section of a corn leaf and identify the following parts: cuticle; upper and lower epidermis; stomata; mesophyll, consisting of chlorenchyma cells with

chloroplasts; and bundle sheath cells surrounding the xylem and phloem cells in a vein. Look closely at the chlorenchyma. Is it organized into palisade and spongy layers as in dicots?_____ Look closely at the upper and lower epidermis. Are stomata more common on one surface compared to the other?_____

Now carefully examine the upper epidermis and find the columnar **bulliform cells.** These thin-walled cells readily lose water on hot days and collapse. When they do, the corn leaf folds inward along its long axis. What purpose do you think this mechanism serves?

Photosynthetic Pigments

The light-absorbing properties of a chlorophyll solution will be studied using a spectrophotometer.

Extraction Procedure

Chlorophyll can be extracted from plant tissue. You or your instructor will use a kitchen blender to homogenize 40 g of frozen spinach leaves in 50 ml of cold water containing 0.1 g of MgO powder. The resulting slurry should be filtered through six layers of cheesecloth into a beaker. Add acetone to the filtrate to bring the volume to 100 ml. This extract is a complex mixture of fats, sugars, orange carotenoid pigments, and chlorophyll.

Contaminating compounds can be partially removed by solvent partitioning. This procedure is based on the solu-

Figure 25.7 A separatory funnel, used to separate a two-phase liquid mixture. The stopper must be removed before the stopcock is opened.

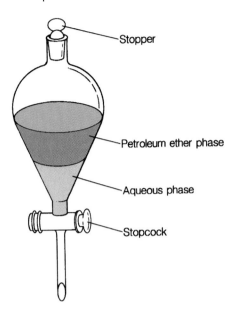

Stopper

Petroleum ether phase

Aqueous phase

Stopcock

TABLE 25.1   Light absorption characteristics of chlorophyll

Wavelength	Light Color	Absorbance
420	_____	___
440		___
460	_____	___
480		___
500	_____	___
520		___
540	_____	___
560		___
580	_____	___
600		___
620	_____	___
640		___
660	_____	___
680		___

bility of chlorophyll only in nonpolar solvents, whereas contaminating compounds are soluble in water or alcohol which are polar.

CAUTION

Petroleum ether, methanol, and acetone are highly flammable. No flames or smoking are permitted in the laboratory during this procedure. Have absorbent material handy to soak up any accidental spills.

▶ Add the acetone extract to a separatory funnel along with 100 ml of petroleum ether and shake vigorously. *Be careful!* Pressure will build up in the separatory funnel and should be released periodically by loosening the stopper. The water-acetone mixture should separate from the less-dense petroleum ether. If it does not, add 25 ml of 10% NaCl. Layers of the two solvent systems should now form, but may require up to 15 minutes for complete separation.

▶ Drain and discard the lower aqueous phase with any particulate interface (fig. 25.7). Now add 60 ml of 80% methanol to the petroleum ether, shake, and then add just enough 10% NaCl to form two layers. Drain the yellow lower aqueous phase containing carotenoid and xanthophyll pigments and discard.

▶ Wash the green petroleum ether extract twice with 10 ml of 10% NaCl to remove residual acetone and methanol and discard the aqueous phase.

▶ Remove traces of water from the petroleum ether by adding 3 g of anhydrous Na_2SO_4 powder to the extract. You should now have about 80 ml of clear green fluid that can

now be used to obtain an absorption spectrum by the entire class. (It should have an absorbance of about 1 at 420 nm.)

The solvents used in this extraction are hazardous wastes and should be disposed of according to directions given to you by your instructor.

Absorption Spectrum

▶ Adjust a spectrophotometer to zero absorbance using pure petroleum ether as a blank. (Review fig. 5.5 for operating instructions.)

Add chlorophyll extract to a spectrophotometer tube and read its absorbance across the visible spectrum at 20 nm wavelength intervals. If the solution is too concentrated (absorbance greater than 2 at 425 nm), dilute it with petroleum ether. Remember to zero the instrument with petroleum ether at each new wavelength. Record your readings at each wavelength in table 25.1.

▶ Observations of the color of light requested in table 25.1 can be obtained by placing a strip of paper in a clean, dry cuvette and inserting it in the spectrophotometer. If you look down into the cuvette with your hands cupped around the cuvette chamber opening, you should see the reflected light as it passes through the cuvette. Change the wavelength and record the colors in the table.

After the laboratory, when you are writing your lab report, you will graph this data. (See appendix B for graphing instructions.) Remember the dependent variable should be on the ordinate. Is wavelength or absorbance the dependent variable?_____ Label all axes and be sure to create a one-sentence figure legend that describes that graph and the treatment of the sample.

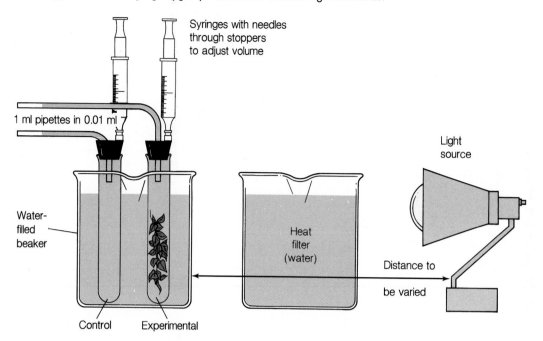

Looking at your data in the graph that you just made, make a hypothesis that relates rate of photosynthesis to wavelength of light striking a plant. Would you predict that photosynthesis would be greatest or least when the wavelength of light striking a plant corresponds to the wavelength of light maximally absorbed by chlorophyll? Test your hypothesis by consulting your text to determine what wavelengths of light are effective in photosynthesis (look up action spectrum).

Light Intensity and Photosynthetic Rate

In this experiment you will test the hypothesis that, as the intensity of light increases, the rate of photosynthesis will also increase. The alternative hypothesis is that the rate will not change or may even decrease. Photosynthetic rate can be measured by determining the rate of oxygen production. Once you have determined which hypothesis is false, you can test another idea.

Some of the oxygen produced in photosynthesis is not released from the plant; instead it is consumed by aerobic respiration in the plant cells. The light intensity at which the rate of oxygen production equals the rate of oxygen consumption is called the **light compensation point.** Using the data collected to test the previous hypothesis, you will graphically determine the light compensation point for the aquatic plant *Elodea* and then experimentally test it.

Procedure

▶ Obtain a piece of *Elodea* 5 to 6 inches long. Choose a healthy specimen with an actively growing leaf bud at the apex. Make a fresh cut across the basal end and place the sprig in a test tube with the cut end up. Fill the test tube with 0.1 M $NaHCO_3$ solution and add a rubber stopper containing a bent glass pipette and syringe, as shown in figure 25.8. The joint between the rubber stopper and the test tube must be dry to get a good seal and fluid must flow into the pipette as you push the stopper in. You must get a good seal between the tube and the stopper. If the stopper "creeps," it will give you erroneous readings. The position of the fluid in the pipette can be adjusted by raising or lowering the plunger on the syringe once the stopper is in place.

Prepare a second tube without *Elodea* to act as a control for temperature and atmospheric pressure fluctuations.

▶ Place both tubes in a beaker of water at room temperature. Set up a heat filter and a lamp as shown in figure 25.8. Move the lamp so it is 75 cm away from the tubes with the heat filter in between. Dim the room lights and wait five minutes for equilibration.

What hypothesis are you testing in this experiment?

TABLE 25.2   Oxygen production readings at 75 cm distance

	Time (Minutes)					
	0	2	4	6	8	10
Elodea						
Control						
Elodea—control						

TABLE 25.3   Oxygen production readings at 50 cm distance

	Time (Minutes)					
	0	2	4	6	8	10
Elodea						
Control						
Elodea—control						

TABLE 25.4   Oxygen production readings at 25 cm distance

	Time (Minutes)					
	0	2	4	6	8	10
Elodea						
Control						
Elodea—control						

Adjust the fluids in the pipettes of both tubes so that they are at the 0.2 ml mark. Read the position of the fluid in the pipettes at two-minute intervals for ten minutes. Record your data in table 25.2. If you do not get a change of about 0.15 ml in ten minutes, you should extend the reading intervals to three to five minutes. If you change the time interval, be sure to change the times printed in table 25.2. At higher light intensities (shorter plant-to-light distances), much greater movement will be seen.

Oxygen is not very soluble in water. As *Elodea* produces oxygen in photosynthesis, the newly produced gas in the tube will force water to move into the pipette. Therefore, fluid movement in the pipette is a measure of photosynthesis. The changes you observed in the tube containing *Elodea* are the result of two simultaneous happenings: (1) the production of oxygen by *Elodea*, which always re-

sults in an increase in volume, and (2) fluctuations in temperature and pressure, which may result in either positive or negative changes in volume. To obtain a "true" reading of the volume change, you must subtract the control reading from the *Elodea* reading. Do this and enter the difference on the last line of table 25.2.

Move the lamp forward 25 cm to the 50 cm mark and let it sit for five minutes of equilibration. Adjust the fluid in the pipettes to 0.2 ml and repeat the experiment, recording the results in table 25.3. Calculate the difference between the experimental and control for this series.

Move the lamp forward to the 25 cm mark and let it sit for five minutes of equilibration. Adjust the fluid in the pipettes to 0.2 ml and repeat the experiment, recording the results in table 25.4. Calculate the difference between the experimental and control for this series.

Analysis

▶ On one panel of the graph paper at the end of this lab topic, plot all three corrected cumulative movements of fluid as a function of time with different plotting symbols for each distance. Using a ruler, draw three straight lines that best fit the points for each treatment. Calculate the slopes at each of the light intensities. (Slope = change in oxygen production per unit of time.)

▶ On a second panel of the graph paper, plot these three slope values as a function of the distance of the plant from the light source. For point light sources, the intensity of the light at a given distance is inversely proportional to the square of the distance, that is $I = 1/d^2$. For an ordinary bulb at the distances in this experiment, the relationship is more nearly linear: $I = 1/d$. The place where the straight line joining the three points intersects the x-axis represents the predicted **light compensation point.**

Testing a Prediction

▶ When the light is placed at the distance from the plants corresponding to the compensation point, the light intensity is such that there should be neither net consumption nor production of oxygen by the plant. Test this prediction (hypothesis) by placing the *Elodea* and control tubes at the compensation distance and measuring any changes that occur.

❖ ❓ Did you see any change? Does this mean that the plant is neither respiring nor photosynthesizing? Explain.

Learning Biology by Writing

In this lab, you explored (1) the properties of chlorophyll and (2) the effects of light on the rate of photosynthesis. Your instructor will tell you which factors to include in your laboratory report.

To begin your report, outline the general purposes of your observations and experiments in a few sentences. Briefly describe the techniques you used. Include your results in the form of half-page graphs of any quantitative data that are to be included in the report. In the discussion, answer such questions as:

1. How does the chlorophyll absorbance spectrum relate to the expected action spectrum for photosynthesis?
2. What is the practical significance if the photosynthetic rate never rises above the compensation point?
3. How might you improve the experiments to gain more information?

As an alternative assignment, your instructor may ask you to answer the following Critical Thinking and Lab Summary Questions.

Lab Summary Questions

1. Draw the anatomy of a leaf in cross section. Indicate the functions of the different cell layers found in a leaf.
2. Plot the data in table 25.1 and turn it in with this summary. Indicate with arrows on the x-axis the wavelengths of light most strongly absorbed by chlorophyll. See appendix B for directions on making graphs.

Internet Sources

As this lab demonstrates, chlorophyll is an essential molecule in photosynthesis. Despite its importance, there are many things that we do not know about it, and much research is being conducted to learn more about it. One type of research has to do with how chlorophyll is synthesized in plants, including what genes are involved. Search the World Wide Web for information about chlorophyll synthesis and review the information available on a few of the sites. Write a short paragraph describing what type of research is being performed today on this topic.

3. Plot the data on the last line of tables 25.2, 25.3, and 25.4 on one sheet of graph paper and turn it in with this summary.

4. Calculate the slopes of the three lines plotted in question 2 and graph the slopes as a function of the distance between the light source and the *Elodea*. Indicate the distance (really a light intensity) at which oxygen production equals consumption. Turn in the plot with this summary.

5. What happened when you placed *Elodea* at the light compensation distance from the light source?

6. Explain why a control tube was included in the light intensity experiments. What did it control for?

Critical Thinking Questions

1. During the chlorophyll extraction procedure, xanthophyll and carotene pigments were extracted and discarded. What are the advantages to the plant of having these accessory pigments?

2. Deep, clear tropical waters are dominated by Rhodophytes (red algae). Their color is derived from red pigments, phycobilins, that mask the chlorophyll. Why does this particular arrangement make them suited to their environment?

3. Discuss the adaptive differences as well as similarities you would expect to find in the structure of leaves taken from plants that lived in hot, relatively dry environments to those from wet, humid environments.

4. Ice and snow on lake surfaces in winter often reduces light intensity in the water below. If light intensity falls below the light compensation point, what is the implication for the ecosystem?

5. At night do green plants produce or consume oxygen? Why?

Investigating Leaf Structure and Photosynthesis

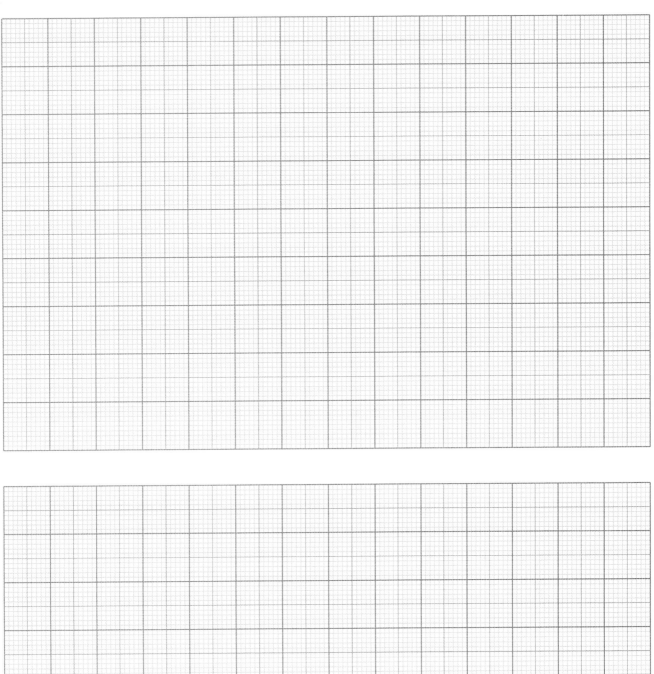

LAB TOPIC 26

Angiosperm: Reproduction, Germination, and Development

Supplies

Preparator's guide available on WWW at
http://www.mhhe.com/dolphin

Equipment

Compound microscopes
Dissecting microscopes

Materials

Living flowers
  Vinca shedding pollen
  Fresh dicots
  Six dicots varying in completeness and perfection
  Gladiolus or other large monocots
Tobacco hormone kit
Wisconsin Fast Plants
  Wild-type seeds
  Rosette seeds
  Growing system kit
Preserved flowers
  Oat inflorescences
  Corn tassel
Fresh fruits
  Apple, tomato, peach, bean, pepper, etc.
Soaked lima beans and corn kernels
Prepared slides
  Corn pistillate flower, longitudinal section
    (demonstration)
  Lily anther, cross section
  Lily ovule, megasporogenesis series
  Lily ovary, cross section
  Capsella embryo development
Miscellaneous
  Slides and coverslips
  Razor blades
  Dissecting needles
  10% raw sugar solution (do not use white
    or brown sugar)

Prelab Preparation

Before doing this lab, you should read the introduction
and sections of the lab topic that have been scheduled
by the instructor.

You should use your textbook to review the
definitions of the following terms:

  anther
  antipodals
  carpel
  cotyledons
  dicot
  embryo sac
  ethylene
  gibberellins
  hypocotyl
  integuments
  megaspore
  meristem
  microspore
  monocot
  ovary
  ovule
  pistil
  polar nuclei
  pollen
  radicle
  seed coat
  synergids

You should be able to describe in your own words
the following concepts:

  How eggs are produced in plants
  How sperm are produced in plants
  How fertilization happens in plants
  Double fertilization
  Formation of embryo in seeds
  Roles of plant hormones in regulating growth

As a result of this review, you most likely have
questions about terms, concepts, or how you will do
the experiments included in this lab. Write these
questions in the space below or in the margins of the
pages of this lab topic. The lab experiments should
help you answer these questions, or you can ask your
instructor for help during the lab.

Figure 26.1   Asexual reproduction occurs by runner (stolon) formation in strawberries.

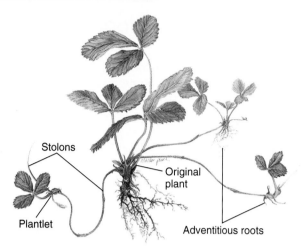

Stolons

Original plant

Plantlet

Adventitious roots

Objectives

1. To observe diversity in flower structure
2. To illustrate gamete formation, fertilization, in flowering plants
3. To examine seed and fruit anatomy
4. To determine experimentally the role of hormones in regulating plant development and differentiation

Background

Flowering plants reproduce by two modes: asexually (vegetatively) by producing new individuals from an older individual's leaves, stems, or roots; and sexually with gametes, which give rise to an embryo contained in a seed. Both modes have their advantages.

Asexual, also called vegetative, reproduction produces robust progeny, which use energy reserves from their parents and thus may increase the chances for survival over those of delicate, sexually produced seedlings. Because asexually produced progeny are genetically identical to the parent, they also have the same genetic advantages that allowed the parent to reproduce asexually. Thus, a genetically superior individual may spread and multiply rapidly without risking genetic recombination inherent in sexual reproduction. Asexual reproduction can yield impressive results: a single aspen tree in the Rocky Mountains is estimated to have produced by repeated root sprouting 47,000 individual trees covering 200 acres. Cattails, prairie grasses, blueberries, blackberries, cherries, and strawberries (fig. 26.1) are examples of other species that are asexually active reproducers. Few annual plants reproduce asexually.

Many named cultivars of fruits and roses are propagated only vegetatively. For example, all McIntosh apple trees in the world today are traceable back to one tree that had superior fruit qualities. These offspring trees were first propagated by cutting branches from the original tree and

grafting them onto roots of crab apple trees. Cuttings from subsequent generations have been used to produce the present number of trees. If you planted the seeds of McIntosh apples, would you expect the tree that grew to bear exactly the same kind of fruit as its parent? Why?

While the principal advantage of asexual reproduction is genetic constancy, the main advantage of sexual reproduction is genetic variability. It results from the meiotic production of sperm and egg, and the combination of genes from two parents at fertilization. No two individuals resulting from the fusion of gametes from the same parents are genetically alike. Each represents a unique combination of genes, which are tested by natural selection, thus increasing the probability of continued survival of the species over time during environmental changes. Many sexually propagating species also produce specialized seeds and fruits, which increases the possibility of dispersal to new locations.

The basic mechanisms in asexual reproduction are simple: cellular multiplication by mitosis and an ability of the new cells to differentiate into the tissues characteristic of roots, shoots, leaves, and other plant organs. Sexual reproduction is more complex. Spores are produced by meiosis. The spores develop into male or female gametophytes within the male or female parts of the flower. Pollen must mature and be released to travel to the egg in the ovule for fertilization. Following fertilization, the development of the zygote into an embryo must be supported by the parent, including the development of the seed coat, which surrounds the embryo during dormancy. The seed containing the embryo must be dispersed and the seed must germinate. Moreover, all these stages of sexual reproduction must be coordinated with the seasons.

Sexual reproduction in flowering plants, in comparison to mosses and ferns (lab topic 15), is highly adapted to the terrestrial environment. Unlike the lower plants, water is not necessary for the male gametes to reach the eggs. Following fertilization, zygotes develop surrounded by maternal tissue and are protected from drying. Mature embryos are encased in a seed coat and often remain viable for years in relatively dry environments. Some desert annuals, for example, grow only during intermittent wet periods. When it rains, the seeds germinate, grow, flower, and produce new seeds in the short period of time that life-sustaining moisture is available. During the dry periods, no mature plants are present and the species survives as seeds that remain dormant until the next rain.

TABLE 26.1 Differences between monocots and dicots

Characteristic	Monocot	Dicot
Characteristic root	Fibrous	Tap
Arrangement of stem vascular tissue	Scattered bundles	Bundles in ring
Presence/absence of wood	Absent	Present or absent
Pattern of leaf veins	Parallel as in grass	Net as in oak or maple
Number of floral petals/sepals	Multiples of 3	Multiples of 4 or 5
Number of cotyledons in seed	One	Two
Pattern of growth	Primary only	Primary and secondary

LAB INSTRUCTIONS

You will dissect a flower, study gamete production, observe plant embryos, and study seeds and fruits. You will grow plants from tissue cultures and investigate the effect of hormones on plant growth.

Sexual Reproduction

Flower Structure

Flowers contain reproductive gametophyte stages (pollen and embryo sac) of plants in the division Anthophyta, about 235,000 species. These are divided into two classes: the monocotyledons (65,000 species) and the dicotyledons (170,000 species). Table 26.1 summarizes the differences between these two groups.

Review the life cycle of an anthophyte shown in figure 16.9 before you start your lab work.

A flower is essentially a short modified stem consisting of several nodes with very short internodes. Modified leaves bourne at the nodes make up the floral parts.

Gladiolus

Obtain a fresh *Gladiolus* flower and observe it through your dissecting microscope. Using figure 26.2 as a guide, identify the following major parts:

1. **Calyx**—the base of the flower made up of whorls of individual modified leaves, the **sepals.**

2. **Corolla**—the usually colored part of the flower made up of whorls of modified leaves called **petals.** Sometimes the petals are fused as in the petunia. The calyx and corolla are collectively known as the **perianth.** In plants pollinated by animals (insects, birds or mammals) the bright colors are an attractant and are a good example of coevolution.

3. **Stamen**—modified leaves composed of an **anther** borne on the end of a thin **filament.** Pollen develops in anther.

4. **Pistil**—sometimes called carpel; a modified leaf or fused group of leaves that consist(s) of the terminal **stigma** where pollen attaches, the supporting **style,** and a basal **ovary** containing **ovules** in which eggs develop and are fertilized.

5. **Receptacle**—expanded tip of flower stalk from which all four of the other structures develop.

Examine a stamen closely. Each anther consists of a **pollen sac** with a line of dehiscence (splitting) separating the lobes. Open a pollen sac by running the tip of a needle along this line and squeezing out the contents into a drop of water on a microscope slide. What does it contain?_____

Take a razor blade and make a longitudinal section of the pistil. Are ovules visible in the ovary?_____

The flowers of dicots tend to have their parts arranged in multiples of fours or fives, such as five sepals, five petals, ten stamens, and five carpels. The flowers of monocots are based on multiples of three. Is a Gladiolus a monocot or dicot? _____

Diversity of Flower Structure

This section is designed to introduce you to diversity in flower structure.

A flower, such as the one you just studied, that has all four types of modified leaves (sepals, petals, stamen, and pistil) is known as a **complete** flower. Flowers that lack one or more of the four parts are **incomplete.**

Perfect flowers contain both stamens and pistils. **Imperfect** flowers are a special case of incomplete flowers and lack either pistils or stamens. Flowers that contain only stamens are **staminate,** or male flowers; flowers that contain only pistils are **pistillate,** or female flowers.

Some plants will have individuals that bear only staminate flowers and others that bear only pistillate flowers. These species are said to be **dioecious** (of two households). Species having both staminate and pistillate flowers on one individual are said to be **monecious** (of one household).

Six flowers from different species are available in the lab. Each type is numbered. Examine these flowers and indicate in table 26.2 whether they are complete or

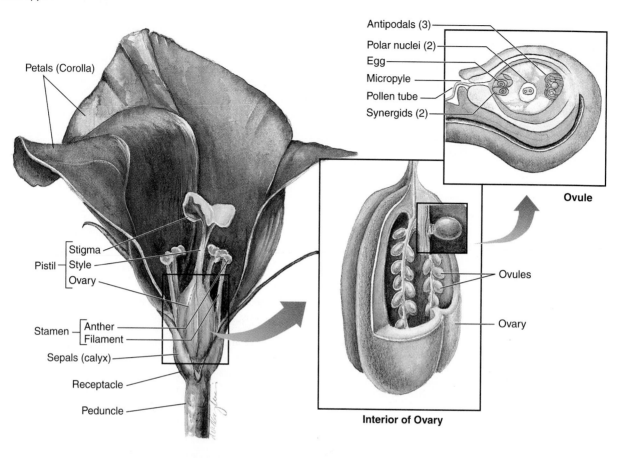

Figure 26.2 Longitudinal section of a flower showing reproductive structures. Eggs are contained in ovules inside of ovary. Fertilization occurs when a pollen tube grows (from pollen landing on stigma) through style tissue to where it enters the ovule through the micropyle.

Petals (Corolla)

Antipodals (3)
Polar nuclei (2)
Egg
Micropyle
Pollen tube
Synergids (2)

Ovule

Pistil — Stigma
Style
Ovary

Ovules

Ovary

Stamen — Anther
Filament

Sepals (calyx)

Receptacle

Peduncle

Interior of Ovary

	Completeness	Perfection	Monecious or Dioecious	Name
1				
2				
3				
4				
5				
6				

TABLE 26.2 Diversity in dicot flowers

incomplete and perfect or imperfect. Your instructor will give you the names of the flowers.

Grass Flowers
The flowers of grasses are highly specialized. Most grasses have flowers arranged in clusters called **inflorescences**

(fig. 26.3). Two arrangements are found: the **spike,** in which the main axis is elongated and the flowers are directly attached; and the **panicle,** in which the main axis is branched with the branches bearing flower clusters. Because grasses are wind-pollinated, they lack the showy petals that attract pollinators. The grasses known as cereals

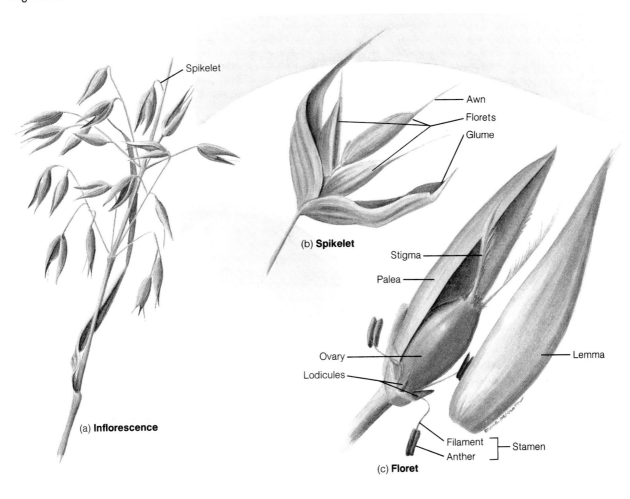

(a) **Inflorescence**

(b) **Spikelet**

Awn
Florets
Glume

Stigma
Palea
Ovary
Lodicules
Lemma
Filament
Anther
Stamen

(c) **Floret**

and grains are a chief part of the human food supply and include corn, wheat, rye, and oats as well as sorghum and sugarcane.

Obtain an inflorescence of an oat from the supply area. What type of inflorescence do oats have?_____

In oats, sepals and petals are missing and the flower is encased in modified leaves, generally known as **bracts.**

Individual flowers are called **florets** and are arranged in groups on a spikelet. Two bracts, called **glumes,** are found at the base of each oat spikelet and enclose several developing florets.

Look at a single floret with your dissecting microscope. It is enveloped by two other bracts. The smaller is called the **palea.** The larger one is called the **lemma.** It has a long extension called the **awn.** At the base of the lemma are two protuberances, the **lodicules.** They swell and open the protecting bracts, as the floret matures.

Spread the bracts and find the three stamens. They are arranged around a central pistil, which bears two feathery **stigmas.** Cut off the stigmas from a pistil and make a wet-

mount slide. Examine the stigma with your microscope and draw it below.

Figure 26.4 Photomicrograph of a cross section of microsporangia in an anther where pollen develops.

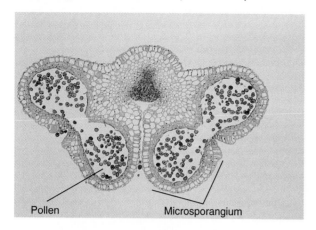

Pollen   Microsporangium

cross section of the anther in a drop of water on a microscope slide. Observe the section under low power with your compound microscope (fig. 26.4).

The four cavities you see are part of the four lengthwise chambers of the anther and are often called pollen sacs, or **microsporangia** (organs producing microspores). Now get a prepared slide of a cross section of an anther. Observe it with the high-power objective and note the pollen grains. Sketch the section and pollen.

Pollen released by the stamens is air blown. The feathery stigmas entrap pollen, and a pollen tube grows down to the single **ovule** in the ovary at the base of the pistil. The seed develops in the ovary and is harvested as grain.

Corn differs from other grains because its flowers are imperfect. A single plant bears both male and female flowers. The staminate flowers are borne at the top of the plant as the **tassel** and release their pollen in the wind. The pistillate flowers are located lower on the plants at nodes and eventually develop into the **ears.** The **silk** found in ears is what remains of the styles that extended from every egg. Pollen that fell on the ends of these styles had to develop pollen tubes that were several centimeters long. Look at the demonstration material of a corn tassel and a longitudinally sectioned pistillate inflorescence.

Gametogenesis in Flowering Plants

As in all plants, flowering plants have an alternation of generations. The gametophyte stage in flowering plants is much reduced and is dependent on the sporophyte stage for protection and nutrition. Consequently, gamete formation does not immediately follow meiosis, as it does in animals. Instead, meiosis gives rise to a haploid **microspore** in the anther and a haploid **megaspore** in the ovule. These haploid cells divide by mitosis to produce a small number of cells, which become either the male or female of the **gametophyte generation.** A pollen grain represents the male gametophyte. The female gametophyte is the embryo sac enclosed by the ovule of the parent. The parent plant is the **sporophyte generation.**

Pollen Formation

Stamens are sometimes refered to as male organs in flowers. This is not the case. The anther and filament are sporophyte tissue and without sex. The male stage, the gametophyte, develops inside the pollen grain. Remove a stamen from your *Gladiolus* flower and use a razor blade to make a

In the young microsporangia, **microspore mother cells** undergo meiosis to produce tetrads of haploid **microspores,** which mature into pollen, the male gametophyte stage of the flowering plant. *Lilium* has a diploid (2N) chromosome number of 24. How many chromosomes are in a microspore?_____

The microspores develop heavy walls as they mature into pollen. Maturation involves the synthesis of a heavy outer wall containing the polymer sporopollenin that protects the cells inside from drying. The haploid nucleus in each developing pollen grain divides by mitosis without cytokinesis to produce two nuclei, the **tube nucleus** and the **generative nucleus.** The generative nucleus divides again by mitosis to produce two male gametes or sperm nuclei. These events may occur in the anther or after the anther splits releasing pollen.

Pollen grains are borne by either wind or animals to the stigma of a pistil. There a pollen grain germinates, producing a pollen tube, which passes through the tissues of the pistil to the ovule and fertilizes the egg (fig. 26.5).

Obtain some pollen from fresh flowers and sprinkle the pollen onto a small drop of 10% raw sugar on a microscope slide. Add a coverslip and observe the specimen at intervals over the next 60 minutes. If pollen tubes do not appear, go to the supply table and get a prepared slide of pollen tubes. Two haploid nuclei will travel down this tube just before

Figure 26.5 Germinating pollen grain showing nuclei in emerging pollen tube.

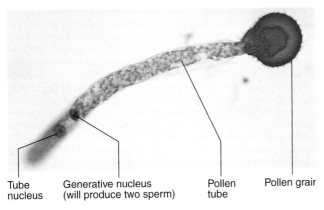

Tube nucleus | Generative nucleus (will produce two sperm) | Pollen tube | Pollen grair

Figure 26.6 Structure of a lily's ovary: (a) photomicrograph of a cross section of an ovary; (b) structure of an ovule, containing the megasporangium in which the female gametophyte develops, producing one egg.

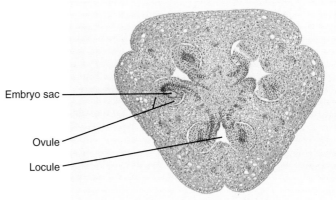

Embryo sac
Ovule
Locule

(a)

fertilization. They are the male gametes of the flower. Sketch a pollen grain and its tube below.

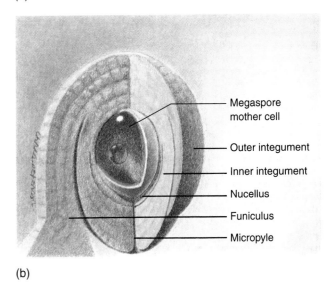

Megaspore mother cell
Outer integument
Inner integument
Nucellus
Funiculus
Micropyle

(b)

Egg Formation

The pistil is sometimes refered to as the female organ in a flower. This is not the case. The pistil is sporophyte tissue and is without sex. The female gametophyte develops within the ovules contained in the ovary. Return to the *Gladiolus* flower bud and cut a thin cross section of the pistil at the level of the ovary. Make a microscope slide by mounting the slice in a drop of water and adding a coverslip. Observe the slice using the scanning objective. Note that the specimen is a compound pistil made up of three fused **carpels.** Sketch the general structure of the pistil in cross section.

Identify the three **locules,** or cavities of the ovary, and the two rows of ovules in each locule. The female gametophyte develops inside the ovule (fig. 26.6). During the evolution of flowering plants, natural selection favored plants in which the carpels folded inward to form the sealed ovarian structure now found in flowering plants. What is the advantage of an internally located ovule?

Depending on the species, an ovary may contain one or more ovules. Each ovule contains a **nucellus** or **megasporangium.** Two integuments grow from the ovarian wall to surround the megasporangium except at a small region known as the **micropyle** (fig. 26.6). The ovule is attached to the ovarian wall by a short stalk, the **funiculus.**

Within the megasporangium is a diploid **megaspore mother cell.** It divides by meiosis to produce four haploid

Figure 26.7 Events in formation of a mature embryo sac, the female gametophyte: (a) through (e) show stages in development of female gametophyte; (f) events of fertilization in mature ovule; (g) embryo develops from zygote formed by fusion of egg and sperm nuclei and endosperm from fusion of two polar nuclei and second sperm nucleus.

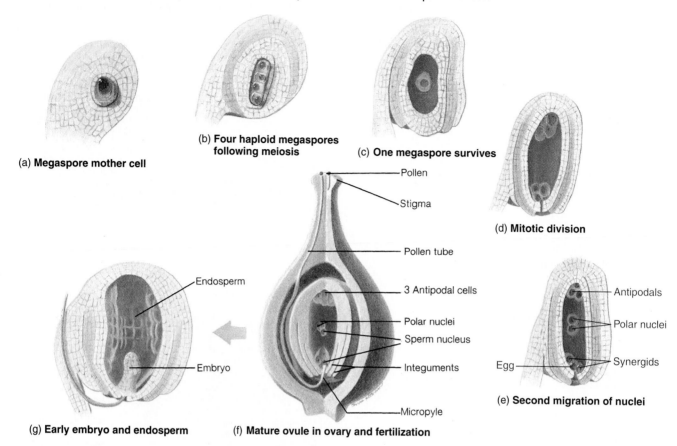

(a) **Megaspore mother cell**

(b) **Four haploid megaspores following meiosis**

(c) **One megaspore survives**

(d) **Mitotic division**

(e) **Second migration of nuclei**

(f) **Mature ovule in ovary and fertilization**

(g) **Early embryo and endosperm**

Labels in (f): Pollen, Stigma, Pollen tube, 3 Antipodal cells, Polar nuclei, Sperm nucleus, Integuments, Micropyle

Labels in (g): Endosperm, Embryo

Labels in (e): Antipodals, Polar nuclei, Egg, Synergids

megaspores. In most plants, three of these die, and the fourth increases greatly in size. Its nucleus divides three times by mitosis to produce eight haploid nuclei. This large multinucleate cell is called the **embryo sac** and is the **female gametophtye** stage of the flowering plant.

Obtain a prepared slide of a cross section of a lily ovary and look at it through your compound microscope. Identify an ovule and its embryo sac. Sketch them below.

While there is no one pattern of embryo sac (female gametophyte) maturation among all flowering plants, the following provides a reasonably general description of events in most flowers (fig. 26.7). The eight haploid nuclei in the embryo sac are arranged in two groups of four, one group near the micropyle and the other at the opposite placental or **chalazal** end. One nucleus from each group of four migrates to the center of the cell. They are termed **polar nuclei.** Cell membranes now form around each of the three remaining nuclei in both groups, yielding a seven-cell stage. The **central cell** is a large binucleate cell, and the remaining six are small mononucleate cells at each end of the embryo sac. Those at the chalazal end are called **antipodals** and soon disintegrate. Of those at the micropylar end, one becomes the functional **egg** and the other two, called **synergids,** disintegrate. Look at the demonstration microscope slide of a seven-celled embryo sac and compare to figure 26.7.

Fertilization

When pollen falls on the surface of the stigma, a sugary secretion stimulates it to germinate. The pollen tubes penetrate the loosely arranged tissue of the style and grow to the ovule. The micropyle allows the pollen tube to enter the ovule. Two haploid nuclei, termed **sperm,** pass down the pollen tube. One enters the egg and fuses with its nucleus to form a diploid zygote. The other enters the central cell where it eventually fuses with the two polar nuclei to form a triploid central cell. It develops into **endosperm,** a tissue that supplies nutrients to the developing embryo.

Flowering plants are unique in having **double fertilization;** both embryo and endosperm receive genetic material from two parents. The developing complex of embryo, endosperm, and embryo sac become the seed of the plant. The ovarian wall becomes the flesh of the plant's fruit.

▶ Obtain a prepared slide showing a pollen tube entering an ovule. Study it with your compound microscope and sketch what you see below.

Embryo Development

Following fertilization, both zygote and endosperm grow and divide. The zygote divides by mitosis to produce a filament of cells called the **proembryo.** Cells at the base of the filament, located at the micropylar end, become **suspensor cells.** Their growth pushes the filament into the nourishing endosperm. Cells at the opposite (chalazal) end of the filament divide to produce at first a **globular** embryo stage.

Continued differential growth and division of the embryo lead to the appearance of characteristic regions. The globular mass of the embryo becomes **heart-shaped** as the seed leaves or **cotyledons** develop. The name *dicot* refers to the fact that these plants have two cotyledons. Monocots have only one. The cotyledons may accumulate food materials from the endosperm and become thick and fleshy as in peas or beans where the cotyledons represent the bulk of the mature embryo. In other species, such as wheat and corn, the cotyledons may be small and thin and the endosperm persists and stores starch, proteins, and lipids.

In a mature dicot embryo, the **shoot apex** (= plumule) is seen as a small node of tissue between the cotyledons. The meristematic regions located here produce the shoot. Beneath the shoot apex and cotyledons is a short embryonic stem, the **hypocotyl.** In many plants, but not all, the hypocotyl elongates at germination to carry the cotyledons

to the soil surface where the new shoot begins to grow from the shoot apex. It connects with an embryonic root, the **radicle,** attached to the suspensor cells. A root apical meristem located at the tip of the radicle will divide to produce new root cells at germination.

▶ Obtain a slide of the fruit of the common weed shepherd's purse *(Capsella)* and look at it through your compound microscope. Several seeds are contained in the fruit. Within each seed, the developing embryo occupies most of the volume. Figure 26.8 shows some of the developmental stages of the embryo in the seed. Depending on its stage of development, your slide may contain only some of these stages.

▶ In a mature seed, the embryo will be bent back on itself and be surrounded by the seed coat, derived from the integuments of the ovule. Identify a mature embryo on your slide and sketch it below, labeling the major tissue regions.

Seeds

▶ Obtain a lima bean that has been soaked overnight to soften the seed coat and hydrate the tissues. Find the **hilum** where the ovule attached to the placenta (fig. 26.9). At one side of the hilum, the micropyle should be visible as a small distinct aperture. Peel away the seed coat to find the two **cotyledons** of the embryo. Separate the cotyledons and examine each half to see the other parts of the embryo. You should see the **shoot apex, radicle,** and the **hypocotyl.** Is the lima bean a monocot or dicot?_____

▶ Obtain a soaked corn kernel and cut it lengthwise at right angles to the broad side of the seed (fig. 26.10). Examine the halves to find the embryo in the base and on one side of the kernel. Note the large amount of endosperm in the kernel. A corn kernel is really a fruit consisting of a single seed to which the ovary wall has grown fast. The covering is called the **pericarp.**

Figure 26.8 Photos and artist's interpretation of stages in the development of a *Capsella* (shepherd's purse) embryo in embryo sac covered by developing seed coats.

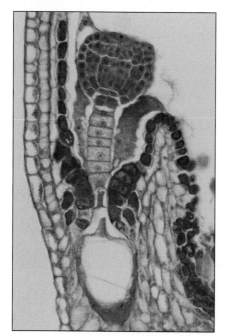
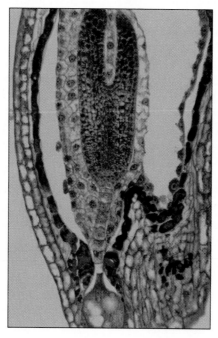
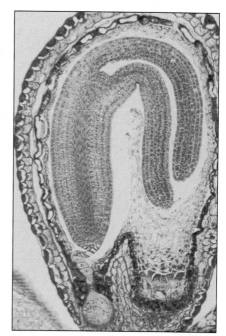

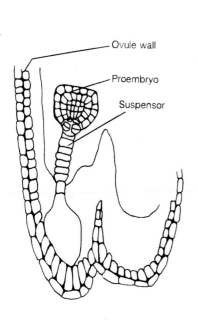

- Ovule wall
- Proembryo
- Suspensor

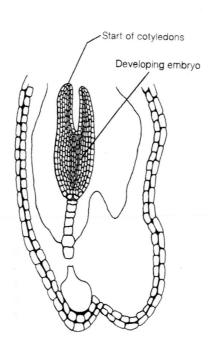

- Start of cotyledons
- Developing embryo

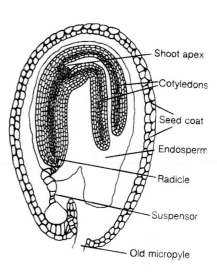

- Shoot apex
- Cotyledons
- Seed coat
- Endosperm
- Radicle
- Suspensor
- Old micropyle

Figure 26.9 Structure of a bean seed.

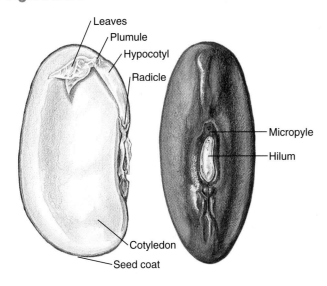

- Leaves
- Plumule
- Hypocotyl
- Radicle
- Micropyle
- Hilum
- Cotyledon
- Seed coat

Figure 26.10 Structure of a corn seed.

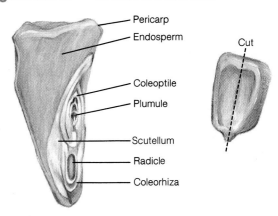

- Pericarp
- Endosperm
- Coleoptile
- Plumule
- Scutellum
- Radicle
- Coleorhiza
- Cut

▶ Draw the cut surface of the corn. Label the shoot apex, the surrounding sheath (coleoptile), the single cotyledon, which in grains is called the **scutellum,** the radicle and its enclosing sheath, the **coleorhiza,** the endosperm, and pericarp.

the **mesocarp,** and the papery part is the **endocarp.** Inside the endocarp is the seed, which develops from the ovule of the flower.

One of the factors that has allowed flowering plants to be the dominant, most diverse group of land plants is their capacity for wide dispersal. Fruits promote dispersal. Animals eat many fleshy fruits and seeds. The tough seed coat protects the embryo as it travels through the digestive tract. When the animal defecates, the seeds have been transported to a new location.

Some fruits, such as those of maples and ash, are winged and are transported by the wind. Others are spiny, such as cocklebur and beggar tick, and stick to the coats of mammals and birds. The variety of fruit types can be overwhelming (fig. 26.13). Botanists have devised a classification system that is outlined in table 26.3.

▶ Dissect several of the fruits available in the lab and assign them to the appropriate type. Make a list below of the fruits studied.

Fruits

Corn, beans, berries, watermelons, and nuts are all examples of fruits because they develop from the ovary of a flower and surround seeds. Vegetables, such as carrots, cabbages, and celery, differ from fruits in that they are not derived from the flower but develop from other parts of the plant, such as the roots, leaves, or stems.

The flesh of fruits develops from the ovary wall, the **pericarp,** which may thicken and differentiate into layers (fig. 26.11). Depending on the species, these layers may or may not be easy to distinguish.

▶ Examine a fresh apple that has been cut in half (fig. 26.12). The "skin" is the **exocarp,** the fleshy part is

Figure 26.11   Development of a pear: (a) flowers; (b) early stage of fruit development showing ovarian wall enlargement; (c) ripe pears.

(a)

(b)

(c)

Figure 26.12   Layers in an apple.

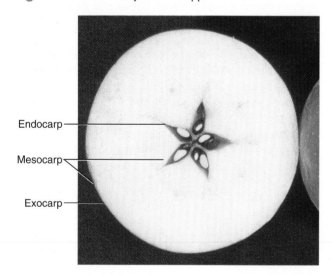

Endocarp

Mesocarp

Exocarp

Investigating the Effects of Hormones on Plant Development

Plant development is influenced by external factors such as water, mineral and light availability, and by internal factors, especially plant hormones. These relatively small, organic molecules play a major role in the regulation of plant growth by altering the expression of genes that lead to a change in the structure and function of the affected target cells. Five classes of plant hormones have been described: auxins, cytokinins, ethylene, abscisic acid, and gibberellins. **Differentiation,** the developmental process by which a relatively unspecialized cell undergoes a progressive change into a more specialized cell with specialized functions, is

known to result from the influence of the ratio of various hormones. In this section of the lab, you will experimentally determine some of the effects of hormones on plant growth.

Differentiation in Plant Tissue Culture

The first experiment is designed to demonstrate the totipotency of plant cells: the capability of plant cells to develop into any structure found in the plant. In testing for totipotency, you will determine that hormones play an important role. The experimental material will be tobacco cells that have been raised in culture producing a mass of undifferentiated cells called a **callus.** Callus cultures can be prepared in the lab or purchased from biological supply houses that grew the cells by taking a small sample of parenchyma cells from the stem of a tobacco plant and putting them into a growth medium containing vitamins, various salts, hormones, and sucrose as an energy source. The growth process takes weeks.

Your instructor will describe to you how to handle the tobacco callus. Strict sterile technique must be observed because fungal spores or bacteria can enter the culture and overgrow the callus destroying it. You will have to use sterile instruments, glassware, culture media, and work in a sterile transfer hood in order to be successful.

You will have available to you three variations of a growth media formula made up in agar. The three tubes will be identical except for the ratio of the concentrations of two types of hormones, auxins, and cytokinins. One medium will have more auxin than cytokinin, another more cytokinin than auxin, and the last will have equivalent amounts. These media will be in tubes labeled *A, B,* and *C,* but you will not know which tube contains which hormone. Read the section in your textbook that describes the effects

Figure 26.13   Examples of different types of fruits.

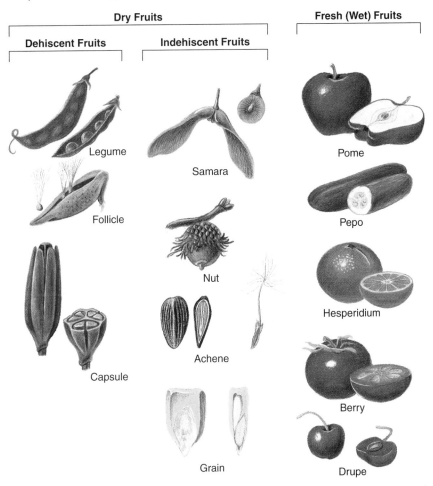

| | Dry Fruits | | Fresh (Wet) Fruits |

Dehiscent Fruits

Legume

Follicle

Capsule

Indehiscent Fruits

Samara

Nut

Achene

Grain

Pome

Pepo

Hesperidium

Berry

Drupe

TABLE 26.3   Classification of fruits by types

Kind of Fruit	Description	Examples
Simple	Develop from the ovary of a flower with single pistil	
Dry		
Achene	Fruit does not split open at maturity; seed remains attached to ovarian wall by stalk	Sunflower
Follicle	Fruit splits along one edge at maturity to release seeds	Milkweed, peony
Legume	Fruit splits along both edges at maturity to release seeds	Bean, carob, lentil, pea, peanut
Nut*	Hard ovarian wall encases a single seed and does not split open at maturity	Hazelnut, hickory nut
Grain	Seed coat is permanently united with ovarian wall; does not split open at maturity	Barley, oats, rice, wheat
Samara	Ovarian wall forms "wings" that aid in dispersal	Maple, elm, ash
Fleshy		
Berry	Entire ovarian wall is fleshy	Tomato, grape, date, avocado
Drupe	Outer zones of ovarian wall become fleshy, while inner zone is hard	Cherry, peach, plum, olive
Pome	Inner zone of ovarian wall is leathery and outer is fleshy incorporating accessory structures	Apple, pear
Aggregate	Develop from single flowers with multiple pistils	Raspberry, strawberry

*Many so-called nuts are not true nuts; the peanut is a legume and walnuts, pecans, and cashews are drupes which lose the fleshy part during maturation.

of these hormones on cellular differentiation. Hypothesize what will happen if you place callus tissue on each of the three media. State your hypotheses below.

Set up the experimental cultures using sterile technique. Divide the callus tissue you are given into thirds. Place one piece in each of three culture tubes containing different growth media formulations. These cultures will now be incubated for four to five weeks and you should observe them weekly during this time and make notes describing any differences that you see among them. At the end of the growth period, tell your instructor which tube had the high auxin concentration relative to cytokinin, which was low auxin, and which was equal. Support your answer with your experimental observations and cite your textbook indicating the known effects of these ratios of hormones.

Effects of Hormones on Seedling Growth Patterns

Wisconsin Fast Plants, developed by the plant breeder Paul Williams at the University of Wisconsin, go through their life cycle in about 40 days, from seed germination to production of plantable seeds, making them ideal for classroom use. These plants are in the genus *Brassica,* which is in the mustard family. Cauliflower, broccoli, rape seed, and turnips are also in this genus. You will use these fast plants to investigate the effects of hormones on plant development.

Brassica seeds with different genetic backgrounds and phenotypes can be ordered from biological supply houses. Wild-type plants have normal phenotypes and the plants grow several inches tall. A mutant strain of *Brassica* produces rosette plants that are very short because there is little elongation of the cells in the internodes. The nodes of the stem are adjacent to each other and the leaves that grow from the nodes give the plant a rosette appearance.

Plant breeders who first observed this hypothesized that the rosette plants were mutants that could not produce a class of plant hormones that cause stem elongation, the gibberellins. Your assignment is to conduct an experiment that tests this hypothesis. To do so, you should know that gibberellins can be absorbed by the leaves and the shoot apex. The gibberellins will be conducted by the vascular tissues throughout the plant.

The experiment may be started in one of two ways. One week before you are to administer the gibberellin you may be asked to plant seeds of wild-type and rosette *Brassica* in special quad pots so that plants will be available at the start of the hormone lab. Alternatively, your instructor may plant the seeds of wild-type and rosette *Brassica* eight days before the hormone lab.

When plants are available, carefully observe both the wild and mutant forms and describe them below. Explain why the mutant is called rosette.

Wild-type

Mutant type (rosette)

Your instructor will assign to you a quad of four plants which will include two wild-type and two rosette mutants. Measure the height of all plants in millimeters from the cotyledons to the very tip of the plant and record in table 26.4.

TABLE 26.4   Height in millimeters of plants in hormone experiment

| | Wild-Type Plants: | | Rosette Plants: | |
	Water Treated	GA Treated	Water Treated	GA treated
Your data				
Start				
End				
Class data summary				
Average height change				
% change (over time)				

If this effect is due to a deficiency of gibberellic acid, what would you hypothesize would happen if you administered gibberellic acid to the plants?

Gibberellic acid can be administered by dissolving it in water and applying a drop of this solution to the leaves of the plant every other day. The gibberellic acid will be transported throughout the plant. Using your plants, create an experimental design with controls that investigates the effects of gibberellic acid on normal and mutant plants.

Your instructor will give you distilled water and a solution of gibberellic acid. Using a different pipette for each solution, apply single drops to all of the leaves on your plants. This procedure will have to be repeated every other day for the next week.

After at least seven days, measure the length of the stem between the cotyledons and shoot tip and record in table 26.4.

When all students have completed their individual measurements, the data should be put on the blackboard and an average calculated for each treatment. Record these in the bottom half of table 26.4. Now calculate the percent change in height during the week of the experiment for each of the treatments.

Restate your hypothesis and decide whether to accept or reject it based on your data.

Learning Biology by Writing

Because the activities in this lab were both observational and experimental, two alternative assignments are described. Your instructor will indicate which one to complete. To summarize your plant anatomy observations, write a brief essay describing the male and female gametophytes of a flowering plant. Indicate how double fertilization occurs and the advantages of seed as a dispersal mechanism. To summarize your experimental work with either tissue culture or whole plants, write a lab report in which you state the hypothesis that was investigated, describe your techniques, present your data, and come to a conclusion regarding your hypothesis.

Internet Sources

Plant tissue culture is becoming a standard technique in the propagation of plants because it is possible to clone individuals with desirable traits or, in some cases, it is possible to insert desirable genes into species of plants that never before had those genes. Search the World Wide Web for information on "plant tissue culture," using the quotes to limit your search to this exact phrase. Read the information available at a few of the sites found and summarize what is happening in this exciting field of research. Include the URLs for future reference.

As an alternative assignment, your instructor may ask you to turn in answers to the Lab Summary or Critical Thinking Questions that follow.

Lab Summary Questions

1. Describe fertilization in a flowering plant.
2. Outline the steps in endosperm formation.
3. A seed has been described as a mature ovule. How can this be true? Explain you answer.
4. Why do some plants have colorful flowers and edible fruits?
5. What evidence do you have that supports the hypothesis that plant hormones influence differentiation in plants?

Critical Thinking Questions

1. Make a list of several vegetables that are actually fruits. In making this list, what was your definition of a fruit?
2. If wind and insects carry pollen from several flower species at the same time, why doesn't the pollen of one species fertilize the eggs of a second species, thus forming hybrids?
3. The pistil is often described as the female part of the flower and the stamens the male part. Which structures actually produce the egg and sperm?
4. Why do you suppose flowers are fragrant and colorful?
5. Why are fruits often colorful and sweet?
6. Many fruits are harvested before they ripen and then are treated with ethylene gas just before they are shipped to stores. Why?

INTERCHAPTER

Investigating Animal Form and Function

In this section of the lab manual are a series of topics on the organ systems of animals. In each topic you are asked to dissect a fetal pig as a representative mammal. Some general instructions are given here to orient you to proper dissection technique.

Vertebrates

General Dissection Information

Fetal pigs are unborn fetuses taken from a sow's uterus when she is slaughtered. They are a by-product of meat preparation and are used in teaching basic mammalian anatomy. Often the circulatory system has been injected with latex so that the veins will appear blue and the arteries red.

Strong preservatives are used and can irritate your skin, eyes, and nose. Rinse your pig with tap water to remove some of the preservative to lessen irritation. If you wear contact lenses, you may want to remove them during dissection, since the preservative vapor can collect in the water behind the lens and can be very irritating. If a lanolin-based hand cream is available, use it on your hands before and after dissection to prevent drying and cracking of the skin. Alternatively, your instructor may ask you to purchase rubber globes and use them during this and subsequent dissections.

You will use this same fetal pig in several future labs to study the circulatory, excretory, and muscular systems. This means that you must do a careful dissection each time so that as many structures as possible are left undamaged and in their natural positions. Two good rules to keep in mind as you dissect your animal are: *cut as little as possible and never remove an organ unless you are told to do so.* If you indiscriminately cut into the pig to find a single structure without regard to other organs, you will undoubtedly ruin your animal for use in future laboratories.

Many of the instructions for dissection in this lab and later ones use anatomical terms to indicate direction and spatial relationships when the animal is alive in normal orientation. You should know the meaning of such terms as:

Anterior—situated near head or, in animals without heads, the end that moves forward.

Caudal—extending toward or located near tail.

Cephalic—extending toward or located on or near head (also cranial).

Distal—located away from the center of the body, the origin, or the point of attachment.

Dorsal—pertaining to the back as opposed to **ventral,** which pertains to the belly or lower surface.

Median—a plane passing through a bilaterally symmetrical animal that divides it into right and left halves.

Posterior—toward the animal's hind end: opposite of anterior.

Proximal—opposite of distal.

Right-left—always in relation to the animal's right and left, not yours.

Sagittal—planes dividing an animal along the median line or parallel to the median.

Obtain a fetal pig and place it in a dissecting tray, ventral side up. Take two pieces of string and tie them tightly to the ankles of both right legs (the animal's right legs, not the legs to your right when the pig is lying on its back). Run the strings under the pan and tie each to the corresponding left leg. Stretch the legs to spread internal organs for easier dissection (fig. I.1). Do not pull so hard that you break internal blood vessels.

External Anatomy of Fetal Pig

Rows of **mammary glands** and the **umbilical cord** should be prominent on the ventral surface of your pig. Mammary glands and hair are two of the diagnostic characteristics of the class *Mammalia.* The umbilical cord attached the fetal pig to the placenta in the sow's uterus. Look at the cut end of the cord and note the blood vessels that carried nutrients, wastes, and dissolved gases from the fetus to the placenta, where they were exchanged by diffusion with the maternal circulatory system.

Determine the sex of your pig. Identify the **anus.** In females, there will be a second opening, the **urogenital opening,** ventral to the anus. In males, the urogenital opening is located just posterior to the umbilical cord. **Scrotal sacs** will be visible just ventral to the anus in males.

Figure I.1 shows the sequence of cuts that should be made to expose the internal organs. Make the cuts in the sequence indicated.

Figure I.1 External anatomy of fetal pig. Numbered lines indicate incisions to be made. Numbers show sequence of incisions.

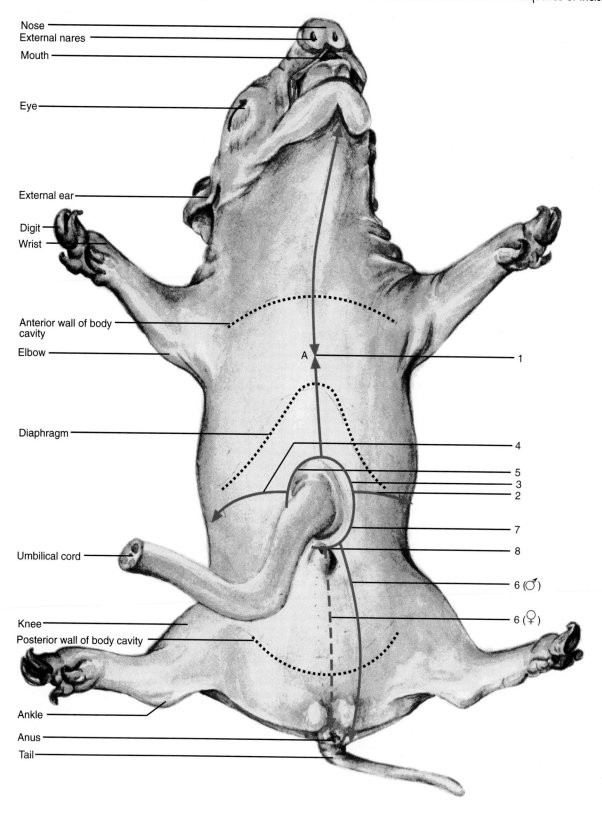

Nose
External nares
Mouth

Eye

External ear

Digit
Wrist

Anterior wall of body cavity

Elbow

Diaphragm

Umbilical cord

Knee
Posterior wall of body cavity

Ankle

Anus
Tail

A

1
4
5
3
2
7
8
6 (♂)
6 (♀)

LAB TOPIC 27

Investigating Digestive and Gas Exchange Systems

Supplies

Preparator's guide available on WWW at
    http://www.mhhe.com/dolphin

Equipment

Compound microscope
Dissecting microscopes

Materials

Fetal pigs
Dissecting trays
Dissecting instruments: scissors, forceps, blunt probe,
    razor blade or scalpel
Thread
Prepared slides
    Mammalian small intestine, cross section
    Mammalian lung section
Hydra culture
Daphnia cultures
Slides and coverslips
Pasteur pipettes with bulbs
Live crickets, grasshoppers, or roaches
Petri plates filled with wax
Demonstration dissection of crayfish gills and gill bailer
Live crayfish
India ink

Solutions

Methyl cellulose
10 and 100 mM glutathione
70% ethanol
Saline (0.7% NaCl)
Ice

Prelab Preparation

Before doing this lab, you should read the introduction
and sections of the lab topic that have been scheduled
by the instructor.

You should use your textbook to review the
definitions of the following terms:

alveolus
circular muscle
gill
longitudinal muscle
mucosa
pharynx
smooth muscle
sphincter
tracheole

You should be able to describe in your own words
the following concepts:

Incomplete and complete digestive tracts
Pathway of food though mammalian digestive tract
Anatomy of *Hydra* (see fig. 19.5)
Pathway of air to mammalian lungs
Mechanics of mammalian breathing

As a result of this review, you most likely have
questions about terms, concepts, or how you will do
the experiments included in this lab. Write these
questions in the space below or in the margins of the
pages of this lab topic. The lab experiments should
help you answer these questions, or you can ask your
instructor for help during the lab.

Objectives

1. To investigate the feeding behavior of *Hydra*
2. To dissect the digestive and respiratory systems in a fetal pig
3. To observe the histology of small intestine and lung
4. To observe the anatomy of gill and tracheole respiratory systems

Background

Digestion involves the chemical breakdown of complex food materials through the actions of enzymes. Chemically, it requires the hydrolysis of the covalent bonds holding together large, polymeric molecules. Proteins are hydrolyzed into amino acids and starches into sugars. These small molecules are used by the animal either as sources of energy or as building units to make other types of molecules, such as proteins, starches, fats, and nucleic acids.

Digestion can occur **intracellularly** or **extracellularly.** In simple organisms, such as protozoa and sponges, the individual cells of the organism ingest food materials by pinocytosis and phagocytosis and digestion occurs in food vacuoles inside of cells. Other organisms have extracellular digestion in special digestive organs that are either sac-like or tubular. Multicellular organisms, such as cnidarians and flatworms, have **incomplete** digestive tracts that are blind sacs; that is, food enters the mouth, passing into a chamber where enzymatic digestion and absorption occur, and nondigestible material must be expelled through the same opening. This seems to be a somewhat inefficient system because, when nondigestible material is expelled, recently ingested food may also be lost.

Higher animals have a **complete** digestive system, a continuous tube from the mouth to the anus in which food is sequentially broken down and absorbed. Different enzymes are secreted by glands at various points along the digestive tube, so that the digestion of different types of molecules occurs as food passes through the system.

Mere digestion, however, is not sufficient for processing food material. To be of value to the animal, absorption must also occur. In vertebrates, specialized regions of the digestive tube absorb nutrients, salts, and water from the digested material, or chyme. The undigested or nondigestible residues pass on to temporary storage areas before being defecated.

The length of the continuous digestive tube in comparison to the length of the organism seems to be related to evolutionary development. Arthropods and annelids have a tube that passes more or less straight through the body from head to tail. The digestive tubes in such organisms include storage areas, grinding areas, digestive areas, and absorptive areas, but the tube is not longer than the organism. In vertebrates, on the other hand, the digestive tube is many times longer than the animal's length, with much of this length devoted to absorption. After absorption, the circulatory system carries food materials to cells throughout the body.

The cells of most animals are capable of aerobic metabolism. It requires an organism to have some mechanism for exchanging gases with its environment. Oxygen must move from the environment to every cell in the body, where it ultimately functions as a terminal electron (hydrogen) acceptor. Without oxygen, the mitochondrial cytochrome system would not operate, and cells could not carry out glycolysis, the Krebs cycle, and other oxidations of food materials. Furthermore, carbon dioxide, which is produced during the Krebs cycle and in other reactions, must pass from cells to the environment to maintain a consistent intracellular acid-base balance.

In small organisms with relatively low metabolic rates and large surface areas relative to body mass, free diffusion of these gases satisfies an animal's needs. Protozoa, sponges, cnidarians, flatworms, and roundworms have no anatomical specializations for respiration. Gases simply pass to and from the environment through their surface layers.

Larger aquatic animals, such as lobsters, clams, and fish, have developed specialized respiratory surfaces called gills. These featherlike surfaces allow the body fluids to circulate in a closed network that is separated by only a cell layer from the water in the environment. As the blood flows through the gills in one direction, water passes over the external gill surface in the opposite direction, allowing a very efficient exchange of respiratory gases.

Because gill surfaces must remain moist to function properly, they are not suited to gas exchange in terrestrial environments. Terrestrial organisms have **tracheolar systems,** as are found in insects, or **lung systems,** as are found in many vertebrates. In these systems, oxygen-rich air enters the animal by a system of tubes and is brought in close proximity to the blood or the tissues where gas exchange occurs. There are exceptions to these generalizations, of course. Many adult amphibians have lungs, but also exchange gases directly through the skin and mouth surface, but as larvae they have gills. One family of salamanders commonly found in North America, the *Plethodontidae,* lacks lungs and depends entirely on cutaneous respiration.

The oxygen-carrying capacity of blood is increased by the presence of a respiratory pigment, such as **hemocyanin** (usually dissolved in hemolymph) or **hemoglobin** (usually in blood cells). These pigments bind loosely and reversibly with oxygen to facilitate oxygen transport from the gills to the tissues. Blood is oxygen deficient when it enters the gills or lungs. Oxygen diffuses into the aqueous portion of the blood and combines with the pigments. As the blood flows to areas where there is little oxygen, the diffusion gradient causes the oxygen to move into the tissues, where it is used by mitochondria. Conversely, a high concentration of carbon dioxide is usually present in active tissues. The CO_2 moves down the diffusion gradient into the blood where it simply may dissolve or may combine chemically with the pigments. The carbon dioxide is, in turn, carried back to the respiratory organs where it is exchanged with the environment.

LAB INSTRUCTIONS

You will study the digestive and respiratory systems of several animals.

Invertebrate Feeding Behavior in *Hydra*

Hydra is a small, freshwater cnidarian related to the jellyfish and sea anemones (see fig. 19.5). It lives attached to submerged rocks, leaves, and twigs. Hydra's body is organized simply, consisting of only two layers of cells surrounding a hollow cavity. However, the organism is highly specialized for food gathering; it uses tentacles to capture food and transfer it into the digestive (**gastrovascular**) cavity. The food is then digested by enzymes that are secreted by cells lining the cavity. Because of the organism's small size, the cells of *Hydra* obtain nutrients from the digestive cavity by simple diffusion. Each body cell also exchanges oxygen and carbon dioxide directly with the surrounding water and releases cellular metabolic wastes directly into the water as well.

▶ Obtain a living *Hydra* from the supply table and place it in a small dish that has been thoroughly washed so it has no chemical residues from other uses. Observe the *Hydra* with a dissecting microscope against a dark background. Sometimes a *Hydra* contracts into a small ball after it is transferred, but it should relax, attach to the glass, and resume its normal functions before long. If the *Hydra* attaches to the surface water film, sink it by dropping water on it from a dropper.

▶ Locate the **basal disc** (the animal's point of attachment), the body **column,** the **tentacles,** and the **hypostome,** a small raised region surrounded by the bases of the tentacles (see fig. 19.7). The **mouth** is located in the center of the hypostome but is difficult to see.

▶ Obtain some *Daphnia,* a freshwater crustacean, that *Hydra* will feed on. Small pieces of aquatic annelids such as California black worms also elicit a strong feeding response. Transfer some food organisms immediately next to your *Hydra* with a Pasteur pipette and watch closely through a dissecting microscope. *Hydra,* like other cnidarians, is a carnivore. It traps food by stinging and paralyzing other animals with specialized epidermal cells called **cnidocytes** located on its tentacles. Each cnidocyte has a special organelle, the nematocyst, which contains a coiled tube. When stimulated, a nematocyst shoots out its thread, which either entangles or pierces and poisons the prey. Seeing the *Hydra* capture food will take patience, but the time is well spent.

Record below how *Hydra* feeds. Include drawings showing how prey are transferred from the tentacles to mouth.

▶ Observe the feeding response carefully. The response is stimulated by the presence of chemicals released by wounded prey. Glutathione is a tripeptide found in the body fluids of many invertebrates, and *Hydra* respond vigorously to it. If a high concentration of reduced glutathione is introduced experimentally into the water around a *Hydra,* the organism may respond by opening its mouth so wide that it virtually turns itself inside out.

Use the glutathione available in the lab to evoke this response. Add a drop or two of 10 mM glutathione next to the *Hydra*. If there is no response, add a drop or two of 100 mM glutathione. Why should you try the 10 mM concentration first? Why not start with the 100 mM glutathione? Record the reponses of the Hydra below.

Once food is in *Hydra*'s gastrovascular cavity, digestion begins. Enzymes secreted by certain cells lining the cavity's begin **extracellular digestion.** Partially digested bits of food material are later taken up by phagocytic cells in the cavity lining, and further digestion occurs inside the food vacuoles in these cells. This is **intracellular digestion.** Food absorbed by the cells lining the gastrovascular cavity supplies all cells of the body.

Study the longitudinal section of *Hydra* in figure 19.5. Note that the body consists of only two layers—the outer layer of the body, the **epidermis,** and the inner **gastrodermis,** which lines the **gastrovascular cavity.** This cavity extends from the body cavity into each of the tentacles. There are several specialized cell types in each of the body layers. Some of these specialized cells are the cnidocytes on the tentacles and enzyme-secreting cells and flagellated cells in the gastrodermis. How many cell layers thick is the body wall?_____. Guess how many cell layers thick the body wall of a mammal is?_____.

When animals increase in size and complexity, diffusion from the digestive cavity to the cells, as occurs in *Hydra,* is no longer adequate to supply the tissues' demands. Increased complexity means special systems for the transport of material from cell to cell. The mammalian digestive and circulatory systems represent the height of this development from physiological and evolutionary perspectives.

Mammalian Digestive System

If you have not already done so, get a fetal pig and dissection tray. Read the general instructions on page 329.

Anatomy of the Mouth

Located in the upper neck region beneath the skin are three pairs of **salivary glands: parotid, submaxillary,** and the **sublingual,** which is difficult to find. They produce saliva containing the enzyme salivary amylase, which hydrolyzes starch during chewing.

Remove the skin and muscle layer from one side of the face and neck, as in figure 27.1 to see the rather dark, triangular-shaped parotid gland. Note the difference in appearance between muscle tissue and glandular tissue. If you dissected carefully, you should find the duct that drains the gland into the mouth near the upper premolar teeth. Try to trace the duct. The other salivary glands lie beneath and below the parotid gland.

With heavy scissors, a razor blade, or scalpel, cut through the corners of the mouth and extend the cut to a point below and caudal to the eye.

Open the mouth, as in figure 27.2 and observe the **hard palate,** composed of bone covered with mucous membrane, and the **soft palate,** which is a caudal continuation of the soft tissue covering the hard palate. The oral cavity ends and the **pharynx** starts at the base of the tongue.

The pharynx is a common passageway for the digestive and respiratory tracts, as seen in figure 27.3. The opening to the **esophagus** may be found by passing a blunt probe down along the back of the pharynx on the midline. This collapsible tube connects the pharynx with the stomach. The **glottis** is the opening into the **trachea** or windpipe and lies ventral to the esophagus. It is covered by a small white tab of cartilage, the **epiglottis.** The epiglottis may be hidden from view in the throat; if so, you will have to pull it forward with forceps or a probe to see it.

Alimentary Canal Anatomy

You will see the path of the esophagus to the stomach when you dissect the respiratory system. To view the rest of the alimentary tract and associated glands, use a scalpel or pair of scissors to make incisions into the **abdominal** cavity as indicated in fig. I.1, page 330. Cut carefully through only the skin and muscles to avoid damaging the internal organs.

As you lay the tissue flaps back, you will see the organs of the abdominal cavity covered by a translucent **peritoneal** membrane. The flap containing the umbilical cord will be held in place by blood vessels. Tie both ends of a 15-cm piece of thread to the blood vessel about 1 cm apart. Cut the vessels between the two knots and lay this tissue flap back. Leave the thread in place so you later can trace the circulatory system.

Figure 27.1   Dissected fetal pig's head showing salivary glands.

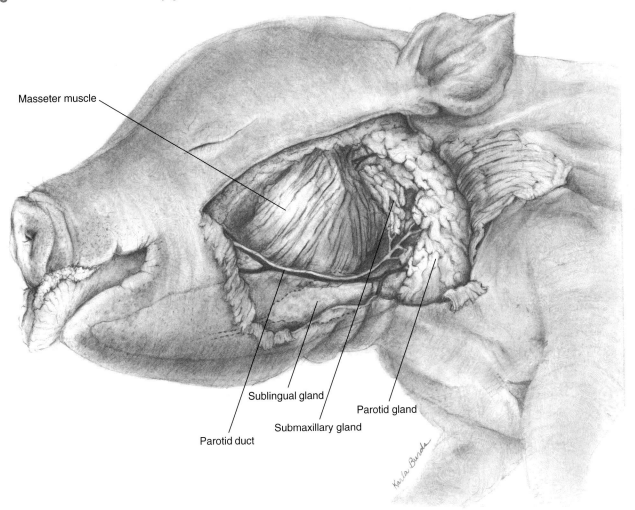

Masseter muscle

Sublingual gland

Parotid duct

Submaxillary gland

Parotid gland

Find the thin, transparent membranes, the **mesenteries,** which suspend and support the internal organs (**viscera**) in the body cavity. The dark brown, multilobed **liver** should be visible caudal to the **diaphragm** (fig. 27.4). If you trace the umbilical vein from the thread to the liver, you will see a green-colored sac, the **gallbladder,** located just below the entrance of the vein into the liver. It stores bile produced in the liver. Bile travels from the gallbladder to the small intestine via the bile duct. Bile is an emulsifying agent that aids in digestion of fats.

Under the liver on the left side is the **stomach.** Locate the point where the esophagus enters the **cardiac region** of the stomach. Gastric glands in the wall of the stomach secrete pepsinogen, hydrochloric acid, and rennin. Pepsinogen is activated by hydrochloric acid to become pepsin, which digests proteins. Rennin is an enzyme that hydrolyzes milk protein. Food leaves the stomach as a fluid suspension, chyme. It enters the **duodenum,** the first part of the small intestine.

Figure 27.2   Anatomy of the fetal pig's mouth.

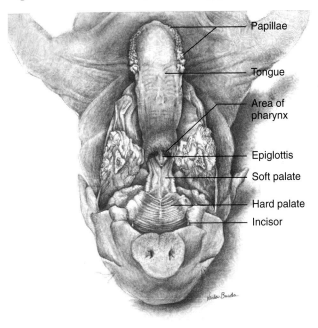

Papillae

Tongue

Area of pharynx

Epiglottis

Soft palate

Hard palate

Incisor

Figure 27.3 Air and food passages across the pharynx.

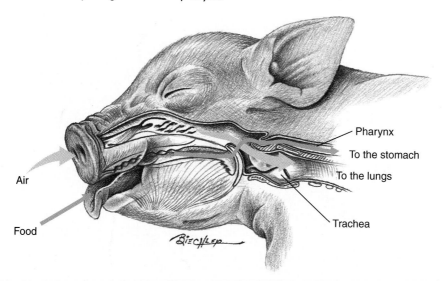

Pharynx
To the stomach
To the lungs
Trachea
Air
Food

Find the **pancreas,** a glandular mass lying in the angle between the curve of the stomach and duodenum. It secretes several enzymes into the duodenum that digest proteins, lipids, carbohydrates, and nucleic acids. Certain cells in the pancreas act as endocrine cells and secrete the hormones insulin and glucagon. In fact, insulin used in human diabetes therapy can be extracted from the pancreases of pigs collected at slaughterhouses.

Remove the stomach by cutting the esophagus and duodenum. Slit the stomach lengthwise, cutting through the cardiac and pyloric **sphincters,** muscles that regulate passage of material into and out of the stomach. The internal surface of the stomach is covered by gastric mucosal cells, which secrete mucus that prevents the stomach from digesting itself. When this protection fails, a peptic ulcer develops.

The small intestine is made up of three sequentially arranged regions: **duodenum, jejunum,** and **ileum.** These areas are difficult to differentiate from each other. Cut out a 2-cm section of the small intestine about 10 cm posterior from the stomach, slit it open, and place it under water in a dish. Use your dissecting microscope to observe the velvety internal lining made up of numerous fingerlike projections called **villi.** The villi are highly vascularized, containing capillaries and lymphatics that transport the products of digestion to other parts of the body, especially the liver.

The ileum opens into the large intestine, or **colon.** They join at an angle, forming a blind pouch, the **cecum,** which in primates and some other mammals often ends in a slender appendage, the **appendix.** In herbivores, the cecum is very large and contains microorganisms that aid digestion by breaking down cellulose.

The **rectum** is the caudal part of the large intestine, where compacted, undigested food material is temporarily stored before being released through the **anus.** The colon of vertebrates contains large numbers of symbiotic bacteria, es-

pecially *Escherichia coli.* These bacteria produce vitamin K, which is absorbed and plays a vital role in blood clotting.

Histology of Small Intestine

Put your fetal pig aside for the moment and obtain a prepared slide of a cross section of a mammalian small intestine. Examine it under scanning power with the compound microscope. Compare what you see to figure 27.5.

The central opening is called the **lumen** and is the space through which food passes as chyme during digestion. Switch to low power and observe the small fingerlike projections of the intestine's inner surface. These are villi and are covered by a layer of cells called the **mucosa.** You should be able to distinguish two cell types in the intestinal mucosa: **goblet cells** and **columnar epithelial cells.** Examine them with the high-power objective. The goblet cells secrete mucus into the small intestine, serving as a lubricant for the passage of chyme. Epithelial cells are involved in absorption.

Return to the low-power objective and observe the **submucosa,** a layer of connective tissue that underlies the mucosa. Look for the blood vessels and lymphatic vessels that ramify through this layer. Sugars, amino acids, glycerides, and other components of digested food must move through the mucosal cells into the submucosa before they can enter the circulatory system and be distributed throughout the body.

To the outside of the submucosa are two smooth muscle layers: an **inner circular layer** and **outer longitudinal layer.** The inner circular muscles change the diameter of the intestine, and the outer muscles alter its length. These muscles contract in a wavelike motion called **peristalsis,** which pushes chyme through the digestive tract. The small intestine is covered by a layer of peritoneal cells, that together with underlying connective tissue, is called the **serosa.**

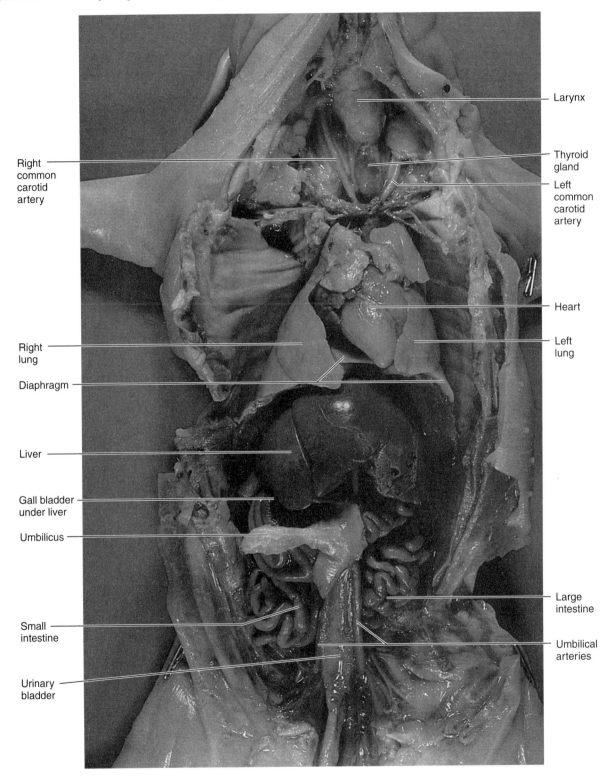

Larynx

Right common carotid artery

Thyroid gland

Left common carotid artery

Heart

Right lung

Left lung

Diaphragm

Liver

Gall bladder under liver

Umbilicus

Large intestine

Small intestine

Umbilical arteries

Urinary bladder

Figure 27.5 Histology of the mammalian small intestine: (*a*) cross section of small intestine; (*b*) magnified view of cell layers.

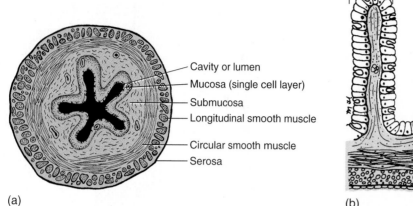

Cavity or lumen
Mucosa (single cell layer)
Submucosa
Longitudinal smooth muscle
Circular smooth muscle
Serosa

(a)

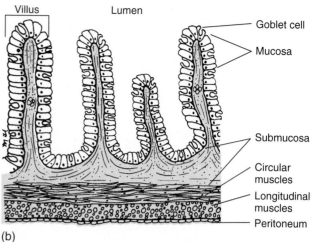

Villus   Lumen

Goblet cell
Mucosa

Submucosa

Circular muscles

Longitudinal muscles

Peritoneum

(b)

Figure 27.6   (*a*) Scanning electron micrograph of cross section of small intestine showing villi (Vi), lumen (Lu), submucosa layer (Su), and muscle layers (Mu). The epithelial cells on the surface of the villi have highly folded membranes, microvilli (Mv), which greatly increase the absorptive surface area of the cell layer, as seen in (*b*) a transmission electron micrograph. (*a*) From R. G. Kessel and R. H. Kardon. *Tissues and Organs: A Text-Atlas of Scanning Electron Microscopy.* 1979. W. H. Freeman and Company.

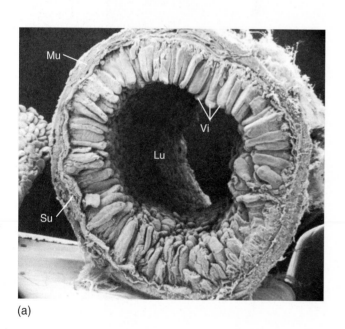

(a)

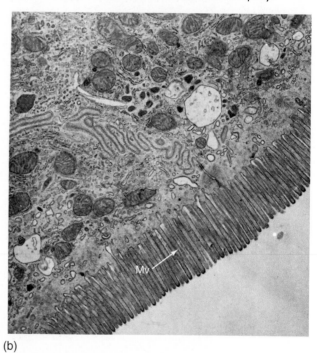

(b)

Figure 27.6 shows scanning electron micrographs of the three-dimensional arrangement of the small intestine. Note how the villi and microvilli increase the surface area. What important process following digestion is facilitated by this increased surface area?

Invertebrate Respiratory Systems

Insect Tracheolar System

The mammalian respiratory system represents only one method of gas exchange in terrestrial animals. Insects and other terrestrial arthropods make use of another, a rather remarkable **tracheole system,** which allows the exchange of gases to occur independent of the circulatory system.

Figure 27.7   Tracheal systems. (*a*) Tracheal system of a flea penetrates all parts of its body, allowing direct gas exchange. (*b*) Large insects pump air through their tracheal systems by contracting muscles that press against air sacs. (*c*) Photomicrograph of tracheae shows ringed appearance due to reinforcement by rings of chitin.

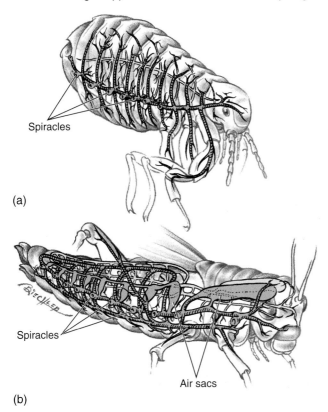

Spiracles

(a)

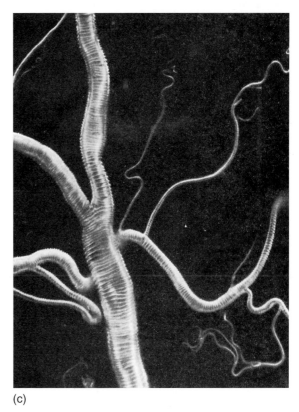

Spiracles

Air sacs

(b)

(c)

In a tracheole system, air enters through several small, lateral openings called **spiracles** and passes into a branching system of tubules called **tracheae** and **tracheoles.** The tracheoles continue branching into a system of fluid-filled tubes where gas exchange occurs (fig. 27.7). The tracheoles branch so extensively that they are never very far from any cell in the body. Gas exchange occurs by diffusion from the cell to the tracheole with little participation of the circulatory system. There are apparently no special muscles that circulate air in the tracheole system, but contraction of the abdominal and flight muscles aid the exchange of air.

Large insects often have **air sacs.** When compressed by movement of surrounding organs, they force large volumes of air through the system.

If your instructor has not prepared a demonstration dissection of a tracheole system, obtain a live, large insect (cricket, grasshopper, or roach) and a petri dish filled with wax to be used as a dissecting pan. Anesthetize the insect with carbon dioxide or by placing it in a freezer. Remove the wings and legs and locate the spiracles along the side. With fine scissors, cut through the dorsal surface close to the lateral margin but do not cut through the spiracles. (Keep the points of your scissors against the inside of the exoskeleton and take small snips so that you do not damage any interior organs.) Run the cuts along the full length of the insect and join them together at the anterior and posterior ends. Remove the dorsal strip of cuticle and pin the insect in

the dish, dorsal side up. Next flood the insect with saline solution to prevent drying and to float tissues for observation. Place the dish under a dissecting microscope and observe.

Gently remove the glistening yellow fat bodies. Vary the lighting angle while looking for a silver-colored tube originating from a spiracle. Remove this trachea with a forceps, mount it in a drop of saline on a microscope slide, and add a coverslip. Observe the trachea with your compound microscope and sketch it below. Compare your observations to figure 27.7c.

Investigating Digestive and Gas Exchange Systems   **339**

Figure 27.8 Direction of water movement over the gills of two animals: (a) crayfish; (b) fish

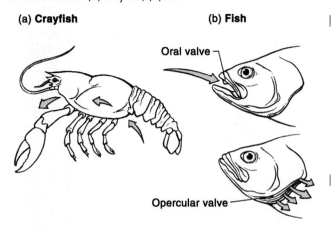

(a) **Crayfish**   (b) **Fish**

Oral valve

Opercular valve

Gills

In aquatic organisms, gills serve in respiratory exchange and ion regulation (fig. 27.8). Gills are found in many different kinds of animals, including larval amphibians, fish, arthropods, and molluscs (see figs. 19.9 and 19.10). Except in molluscs, gills originate in the embryo as featherlike pocketings of the body wall that are amply supplied with blood. In most cases, a special mechanism exists to ventilate the gills, that is, to move water over them. In crayfish, this involves a specialized mouthpart, the **gill bailer,** which moves water forward over the gills in a sculling motion.

In fish, the sides of the head are modified to form gill covers called **opercula.** When the opercula are raised, water is drawn in through the open mouth. The mouth is then closed and depression of the opercula forces water over the gills and out through the external gill openings.

In some fish, the pumping mechanism is not sufficient to satisfy the metabolic demands for oxygen, so that such fish must constantly swim with its mouth open to improve the rate of water flow over the gills. In clams, the gills are not feathery appendages but instead are netlike. Not only do they function in gas exchange but they also are involved in filtering food particles from the water stream that passes over and through them.

In your lab, you may have some living crayfish in an aquarium. Place one in a bowl of water and give it time to resume normal gill ventilation. Take a Pasteur pipette and place a drop of india ink next to the side of the crayfish just behind the thorax. Watch the direction of water flow using the ink as a tracer. Record your observations below.

Mammalian Respiratory System

Respiratory Tree Anatomy

Return to your fetal pig and examine the external openings (**external nares**) on the snout. Cut across the snout with a scalpel about 1 to 2 cm from the end and remove the tip. The nasal passages are separated from each other by the **nasal septum.** The curved **turbinate bones** in the sinus area increase the surface area of the passageways, creating eddy currents that, along with hairs, cilia, and mucus, help remove dust in the inhaled air and humidify it. The floor of the nasal passages is made up of the hard palate and the soft palate (posterior to hard palate).

Look into the pig's mouth. Behind and above the soft palate is the **nasopharynx.** It may be necessary to slit the soft palate to observe this. Air enters the nasopharynx from the posterior end of the nasal passages, then passes into the pharynx, through the glottis, and into the larynx and ultimately the trachea. If food accidentally enters the glottis, choking results. The Heimlich maneuver can often "save" a person who has food wedged in the glottis. Your instructor will demonstrate this maneuver for you.

In the nasopharyngeal area, look for the openings of the **eustachian tubes.** They are very difficult to find. These tubes allow air pressure to equilibrate between the middle ear chamber and the atmosphere. (This is why changes in altitude cause the ears to "pop.") Throat infections often spread to the ears through the eustachian tubes.

Run your fingers over the pig's throat and locate the hard, round **larynx.** Make a **medial** incision in the skin of the throat and extend the end cuts laterally, folding back the skin flaps. Repeat this procedure for the muscle layers.

Note the large mass of glandular material, the **thymus,** in this area. In the young pig, the thymus produces **lymphocytes,** an important component of the immune system. Later in life, the thymus atrophies and is of little consequence.

As you approach the larynx and trachea in your dissection, use a blunt probe to separate the muscles and expose these structures. Ventral to the trachea, observe the brownish-colored **thyroid gland.** Note how both the trachea and larynx are supported by rings of cartilage. The **hyoid apparatus** is anterior to the larynx and is a supporting frame for the tongue extensor muscles.

The larynx, or voice box, contains folds of elastic tissue, which are stretched across the cavity. These **vocal cords** vibrate when air passes over them, producing sound, and attached muscles vary the cord tension, allowing variations in pitch. Slit the larynx longitudinally and observe the vocal cords. Continue the slit posteriorly into the trachea and observe its lining. The esophagus is located behind the trachea. Pass a blunt probe into the esophagus from the mouth to help identify it.

If the **thorax** of your animal is not already opened, make a longitudinal cut with heavy scissors through the ribs just to the right of the **sternum,** or breastbone. Always keep the lower scissor tip pointed upward against the inside of the sternum to avoid catching and cutting internal struc-

Figure 27.9   Inspiration and expiration mechanics. (*a*) Contraction of intercostal muscles lifts rib cage and contraction of diaphragm muscles flattens diaphragm: both enlarge chest cavity and cause its pressure to drop. During maximum inspiration, both sternocleidomastoid and pectoralis minor muscles help elevate rib cage. (*b*) During resting expiration, elastic recoil of lung tissues, together with dropping of the rib cage and recoil of diaphragm because of organ pressure, decrease size of chest cavity. (*c*) Forced expiration involves compressing abdominal organs by contraction of the abdominal wall muscles that pull rib cage down.

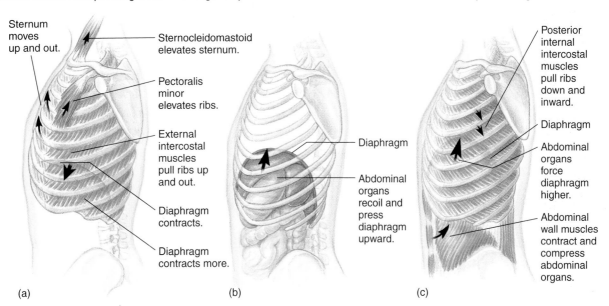

tures. Be careful! Several major blood vessels are under the sternum and should not be damaged.

The **diaphragm** is a sheet of muscle that separates the **abdominal cavity** from the **thoracic cavity.** The thoracic cavity is divided into three areas by membranes: the right and left **pleural cavities,** which surround the lungs, and the **pericardial cavity** where the heart is located.

If the pleural membranes are removed, the **lung** structure can be seen. The trachea, when it enters the thorax, divides into two **bronchi,** which are hidden from direct view beneath the heart and blood vessels. These bronchi, in turn, divide into progressively smaller **bronchioles,** which finally end in microscopic air sacs called **alveoli.** Alveoli have walls only a single cell layer thick and they are covered by capillaries. In these air sacs, oxygen and carbon dioxide are exchanged between the blood and the inhaled air.

Lung Ventilation Mechanism

Air enters the lungs as a result of the combined effects of the contraction of the diaphragm and three sets of muscles: **sternocleidomastoid, pectoralis minor,** and **intercostal** (between the ribs). When these muscles contract, the volume of the thoracic cavity increases as the rib cage elevates and the diaphragm depresses, causing the air pressure in the cavity to decrease. Air rushes in through the respiratory passageways and expands the alveoli. This causes the pressure between the atmosphere and the pleural cavities to equilibrate. When the diaphragm and intercostal muscles relax, the rib cage drops and the diaphragm rises decreasing the volume of the thoracic cavity. Consequently, the pres-

sure in the pleural cavity increases and collapses the alveoli, driving air out of the "respiratory tree." Figure 27.9 illustrates these mechanics in humans.

Note the convex shape of the diaphragm in your pig and imagine how contraction of the diaphragm increases the volume of the thoracic cavity. Push the sternum up to mimic the contraction of the involved muscles and note the expansion of the chest cavity.

Microscopic Examination of Mammalian Lung

Remove a piece of the lung and put it in a small bowl of water. Observe it with a dissecting microscope and find the alveoli and bronchioles. Also look at the demonstration slide of a section across several alveoli. How many cell thicknesses separate the air in a mammalian lung from the red blood cells in the capillaries?_____

Figure 27.10 contains scanning electron micrographs of an alveolus and surrounding capillaries. The alveoli form an interconnecting system of chambers, and macrophages in the alveoli scavenge for microorganisms and particulate material.

You are now finished with the fetal pig for this lab. Return it to the storage area according to the directions given by your lab instructor.

Comparative Summary

To gain an understanding of comparative anatomy, fill in tables 27.1–27.6. Write a brief description of each structure and its function. Include a summary statement of the organ system for each animal.

Figure 27.10 Scanning electron micrographs of a mammalian lung: (*a*) alveolar sac; (*b*) lung capillaries after lung tissue was removed. Note how capillaries surround each alveolus.

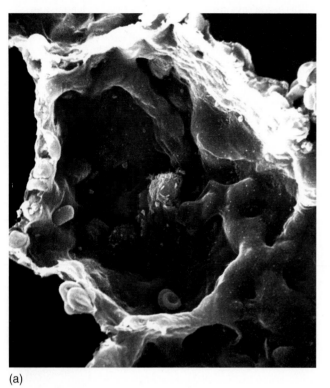

(a)

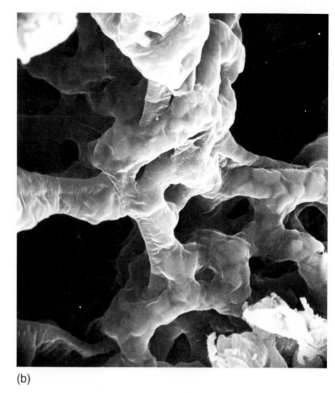

(b)

TABLE 27.1	Hydra—digestive system summary

Cnidocytes

Mouth/anus

Gastrovascular cavity

Mesoglea

Description of digestive process

TABLE 27.2	Fetal pig—mammal—digestive system summary
Mouth	
Salivary glands	
Esophagus	
Stomach	
Small intestine	
Liver	
Pancreas	
Large intestine	
Rectum	
Anus	

Description of digestive process

TABLE 27.3	Crayfish—gas exchange system summary
Gill bailer	
Gills	
Haemolymph	

Description of gas exchange

blood rapidly from one location to another within an animal. The circulating fluid often contains a respiratory pigment, a protein that aids in transporting oxygen and carbon dioxide between the tissues and the respiratory surface. Hemocyanin and hemoglobin (which in mammals occurs in red blood cells) are common pigments. Blood also contains cells or proteins that protect against invasion by microorganisms and proteins that are involved in clotting, the sealing of leaks. The blood-vessel system often has anatomical provisions so that the blood is brought into close contact with three other physiological systems: lung or gill, where gas exchange occurs; excretory, where waste, salt, and water exchange occur; and digestive, where nutrients are absorbed.

Circulatory systems may be either open or closed. In **open circulatory systems,** found in molluscs and arthropods, the **arterial system** is not connected to the **venous system** through **capillary beds.** Instead, the small arteries simply terminate, emptying their contents into the tissue spaces, and the blood (properly called hemolymph) directly bathes the tissues, eventually finding its way back to the heart.

In **closed circulatory systems,** found in some invertebrates and all vertebrates, the flow of blood is always within blood vessels. The arterial system is connected to the venous system by means of capillaries which have very thin walls only one cell thick. Blood entering the capillaries is under relatively high pressure, and part of the fluid portion is filtered through the capillary walls, entering the tissue spaces. On the venous side of the capillary bed, most of this fluid flows back into the capillaries due to osmotic force. Gaseous, waste, and nutrient exchanges between the blood and tissues occurs by way of this fluid exchange as well as by diffusion. Blood flow in each capillary bed is regulated by the opening and closing of the **precapillary sphincter,** as seen in figure 28.1.

The capillary bed, and only the capillary bed, is the functional site of the closed circulatory system where all exchanges take place. The **lymphatic system** consists of small open-ended lymphatic capillaries that conduct fluid into larger lymphatic ducts. Fluid that does not return to the blood capillaries enters the lymphatic capillaries. This fluid is collected in lymphatic ducts and is returned to the venous system near the heart.

Vertebrate circulation is summarized in figure 28.2. Consider the changes that occur in the blood as it passes through the various circuits. It is more important to understand the purpose of the circulatory system than to know a

LAB INSTRUCTIONS

You will observe the gross and microscopic features of the mammalian circulatory system and circulation in the capillary beds of a living vertebrate. Finally, you will observe a living arthropod heart in order to compare open and closed circulatory systems.

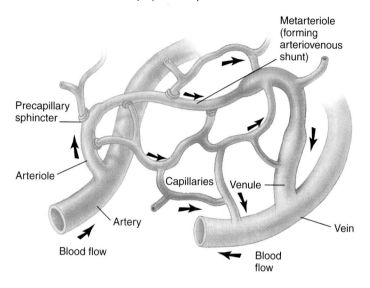

Figure 28.1   Anatomy of a capillary bed. Fluid leaves the capillaries because of the pumping force of the heart, raising the osmotic pressure of the blood. Pumping pressure falls across the capillary bed due to drag and volume loss. On the venous side, fluids are drawn into the capillary by osmosis. Excess fluids enter the lymphatic capillaries.

Figure 28.2   Schematic of major mammalian blood vessels and their relationship to one another.

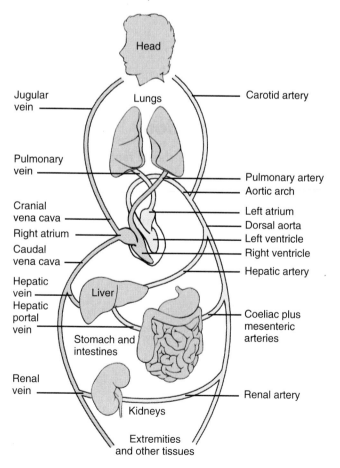

long list of the names of blood vessels.

Invertebrate Circulatory System

Most molluscs, arthropods, and many other invertebrates (but not all) have open circulatory systems. Of these, the crayfish's is most easily observed. See figure 21.7 for a diagram showing the crayfish's circulatory system.

Demonstration of Open Circulatory Systems

▶ To observe the heart of a crayfish, immobilize the animal by covering it with wet cotton and pinning it down in a dissecting tray. Part of the carapace, the exoskeleton covering the thorax, should be removed by inserting scissors under the posterior edge 1 cm to the left of the midline and cutting forward to the region of the eye. This procedure should be repeated on the right side, and the strip of exoskeleton should be carefully removed, so that none of the underlying membranes are torn.

The heart can now be seen beating in the **pericardial sinus** covered by the **epidermal** and **pericardium membranes.** These membranes can be removed to expose the heart, which should be bathed in Ringer's solution to keep it from drying.

▶ In an open circulatory system such as this one, **hemolymph,** the circulating fluid, leaves the heart in arteries but returns in open **sinuses** instead of veins. Under a dissecting microscope, you will be able to see the fluid surrounding the heart enter it through three pairs of slitlike openings called **ostia.** These ostia open when the heart relaxes and allow hemolymph to flow in. When the heart contracts, flaps of tissue inside the heart close the ostia, and hemolymph is forced out of the heart through the arteries.

▶ You should be able to see the **dorsal abdominal artery,** which carries hemolymph to the tail, and the **opthalmic** and **antennary arteries,** which carry hemolymph to the head region. Other arteries lie beneath the heart. Make a diagram of how the crayfish heart works.

Mammalian Circulatory System

▶ If your fetal pig has not previously been opened, make a series of cuts as diagramed in the figure I.1, page 330. If you have followed the lab sequence in this manual, complete the opening as follows:

1. Make a longitudinal cut 1 cm to the left of the sternum from the lower ribs to the region of the forelimbs and parallel to the previous cut. Sever all ribs.

2. Lift up the center section, labeled (*A*) in figure I.1 (page 330), and cut any tissues adhering underneath. A transverse cut at the anterior end will detach this center piece, which should be discarded.

3. The heart and lungs will be easier to observe if the diaphragm is cut away from the rib cage on the animal's left side only. Cut close to the ribs.

4. In the region of the throat, remove the thymus glands, thyroid, and muscle bands, but do not cut or tear any major blood vessels.

The Heart and Its Vessels

▶ Find the heart encased in the **pericardial sac.** Remove the sac and identify the four heart chambers. The paired **atria,** thin-walled, distensible sacs, collect blood as it returns to the heart. The two **ventricles** are the large, muscular pumping chambers of the heart.

Blood returning from the systemic circulation enters the right atrium from the cranial and caudal **vena cavae.** After passing into the right ventricle, it is pumped to the lungs through the **pulmonary trunk,** which divides into the left and right **pulmonary arteries.** The trunk is visible passing from bottom right to upper left over the front of the heart and passing between the two atria (fig. 28.3). Trace the pulmonary arteries to the lungs.

Following gas exchange in the capillaries of the lungs, blood collects in the **pulmonary veins** and flows to the left atrium. These veins enter on the dorsal side of the heart and will be difficult to find. If you remove some of the lung tissue from the left side, you may be able to locate these vessels.

From the left ventricle, blood is pumped at high pressure through the **aorta** to the systemic circulation. Find the ▶ aorta. It will be partially covered by the pulmonary trunk but can be identified as the major vessel that curves 180° to the left, forming the **aortic arch** (fig. 28.3). In the mammalian fetus, the pulmonary trunk and aorta are connected by a short, shunting vessel, the **ductus arteriosus.** Find this vessel in your animal. During the intrauterine life, when the lungs are not functional, blood entering the pulmonary circuit does not pass to the lungs. Instead, it is shunted to the aorta. At birth, the shunting vessel constricts so that blood enters the lungs. The constricted vessel fills

Figure 28.3 External ventral and dorsal views of the fetal pig's heart.

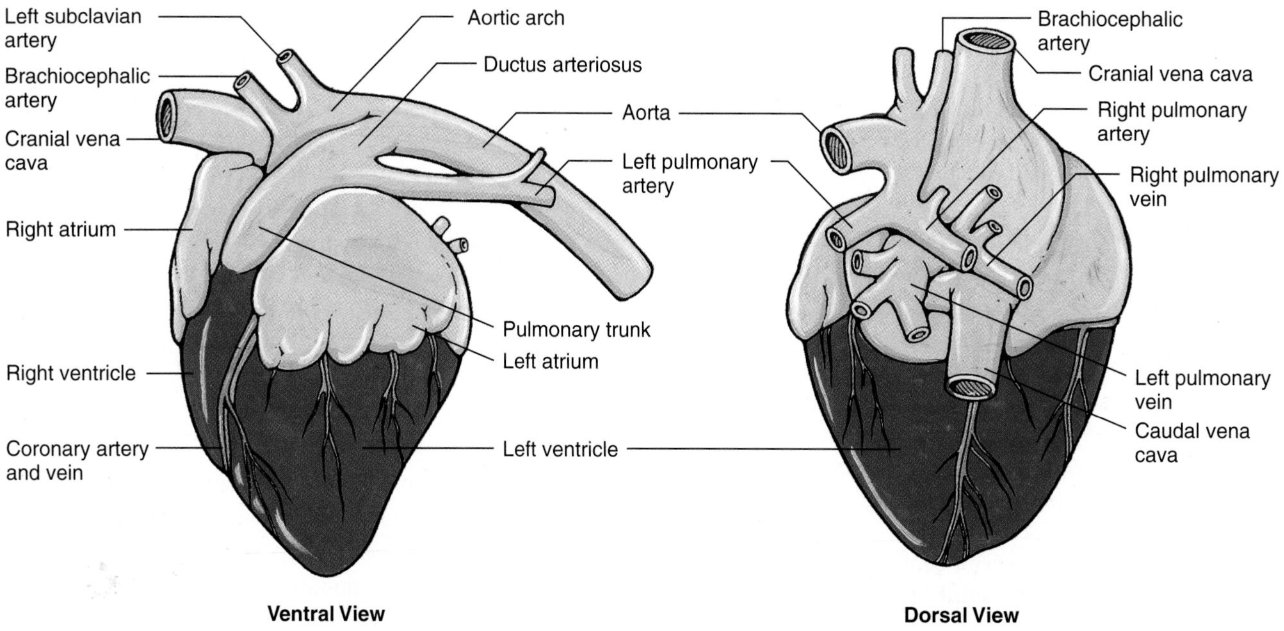

Left subclavian artery
Brachiocephalic artery
Cranial vena cava
Right atrium
Right ventricle
Coronary artery and vein
Aortic arch
Ductus arteriosus
Aorta
Left pulmonary artery
Pulmonary trunk
Left atrium
Left ventricle
Brachiocephalic artery
Cranial vena cava
Right pulmonary artery
Right pulmonary vein
Left pulmonary vein
Caudal vena cava
Ventral View
Dorsal View

with connective tissue to become a solid cord seen in adults as the arterial ligament.

Vessels Cranial to the Heart

Veins

Because the venous system is generally ventral to the arterial system, it will be studied first in the congested region of the heart. Refer to figure 28.4 and place a check next to each vein identified.

Trace the **cranial vena cava** forward from the heart to where it is formed by the union of the two very short **brachiocephalic veins.** Each of these in turn is formed by the union of the five major veins: the **internal** and **external jugular veins,** which drain the head and neck; the ~~**cephalic vein,**~~ which lies beneath the skin anterior to the upper forelimb and typically enters at the base of the external jugular; the ~~**subscapular vein**~~ from the dorsal aspect of the shoulder; and the **subclavian vein** from the shoulder and forelimb. As the latter passes into the forelimb, it is known as the **axillary vein** in the armpit and the **brachial vein** in the upper forelimb. Caudal to the union of the brachiocephalic veins, find the pair of **internal thoracic veins** entering the ventral surface of the vena cava. They extend along the sternum and drain the chest wall. These veins were most likely cut when you opened the animal.

Arteries

Find the aortic arch and trace it back to the heart. Note the several arteries that branch off to supply the anterior region of the animal. Refer to figures 28.4 and 28.5 to identify the vessels. Check off the arteries as they are identified.

Find the small **coronary arteries** that arise from the base of the aorta behind the pulmonary trunk. They pass to the groove between the ventricles on the ventral surface of the heart and branch to supply the muscles of the heart. The first major artery to branch from the aorta is the **brachiocephalic artery.** It gives rise to the two **carotid arteries,** which pass anteriorly to supply the head, and the **right subclavian artery,** which passes to the right forelimb. Just to the left of the brachiocephalic artery, find the **left subclavian artery** arising as a separate branch from the arotic arch. Blood in this vessel goes to which region of the body?

Once the aorta runs posteriorward along the dorsal wall of the thorax, it gives rise to intercostal arteries, which supply the walls of the chest. The aorta then passes through the diaphragm to become the **abdominal aorta.**

Return to the left subclavian artery and trace it into the forelimb, removing skin and separating muscles as necessary. In the armpit it is known as the **axillary artery,** and in the upper forelimb as the **brachial artery.** The **subscapular artery** branches from the axillary artery and supplies the shoulder muscles. The brachial artery divides in the lower forelimb to give rise to the **radial** and **ulnar arteries.**

Find another student in the lab who is at the same stage in the dissection as you are. Quiz one another about the path blood takes as it flows from the forelimb through the heart to the head.

Figure 28.4 Ventral view of the fetal pig's major arteries (a) and veins (v).

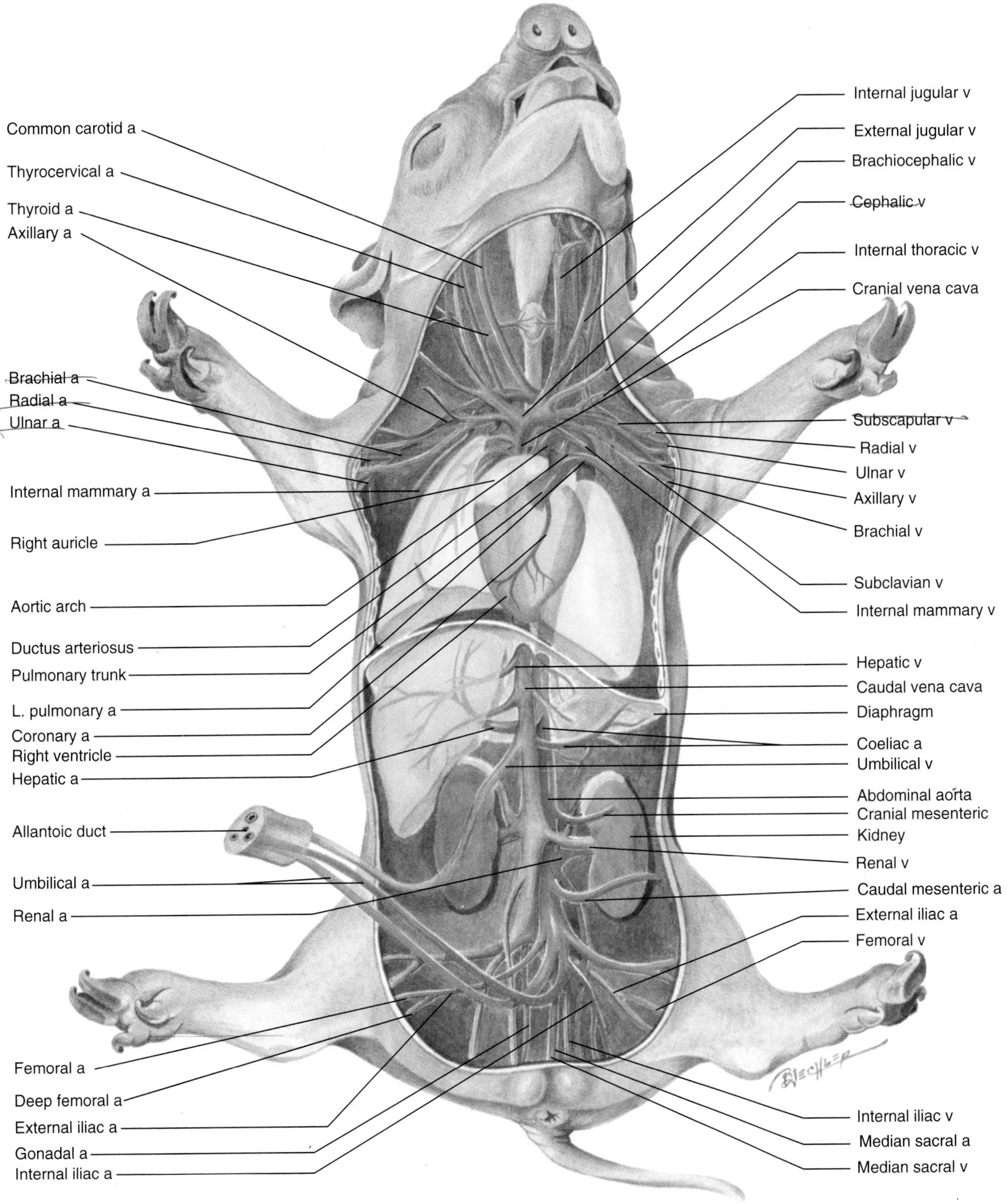

Common carotid a
Thyrocervical a
Thyroid a
Axillary a

Brachial a
Radial a
Ulnar a

Internal mammary a

Right auricle

Aortic arch

Ductus arteriosus
Pulmonary trunk
L. pulmonary a
Coronary a
Right ventricle
Hepatic a

Allantoic duct

Umbilical a
Renal a

Femoral a
Deep femoral a
External iliac a
Gonadal a
Internal iliac a

Internal jugular v
External jugular v
Brachiocephalic v
Cephalic v
Internal thoracic v
Cranial vena cava

Subscapular v
Radial v
Ulnar v
Axillary v
Brachial v

Subclavian v
Internal mammary v

Hepatic v
Caudal vena cava
Diaphragm
Coeliac a
Umbilical v

Abdominal aorta
Cranial mesenteric
Kidney
Renal v
Caudal mesenteric a
External iliac a
Femoral v

Internal iliac v
Median sacral a
Median sacral v

Figure 28.5   Major arteries in the region of the fetal pig's heart.

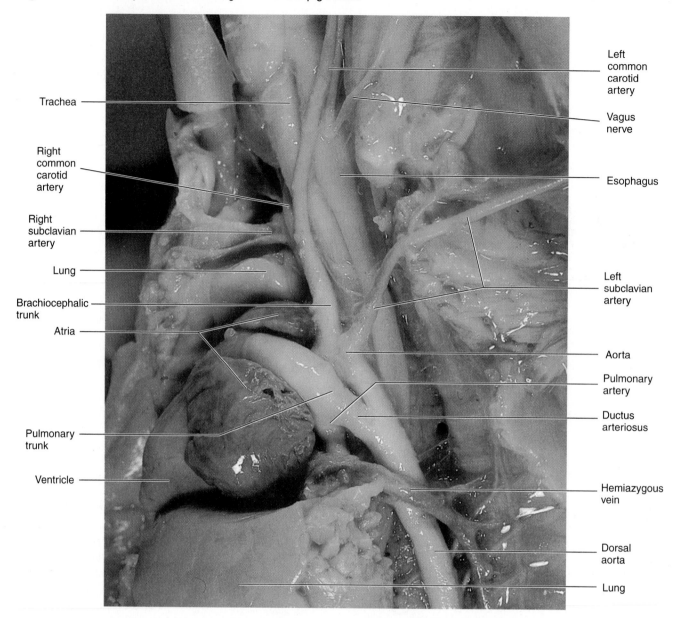

Trachea

Right
common
carotid
artery

Right
subclavian
artery

Lung

Brachiocephalic
trunk

Atria

Pulmonary
trunk

Ventricle

Left
common
carotid
artery

Vagus
nerve

Esophagus

Left
subclavian
artery

Aorta

Pulmonary
artery

Ductus
arteriosus

Hemiazygous
vein

Dorsal
aorta

Lung

Vessels Caudal to the Heart

Veins

If the heart is lifted and tilted forward, the **caudal vena cava** can be viewed at the point where it enters the right atrium. As this vein is traced caudally, several veins will be found flowing into it. After it passes through the diaphragm, the paired **hepatic veins** and single **umbilical vein** enter first. The umbilical vein carries oxygenated, nutrient-laden blood from the placenta. This vein passes through the liver where it is known as the ductus venosus.

While studying the major veins of the body cavity, note the **hepatic portal system.** It consists of veins that flow from the digestive tract and spleen to the liver where they again divide into a system of capillaries. The term **portal system** refers to a system of veins that arises from a

capillary bed and carries blood to a second location. There the veins again fan out to form capillaries, which again flow into veins before the blood returns to the heart.

The hepatic portal system can be traced partway from the liver to the digestive system. The **portal vein** runs next to the common bile duct in the **hepatoduodenal** ligament under the lobes of the liver. Nutrient-laden blood flows up this vein to the liver where exchange occurs between the blood and the liver across the walls of the portal system capillaries. The blood then flows into the **hepatic veins,** which enter the caudal vena cava.

Follow the vena cava caudally to where the **renal veins** enter from the kidneys. In the male, the **spermatic veins,** and in the female, the **ovarian veins,** enter next. On the left side, these veins may enter the renal vein first. Below the

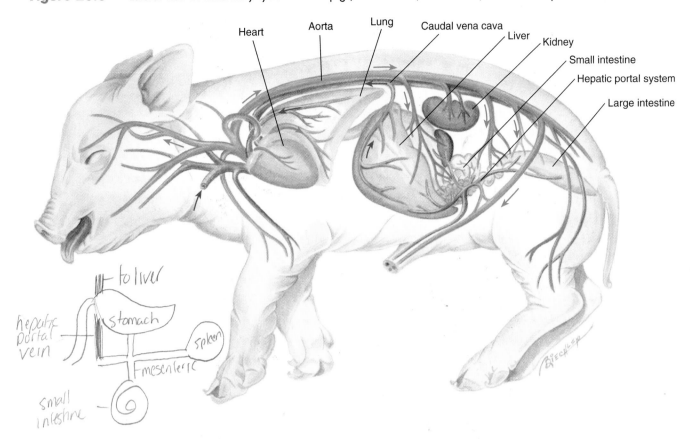

Internal Heart Structure

kidneys, the vena cava splits into the **internal** and **external iliac veins** and the **median sacral vein,** a small vein that comes from the tail. The external iliac veins collect blood from the **femoral veins** in the hind legs, whereas the internal iliacs collect blood from the pelvic area.

Arteries

▶ After the aorta enters the abdominal cavity, a large, single **coeliac artery** arises from it at the cranial end of the kidneys. You will have to remove some connective tissues to obtain a full view of this artery. The coeliac artery eventually divides into three arteries supplying the stomach, spleen, and liver. The **mesenteric artery** next arises from the aorta and supplies the pancreas, small intestine, and large intestine. The **renal arteries** are short, paired arteries supplying the kidneys. The next large arteries arising from the aorta are the **external iliacs,** which supply the hind legs with an **iliolumbar** branch to the lower back. In the fetus, the **umbilical arteries** branch from the caudal end of the abdominal aorta and pass out through the umbilical cord. They form a capillary bed in the placenta for nutrient, gas, and waste exchange with the maternal circulatory system.

▶ Find another student in the lab who is at the same stage in the dissection as you are. Quiz one another about the circulation paths to the major organs and hind limbs.

▶ Look at the diagram in figure 28.6 and add labels to the major arteries and veins that you identified in your dissection.

▶ Study the orientation of the heart so that you can later identify it in isolation. Now, free the heart from the body by cutting through all the vessels holding it in place. Be careful to leave enough of each vessel so that they can be identified in the isolated heart. Alternative to removing the fetal pig's heart, your instructor may have a demonstration dissection of a beef heart for your study. Whichever specimen you are using, orient yourself by identifying the **aorta, pulmonary artery, pulmonary vein,** and **vena cava.**

▶ Place the heart in your dissecting pan, ventral side up. Make a razor cut along the pulmonary trunk down through the right ventricle. Spread the tissue open, pin it down, and remove the latex. You may wish to use a dissecting microscope to observe the open ventricle.

▶ Identify the **semilunar valves** at the junction of the artery and ventricle. Consider how these valves work. The open flaps face into the **pulmonary trunk,** and any backflow in the pulmonary trunk fills the valve flaps with blood and closes the valve. You may have to add water to the heart to float the valve flaps so that you can see them (fig. 28.7).

▶ Now cut through the right atrium and remove the latex and coagulated blood. The **tricuspid valves** are between the atrium and ventricle. These valves also work on the backflow principle, allowing blood to flow only one way from the atrium into the ventricle. In the ventricle, fine fibers called **chordae tendinae** are attached to the valve flaps.

Figure 28.7 Human heart: (a) the path of blood flow through the major chambers and vessels; (b) valves of the heart viewed from above.

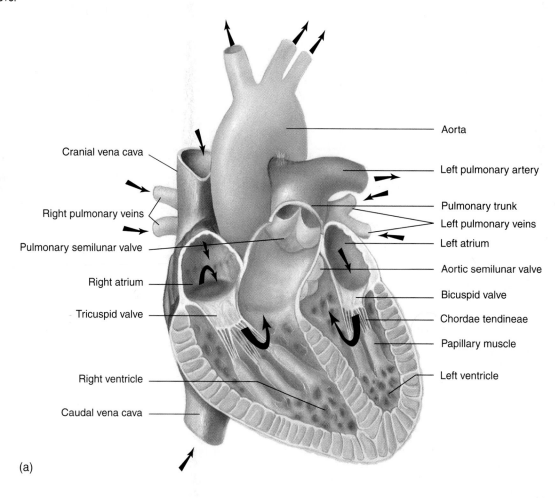

Aorta

Left pulmonary artery

Cranial vena cava

Pulmonary trunk

Right pulmonary veins

Left pulmonary veins

Pulmonary semilunar valve

Left atrium

Right atrium

Aortic semilunar valve

Tricuspid valve

Bicuspid valve

Chordae tendineae

Papillary muscle

Left ventricle

Right ventricle

Caudal vena cava

(a)

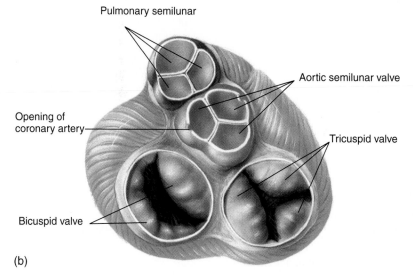

Pulmonary semilunar

Aortic semilunar valve

Opening of
coronary artery

Tricuspid valve

Bicuspid valve

(b)

Figure 28.8 Histology of arteries and veins: (*a*) tissues in the wall of an artery; (*b*) tissues in the wall of a vein; valves in veins prevent backflow of blood in the venous sytem.

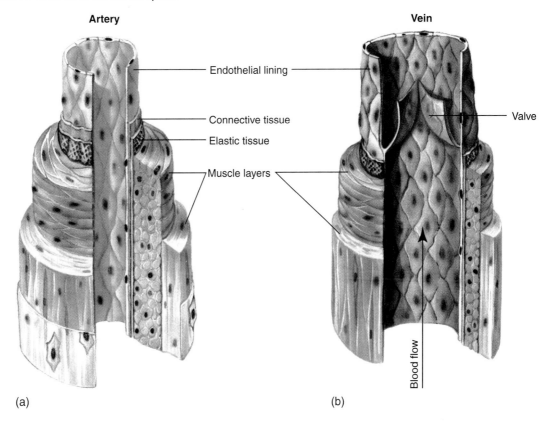

Artery

Endothelial lining

Connective tissue

Elastic tissue

Muscle layers

(a)

Vein

Valve

Blood flow

(b)

These cords prevent the flaps from "blowing back" from high pressures developed when the ventricle contracts.

Cut into the left atrium and ventricle as you did on the right side. Identify the **bicuspid** or **mitral valve** with its associated chordae tendinae, between the atrium and ventricle. After cutting into the ventricle and cleaning it, find the **aortic semilunar valve.** Blood flow through the human heart is shown in figure 28.7.

Pair off with another student and describe how blood returning to the heart from the foreleg travels to the lungs and then to the hindleg. Describe the operation of the heart valves.

Clean your dissecting instruments and tray and return your fetal pig to the storage area.

Histology of Vessels

Obtain prepared slides of cross sections of arteries and veins and observe them under low power with a compound microscope. Note that arteries have thicker walls than veins. Most of the difference in thickness is due to the increased amounts of muscle and connective tissue in the artery. Since arteries carry blood from the heart, they operate under relatively high pressure (average 120 mm of mercury equivalent). Veins experience only a tenth as much pressure.

Blood flows through veins because skeletal muscles press on them and move the blood along. **Valves** in the

veins prevent backflow and make the passage of blood unidirectional (fig. 28.8).

Observe the tissues of a blood vessel under high power. **Endothelial cells** are epithelial cells that line both arteries and veins; capillary walls consist of only endothelial cells. When the muscle layers are contracted in the smallest arteries, the total volume of the vascular system is reduced and the blood pressure rises.

The complexity of the microvasculature is evident in scanning electron micrographs of casts of the circulatory system. Note the capillaries and their relationship to arterioles. Figure 28.9 shows the nature of the endothelial lining of blood vessels and red blood cells in an arteriole. The nature of a capillary bed is shown in figure 28.10.

Blood

Human blood consists of 55% plasma and 45% cells by volume. Plasma is the fluid portion of the blood containing dissolved proteins, salts, nutrients, and waste products. Several different types of cells and cell fragments are contained in blood. By far the most common (about 95% of the cells) are erythrocytes (red blood cells) which are red because they contain hemoglobin. The other 5% are collectively called **leukocytes** (white blood cells) and **platelets** that are important in blood clotting. There are several types of leukocytes and the most common—neutrophils and lymphocytes—are shown in figure 28.11.

Investigating Circulatory Systems **355**

Figure 28.9 Scanning electron micrograph of a broken blood vessel, showing the smooth endothelial lining and several red blood cells (RBC).

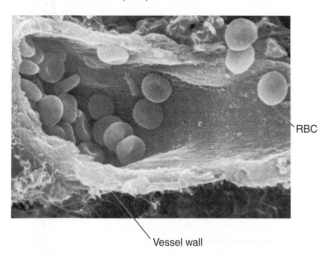

RBC

Vessel wall

Figure 28.10 Scanning electron micrograph of a plastic cast of a capillary bed from skeletal muscle in which individual arterioles can be traced to capillaries. From R. G. Kessel and R. H. Kardon. *Tissues and Organs: A Text-Atlas of Scanning Electron Microscopy.* 1979. W. H. Freeman and Company.

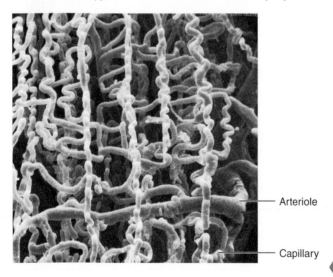

Arteriole

Capillary

Get a prepared slide of a Wright-stained human blood smear from the supply area and look at it with your compound microscope under medium power. Note the large number of red blood cells. Can you see a nucleus in these cells?_____ Why do you think that the cells are lighter in color in the center and darker at the periphery?

Figure 28.11 Human blood stained with Wright's stain shows red blood cells and different types of white blood cells; (*a*) neutrophil; (*b*) lymphocyte.

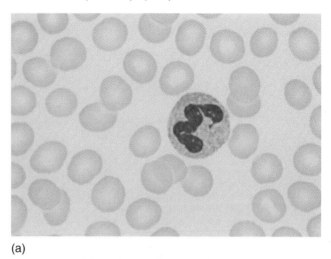

(a)

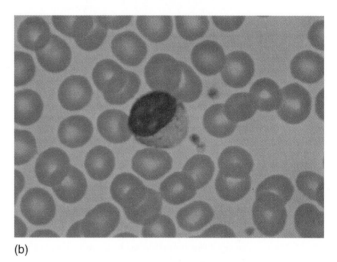

(b)

As you look carefully at the slide, you will see occasional cells that look different. These are leukocytes and because of the staining they have a blue/purple color. Center a leukocyte in the field of view and observe it with high power. What structure in the cell is stained?_____
Return to medium power and scan the slide to locate other leukocytes. Try to find examples of each of the types shown in figure 28.11.

Neutrophils leave the blood early in the inflammation process and become phagocytic cells consuming cell debris and bacteria. **Lymphocytes** are important in the immune response. Some are involved in cellular immunity and others secrete antibodies that neutralize foreign proteins and other macromolecules.

After you have observed these four cell types, return your slide to the supply area.

Figure 28.12   Method for observing microcirculation in a fish tail.

(1) Wrap the fish (except for the head and tail) with dripping wet cotton. Place fish in half of a petri dish. Place coverslip over tail.

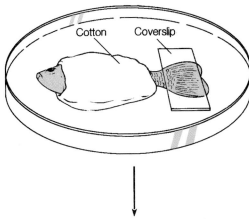

Cotton   Coverslip

(2) Place dish on microscope so that fish's tail is over hole in stage.

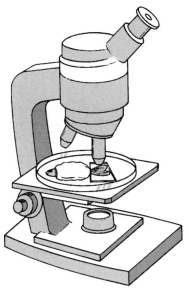

(3) Examine with low- and medium-power objectives of your microscope.

Circulation in capillaries

To observe circulation in capillaries, net a small fish or tadpole from an aquarium and wrap it in dripping wet cotton, as shown in figure 28.12, being careful not to cover the head or the tail. Lay the wrapped fish in an open petri dish. Place a few drops of water on the tail and add a coverslip over the tail.

Place the dish on a compound microscope stage and observe the tail under scanning power. Sketch your observations, answering the following questions:

1. Can you identify capillaries, venules, and arterioles?

2. Is blood flow faster in certain vessels compared to others?

3. Is blood flow continuous in all vessels? What might control this?

Dilute solutions of nicotine, caffeine, and adrenalin are available in the lab in dropper bottles. You can see the effects of these chemicals by placing a drop on the tail.

Learning Biology by Writing

Your instructor may ask you to answer the Lab Summary and Critical Thinking questions that follow.

To gain an understanding of comparative anatomy, fill in tables 28.1 and 28.2. Write a brief description of each structure and its function. Include a summary statement of the organ system for each animal.

Internet Sources

Many medical schools have extensive collections of pictures of pathological conditions. These collections are available over the WWW. Use your browser program and a search engine to locate pictures of a blood vessel with arteriosclerosis. Compare this picture to your observations of a normal artery in the lab. Describe the differences.

TABLE 28.1	Fetal pig—circulatory system

Blood

Hemoglobin

Red blood cells

Heart

Pulmonary circulation

Systemic circulation

Aorta

Arteries

Arterioles

Capillaries

Venules

Vena cavae

Portal system

Description of circulation

Lab Summary Questions

1. Create a flowchart showing the major vessels and heart chambers that a drop of blood passes through as it returns from the arm and passes to the back leg.
2. Create a flowchart showing the major vessels and heart chambers that a drop of blood passes through as it returns from the small intestine and passes to the brain.
3. List the valves of the heart and describe how they operate.
4. Describe microcirculation in a capillary network. What controls whether blood enters a capillary?
5. How does the open circulatory system of a crayfish differ from the closed system of a mammal? Describe how blood returns to and enters the heart of both types of animals.
6. Trace a molecule of oxygen from when it enters your nostril as it passes to your leg and then a carbon dioxide molecule as it passes from your leg back to your nostril.

TABLE 28.2   Crayfish—Circulatory system

Hemolymph

Hemocyanin

Heart

Dorsal abdominal artery

Ophthalmic/antennary arteries

Sinuses

Ostia

Description of circulation

Critical Thinking Questions

1. What are the roles of the lymphatic system and the venous system in returning fluid filtered through the microcirculation?
2. Since a giraffe's head is 15 feet above the ground, what circulation adaptations are necessary to allow adequate blood supply to the head?
3. What circulatory system adaptations occur in response to environmental demands in the animal kingdom?
4. If arthropods such as insects have open circulatory systems that lack veins, how does hemolymph return from the tissues to the heart?

Know both egg/sperm pathways
from development to exit

Know urine from pyramids to exist

LAB TOPIC 29

Investigating the Excretory and Reproductive Systems

Supplies

Preparator's guide available on WWW at
http://www.mhhe.com/dolphin

Equipment

Compound microscopes
Dissecting microscopes

Materials

Fetal pig
Dissecting pan and instruments
Prepared slides
　　Mammalian kidney cortex
　　Mammalian ovary with Graafian follicles
　　Mammalian seminiferous tubules, cross section
Microscope slides and coverslips
Live earthworm

Solutions

0.5% methylene blue in 0.7% NaCl
0.7% NaCl
10% ethanol

Prelab Preparation

Before doing this lab, you should read the introduction and sections of the lab topic that have been scheduled by the instructor.

　　You should use your textbook to review the definitions of the following terms:

　　bladder
　　Bowman's capsule
　　collecting duct
　　distal tubule
　　epididymis
　　fallopian tube
　　glomerulus
　　loop of Henle
　　ovary
　　oviduct
　　proximal tubule
　　seminiferous tubule
　　testis
　　ureter
　　urethra
　　uterus
　　vas deferens

You should be able to describe in your own words the following concepts:

　　Glomerular filtration in kidney
　　Tubule reabsorption in kidney

　　As a result of this review, you most likely have questions about terms, concepts, or how you will do the experiments included in this lab. Write these questions in the space below or in the margins of the pages of this lab topic. The lab experiments should help you answer these questions, or you can ask your instructor for help during the lab.

Objectives

1. To make living preparations of nephridia from live earthworms

2. To observe the gross and microscopic anatomy of the mammalian excretory and reproductive systems in a fetal pig by dissection

3. To be able to trace a drop of water leaving the blood in the kidney until it shows up in urine being voided

Background

The excretory systems of animals maintain a constant chemical state in the internal body fluids. The name *excretory system,* with its implicit emphasis on waste removal, does not suggest the three other important functions of these systems: (1) controlling water volume, (2) regulating salt concentrations, and (3) eliminating nonmetabolizable compounds absorbed from food.

　　Excretion involves the elimination of the waste products from the metabolism of nitrogen-containing compounds. **Ammonia,** a toxic compound at high concentrations, is produced by all animals when they metabolize amino acids and nucleotides. Depending on whether an animal inhabits an aquatic or terrestrial environment, this toxic product is handled in different ways. Aquatic organisms generally excrete the NH_3 directly into the surrounding water, often through their gills. Terrestrial animals convert it into less toxic compounds, such as **urea** in mammals or **uric acid** in insects, reptiles, and birds. These products are collected by excretory organs and periodically voided.

Marine, freshwater, and terrestrial environments present quite different challenges to animals in terms of internal water volume regulation and regulation of salt concentrations in body fluids. The following paragraphs outline these challenges.

Many marine invertebrates lack well-developed excretory systems. They have body fluids that resemble seawater in terms of osmotic concentration and ionic composition. Consequently, diffusion and active transport are sufficient to maintain the minor differences that exist. Furthermore, such animals usually have little tissue volume relative to total surface area. Ammonia diffuses from the body into the virtually infinite reservoir of the ocean at rates that are sufficient to prevent internal toxic levels.

In more advanced marine invertebrates and fish, the mass of tissues dictates that there be mechanisms other than diffusion to get rid of nitrogenous wastes. The active excretion of ammonia or urea requires water; that is, these compounds must be dissolved in a certain volume of water when voided. Obviously, water will be lost and must be continually replaced. Water is also lost across the gill surface by osmosis, as water flows from the blood to the sea down an osmotic gradient. If salt water is used to replace the water lost in the urine and through the gills, salts accumulate in the body, and the animal must have some means of excreting this salt. Marine fish produce urine that contains ammonia, water, and some salts. Additional salts are actively transported across the gills from the blood into the sea. Acting together, the kidneys and gills maintain the homeostasis (a constant state) of the marine fish's internal environment by regulating water, salt, and waste concentrations.

In freshwater fish, the situation is different. Water and salts are still excreted along with ammonia in the urine, but water is replaced by water that flows osmotically across the gill surface from the environment into the blood. Salts would be gradually depleted in body fluids if excretion and the osmotic replacement of water were the only processes operating. The gills of freshwater fish accumulate salts from the environment by active transport. In amphibians, similar "pumps" for ion accumulation are located in the skin.

Terrestrial animals face different problems. They do not gain water or salt from their environment. Both materials are in short supply and are conserved. Two different conservation mechanisms seem to have evolved. Insects, reptiles, and birds convert waste ammonia into uric acid, which is practically insoluble and requires little accompanying water when it is excreted. Mammals, on the other hand, have well-developed kidneys that can recover both water and salt from the urine after it is produced, while allowing urea to be excreted.

Mammalian kidneys function by filtering wastes and then selectively reabsorbing materials from the filtrate. The mammalian kidney consists of millions of capillary tufts, the **glomeruli.** They filter blood so that the fluid portion (that is, the portion with no blood cells or proteins) enters the tubules of the urinary system. The concentrations of water, salts, urea, sugars, amino acids, and so on, in the filtrate are very similar to those in the blood. As the filtrate passes through the tubular network of the **nephrons** toward the **collecting ducts** of the urinary system, many of these materials are reabsorbed by active and passive transport, leaving wastes, excess salts, and water. By the time the filtrate reaches the collecting ducts, it has been changed by the reabsorption of water and salts, but not waste products.

In the adult human, 180 liters of filtrate are produced each day, but the daily urine volume is only 0.6 to 2.0 liters. If we assume that the plasma volume is 3 liters, the production of 180 liters of filtrate means that all of the blood plasma must be filtered through the glomeruli 60 times in 24 hours. A volume of fluid equal to 178 to 179 liters is reabsorbed along with many salts. The reabsorption process is obviously of some magnitude.

LAB INSTRUCTIONS

You will observe the excretory system of the earthworm and mammal. Keep in mind that these systems are involved in water, salt, and waste balance.

Because of this close anatomical location of the excretory and reproductive systems, you will also study the anatomy of the mammalian reproductive system.

Invertebrate Excretory Systems

Nephridial Systems

Invertebrates in several phyla have excretory organs called **nephridia.** The earthworm is one example of an animal with a nephridial system (refer to fig. 20.7). In each segment, except for a few anterior ones, is a pair of nephridia, which open independently to the outside. Figure 29.1 diagrams the anatomy of a nephridium. Body fluids enter the nephridium via a ciliated, funnel-like structure, the **nephrostome,** and pass through a convoluted **tubule** to a distensible **bladder.** As fluids pass along the tubule, materials such as water and salts exchange between the tubule and the capillary network surrounding it. The contents of the bladder are voided to the outside through the **nephridiopore.**

If live earthworms are available in class, they can be used to view the nephridial system directly. Anesthetize a worm by placing it in 10% ethanol until it does not respond. Remove the worm and place it in a clean dissecting dish with a thick layer of wax on the bottom. (Dirty dishes

Figure 29.1 The anatomy of a nephridium in an earthworm segment.

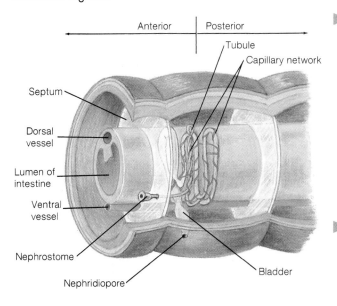

may contain formalin, which will kill the nephridial cells.) Make a careful longitudinal cut into the posterior half of the worm and pin back the tissue flaps. Place the worm under a dissecting microscope, flood the body cavity with 0.7% saline, and observe it with reflected light. In areas where the intersegmental membranes, the **septa,** are not stretched too tightly, you should be able to distinguish the parts of the nephridium. Surgically remove part of a septum containing a nephridium. Place the septum in a drop of 0.5% methylene blue dye made up in 0.7% saline for one minute, then make a wet-mount slide using 0.7% saline and look at it with your compound microscope.

Observe the relationship between the capillaries and the tubules. Parasitic roundworms are often found in the bladder. Sketch your observations below.

Mammalian Excretory System

Anatomy of Fetal Pig Excretory System

▶ Use your fetal pig to study the urogenital system (figures 29.2, 29.3, 29.7, and 29.8). Refer to these as you locate each of the structures indicated in boldface type. Be sure to look at a pig of the opposite sex before you leave lab.

The kidneys are located on the dorsal wall of the abdominal cavity. Remove the **peritoneum,** the connective tissue membrane that holds them in place. Note the kidneys' proximity to the **descending aorta.** The blood pressure drops very little as blood passes from the aorta into the kidneys via the **renal arteries.** A high blood pressure is essential to force-filter the blood through the walls of the **glomerular capillaries. Renal veins** drain the kidney, returning blood to the caudal vena cava.

▶ Find the **ureter,** which originates from the medial face of the kidney. Trace it caudally to where it empties into the **urinary bladder.** In the fetal pig, part of the urinary bladder extends between the two **umbilical arteries** and continues into the umbilical cord, where it is called the **allantoic duct.** After birth, the duct atrophies, becoming nonfunctional.

The **urethra** proceeds caudally from the bladder to its opening; its location depends on the sex of your pig. To observe the urethra, cut through the cartilage and muscles of the **pelvic girdle** on the medial line and press the hind legs down against the dissecting tray.

▶ Remove one kidney. Make a longitudinal cut through the kidney with a sharp razor blade in a plane parallel with the front and back of the kidney. Study the cut surface with the dissecting microscope (fig. 29.4). The concave side of the kidney is the **hilum** and the outer, convex surface, the **cortex.** The **medulla** is the compact tissue containing collecting ducts and blood vessels.

Follow the ureter into the kidney. It expands into a structure called the **renal pelvis,** which subdivides into funnel-shaped **calyxes.** Urine is produced by the thousands of filtration-reabsorption units called **nephrons.** It flows through collecting ducts to the calyxes and passes by way of the ureter to the bladder for storage.

▶ Though they are not involved in excretion, note the **adrenal glands,** which are located cranially and medially to the kidneys in the abdominal cavity. Adrenaline and noradrenaline, hormones produced by these glands, regulate a number of body functions, including heart rate, blood sugar levels, and arteriole constriction.

Histology of the Kidney

▶ Obtain a prepared slide of a section from the cortex of the kidney and observe it under a 10× objective.

In the cortex are tufts of capillaries each called the **glomerulus** surrounded by a cuplike **Bowman's capsule** that function as a filter unit. You will have to search on the slide to find these structures because the plane of the slice does not pass directly through each nephron.

Figure 29.2 Ventral view of a male fetal pig's urogenital system.

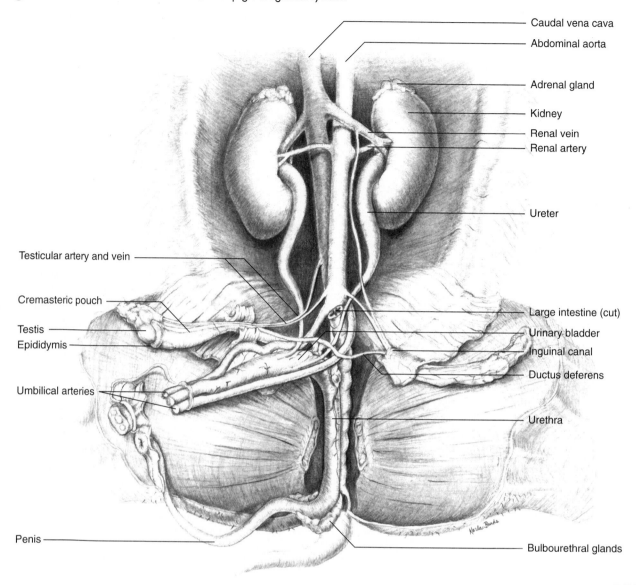

Caudal vena cava
Abdominal aorta
Adrenal gland
Kidney
Renal vein
Renal artery
Ureter
Large intestine (cut)
Urinary bladder
Inguinal canal
Ductus deferens
Urethra
Bulbourethral glands

Testicular artery and vein
Cremasteric pouch
Testis
Epididymis
Umbilical arteries
Penis

Figure 29.5 shows scanning electron micrographs of the glomerulus. Afferent arterioles entering the glomeruli are large, while efferent arterioles exiting the glomeruli are small. This causes a high filtration pressure in the glomerulus. The Bowman's capsule is a funnel-like device that surrounds and encloses the glomerulus (fig 29.6).

Blood from the renal arteries enters the glomeruli at high pressure and is filtered through the capillary walls and into the Bowman's capsule. Blood cells and proteins remain in the capillaries along with some liquid. Water and dissolved, low molecular-weight substances, such as sugars, amino acids, urea, and salts, pass into the Bowman's capsule, which is the start of the **nephron.**

Nephron structure is difficult to discern in a prepared slide, since the plane of the section rarely coincides with the plane of the nephron. Figure 29.6 shows the anatomy of the nephron. As the filtrate passes through the tubules and

the **loop of Henle,** water, salts, and nutrients are reabsorbed and pass into the capillaries surrounding the tubules. These absorption processes are controlled by hormones. Some water, urea, and other waste products are not reabsorbed and consequently flow through the collecting ducts to become urine.

To gain an understanding of comparative anatomy, fill in tables 29.1 and 29.2. Write a brief description of each structure and how it functions in filtration, reabsorption, excretion, or storage. Include a summary statement of the organ system for each animal.

Mammalian Reproductive System

Though not part of the excretory system, the organs of the reproductive system are located so close to the excretory system that they warrant a brief discussion here.

Figure 29.3   Ventral view of a female pig's urogenital system.

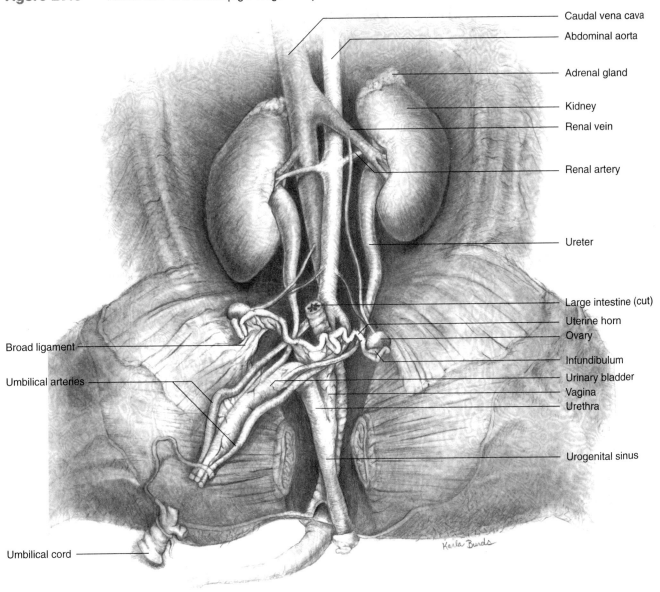

Caudal vena cava
Abdominal aorta
Adrenal gland
Kidney
Renal vein
Renal artery
Ureter
Large intestine (cut)
Uterine horn
Ovary
Infundibulum
Urinary bladder
Vagina
Urethra
Urogenital sinus

Broad ligament
Umbilical arteries
Umbilical cord

Karla Burds

Male System

During embryonic development, the **testes** originate in the coelom caudal to the kidneys, and then descend to lie in an external pouch, the **scrotum,** in adult pigs.

To find the testes in a fetal pig, locate the testicular artery and vein on one side of the animal and trace them to the **inguinal canal** (see fig. 29.2). The testis passed through this canal during its descent guided by the gubernaculum, a mass of smooth muscle connecting the testis to the scrotal sac, which shortens during growth. Pass a blunt probe through the inguinal canal.

The probe should help you locate a thin-walled, elongated sac extending across the ventral surface of the thighs. Free the sac from the surrounding tissue. This is the **cremasteric pouch** and represents an outpocketing of the con-

nective and muscle tissues of the abdominal wall. Cut it open to view the testis.

Sperm are produced in the microscopic seminiferous tubules of the testis. They in turn empty into the **epididymis,** visible as a tightly coiled mass of tubules along one side of the testis. The epididymis flows into the **vas deferens,** which passes from the scrotal area through the abdominal wall and into the body cavity via the inguinal canal. After entering the body cavity, the vas deferens on each side loops over the ureter and enters the **urethra.** The paired **seminal vesicles** and single, bilobed **prostate gland** are also located at this juncture. The large **bulbourethral glands** lie on either side of the junction of the urethra with the **penis.** These three glands secrete additional fluids that carry the sperm during an ejaculation. The penis of a fetal

Investigating the Excretory and Reproductive Systems   **365**

Figure 29.4
Longitudinal section of a mammal's kidney.

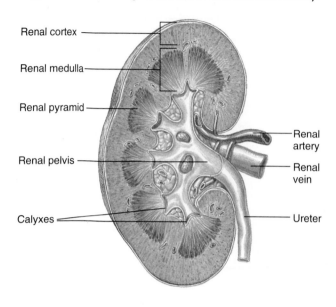

Renal cortex

Renal medulla

Renal pyramid

Renal pelvis

Calyxes

Renal artery

Renal vein

Ureter

Figure 29.5
Scanning electron micrographs of the glomerulus. Plastic is injected into the circulatory system and then tissue is digested away, leaving the glomerulus (G) and arterioles. Note the difference in the size of the afferent (AA) and efferent (EA) arterioles.

Figure 29.6
Diagram of a nephron.

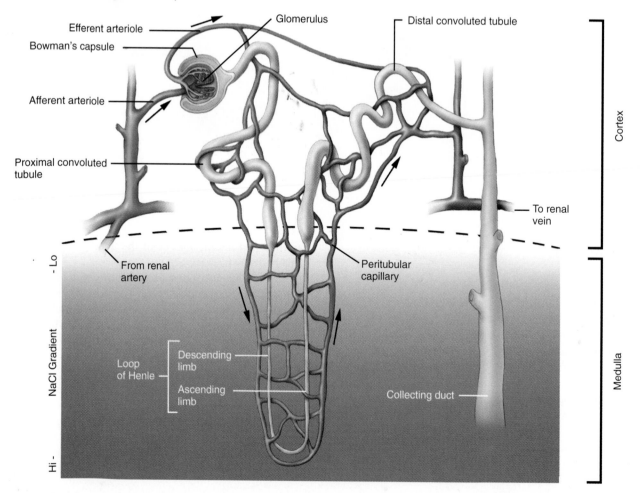

Efferent arteriole

Bowman's capsule

Afferent arteriole

Proximal convoluted tubule

From renal artery

Glomerulus

Distal convoluted tubule

Cortex

To renal vein

Peritubular capillary

Medulla

NaCl Gradient

- Lo

Loop of Henle

Descending limb

Ascending limb

Collecting duct

Hi -

TABLE 29.1	Earthworm—excretory system summary

Body surface

Nephrostome

Septum

Tubule

Capillary network

Bladder

Nephridiopore

Description of excretory process

TABLE 29.2	Fetal pig—excretory system summary

Glomerulus

Bowman's capsule

Descending and ascending tubules

Loop of Henle

Collecting duct

Ureters

Bladder

Urethra

Description of excretory process

Figure 29.7   Male fetal pig's urogenital system in ventral view.

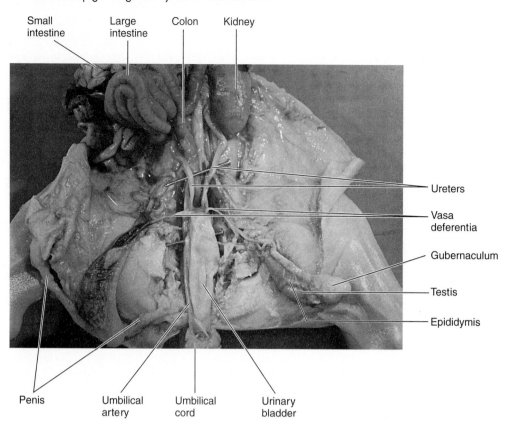

Small intestine   Large intestine   Colon   Kidney

Ureters

Vasa deferentia

Gubernaculum

Testis

Epididymis

Penis   Umbilical artery   Umbilical cord   Urinary bladder

Figure 29.8   Male fetal pig's urogenital system in sagittal section view.

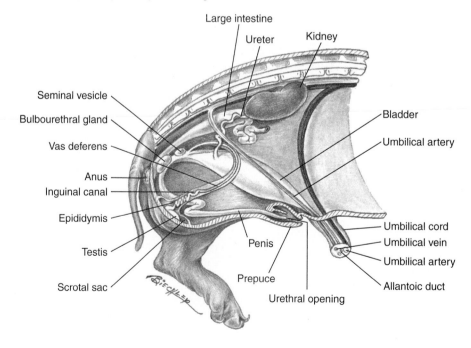

Large intestine

Ureter   Kidney

Seminal vesicle

Bulbourethral gland

Vas deferens

Anus

Inguinal canal

Epididymis

Testis

Scrotal sac

Bladder

Umbilical artery

Penis

Prepuce

Urethral opening

Umbilical cord

Umbilical vein

Umbilical artery

Allantoic duct

Figure 29.9 Cross section of a seminiferous tubule, showing stages in sperm development as you pass from outer wall to lumen of tubule.

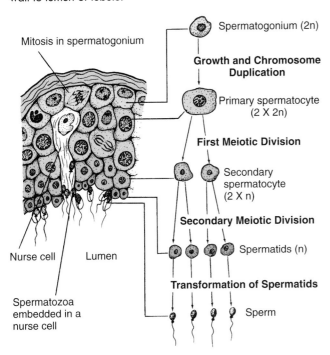

Mitosis in spermatogonium

Spermatogonium (2n)

Growth and Chromosome Duplication

Primary spermatocyte (2 X 2n)

First Meiotic Division

Secondary spermatocyte (2 X n)

Secondary Meiotic Division

Nurse cell   Lumen

Spermatids (n)

Transformation of Spermatids

Spermatozoa embedded in a nurse cell

Sperm

pig is retracted and lies deep in the muscle layers of the groin area. The location of these structures is shown in sagittal section in figure 29.8

Microanatomy of Seminiferous Tubules

Obtain a slide of a cross section of a mammalian testis containing **seminiferous tubules.** Using the compound microscope, look at it first with low power. Center a tubule in the field of view and then switch to high power. Compare the specimen to figure 29.9.

Note that the cells grow smaller as you scan from the outer tubule wall to the inside, where mature sperm may be present in the **lumen.** The largest outer cells, the **spermatogonia,** divide by mitosis to produce other spermatogonia. Half of the spermatogonia produced undergo meiosis and become sperm, while the other half again divide by mitosis to replenish the spermatogonia population.

Primary spermatocytes are spermatogonia destined to enter meiosis. When these cells finish meiosis I, two **secondary spermatocytes** are produced. They rapidly undergo meiosis II to produce spermatids that mature into functional sperm. **Nurse cells** (or **Sertoli cells**) located in the seminiferous tubule walls aid in this process. Between the seminiferous tubules, **interstitial cells** secrete the male hormone **testosterone.**

Female System

The **ovaries** are small, bean-shaped organs on the dorsal wall near the caudal end of the kidneys (fig. 29.10). The ovaries are suspended by two sheets of connective tissue, the **broad ligament** (see fig. 29.3). On the dorsal side of each ovary is a **fallopian tube** (oviduct) ending in a hood-like structure, the **infundibulum.** Eggs bursting from the ovary enter the openings, or **ostia,** of the infundibulum and pass down to the **uterine horns,** where, if fertilized, they implant and develop. The uterine horns unite to form the **uterus,** which opens to the outside through a muscular tube, the **vagina.** The vagina and the urethra open into a common area.

You are now finished with the fetal pig for this laboratory. Return it to the storage area.

Microanatomy of the Ovary

Obtain a slide of a section of a mammalian ovary and look at it under low power with the compound microscope. Large, clear areas will be visible in the section, as in figure 29.11. These are **Graafian follicles,** structures in which egg maturation occurs. There are several hundred thousand primordial follicles in each ovary of a female at the time of her birth. The ova in these follicles are in an arrested prophase of meiosis I. At sexual maturity, hormones from the pituitary gland cause the follicles to secrete the female hormones **estrogen** and **progesterone.**

If you are lucky, your slide will show such an enlarged follicle containing an ovum and surrounded by follicle cells. During ovulation, the follicle swells until it finally bursts, releasing the egg in a rush of fluid. Fingerlike projections of the infundibulum surround the ovary and collect the egg. The egg moves down the fallopian tube by the action of cilia on the tube surface.

Sperm usually fertilize the egg as it passes down the fallopian tube. A second meiotic division will take place only if a sperm fertilizes the egg.

The collapsed follicle left in the ovary following ovulation assumes a star-shaped appearance and fills with a yellow fluid. At this stage, it is called a **corpus luteum** and secretes **progesterone.** If fertilization occurs and the ovum implants in the uterus, a hormone produced by the developing placenta will sustain the corpus luteum and progesterone production. If implantation does not occur, the corpus luteum will degenerate after several days and progesterone production will decrease. Falling levels of progesterone trigger the pituitary to produce follicle-stimulating hormone (FSH), which spurs another Graafian follicle toward ovulation. The total time course for these events in a human is approximately 28 days.

Figure 29.10   Ventral view of female urogenital organs. Intestines have been removed.

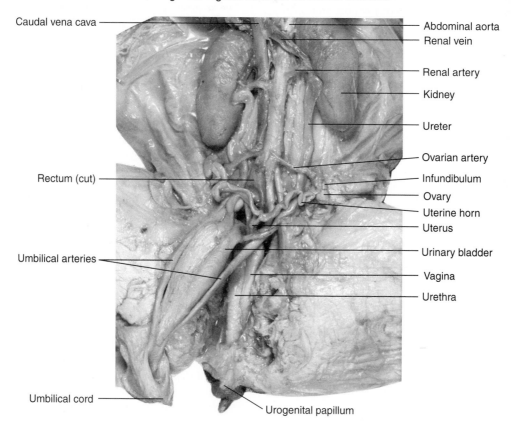

Caudal vena cava

Rectum (cut)

Umbilical arteries

Umbilical cord

Abdominal aorta
Renal vein
Renal artery
Kidney
Ureter
Ovarian artery
Infundibulum
Ovary
Uterine horn
Uterus
Urinary bladder
Vagina
Urethra

Urogenital papillum

Figure 29.11   Photomicrograph of a section from a rhesus monkey's ovary.

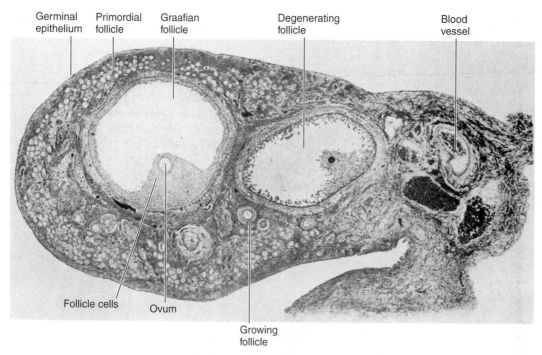

Germinal epithelium   Primordial follicle   Graafian follicle   Degenerating follicle   Blood vessel

Follicle cells   Ovum

Growing follicle

Learning Biology by Writing

The nephron is the basic functional unit of the kidney. Briefly describe the parts of the nephron and their functions, describing how water and small dissolved molecules pass from the blood and become urine. Describe the pathway that urine follows as it flows from the nephron to the time it is voided.

As an alternative assignment, your instructor may ask you to answer the Lab Summary and Critical Thinking Questions that follow.

Internet Sources

Search the World Wide Web for information on kidney stones. What are they and how are they formed? Be sure to record **URLs** for future reference.

Lab Summary Questions

1. How does the nephridial system of an earthworm work?
2. Create a flowchart showing how urine is formed in a mammal from fluids in the blood and is eventually voided. To the right of the structures named, briefly describe their function.
3. Discuss why filtration and reabsorption are both important processes in the kidney.
4. Name the structures that a sperm passes through from the time it is formed until it is ejaculated.
5. Name the structures an egg passes through from the time it is formed until it reaches the uterus.
6. Create a flowchart that traces a drop of water from the time it enters the mouth until it is voided in the urine of a mammal.

Critical Thinking Questions

1. During embryonic development, the testes develop within the abdominal cavity. Shortly before birth, or shortly thereafter, the testes descend through the inguinal canal to lie outside the abdominal cavity within the scrotal sacs. Why is this important?
2. Some mammals, such as bears, badgers, and raccoons, enter a period of prolonged sleep during the winter months when temperatures are extreme and food is scarce. This is not considered "hibernation" because there is little if any drop in body temperature. Also, the heart rate in bears may drop only from 40 to 10 beats per minute. On the other hand, some mammals, such as woodchucks and ground squirrels, enter a period of true hibernation. Their body temperature cools to within a degree or two of the ambient temperature. Their respiratory rate decreases from 200 to around 5 breaths per minute, and heart rate drops from 150 to 5 beats per minute.

   What excretory problems do bears face during their long sleep? Would woodchucks have the same challenge? Suggest possible adaptations that these mammals might have evolved to deal with nitrogenous wastes during winter.
3. Is internal fertilization superior fertilization? Why do you say so?
4. Researchers interested in the physiology of the kidney made the measurements in the following table. Based on your understanding of how the kidney works, explain why some substances are found at the same concentration in the plasma and glomerular filtrate and why others are not. Also explain why most substances are found at different concentrations when the glomerular filtrate is compared to the urine.

TABLE 29.3 Concentrations of selected substances in mg per 100 ml of fluid

Substance	Plasma	Glomerular Filtrate	Urine
Albumin	4500	0	0
Glucose	100	0	0
Urea	26	26g	1820
Uric acid	4	4	53
Na	330	330	297
K	16	16	192
Cl	350	350	455

Assume that a normal adult has 31 liters of blood plasma; glomerular filtration rate is 180 liters per day and normal urine output is about 1.5 liters per day.

Investigating the Excretory and Reproductive Systems

LAB TOPIC 30

Investigating the Properties of Muscle and Skeletal Systems

Supplies

Preparator's guide available on WWW at
http://www.mhhe.com/dolphin

Equipment

Compound microscopes
Intellitool Setups (Batavia, IL)
   Physiogrip and software
   Isolated, square wave stimulator
   Computer

Materials

Fetal pig
Dissection pans and instruments
Millipedes, spiders, crabs, or insects
Bird skeleton
Human skeleton
Frog skeleton
Fresh beef long bones split longitudinally
Dry long bones
Prepared slides
   Skeletal muscle, longitudinal
   Smooth muscle section
   Cardiac muscle, longitudinal
   Bone
   Hyaline cartilage
   Tendon
   Earthworm, cross section
   Neuromuscular junction (demonstration)

Prelab Preparation

Before doing this lab, you should read the introduction and sections of the lab topic that have been scheduled by the instructor.
   You should use your textbook to review the definitions of the following terms:

   antagonistic muscles
   appendicular skeleton
   axial skeleton
   endoskeleton
   exoskeleton
   extensor
   flexor
   hydrostatic skeleton
   pelvic and shoulder girdles
   sarcomere

   smooth, cardiac, and skeletal muscle
   summation
   tetany
   threshold effect
   twitch

   You should be able to describe in your own words the following concepts:

   Sliding filament theory of muscle contraction
   How a nerve impulse causes a muscle to contract
   Basic structure of bone

   As a result of this review, you most likely have questions about terms, concepts, or how you will do the experiments included in this lab. Write these questions in the space below or in the margins of the pages of this lab topic. The lab experiments should help you answer these questions, or you can ask your instructor for help during the lab.

Objectives

1. To observe the microanatomy of skeletal muscle
2. To observe the surface muscles in the fetal pig
3. To investigate the properties of threshold, recruitment, and temporal summation during muscle contraction
4. To observe examples of hydrostatic skeletons and exoskeletons
5. To observe the microanatomy of bone
6. To compare the endoskeletons of several vertebrates

Background

Simple small organisms, such as bacteria, algae, protozoa, and sponges, are capable of movement without muscle systems. They use cilia or flagella to move through their aqueous environments. All higher animals, whether aquatic or terrestrial, depend on muscle systems for movement and on nervous systems to control that movement. These muscle systems are also intimately associated with

Figure 30.1 Antagonistic muscle arrangements in human limb. Contraction of human biceps flexes the forearm, while contraction of the triceps extends the forearm.

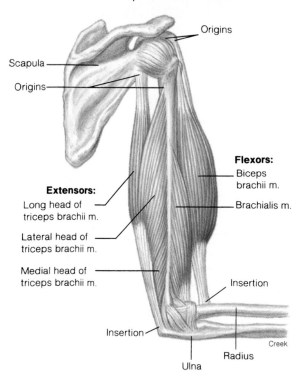

the skeletal system, which converts muscular contraction into effective movement and locomotion.

Muscles are nearly always found arranged as **antagonistic** pairs attached to skeletons. As a result, contraction of one member of the pair causes an action, while contraction of the second member restores the body to its original position. In earthworms, the longitudinal and circular muscles in the body wall are antagonists. Contraction of the circular muscles elongates the segments by forcing the fluid of the hydrostatic skeleton forward and back in each segment, but contraction of the longitudinal muscles shortens the segments, leading to an increase in girth. In antagonistic pairs, neuronal activity is such that the contraction of one antagonist is usually accompanied by the relaxation of the second.

Figure 30.1 shows the antagonistic arrangement of human muscles associated with an internal skeleton. Muscles that increase the angle between two bones at a joint are called **extensors. Flexor** muscles are antagonistic to extensors and decrease the angle between two bones. Straightening your elbow is extension and bending it is flexion.

About 80% of the mammalian body mass is muscle. Most of this is **skeletal muscle** (sometimes called **striated** or **voluntary muscle**). Skeletal muscle is characterized by a high degree of cellular organization. Muscle proteins, consisting mostly of **actin** and **myosin**, occupy most of the cytoplasm of the cell. These proteins are arranged in a series of parallel filaments. Many cylindrical muscle cells make up a muscle, and these cells are arranged in register, giving a striated pattern to the muscle.

Two types of skeletal muscle are found in vertebrates and many invertebrates: red muscle and white muscle. **Red muscle** is a relatively slow-acting and slow-fatiguing muscle. It is usually characterized by a good blood supply and has many mitochondria and much myoglobin, a compound similar to hemoglobin, which stores oxygen in the muscle. Red muscle cells aerobically oxidize fatty acids for energy. **White muscle** has a less well-developed blood supply, fewer mitochondria, and less myoglobin. Cells of this type preferentially metabolize stored glycogen to produce lactic acid anaerobically. They can contract rapidly and develop high tension but they also fatigue rapidly. Muscles consist of both types of cells, and the relative composition of each may change with use or disuse.

The contractile mechanism in both types of muscle is the same. Actin and myosin fibers overlap and slide across each other during contraction, shortening the muscle in much the same manner as a deck of cards is aligned after it is shuffled (see fig. 30.2). The force for movement comes from the interaction of the filaments themselves rather than from a shortening of the filaments.

Along with skeletal muscle, two other types of muscle tissue are found in animals. **Smooth muscle,** also called **involuntary muscle,** is found in the gut, blood vessels, pupil, and reproductive system. These muscle cells are spindle shaped, and their actin and myosin are not organized in parallel arrays. Smooth muscle is innervated by the autonomic nervous system and responds to various hormones. Its response is characterized by slow, rhythmic contractions. In some cases, contractions arise spontaneously in smooth muscles and are propagated along the length of the organ.

Cardiac muscle is found in the heart. The actin and myosin components in cardiac muscle are found in interdigitating linear arrays, as in skeletal muscle, but the cells are capable of spontaneous contraction, as in smooth muscle. The muscle cells are shorter than in skeletal muscle, often branch, and are joined end to end by tight junctions, which electrically couple the cells and allow contraction to spread from cell to cell independent of the nervous system.

An integral part of any animal's locomotory system is its skeleton. While the word *skeleton* usually brings to mind the bony internal structure of vertebrates, other types of skeletons occur in the animal kingdom. **Hydrostatic skeletons** are found in earthworms and other invertebrates in which the pressure of internal fluids gives shape and allows the organism to move. **Exoskeletons** are found among such diverse animals as arthropods, molluscs, and corals. This tough outer covering protects the organism and supports it in an upright position while allowing for muscle attachment and movement.

Skeletons show adaptations to an organism's environment and lifestyle. Frogs and birds not only have limbs modified to fit their environments, but also have bones that show either thickening or shortening to fit their forms of locomotion. Consider the skeletal adaptations of a human, horse, bat, and seal to their respective forms of locomotion. Careful study of the skeleton by the trained observer can reveal much

Figure 30.2 The structure of muscle: (*a*) a muscle fiber or cell, showing the outer membrane with tubules that project into the cell from the cell surface, sarcoplasmic reticulum, and the myofibrils inside the cell; (*b*) thin actin filaments and thick myosin filaments slide past one another, shortening the sarcomere.

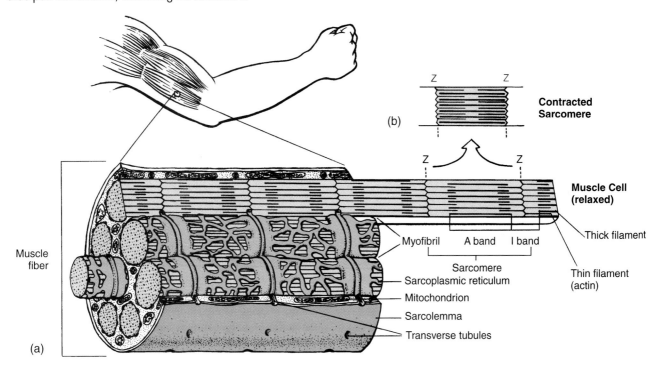

about an animal, because skeletons are excellent examples of the biological principle that form reflects function. Skeletons serve many other functions besides locomotion. They protect soft tissue, serve as a reservoir for calcium and phosphate, and are sites of blood cell formation.

Skeletons are often the only parts of animals that survive from the past as fossils. Because the muscles attach to the skeleton, it is possible to see which muscles might have been highly developed and to speculate about how an extinct animal lived on a day-to-day basis.

LAB INSTRUCTIONS

You will study the properties of skeletal muscle and compare several types of skeletal systems.

Muscular System

Microscopic Anatomy of Muscle

Smooth Muscle

▷ Obtain prepared slides of smooth muscle, cardiac muscle, and a longitudinal section of skeletal muscle. Observe the smooth muscle slide first. Note the shape of individual cells, the presence of nuclei, and the absence of fiber organization in the cytoplasm. Sketch two or three examples of smooth muscle cells.

Skeletal Muscle

Now examine the slide of skeletal muscle under the medium-power objective and compare its organization to that of smooth muscle. Note the substantial differences. Identify the individual cylindrical cells that run the length of the tissue specimen on the slide. Find a very thin area of the section that is only one cell thick and examine it under high power. On the periphery of each cell, you will see several nuclei. Each skeletal muscle cell is **multinucleated** (a **syncytium**) because it is formed by the fusion of several smaller, uninucleated cells during embryonic development. Though probably not visible, several mitochondria are also

Figure 30.3 Transmission electron micrograph of skeletal muscle, showing that each myofibril is made up of sarcomeres joined end to end.

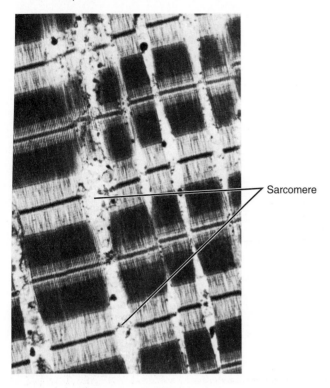

Sarcomere

present in this peripheral cytoplasmic area. The central area of the cell consists of many parallel fibers called **myofilaments** running lengthwise in the cell. These fibers give a cross-banding appearance to the cytoplasm. Draw two or three of these cells.

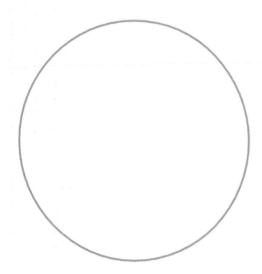

To understand how your drawing relates to the structure of a whole muscle, such as the biceps in your upper arm, look at figure 30.2. What most people call a muscle contains thousands of muscle cells arranged parallel to one

another and surrounded by a sheath of connective tissue. Within a single muscle cell are many **myofibrils,** which consist of contractile units called **sarcomeres** joined end to end (fig. 30.3). Each sarcomere contains two types of filamentous protein. **Actin** is the protein in thin filaments and **myosin** is the protein in thick filaments. The actin filaments at each end are anchored in the Z lines, which mark the ends of the sarcomere (fig. 30.2). The myosin filaments are suspended between and surrounded by actin filaments. It is the interdigitation of these filaments and the areas of overlap that create the alternating light and dark banding patterns that you saw on your slide of skeletal muscle.

During contraction, the thin filaments slide over the thick filaments toward the middle, pulling the Z lines inward and shortening the overall length of the sarcomere. Because this process is repeated simultaneously in each sarcomere, muscle cells can dramatically shorten.

In recent years, research has revealed the molecular mechanisms involved in muscle contractions. When a neural action potential arrives at the **neuromuscular junction,** it causes the release of a chemical called acetylcholine (fig. 30.4). This in turn depolarizes the muscle cell membrane and triggers an action potential in the muscle cell. As the action potential moves along the muscle cell, it penetrates to the interior of the cell via the transverse tubule system (fig. 30.2). Membrane depolarization at the Z lines causes the sarcoplasmic reticulum to release calcium ions into the sarcomere. The calcium ions affect the actin, so that it reacts chemically with the bridges on the myosin filaments, forming momentary cross-links. This involves pulling the actin filament over the surface of the myosin filament. The coordinated effect of this **sliding filament mechanism** is the shortening of the muscle.

Look at the demonstration slide of a neuromuscular junction that is available in the lab. What do you find interesting about this slide?

Cardiac Muscle

As the name implies, cardiac muscle is found in the heart, where it makes up the mass of the heart wall along with connective and nervous tissue. It is capable of rhythmic contraction which can be modulated by the pacemaker of the heart.

Figure 30.4   Scanning electron micrograph showing axon ending at the neuromuscular junction on a muscle cell.

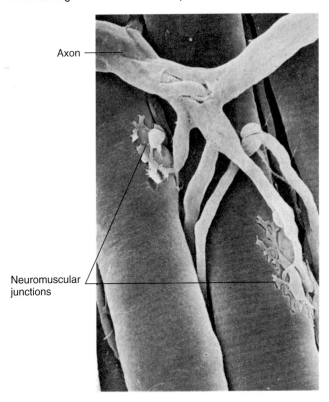

Axon

Neuromuscular junctions

Look at a slide of cardiac muscle through your compound microscope. Would you say that it is similar to smooth muscle or to striated muscle? Why?

Use your 43× objective and look closely at one muscle cell. Move the slide so that you can follow a single cell for some distance. You should find two things in cardiac muscle that are different from what you saw in striated muscle. What are these differences?

Fetal Pig Superficial Muscles

This dissection may be done by students or by the instructor as a demonstration, depending on the time available and the emphasis of the course on gross anatomy.

Three terms are used to describe the anatomical orientations of muscles: origin, belly, and insertion. The **origin** is the end attached to the less mobile portion of the skeleton. The **belly** is the central part of the muscle, and the **insertion** is the end of the muscle attached to the freely moving part of the skeleton. **Tendons** are fibrous connective tissues that connect muscles to the skeleton.

If your fetal pig is not already skinned, skin it—being careful not to tear away muscle as you remove the skin. Use a probe or finger to separate the two. Under the skin, there may be adipose (fat) tissue, which should be removed to reveal the muscles. When the skin is removed, dry the carcass with paper towels to improve viewing. In young fetal pigs, the muscles are not fully developed and are often tightly connected to the skin so that they are torn during skinning. Identification will be difficult.

Starting at the head, identify the major muscles indicated in figure 30.5.

1. The **latissimus dorsi** is a broad muscle running obliquely around the sides of the thoracic region. It originates on the vertebrae and inserts on the proximal end of the humerus. The latissimus dorsi is involved in moving the foreleg.

2. The **trapezius** originates on the occipital bone of the skull and from the first ten vertebrae and inserts on the scapula or shoulder blade. This broad muscle draws the scapula medially. When this muscle contracts, how does the leg move?

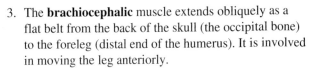

3. The **brachiocephalic** muscle extends obliquely as a flat belt from the back of the skull (the occipital bone) to the foreleg (distal end of the humerus). It is involved in moving the leg anteriorly.

4. The **sternocephalicus** is a long muscle below the brachiocephalic muscle. It controls the flexing of the head. It originates on the sternum and inserts on the mastoid process of the skull by means of a long tendon. Remove the parotid salivary gland to see the tendon.

Investigating the Properties of Muscle and Skeletal Systems   **377**

Figure 30.5 The major muscles of the fetal pig.

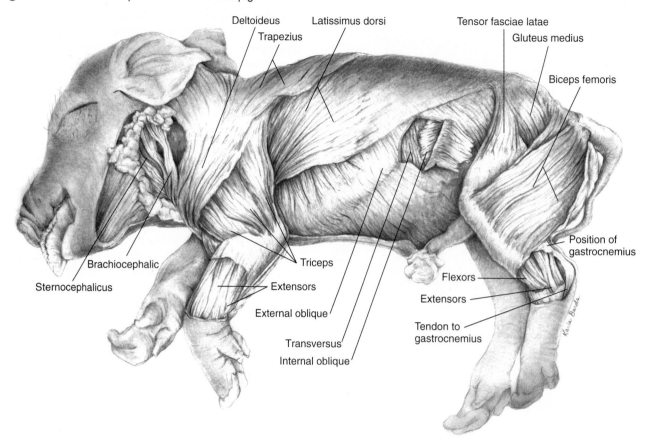

Deltoideus

Trapezius

Latissimus dorsi

Tensor fasciae latae

Gluteus medius

Biceps femoris

Brachiocephalic

Sternocephalicus

Triceps

Extensors

External oblique

Transversus

Internal oblique

Flexors

Extensors

Tendon to gastrocnemius

Position of gastrocnemius

 5. The **brachialis** is a small muscle located in the angle formed by the flexed foreleg. It originates on the humerus and inserts on the ulna. What is its function?

6. The **deltoideus** is a broad shoulder muscle originating on the scapula and inserting on the humerus. It aids in flexing the humerus.

7. The **extensors** are several muscles in the lower foreleg that extend and rotate the wrist and digits.

8. The **triceps** (brachii) is a large muscle making up practically the entire outer surface of the forelimb. It originates on the humerus and inserts on the proximal end of the ulna. When it contracts, the forelimb extends.

9. The **external oblique** muscles make up the outer wall of the abdomen and run obliquely from the ribs to a ventral **longitudinal ligament** along the ventral midline. Contraction of these muscles constricts the abdomen.

10. The **internal oblique** fibers lie just under the external oblique muscle and run almost at right angles to them. Contraction of these fibers also results in abdominal constriction.

11. The **transversus** muscles are a third layer of muscles in the abdominal wall. They run dorsoventrally in the body wall somewhat like the obliques. Carefully scrape away this muscle layer, and you will expose the outside of the peritoneum, a connective tissue membrane that lines the body cavity, or **coelom.**

12. The **tensor fasciae latae** is the most cranial of the thigh muscles. It originates on the pelvis and continues ventrally as a thin, triangular muscle until it becomes a sheet of connective tissue called the **fasciae latae,** which inserts on the kneecap and extends the leg.

13. The **gluteus medius** is a thick muscle covered by the tensor fasciae latae in the hip region. If the overlying muscle is removed, its origin on the hip and insertion on the femur can be seen. What action does it cause?

14. The **biceps femoris** is a large muscle making up most of the back half of the thigh. It originates on the pelvis and inserts on the lower femur and upper part of the tibia.

15. The **gastrocnemius** is the large muscle of the calf, originating on the lower end of the femur and attaching to the heel by means of the Achilles tendon. Its action extends the foot. The **soleus** muscle lies close to the gastrocnemius and has a similar function.

16. The **digital flexors** and **extensors** originate on the tibia and fibula and insert on the metatarsals. What are their functions?

17. The **peroneus** muscles have origins and insertions similar to digital muscles. They are involved in moving the whole foot.

Physiology of Muscle

This part of the lab may be performed by the instructor as a demonstration or by students working in groups of four or more, depending on the time and equipment available.

Overview of Experiment

In this section of the lab, you will investigate four properties of human muscles: threshold stimulus, recruitment, temporal summation, and tetany. You will use special equipment and software that allows a computer to collect and analyze data. The experimental subject(s) will be a volunteer member(s) of your class. Anyone who has heart problems or does not wish to participate directly in this experiment should not be a subject. They can operate the computer or record notes during the experiment.

To do this experiment, you need to make adjustments to four major pieces of equipment—a stimulator, a transducer, an interface unit, and a computer running special software. A **stimulator** is an electronic device that produces a pulse of electricity that can be varied by volts (strength), duration of the pulse (length), and frequency (number of pulses per second). The **transducer** is a device that converts the movement of a trigger in a pistol handle into an electrical resistance that can be recorded. The **interface** is a box that connects various cables together into a single lead that is connected to the serial port of your computer. The **software** allows your computer to record raw data from the transducer and process the data, converting it

into graphical displays that can be analyzed. Before you start investigating the biology of muscle, this equipment must be calibrated.

Equipment Calibration
Stimulator

Look at the stimulator and note the dials on the front. Make sure that the power switch is turned off. The various dials and switches allow you to adjust the characteristics of the stimulus. Set the duration of the pulse to 10 msec and the voltage to 30 volts. Note that there may be a voltage dial and a multiplier dial. If the voltage dial is set at 3.0 and the multiplier at 10, then the pulse will be 30 volts. There should also be a switch that allows you to select between two modes of operation: single stimulus or continuous (multiple stimuli). In multiple mode a dial should allow you to set the number of stimuli per second. To start, set the switch to single mode. Connect a cable from the stimulator to the interface box according to the manufacturer's directions. This will send a signal to the computer each time a stimulus is administered. Remember that red connectors should always be connected to positive terminals and black to negative ground terminals. Now connect wires from the stimulus electrodes to the stimulator output terminals according to the manufacturer's directions. One electrode will have a flat plate ground electrode (–) and the other will be a pencil-like probe electrode (+). Note the button that must be depressed to administer a stimulus. This is a safety feature that prevents shocks from happening. Do not turn on the stimulator power until you are ready to do the experiment.

Transducer

Find the Physiogrip transducer. This device allows you to measure the amount of muscle contraction. When you squeeze the trigger, the electrical resistance of a wire passing through the device changes. The Physiogrip is not a device to measure the strength of your grip, and it should be treated gently. It is a sensitive device that measures how rapidly and forcefully a single muscle can contract.

Computer Calibration

Connect the Physiogrip to the interface box according to the manufacturer's directions. Connect the interface box to the computer. The software should have been previously installed on the computer. Run the Physiogrip software according to directions from your instructor. Click on NEW under the FILE menu and a dialog box will appear to lead you through transducer calibration. Follow the directions that appear on the computer screen.

Trial Use of Software

After calibrating the computer and Physiogrip, take a moment to familiarize yourself with the software. In the Acquiring Data mode, you can record the output from that Physiogrip. Do this by clicking on START on the Acquiring

Figure 30.6 Approximate location of nerve innervating flexor digitorum superficialis.

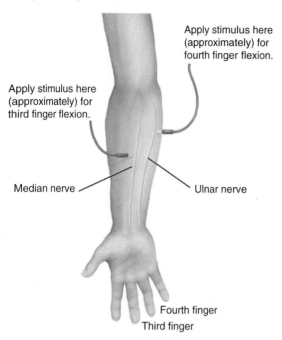

Apply stimulus here (approximately) for fourth finger flexion.

Apply stimulus here (approximately) for third finger flexion.

Median nerve

Ulnar nerve

Fourth finger

Third finger

Figure 30.7 Position of electrode attachment.

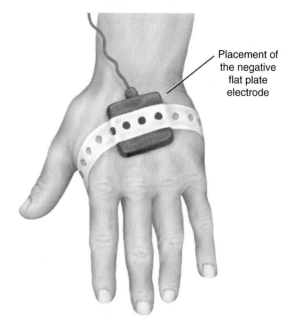

Placement of the negative flat plate electrode

Data screen and then squeeze the trigger of the Physiogrip in and out. You should see a movement of the trace on the computer screen and a record of the contractions. You can stop the trace by clicking AUTOSTOP or by letting it go for the preset time of 50 seconds. The data just collected can be printed by selecting PRINT SCREEN from the FILE menu or it can be saved by choosing SAVE from the FILE menu.

You are now ready to use this equipment in your investigation of muscle physiology.

Determining a Motor Point

Motor points are locations on the body surface where nerves innervating certain skeletal muscles pass close to the skin. If a stimulating electrical current is given at this point, the nerve passing to the muscle is stimulated and the muscle contracts. In this part of the lab, you will determine the motor point for the nerve innervating the **flexor digitorum superficialis,** a muscle involved in the rapid flexing of the third and fourth digits. This nerve is located in the belly area of the underneath surface of the forearm (fig. 30.6).

To start the experiment, one person should volunteer to be the subject. Thin males have less subcutaneous fat than other students and will be the best subjects. Be sure that the stimulator is turned off. Attach the negative plate electrode to the back of the hand **on the arm that will be stimulated,** using a liberal amount of electrode gel between the plate and the skin (fig. 30.7). When handling the electrodes, use caution.

CAUTION

Guard against any situation in which one electrode is held in one hand and the second electrode is held in the other hand. In such a situation, when the stimulator is turned on an electric current could flow across the chest and cause an arrhythmia in the heartbeat.

The three remaining people in the group should each take a different job. One should handle the stimulating electrode, a second should work at the computer, and the third should work the controls on the stimulator and depress the stimulator button to activate the stimulator probe.

When the first electrode has been attached to the back of the hand, have the person rest their arm on the table with the palm up and muscles relaxed. Check that the stimulator is set at 10 msec duration, 30 volts at a frequency of 1 per second. Turn it on. Put a dab of electrode gel on the end of the probe stimulator and press it on a spot near the center of the belly of the forearm muscles (fig. 30.6). Try several different spots in this general vicinity. You may have to press rather hard to get the best results. When you find the motor point, the third and fourth digits will contract involuntarily and almost all discomfort from the stimulus disappears. If you cannot find the spot, increase the strength of the stimulus to 40 volts and try again. When a motor point is identified, mark it with marker or a ball point pen.

After the motor point is identified, have the subject grip the Physiogrip and start the software. Set the controls so that:

Metronome and Autostop are off;

Samples per second is 100;

Seconds per data set is 50.

Apply the stimulus to the previously identified motor point and work out a procedure that allows you to record twitches on the screen. Once you can cause involuntary contractions by stimulating the motor point and record data, you are ready to do the experiment. (*Note:* Changing the position of the subject's arm may require that you locate the motor point again.)

Determining Threshold Stimulus Intensity

Nerve cells are excitable cells. When subjected to an electrical current, ions move across the cell membrane causing a change in the membrane potential that is proportional to the current. If the change in potential exceeds a **threshold value,** an action potential is triggered in the excitable cell. You will determine this threshold for the nerve innervating the flexor digitorum superficialis. What do you hypothesize will happen when the threshold potential is reached?

The person operating the computer should start a new data file by selecting NEW from the FILE menu, with the computer screen controls set as before. The subject should grasp the Physiogrip so that a very slight pressure is put on the trigger, raising the baseline a few millimeters on the computer screen. The pressure should be steady. This takes the slack out of the system and ensures that small twitches will be recorded.

The person applying the stimulus should turn the voltage down to 20, but keep the duration at 10 msec and the frequency at 1 per sec. Turn on the stimulator. Place the stimulating probe at the previously determined motor point and see if there is any response. If not, raise the voltage by 5 volts and repeat the stimulation. Continue doing so until a stimulus intensity is reached that elicits a twitch. Because the threshold could be up to 4 millivolts less than the stimulus just used, decrease the intensity by 4 volts and see if a twitch results. If not, increase the intensity by 2 volts and apply the stimulus. Repeat until the threshold is deter-

 mined. What is the voltage of this threshold?_____ Why is it called a threshold?

Recruitment of Motor Units

Muscles are composed of **motor units,** groups of muscle cells that are separately innervated. Different motor units have different thresholds. Consequently, if the strength of the stimulus is increased, more muscle cells contract, a phenomenon known as **recruitment** or spatial summation. What should happen to the deflection of the trace on the computer screen as recruitment occurs?

Use the same setup as before. Be sure to use NEW from the FILE menu to select a new data file.

Start with the previously determined threshold voltage and record a twitch. Slowly increase the voltage and continue recording twitches. Continue this procedure until a maximum height of contraction is reached and further increments of stimulus intensity do not evoke stronger contractions. What is the voltage at which no further increase in strength of contraction occurs?_____ Turn off the stimulator.

If excitable cells exhibit an all-or-none response when exposed to a threshold stimulus, explain why the height of contraction increased with stimulus strength above the threshold? (*Hint:* Consider the phenomenon of recruitment.)

Temporal Summation and Tetany

So far in your investigations of muscle you have used a low frequency of stimulation. In this section, you will investigate the effects of increasing frequency of contraction on the strength of muscle contraction. With increasing frequency of contraction, there is insufficient time for the muscle to relax before the next stimulating impulse arrives. Consequently, the contractions begin to summate with a new contraction starting in a partially contracted muscle. This phenomenon is called **temporal summation,** and when the muscle is fully contracted and cannot relax between stimuli,

Investigating the Properties of Muscle and Skeletal Systems

tetany occurs. What would you hypothesize is the effect on the strength of muscle contraction? State your hypothesis.

▶ Test your hypothesis, using the following procedure. Choose NEW from the FILE menu to create a new data file. Turn on the stimulator. Have the subject grasp the Physiogrip and put a slight pressure on the trigger to raise the baseline a few millimeters. Set the stimulus parameters to 10 msec duration, frequency to 1 per sec in continuous mode, and a voltage that gives a single contraction.

Once this base is established, gradually increase the frequency of stimulation over a 10-second interval. Can you see individual contractions at low frequencies?_____ At higher frequencies?_____ What happens to the strength of contraction as the frequency of stimulation increases?

If the strength of contraction is so strong that the trigger reaches the end of its travel, stop the experiment and change the spring in the Physiogrip to a stronger one. Erase the previous trace from memory in the computer and repeat the experiment.

When the strength of contraction no longer increases with increasing frequency of stimulation, stop the experiment and turn off the stimulator.

Analyze the data that you have collected in the following way. First, print a copy of the screen by choosing PRINT from the FILE menu. This will provide a record for any report that you may have to write.

From the FILE menu, select ANALYZE and then DISPLACEMENT. You should see a record of the contractions on the screen and at the top should be a series of hash marks indicating how often a stimulus was given. Using the ANALYZE cursor, determine the frequency of stimulation just before temporal summation started.

What is the value in number per second?_____ Now repeat this measurement at the point where tetanic contraction was reached. What is the frequency of stimulation that results in tetany?_____

Measure the height of an individual twitch at the beginning of this experiment. What is it?_____

Now measure the height of contraction reached in tetany. What is it?_____

What is the percent change due to tetany?_____

Skeletal Systems

Skeletal systems support and protect an animal and often amplify the effects of muscle contraction, making locomotion possible. In some animals, especially those with endoskeletons, the skeleton is a storage site for calcium and phosphate ions needed in metabolism. Three types of skeletal systems are found in the animal kingdom: hydrostatic (sometimes called hydraulic), exoskeletons, and endoskeletons.

Invertebrate Skeletons

Hydrostatic Skeletons

Many invertebrates, such as roundworms, cnidarians, and annelids, do not appear to have specialized, differentiated skeletal systems. However, close examination reveals that the body wall muscles act on the incompressible fluids in the body cavity and intracellular spaces to facilitate very effective movement and support. Such skeletal systems are called **hydrostatic skeletons.**

▶ Obtain a microscope slide of a cross section of an earthworm. Look at it first under the low-power objective and note the general anatomy of the animal by comparing the slide to figure 20.7.

The external surface of the earthworm is covered by a highly flexible **cuticle** secreted by the underlying cells of the dermis. The fibrous proteins of the cuticle protect the worm and help maintain its form. Some cells in the dermis secrete mucus, which lubricates the passage of the earthworm through the soil and also maintains a moist surface through which respiratory gases are exchanged.

▶ Note that the body wall is made up of two layers of muscles: an outer circular set and a featherlike inner longitudinal set. Since the longitudinal set runs parallel to the long axis of the body, it will appear in cross section in the slide. Also note the well-developed body cavity, which is filled with fluid in live animals.

What happens to the worm's shape when the circular muscles contract? (Consider the effects of both the muscles and internal fluids.)

What happens to the worm's shape when the longitudinal muscles contract? (Consider the effects of both the muscles and internal fluids.)

Note the spinelike **setae** in the body wall. Sets of muscles attach to each seta and extend or retract them. When extended they project into the soil and anchor the worm in its burrow, allowing the contraction of the longitudinal muscles to quickly move one end of the worm. Explain how the setae and circular muscles allow a worm to crawl forward.

The body cavity of the earthworm is divided into compartments, corresponding to each segment, by cross walls called **septa.** These prevent fluids from "sloshing" from one end of the worm to the other and allow earthworms to make a variety of movements with fine gradations.

Exoskeletons

Exoskeletons are characteristic of several animal phyla. In addition to supporting the animal, exoskeletons prevent water loss, protect the organs, and serve as points of attachment for muscles. Animals in the phylum **Arthropoda,** which includes crayfish, insects, spiders, and millipedes, have a well-developed exoskeleton composed of a complex polysaccharide called **chitin** combined with proteins to give flexibility. Several examples of these animals should be available in the laboratory.

Look at an arthropod through your dissecting microscope and note the segmentation of the exoskeleton. The **head** and **thorax** are composed of several fused body segments. In insects, the thorax bears the walking legs and wings. The segments should be clearly visible in the **abdomen.**

The insect's exoskeleton is made up of hardened plates called **sclerites.** In the head and thorax, the sclerites are rigidly joined to one another along **suture** lines. In the abdomen, the dorsal and ventral sclerites are enlarged and joined to one another by **pleural membranes,** which correspond to thin, reduced vestiges of the lateral sclerites. Similar membranous areas are found between adjacent dorsal or ventral sclerites. These thin, flexible areas allow the animal to expand, contract, and curl its abdomen. Flexible membranous areas are also found between the joints of the various appendages.

Endoskeletons

Defined as an internal supporting system of hardened material, endoskeletons are highly characteristic of vertebrates. Cartilage and bone make up the vertebrate endoskeleton. You will start your study of endoskeletons by examining the structure of bone.

Bone Structure

Examine a fresh beef leg bone that has been cut longitudinally in half. The central shaft of a long bone such as this one is called the bone's **diaphysis,** while the enlarged ends are the bone's **epiphyses** (fig. 30.8). Though difficult to observe in older bones, a narrow zone of cartilage, the **epiphyseal disk** (growth center) of the bone, separates the epiphysis from the diaphysis.

The rounded projection from the upper end of the bone (fig. 30.8a) is called the **head,** and it articulates with the **acetabulum,** a depression in one of the pelvic bones. **Condyles** (rounded, articular prominences) are located on the lateral and medial sides of the distal end of the femur. With which bone(s) do they articulate?

Find the **periosteum,** the tough connective tissue layer surrounding the diaphysis. The surfaces of the condyles are covered by a glassy smooth **articular cartilage,** which, along with synovial fluid, reduces frictional erosion between articulating bones at the joints. Residues from the ligament that held the joint together may also be visible.

Investigating the Properties of Muscle and Skeletal Systems

Figure 30.8 Structure of a bone: (*a*) a bone in partial longitudinal section; (*b*) scanning electron micrograph of cancellous (spongy) bone; (*c*) scanning electron micrograph of Haversian canal (HC) systems in compact bone. Canals allow blood vessels to pass into bone. Bone cells are located in the lacunae (La). From R. G. Kessel and R. H. Kardon. *Tissues and Organs: A Text-Atlas of Scanning Electron Microscopy.* 1979. W. H. Freeman and Co.

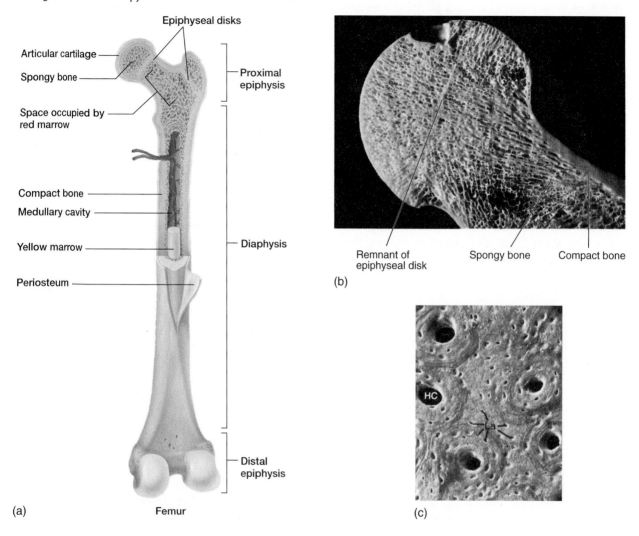

(a) Femur

(b)

(c)

Comparative Vertebrate Endoskeletons

The vertebrate skeleton has two components: the **axial skeleton,** consisting of the **skull** and **vertebral column;** and the **appendicular skeleton,** which includes the **pelvic** and **pectoral girdles** that provide support for the bones of the appendages.

Skeletal Comparisons

In the laboratory, you will observe the skeletons of three vertebrates: a frog, a bird, and a human (figs. 30.9–30.11). Compare these skeletons. Identify the components of the axial and appendicular skeletons first and then go on to identify the bones listed on the next page.

On the outer surface of the bone, find the narrow ridges called **crests** and the small, round, elevated **tubercles,** where muscles attach by tendons to the bones. The surface of the bone is perforated by openings called **foramina** (sing. **foramen**) that allow blood vessels and nerves to penetrate the bone.

The center of the bone is a hollow chamber called the **medullary cavity.** It is filled with **marrow.** Red marrow is where new red blood cells are formed, but in older animals much of the marrow is fatty yellow matter, which no longer produces blood cells.

Compare the structure of the bone in the diaphysis with that in epiphysis. **Compact bone** is very dense, whereas **cancellous bone** is spongy in appearance because of the many interior cross braces.

Figure 30.9 (a) Skeleton of frog (ventral view), including (b) ventral view of pectoral girdle.

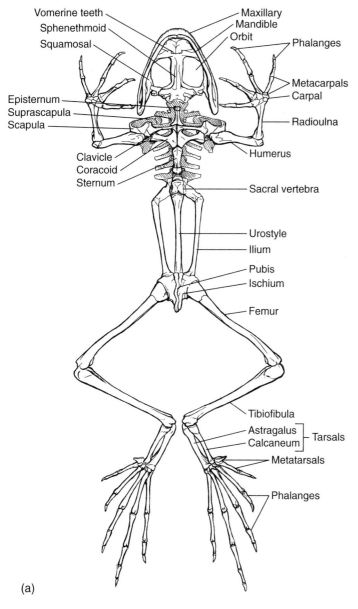

Vomerine teeth
Sphenethmoid
Squamosal
Maxillary
Mandible
Orbit
Phalanges
Metacarpals
Carpal
Episternum
Suprascapula
Scapula
Radioulna
Clavicle
Coracoid
Sternum
Humerus
Sacral vertebra
Urostyle
Ilium
Pubis
Ischium
Femur
Tibiofibula
Astragalus
Calcaneum
Tarsals
Metatarsals
Phalanges

(a)

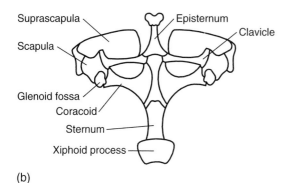

Suprascapula
Scapula
Episternum
Clavicle
Glenoid fossa
Coracoid
Sternum
Xiphoid process

(b)

Axial skeleton

Skull

cranium (fused bones encasing brain and sense organs) consisting of the following major bones:
frontal
parietal
temporal
occipital

facial bones consisting of the following major bones:
zygomatic
maxilla
mandible (lower jaw), the only movable bone in skull

Vertebral column, which supports and protects the spinal cord. Cartilaginous discs are found between the vertebrae in living animals. The vertebral column is divided into five regions:
cervical vertebrae (neck)
thoracic vertebrae (on which the ribs articulate)
lumbar vertebrae (abdominal region)
sacral vertebrae (enclosed by pelvic girdle)
caudal vertebrae or tail (coccyx in human)

Ribs (attached and floating)
Sternum
~~**Xyphoid process**~~
~~**Manubrium**~~
Costal cartilage

Appendicular skeleton

Pectoral girdle consisting of:
clavicle (collarbone)
scapula (shoulder blade)
~~**coracoid**~~

Forelimbs consisting of:
humerus (long bone of upper limb)
radius (long bone of lower limb, which forms a pivot joint with the ulna and is also part of the hinge joint of the elbow)
ulna (other long bone of lower arm, forming hinge joint at elbow)
carpals (small bones of wrist)
metacarpals (bones of palm)
phalanges (finger bones)

Pelvic girdle consisting of:
ilium
ischium
pubis

Hind limbs consisting of:
femur (long bone of thigh)
patella (kneecap)
tibia (larger of two long bones of lower limb)
fibula (smaller long bone of lower limb)
tarsals (bones of ankle)
calcaneus (heel bone)
metatarsals (slender foot bones)
phalanges (toe bones)

Investigating the Properties of Muscle and Skeletal Systems

Figure 30.10 Skeleton of a bird.

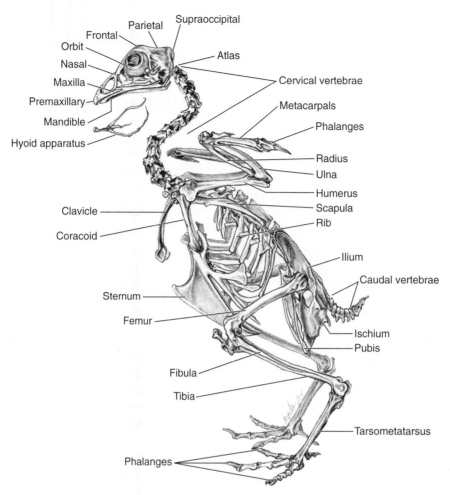

Supraoccipital
Parietal
Frontal
Orbit
Nasal
Maxilla
Premaxillary
Mandible
Hyoid apparatus
Atlas
Cervical vertebrae
Metacarpals
Phalanges
Radius
Ulna
Humerus
Scapula
Rib
Clavicle
Coracoid
Ilium
Caudal vertebrae
Sternum
Femur
Ischium
Pubis
Fibula
Tibia
Tarsometatarsus
Phalanges

◆ Compare a bird skeleton and a human skeleton. List five skeletal differences associated with locomotion and explain how each is involved in locomotory movement. Differences will include fusions, increase in size, reduction in size, but not the appearance of new bones.

◆ List the skeletal differences between a frog and a human that are associated with locomotion.

Skeletal System Summary

Using the information gathered from your laboratory observations, fill in tables 30.1 and 30.2. Describe the role that each of the indicated components has in support, protection, movement, or storage.

Figure 30.11 Human skeleton.

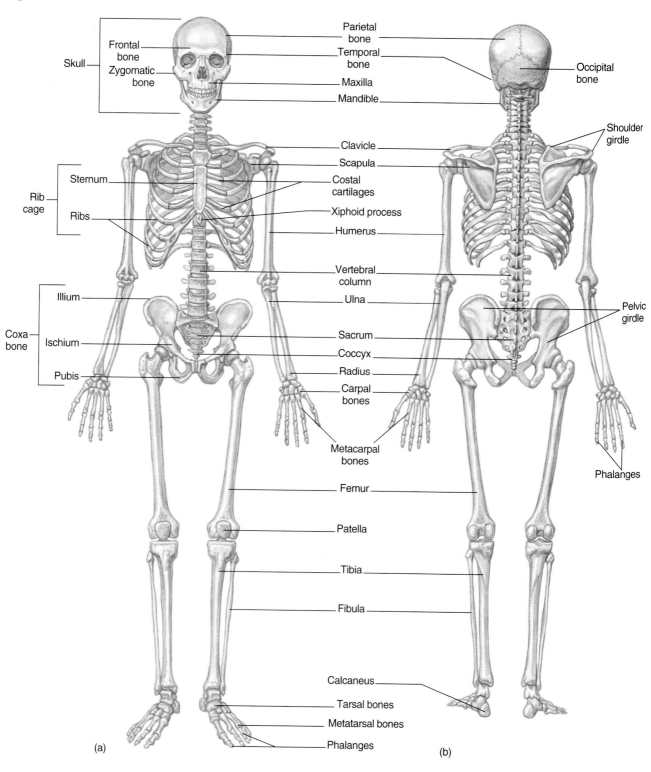

Skull
Frontal bone
Zygomatic bone
Parietal bone
Temporal bone
Maxilla
Mandible
Occipital bone

Clavicle
Scapula
Sternum
Costal cartilages
Xiphoid process
Rib cage
Ribs
Humerus

Shoulder girdle

Vertebral column
Illium
Ulna
Coxa bone
Ischium
Sacrum
Coccyx
Pubis
Radius
Carpal bones

Pelvic girdle

Metacarpal bones

Femur

Patella

Tibia

Fibula

Phalanges

Calcaneus
Tarsal bones
Metatarsal bones
Phalanges

(a)

(b)

TABLE 30.1   Earthworm—skeletal system summary

Cuticle

Body fluids

Circular muscle

Longitudinal muscle

Septa

Description of skeletal system

TABLE 30.2   Vertebrates—skeletal system summary

Bone

Axial skeleton

Skull

Vertebral column

Appendicular skeleton

Pectoral girdle

Pelvic girdle

Description of skeletal system

Learning Biology by Writing

Write a lab report that describes your experiments with human muscles. It should have three sections in the results:

Determining threshold
Demonstrating motor unit recruitment
Demonstrating temporal summation and tetany

The discussion section should describe how recruitment and summation are important in graded responses such as are used in sports or playing a muscial instrument.

Your instructor may ask you to answer the Lab Summary and Critical Thinking Questions that follow.

Lab Summary Questions

1. Describe the differences you observed between smooth and skeletal muscle cells. Give examples of where you would find each in your body.
2. Describe the all-or-none response in muscle cells. How is it possible for us to have graded responses in our muscles if muscle cells respond in an all-or-none manner to stimuli?

Internet Sources

The disease rickets affects the human skeletal system. Search the WWW to learn what this disease is. Where is it common? Is it found in the U.S.? How can it be treated?

3. Both recruitment and temporal summation lead to greater force of muscle contraction. Explain what is happening in both of these phenomena.
4. Explain how a nerve impulse arriving at a muscle causes a muscle to contract.
5. "Form reflects function" is a statement that applies to the skeletal system. Explain what this statement means and give examples from your study of vertebrate endoskeletons.
6. Describe the major common features of skeletons found in all vertebrates.

Critical Thinking Questions

1. About 99% of the body's calcium is found as calcium phosphate salts in bone tissue. The calcium in plasma is in an ionic form, Ca^{++}. Although the level of Ca^{++} in the blood and tissues is closely regulated by parathyroid hormone and calcitonin, imbalances in Ca^{++} can occur. What effect would an elevated Ca^{++} concentration have on muscle activity? What effect would a Ca^{++} deficiency have on muscle activity?
2. As can be seen in figure 30.10, the skeleton of a bird is modified for flight. Aside from the size and arrangement of the bones in the skeleton, what other feature of its bones might affect the ability of a bird to fly?
3. How do scientists deduce the appearance of hominids or dinosaurs when all they have to study are a few bone fragments?

LAB TOPIC 31

Investigating Nervous and Sensory Systems

Supplies

Preparator's guide available on WWW at http://www.mhhe.com/dolphin

Equipment

Compound microscopes
Dissecting microscopes

Materials

Fetal pig
Pig brain, sagittal section
Whole pig brains
Prepared slides
    Mammalian nervous cord with dorsal and ventral roots, cross section
    Neurons or multipolar neurons
    Muscle with nerve end plate (demonstration slide)
    Mammalian retina (demonstration slide)
    Mammalian cochlea, cross section
Models of human eye and ear
Overhead transparencies made from black paper with one inch squares cut in center and covered with red, blue, or green cellophane so that a small square is projected. Alternatively, photographic slides can be made to project one of the colors with a black background.

Prelab Preparation

Before doing this lab, you should read the introduction and sections of the lab topic that have been scheduled by the instructor.

You should use your textbook to review the definitions of the following terms:

axon
cerebellum
cerebrum
cochlea
cornea
cranial nerves
dendrite
gray matter
iris
medulla oblongata
meninges

myelin sheath
organ of corti
pupil
reflex arc
retina
semicircular canals
spinal cord
spinal nerves
white matter

You should be able to describe in your own words the following concepts:

How a nerve cell functions
Types of nervous systems found in animals
Major regions of the mammalian brain
Reflex arc
How the eye "sees" light
How the ear "hears" sound

As a result of this review, you most likely have questions about terms, concepts, or how you will do the experiments included in this lab. Write these questions in the space below or in the margins of the pages of this lab topic. The lab experiments should help you answer these questions, or you can ask your instructor for help during the lab.

Objectives

1. To observe the microscopic structures of nerves, nerve cells, and neuromuscular junctions
2. To learn the gross anatomy of the mammalian central nervous system, eye, and ear
3. To determine experimentally some of the properties of rods and cones

Figure 31.1 Nervous systems in invertebrates: (a) nerve net in *Hydra* lacks central tracts; (b) flatworms have longitudinal nerve cords with cross connectives; (c) ventral central nervous system of earthworm shows cephalization and has integrating ganglia in each segment.

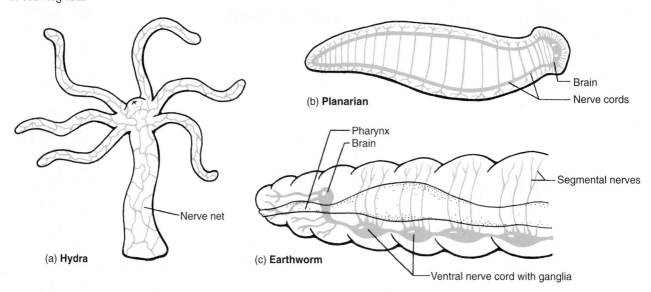

(b) **Planarian**

Brain
Nerve cords

Pharynx
Brain

Segmental nerves

Nerve net

(a) **Hydra**

(c) **Earthworm**

Ventral nerve cord with ganglia

Background

Coordination of the several different types of specialized tissues in multicellular animals is necessary for the organism to operate as an integrated whole. The nervous system coordinates the body's relatively rapid responses to changes in the environment. The endocrine system regulates longer term adaptive responses to changes in body chemistry between meals, as the seasons change, or as developmental changes occur during maturation. Because the functions of these two systems often complement one another, biologists often speak of the **neuroendocrine system.**

The nervous system has three functions: (1) to receive signals from the environment and from within the body through the sense organs, (2) to process the information received, which can involve integration, modulation, learning, and memory, and (3) to cause a response in appropriate muscles or glands. **Receptors** are usually specialized cells outside of the central nervous system that detect physical and chemical changes. There are separate receptor cells for heat, cold, light, and so on. The function of the receptor cells is to convert the environmental signal into a change in the cellular membrane's ion permeability, leading to a voltage change across the cell membrane. If the voltage change is of sufficient magnitude, an **action potential** will be created in an adjacent nerve cell, and a nerve impulse will travel from the sensing zone to the central nervous system.

The central nervous system consists of neuronal networks, which are involved in decision making. A decision may involve the simple passage of nervous impulses from an **afferent sensory fiber** to an **efferent motor (effector) fiber,** assuring that perception results in a response, as in a **reflex arc.** In other cases, many **interneurons** in the nervous system may become involved, tempering the inputs

with memory or reasoning. The response resulting from this interaction of nerve signals may be different according to conditions. For example, if a person is frightened by a loud sound, he or she will gasp, but if this stimulus is repeated when the person is under water, higher nerve center input inhibits the response.

To gain insights into how nervous systems operate, many biologists study the simpler systems of invertebrates. These scientists have found that neurons are similar in form and function no matter what type of animal they are taken from. The **action potential,** the electrical potential changes comprising a nerve impulse, also seems to be universal in animals. The differences between simple and complex nervous systems involve the types of receptors present, the organization of the central nervous system, and the **neurotransmitter** chemicals that function in cell-to-cell impulse transmission.

One of the simplest nervous systems is found in the radially symmetrical cnidarians, such as *Hydra*. The **nerve net** found in these organisms does not appear to be organized into **tracts** or **centers** (fig. 31.1a). Bilaterally symmetrical animals as simple as the Platyhelminthes have tracts running the length of the organism, with the anterior end of the tract serving as a coordinating center (fig. 31.1b). In higher phyla, the coordinating centers become more complex, creating an evolutionary trend referred to as **cephalization,** in which sense organs and the coordinating centers are concentrated at the animal's anterior.

The central nervous systems of annelids and arthropods consist of two **ventral nerve cords** with interspersed **ganglia,** areas of the cord where the cell bodies of neurons are concentrated. The cords rapidly transmit signals, whereas the ganglia serve as processing centers (fig. 31.1c). In arthropods, the ganglia are clustered in the head and thorax regions, where most activity takes place.

Figure 31.2   Three kinds of neurons found in humans.

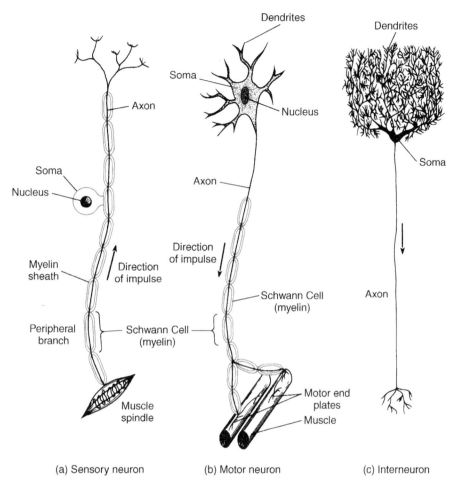

(a) Sensory neuron   (b) Motor neuron   (c) Interneuron

In the vertebrates, it is obvious that the brain dominates the nervous system. It is estimated that the brain contains between 10 and 100 billion neurons, which make over 100 trillion points of contact with each other. Well over 99% of these neurons are interneurons; only 2 to 3 million are motor neurons.

The nervous system receives its information about the environment and the position of an animal in its environment from sensory cells. The sensory cells that receive most environmental signals usually do not occur as isolated units but are part of a larger organized unit, the **sensory organ.** The mammalian eye consists not only of light-sensitive rod and cone cells, but also of a focusable lens, a light-regulating iris, a tear gland, and a covering eyelid. The proper functioning of a sense depends not only on the sensory and neuronal cells, but also on the anatomical structures that collect, amplify, and direct environmental signals to the receptor cells.

These senses are "hard-wired" into the organism. You feel cold because specific cold receptors in the skin are stimulated, and impulses are carried by specific neuronal routes to specific areas of the brain where that information is integrated with other sensory inputs and memory. You see something because light has stimulated individual cells, and these cells each stimulate different neurons that carry nerve impulses to the brain. The brain integrates those signals with memory, and you realize the meaning of that particular pattern of neuronal stimulation. The perception of a face, especially a familiar face, is actually a comparison of new neuronal input with a record of previous neuronal inputs. Sensory cells do not perceive, they receive. Perception is an integrative process in which nerve impulses from receptors, not environmental stimuli, are processed and may result in some action by the animal.

LAB INSTRUCTIONS

You will study the major components of the mammalian central nervous system and the anatomy of nerves, the eyes, and the ear.

Microanatomy of the Nervous System

▶ Obtain a slide of neurons and observe it through a compound microscope, comparing what you see to figure 31.2.

Figure 31.3   Cross section of spinal cord showing reflex arc components.

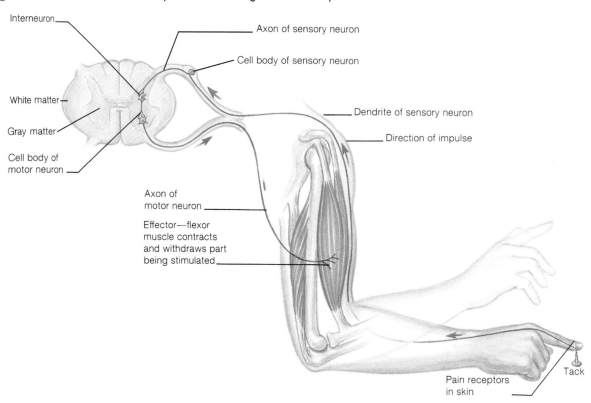

Interneuron

Axon of sensory neuron

Cell body of sensory neuron

White matter

Gray matter

Dendrite of sensory neuron

Direction of impulse

Cell body of
motor neuron

Axon of
motor neuron

Effector—flexor
muscle contracts
and withdraws part
being stimulated

Pain receptors
in skin

Tack

Sketch a neuron in the blue circle. Small cells surrounding the neurons are neuroglial cells which supply nutrients for the neurons and produce chemicals that modulate the functions of neurons.

Label the cell body called the **soma,** and the cytoplasmic extensions called **neurites. Axons** are neurites that conduct impulses away from the soma. **Dendrites** are neurites conducting impulses toward the soma. Can you morphologically distinguish between axons and dendrites? Neurites often extend long distances. In a 7-foot basketball player, axons extend from the base of the spine to the toes, a distance of 3 to 4 feet!

Obtain a slide of a spinal cord cross section through the **dorsal root ganglion.** Observe it at first under low power with a compound microscope or with a dissecting microscope against a white background.

Note the general organization and compare with figure 31.3. Paired nerve roots enter the top and others leave the bottom of the cord. Note that the nerve cord is hollow, containing a **central canal,** a vestige from its embryonic development. The central canal is connected with the ventricles of the brain.

Examine the peripheral white matter surrounding the central gray matter with a high-power objective. Note that the white matter consists primarily of neurites that have been cut in cross section. They represent myelinated nerves that travel up and down the spinal cord. When the spinal cord is severed in an accident, it is impossible to rejoin these thousands of neurites and numbness and paralysis result.

Examine the central gray matter and find the cell bodies in this area. Interneurons here process information and send motor axons out the ventral root to muscles.

Examine the demonstration slide your instructor set up, showing a neuromuscular junction where a motor nerve connects to the muscle fibers it innervates. The axon of the motor nerve divides into fine branches (**terminal arborizations**), which terminate as **motor end plates** on the muscle fiber surface (see fig. 30.4). Nerve impulses arriving at the end plates cause the release of **acetylcholine,** a neurotrans-

Figure 31.4   Divisions of the nervous system in mammals.

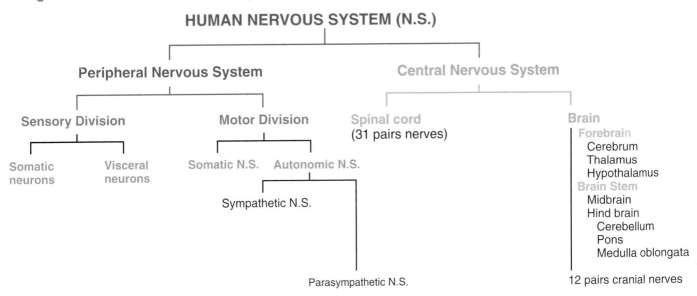

HUMAN NERVOUS SYSTEM (N.S.)

Peripheral Nervous System

Central Nervous System

Sensory Division

Motor Division

Spinal cord
(31 pairs nerves)

Brain
Forebrain
Cerebrum
Thalamus
Hypothalamus
Brain Stem
Midbrain
Hind brain
Cerebellum
Pons
Medulla oblongata

Somatic neurons

Visceral neurons

Somatic N.S.

Autonomic N.S.

Sympathetic N.S.

Parasympathetic N.S.

12 pairs cranial nerves

mitter chemical that diffuses across the neuromuscular junction cleft and triggers an action potential in the muscle cell. The action potential leads to the muscle contracting. Sketch and label a junction below.

Compare your sketch to figure 30.4.

Mammalian Nervous System

Figure 31.4 provides an organizational overview of the mammalian nervous system. It consists of two major parts: the peripheral nervous system and the central nervous system, which in turn can be further subdivided into their component parts. As you work through the dissection of the fetal pig's nervous system, refer to this chart to refresh your memory of the relationships.

If you have not already done so, remove the skin of your fetal pig. Lay the pig on its stomach. Expose the **spine** by removing the muscles that cover and extend on each

side of the vertebrae. Since it is time consuming for each student to expose the entire spine, groups in the laboratory should expose only short sections of about five to eight vertebrae each. Coordinate your efforts so that representative parts along the entire spinal cord are exposed.

Spinal Cord

Once the vertebrae are exposed, take a sharp scalpel and gradually remove the cartilaginous spines and neural arches of several vertebrae, exposing the spinal cord (fig. 31.5). The spinal cord will be surrounded by three membranes, or **meninges.** The tough, outer, transparent connective tissue sheath is the **dura mater.** The next layer, the **arachnoid,** is so thin and fragile that it is hard to find in routine dissection. The **pia mater,** the innermost covering, lies close to the spinal cord. Slit the dura mater to observe the spinal cord.

Note the **spinal nerves** branching from the cord. There are 33 pairs. Each nerve is composed of several small roots or fibers. Sensory neurons carrying impulses into the spinal cord enter by the **dorsal root,** and motor neurons carrying impulses to effectors leave by the **ventral root.** These two roots come together to form the spinal nerves.

The cranial end of the spinal cord gradually enlarges to become the **medulla oblongata.** This region of the brain contains the **cardiac** and **respiratory centers** as well as numerous **sensory** and **motor nerve tracts,** which transmit impulses to and from the higher brain centers.

Autonomic System

The **autonomic nervous system** consists of the **sympathetic** and **parasympathetic systems,** which are involved in the control of glands, smooth muscle, and viscera. These two systems have opposing effects to slow or speed muscle movements in the viscera. This system is

Figure 31.5   The relationship of the spinal cord, spinal nerves, and sympathetic nervous system to the vertebrae and meninges.

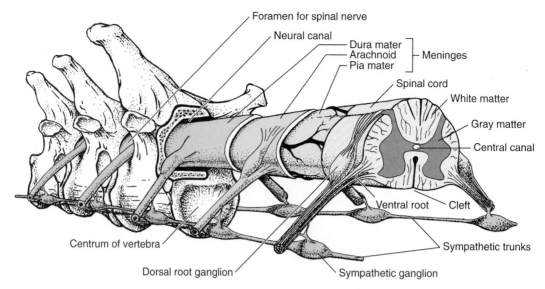

The cerebrum has a convoluted surface. The ridges are called **gyri** and the valleys, **sulci.** The outer part of the cerebrum, the **cortex,** is made up of **gray matter,** which is composed of nerve cell bodies and supporting, but not conducting, **neuroglial** cells. The inner part of the cerebrum is made up of **white matter,** which is composed of nerve-cell extensions or fibers encased in an insulating, fatty **myelin** sheath. In the nerve cord, the relative positions of white and gray matter are reversed, with **myelinated** nerve fibers on the outside and cell bodies on the inside.

part of the **peripheral nervous system** and is connected to the central nervous system by many nerve fibers and ganglia.

The **sympathetic trunks** can be viewed by turning the fetal pig over and looking at the dorsal wall of the body cavity on either side of the spinal column in the thoracic and abdominal cavities. They are thin strands of nerves with small enlargements, the **ganglia,** along their length, running parallel to the spine.

Brain

This dissection takes some time to perform; your instructor may provide a demonstration of the larger brain from an adult pig. If you are to do the dissection, expose the skull by making a longitudinal cut from the base of the snout to the base of the skull. Make lateral cuts from the ends of the first cut to the angle of the jaws at the anterior end and to the level of the ears at the caudal end. Remove the muscle layers. Make a shallow longitudinal cut in the skull and then lateral cuts from this incision at 2 cm intervals. Break off chips of the skull and gradually expose the brain. At the base of the skull, the tough **occipital** bone will have to be dissected out separately. The spinal cord passes through this bone.

The brain is surrounded by three meninges as was the spinal cord. Remove the membranes and observe the gross features of the brain (fig. 31.6). A **longitudinal fissure** separates the right and left **hemispheres** of the **cerebrum.** Higher order functions, such as memory, intelligence, and perception, are associated with this part of the brain. Caudal to the cerebrum is the smaller **cerebellum,** which coordinates motor activity and equilibrium. The **medulla oblongata** lies ventral and caudal to the cerebellum and is continuous with the spinal cord.

The **transverse fissure** separates the cerebrum from the cerebellum. There are three parts to the cerebellum, the two **lateral hemispheres** and a central **vermis.** If the caudal edge of the cerebellum is raised, a thin vascular membrane called the **posterior choroid plexus** may be observed covering the medulla. The space beneath the choroid plexus is the **fourth ventricle** of four ventricles, or cavities, found in the brain. **Cerebrospinal fluid** created by filtration from the capillaries in choroid plexi located in the ventricles passes into the spinal cord spaces. This fluid carries nutrients to the cells and protects the nervous system from mechanical shock. The ventricles are best seen in a sagittal section of the brain (fig. 31.7).

Remove the brain from the skull (or obtain demonstration material) and orient it with the ventral surface up (fig. 31.8). At the anterior end of the cerebrum, find the **olfactory bulbs.** In the midportion of the cerebrum, two large nerve trunks, the **optic tracts,** cross to form the **optic chiasma.** Just posterior to the crossover is the **pituitary,** but this small body is often broken off when the brain is removed from the cranium. Dorsal to the pituitary comprising the ventral surface of the brain is a conspicuous oval area, the **hypothalamus.** It integrates many autonomic functions, such as sleep, body temperature, appetite, and water balance. Posterior to this is a wide transverse group of fibers,

Figure 31.6   Dorsal view of the fetal pig's brain.

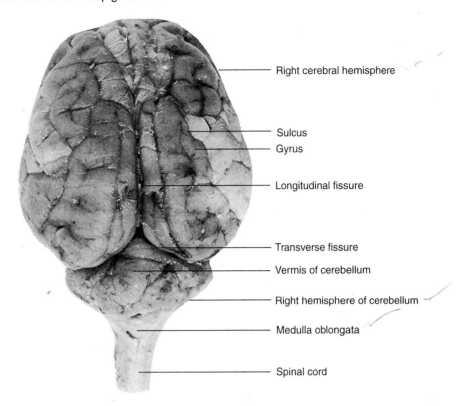

Right cerebral hemisphere

Sulcus
Gyrus

Longitudinal fissure

Transverse fissure
Vermis of cerebellum

Right hemisphere of cerebellum

Medulla oblongata

Spinal cord

Figure 31.7   Sagital section of the fetal pig's brain.

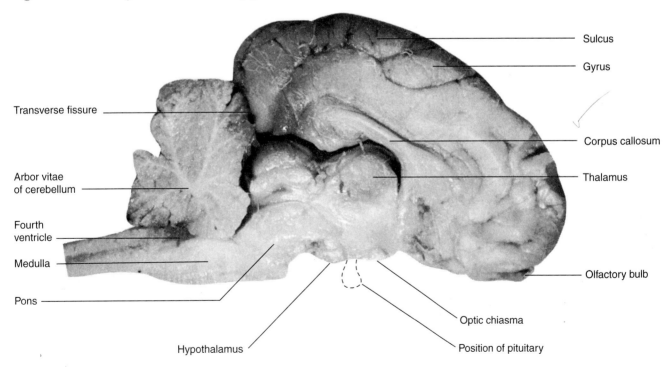

Sulcus

Gyrus

Transverse fissure

Corpus callosum

Arbor vitae
of cerebellum

Thalamus

Fourth
ventricle

Medulla

Olfactory bulb

Pons

Optic chiasma

Hypothalamus

Position of pituitary

Figure 31.8 Ventral view of the fetal pig's brain with cranial nerves indicated by standard numbering system.

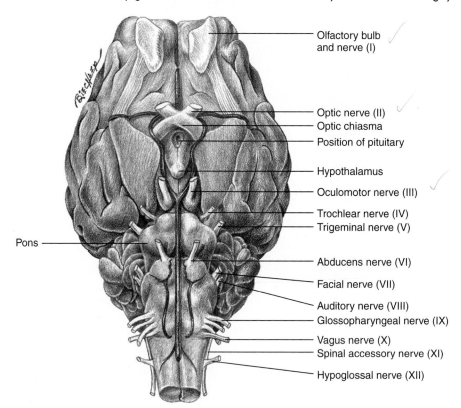

Olfactory bulb and nerve (I)

Optic nerve (II)
Optic chiasma
Position of pituitary

Hypothalamus
Oculomotor nerve (III)

Trochlear nerve (IV)
Trigeminal nerve (V)

Pons

Abducens nerve (VI)

Facial nerve (VII)

Auditory nerve (VIII)
Glossopharyngeal nerve (IX)

Vagus nerve (X)
Spinal accessory nerve (XI)

Hypoglossal nerve (XII)

the **pons,** which serves as a passageway for nerves running from the medulla to higher centers. Twelve pairs of **cranial nerves** directly enter the brain, bringing sensory information concerning sight, sound, taste, equilibrium, and touch. Motor fibers to the head region also exit via these nerves. The nerves are named in figure 31.8.

In table 31.1 write a brief description of each structure and a summary statement of its function.

Sensory Systems

Eye Anatomy

Dissection of beef eyes may be substituted for fetal pig dissection, if desired.

To remove an eye from the fetal pig, make an incision that extends from the external corner of the eye completely around the eye, removing the upper and lower lids. A thin mucous membrane, the **conjunctiva,** covers the eye and folds back to line the undersurface of the lids, preventing foreign material from entering the socket. The **nictitating membrane,** a third eyelid, which can pass across the eye, will appear as a white, semicircular structure in the medial corner of the eye. (This structure is not found in humans.) Now remove the connective tissue, the muscle, and part of the bony orbit surrounding the eye. Push the eye to the side and up and down, noting the seven thin strips of the **ocular** muscles that originate on the skull and insert on the eye.

Cut these muscles as far as possible from the eye. Draw the eye out of the orbit and snip the optic nerve. Once the eye is free, note the muscle mass that surrounds the eyeball and holds it in orbit.

With a sharp razor blade, cut longitudinally through the eye to one side of the **optic nerve** and place the halves in a dish of water. Identify the three layers making up the wall of the eye: the white **sclera,** the dark **choroid,** and the **retina** (fig. 31.9). The front surface of the eye is covered by a transparent layer of connective tissue called the **cornea.** The large white structure within the eye at the front is the **lens.** (It has turned white from the preservative.) The colored material surrounding the lens is the **iris,** and the central opening in the iris is the **pupil.** Muscles in the iris regulate the opening of the pupil, depending on the light intensity in the environment.

The large posterior space of the eye is filled with a gelatinous fluid, the **vitreous humor,** whereas the small chamber anterior to the lens is filled with **aqueous humor.** Where the optic nerve enters the eye, the retina is devoid of rods and cones. This area is called the **blind spot** (optic disk) since light falling on this area does not trigger nerve impulses. The small, yellowish spot in the center of the retina is called the **macula lutea.** The depression (**fovea**) in the center of this spot has a very high density of cones, and is the area producing the greatest visual acuity. Figure 31.9 shows the structure of the eye. Describe how the eye is able to focus on objects at different distances.

Dorsal nerve cord

Spinal nerves

Sensory nerve tracts

Motor nerve tracts

Ganglia

Medulla oblongata

Pons

Cerebellum

Cerebrum

Meninges

Cranial nerves

Figure 31.9   Section of the eye.

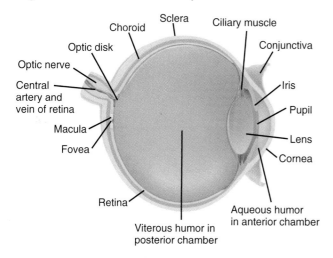

Choroid • Sclera • Ciliary muscle
Optic disk
Conjunctiva
Optic nerve
Central artery and vein of retina
Iris
Pupil
Macula
Lens
Fovea
Cornea
Retina
Aqueous humor in anterior chamber
Viterous humor in posterior chamber

Color Vision

In the human retina, there are two kinds of light-sensitive cells: **rods** and **cones.** These cells are located on the outside of the retina, away from the vitreous humor and against the choroid layer, and are covered by a layer of **bipolar** and **ganglion** cells as well as neurons, as shown in figure 31.10. Light enters the posterior chamber through the lens and must pass through several cell layers before it is absorbed by the rods and cones. Look at the demonstration slide of the mammalian retina and identify rods and cones.

Cones are involved in color vision at daylight intensities. Figure 31.10 shows that each cone is associated with one bipolar cell. Generator potentials in a cone excite one neuron, giving high-level resolution vision. Cones will not cause neuronal firing at low light intensities and are, therefore, nonfunctional at night.

Rods provide black-and-white vision at low light intensities. A bipolar cell associates with several rods, so that generator potentials sum in their effect. This assures neuronal

Figure 31.10 Diagram of the cells in the retina. Note the light direction. Light must pass through the layer of nerve cells before it reaches the rods and cones.

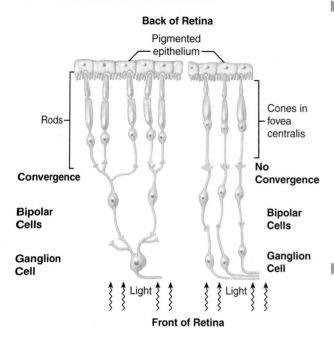

for a long period of time at high light intensity, the cones for that color will become bleached. The afterimage of the object will be "seen" in the complementary color.

Your instructor will demonstrate afterimages to you. A small square of light will be projected on the screen in the dark lab room. After staring intensely at the square for 30 seconds or so without shifting your gaze, the slide will be changed to bright white. Continue staring at the screen as the change is made. What do you see?

The experiment will be repeated using a red square of light. Stare at the red square for one minute and then your instructor will change the slide so the entire screen is white. What do you "see"? What was the color?

firing at low light intensities and provides a highly sensitive vision system but with less resolution. Why?

Repeat this experiment using a blue square and record your results.

When light strikes a receptor cell, it triggers a photochemical reaction that leads to neuronal firing. If a rod or cone receives high-intensity light, the visual pigments will be bleached temporarily and will not absorb any more light. Within a matter of a few seconds, enzyme systems in the cells will restore the visual pigment to its light-absorbing form. This can be easily demonstrated. If, after viewing a bright light, the eye is quickly shut or turned to a dark wall, the bright image will persist as a positive **afterimage** because the high generator potential causes the continued firing of neurons. If the eye is cast on a lighter background after a few seconds, a negative afterimage, or a dark image of the object, will appear. This is because the still bleached cells are not receptive to the light coming from the light background.

According to the **Young-Helmholtz theory of color vision,** there are three types of cone cells, which respond respectively to red, green, and blue light. All other colors are perceived as the brain interprets impulses coming from a mix of these receptors. If an object of one color is viewed

How do you explain the colors seen in these afterimages?

Figure 31.11 Anatomy of the human ear: (*a*) the relationship between the external and inner ears, and a cross section of the cochlea, showing the canal and membrane organization; (*b*) magnified view of the cochlear canal showing organ of Corti and the relationship of hair cells to the tectorial membrane; (*c*) the organ of Corti in magnified view.

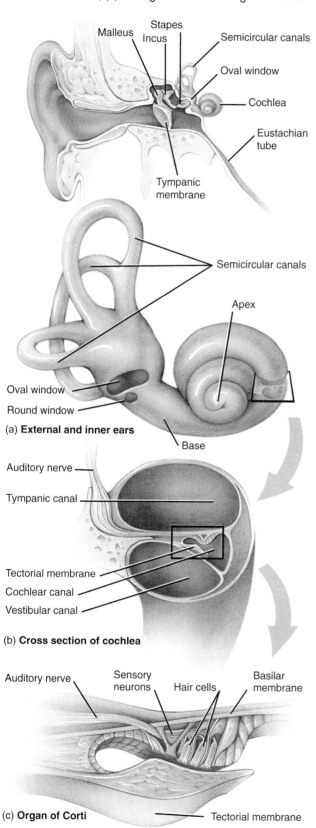

(a) **External and inner ears**

(b) **Cross section of cochlea**

(c) **Organ of Corti**

Ear Anatomy

The mammalian inner ear is difficult to dissect because it is part of the bony skull. However, the anatomy is not difficult to visualize. Study figure 31.11 and the models provided in the laboratory.

Sound waves are collected by the **pinna** and pass down the auditory canal, causing the eardrum (**tympanic membrane**) to vibrate. The movement of the eardrum is transmitted by the three **auditory ossicles** of the middle ear to the flexible membrane covering the **oval window** of the **cochlea** or inner ear. Sound energy is amplified in this process for two reasons: the area of the tympanic membrane is about 30 times larger than the oval window membrane; and the mechanical leverage systems of the ossicles, the **malleus, incus,** and **stapes,** greatly amplify any movement. The back-and-forth movement of the oval window causes the fluid in the cochlea to move. The cochlea is divided lengthwise into three canals by soft-tissue walls composed of the **auditory nerve,** sensory hair cells, and connective tissue.

Pressure waves generated at the oval window travel up the **vestibular canal** to the apex of the cochlea and down the **tympanic canal.** The **round window** is at the base of the tympanic canal. The canal's covering membrane serves as a pressure-release valve, allowing the fluid of the cochlea to move back and forth. Thus, these gross anatomical structures of the ear convert the movement of air molecules, or sound, into the movement of the fluid **perilymph** in the inner ear.

The pressure waves cause the thin, longitudinal **basilar membrane** to vibrate. **Sensory hair cells** which are found on this membrane in an area known as the **organ of Corti** (fig. 31.11), are displaced upward and brush against a stiff, overhanging structure called the **tectorial membrane.** Distortion of the hair cells causes generator potentials, which in turn cause the firing of the cochlear neurons that travel along cranial nerve VIII to the brain.

Loudness is encoded in the neurons by the frequency at which the neurons fire. Loud noises cause a large displacement of the basilar membrane and large generator potentials in the hair cells. Pitch is detected in a different way. High-frequency sounds stimulate cells near the base of the cochlea and low-frequency sounds stimulate cells near the apex. Separate neurons lead from hair cells at different points along the cochlea to the brain. Location of receptors thus determines pitch discrimination.

Obtain a slide of a cross section of the cochlea. First look at the slide under the 10× objective. Identify the structures shown in figure 31.11*b* and *c*. Switch to the high-power objective and observe the hair cells and their relationship to the tectorial membrane in the organ of Corti.

Pair off with another student and quiz one another about how the ear functions. Use the model and explain to one another how sound passes into the inner ear and is transduced into a nerve impulse.

Learning Biology by Writing

Write a 250-word essay on how a sensory nerve impulse travels from a muscle spindle in the calf to the central nervous system and back to a muscle cell in the leg via a motor neuron. Discuss the ways in which action potentials, synapses, and conduction velocities are involved.

As an alternative assignment, your instructor may ask you to answer the Lab Summary and Critical Thinking Questions that follow.

Internet Sources

Neuromuscular junctions are fascinating anatomical sites where the nervous system meets the muscular system. These are still the topics of much research. Search the World Wide Web to discover the type of research that is being conducted now. Jump to at least three of the sites returned in the list and summarize the research that is being done. Include the URLs for future reference.

Lab Summary Questions

1. Describe the organizational structure of nervous systems in the animal kingdom as you proceed from Cnidaria to Vertebrate.

2. How does the structure of a nerve cell reflect its function?
3. What is the difference between white matter and gray matter in the mammalian central nervous system?
4. Name the major anatomical features of the brain as viewed whole from the dorsal perspective.
5. Describe how light entering the eye is regulated and focused on the retina. Describe the response of rods and cones to different colors of light.
6. Explain the nature of afterimages that are seen after looking at a bright light. Why are afterimages a different color than the actual images?
7. Trace the pathway of sound energy as it enters the auditory canal and passes into the inner ear. How are pitch and loudness encoded by the nervous system?

Critical Thinking Questions

1. Hearing impairment can result from mechanical failure of parts of the ear, or from sensory cell damage, or from auditory nerve damage. Which of these might be easily corrected by a hearing aid? Why?
2. If you examined the retinas of various animals with a microscope, what differences would you expect to find between animals that are active at night versus those active during the day?

LAB TOPIC 32

Statistically Analyzing Simple Behaviors

Supplies

Preparator's guide available on WWW at
   http://www.mhhe.com/dolphin

Equipment

Lamp with 100-watt bulb

Materials

Ring stand, two clamps and one ring clamp
Ruler
Paper towel tube. At one end punch two holes
   opposite each other ¼" in from edge. Close that
   end with a #8 or #9 rubber stopper.
Plastic vials that fit into paper towel tube when taped
   together end to end
Tape (clear)
Thermometer
400-ml and 800-ml beakers filled with water
Wooden stick ⅛" diam × 6"
Fruit flies (Drosophila melanogaster)
Marking pencil
Black cloth 24" × 24"

Prelab Preparation

Before doing this lab, you should read the introduction
and sections of the lab topic that have been scheduled
by the instructor.

   You should use your texbook to review the
definitions of the following terms:

   innate
   instinct
   kinesis
   taxis

   You should be able to describe in your own words
the following concepts:

   Nature versus nurture debate
   Chi-square test
   Read Appendix C, page 437

   As a result of this review, you most likely have
questions about terms, concepts, or how you will do
the experiments included in this lab. Write these
questions in the space below or in the margins of the
pages of this lab topic. The lab experiments should

help you answer these questions, or you can ask your
instructor for help during the lab.

Objectives

1. To test hypotheses regarding phototactic and
   geotactic behavior, using fruit flies
2. To use statistical tests to reject or accept
   experimental hypotheses

Background

Behavior can and has been studied from two philosophical
perspectives. The older view involves observing animals
and trying to determine which behaviors are due to **instinct** or **innate behavior patterns** and which are due to
learning. This is the **nature-versus-nurture** perspective.
Behaviorists, such as Konrad Lorenz and Niko Tinbergen,
have developed another approach, called **ethology,** or the
study of the evolution of behavioral characteristics of a
species.

   Although ethology embraces the nature-versus-nurture
viewpoint, it also asks *how* the behavior developed and
why it is an advantage to the species. This viewpoint
should not be confused with an anthropomorphic, or
human-oriented, view of animal behavior. For example, an
anthropomorphic explanation of why a bird sings on a
beautiful spring morning might be because it is happy. An
ethological explanation for the same event might be that
the bird is singing to communicate two things to other
members of the population: availability for mating and ter-
ritorial occupation.

   Like physiological processes, behavior must be considered in terms of survival value. Like other biological
phenomena, complex behavior ultimately depends on
cells: sensory cells, neurons, and effector cells acting in
coordination with one another. The functioning of these

is ultimately related to the genes that dictate the behavioral repertoire of the animal. Experiments with fruit flies have confirmed this idea by showing that some behaviors, such as circadian rhythms and mating behavior, are inherited in accordance with Mendelian genetic mechanisms.

However, biologists are a long way from knowing the molecular mechanisms involved in the development of these behaviors. Neurobiologists are just now beginning to map the neuronal circuits involved in many complex behaviors. Some day research may reveal how innate behavior and memory are stored in the nervous system. When scientists understand the mechanisms of storage and the way in which incoming stimuli are processed, they may be able to understand exactly why an animal behaves as it does in certain situations.

Until that time occurs, the study of behavior is a "black box" approach. This scientific approach to studying a process where the mechanisms are unknown involves observing an animal first under normal conditions and then under experimental conditions with the changes in response noted. In this way, the observer obtains some information about the capabilities of the underlying mechanism and may also gain information that proves useful in managing the system or organism. The black box approach has led to great advances in our understanding of animal communications, orientation, and sociobiology.

The simplest behaviors are **kineses** and **taxes.** Kineses involve a change in the rate of random movement in response to an environmental stimulus. For example, pill bugs, usually found in moist places, get there because of kineses. When in dry conditions, they move quickly in random directions, but in moist conditions, they move slowly or not at all. Thus, they tend to congregate where there is moisture but they do not seek it. In contrast to kineses, taxes are directed movements. An animal that senses and moves toward a stimulus is said to show positive taxis; one that moves away shows negative taxis. Both taxes and kineses are innate behaviors.

LAB INSTRUCTIONS

You will investigate taxes in fruit flies, specifically, their response to light and gravity.

Hypothesis to Be Tested

A statistical test always tests a **null hypothesis (H_o)**, which is based on a scientific model of what the outcomes from an experiment should be. The null hypothesis proposes that there is no significant difference between the results actually obtained and the results predicted by a model for an experiment.

A null hypothesis is always paired with an **alternative hypothesis (H_a)**, which proposes an alternative explanation of the results, not based on the same scientific model as

was the null hypothesis. If the statistical test indicates that you cannot accept H_o, then H_a is true by inference: the variation in your data compared to that predicted by the model is greater than that expected by chance, and some other factor is responsible for the variation between observed and predicted results. A more thorough discussion of scientific hypotheses may be found in appendix C.

For these experiments, we will propose a simple null hypothesis. (Since the response of fruit flies cannot be predicted, the model used will be a random distribution model.)

H_o: Fruit flies randomly distribute themselves in the container regardless of the direction of light and/or gravity.

H_a: Fruit flies orient and move either positively or negatively in response to light or gravity.

Therefore, to show taxis is to reject H_o and infer H_a.

To collect data to test your hypotheses, obtain one small plastic vial, some transparent tape, and 15 fruit flies in a second plastic vial from the supply table. Tap the vial containing the flies on the table, so that all the flies fall to the bottom. Quickly remove the stopper and tape the open end of the empty vial to the open end of the vial containing the flies, so that you have a double vial experimental chamber. The tape should pass completely around the seam between vials (fig. 32.1).

Use a marker to place lines on the double vial, dividing it into thirds. Take a 5-inch piece of tape and turn it back on itself with a 2-inch overlap. Attach this as a pull tab on one end of the double vial. Do the same for the other end. This chamber and the flies it contains will now be used to test hypotheses about behavior.

Phototaxis

You should work in groups of three for all experiments in this lab topic.

To determine if fruit flies respond to light, slide the chamber into a cardboard tube and put stoppers in the ends of the tube so no light enters. Place the tube horizontally on a table. Place a lamp (off) 12 inches from one end of the tube with a 800-ml beaker of water in between the tube and the lamp to serve as a heat filter. Wait for at least five minutes and measure the temperature at the end of the tube nearest the lamp. Record the temperature in table 32.1. Leave the thermometer in front of the tube but elevated off the table. Why is it necessary to measure the temperature before and after the experiment? If temperature rises, how would it affect your analysis of the data?

Figure 32.1 Experimental setup for studying phototaxis.

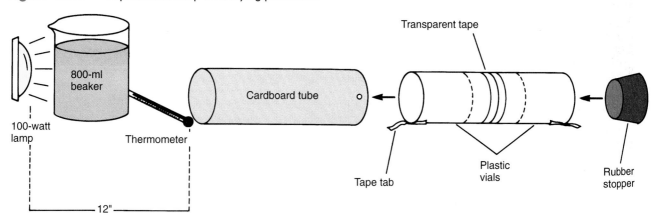

TABLE 32.1 Results from your phototaxis experiments

| | Temperature (°C) | | | Fruit Fly Distribution | | |
	Before	After		Nearest ⅓	Middle ⅓	Farthest ⅓
Trial 1						
Trial 2						
Trial 3						
			Sum across trials			

Now remove the stopper nearest the lamp and turn the light on for five minutes. Record the temperature again and slip the tube off while holding the experimental chamber by its tab to prevent movement. Immediately count the number of flies in each third of the chamber and record your results in table 32.1.

Repeat the experiment twice more, but each time turn the experimental chamber end-over-end before slipping it back into its tube. Why?

Geotaxis

To determine if the fruit flies will react to gravity, slip the chamber into the cardboard tube, stopper both ends, and lay the tube horizontally on the table for three minutes.

Now remove one stopper and stand the tube and the chamber upright on the table with the remaining stopper on the top. After five minutes, quickly slip off the tube and count the flies in each of the three sections. Record your results in table 32.3.

Reverse the vial and repeat the procedure for measuring the response to gravity twice more. Be sure to precede each trial with a three-minute horizontal equilibration period.

Put your sum across trials data for geotactic response on the board and transfer the class data into table 32.4.

Why should you reverse the vials for each new trial?

On the blackboard, record your sums across trials from table 32.1. Copy down the class data in table 32.2. The methods for analyzing this data will be discussed later. Turn off the light used in this part of the experiment.

Statistically Analyzing Simple Behaviors   **405**

TABLE 32.2 Class data from phototaxis experiments

Fruit Fly Distribution

Group	ΔTemp (°C)	Top ⅓	Middle ⅓	Bottom ⅓
1				
2				
3				
4				
5				
6				
7				
8				
Sum across groups (observed)				
Expected by chance				
(obs-exp)				
(obs-exp)²				
$\dfrac{(obs-exp)^2}{exp}$				

$\chi^2 = \dfrac{\Sigma(obs-exp)^2}{exp}$ across chamber areas = _____

Critical value for χ^2 from table C.4, p. 442 _____

d.f. = _____: Probability level _____

State null hypothesis for this experiment _____

_____

_____

_____

Total _____

State the alternative hypothesis _____

_____

_____

Accept or reject null hypothesis _____

TABLE 32.3 Results from your geotaxis experiments

Fruit Fly Distribution

	Top ⅓	Middle ⅓	Bottom ⅓
Trial 1			
Trial 2			
Trial 3			
Sum across trials			

Combined Test of Phototaxis and Geotaxis

You will now test the relative strength of the responses to light and gravity, simultaneously. Place the container of flies in the cardboard tube and stopper both ends. Clamp the tube and container horizontally on a ring stand for a three-minute equilibration period. Clamp a 400-ml beaker full of water above the tube to serve as a heat filter. Measure the temperature at the cardboard tube. Record your results in table 32.5.

Remove the stopper from one end of the tube and rotate the clamp holding the tube so that the open end is upward (fig. 32.2). Turn on the light and place it 12 inches above the open end of the tube for five minutes. Have someone hold a thermometer next to the open end of the tube during this time. Read the temperature after five minutes. Remove the stopper from the bottom of the tube and

carefully slip the container out without jarring it. Count the number of flies in each section of the container and record the results in table 32.5.

Repeat this procedure twice more and record the results in the top section of table 32.5.

The procedure will now be reversed, so that both light and gravity come from the bottom of the tube (fig. 32.3). Put a stopper in both ends and clamp the tube in a horizontal position on the ring stand. Adjust the position of the ring stand on the table, so that the tube hangs over the edge. Clamp a 400-ml beaker of water below the tube. Measure the temperature at the tube. Record the temperature in table 32.5. After three minutes, remove the stopper from the end with the two holes and insert a wooden applicator stick through the holes in the wall of the cardboard tube. Rotate the clamp so the end with the stick is downward.

Turn on the lamp and hold it 12 inches below the open end of the tube for five minutes.

C A U T I O N
If water spills on the bulb, it will explode.

Have someone check the temperature and record it in table 32.5. Now remove the stopper and carefully slip the vial upward, using the tab handle. Note the number of flies in each container section and record it in the lower section of table 32.5. Repeat the procedure twice more.

TABLE 32.4   Results from class geotaxis experiments

Fruit Fly Distribution

Group	Top ⅓	Middle ⅓	Bottom ⅓
1			
2			
3			
4			
5			
6			
7			
8			
Sum across groups (observed)			
Expected by chance			
(obs-exp)			
(obs-exp)²			
$\frac{(obs-exp)^2}{exp}$			

$\chi^2 = \dfrac{\Sigma(obs-exp)^2}{exp}$ across chamber areas = _____

Critical value for χ^2 from table C.4, p. 442 _____

d.f. = _____: Probability level _____

State null hypothesis for this experiment _____

_____

_____

_____

Total _____

State the alternative hypothesis _____

_____

_____

Accept or reject null hypothesis _____

TABLE 32.5   Results from your experiment on combined phototaxis and geotaxis

Light from top

	Temperature (°C)		Fruit Fly Distribution		
	Before	After	Top ⅓	Middle ⅓	Bottom ⅓
Trial 1					
Trial 2					
Trial 3					
Sum across trials					

Light from bottom

	Temperature (°C)		Fruit Fly Distribution		
	Before	After	Top ⅓	Middle ⅓	Bottom ⅓
Trial 1					
Trial 2					
Trial 3					
Sum across trials					

▶ Place your sum across trials data on the blackboard and copy the class data into tables 32.6 and 32.7.

Statistical Analysis of Results

In these four experiments, you collected quantitative data that describe the taxis response of fruit flies. If all fruit flies went to one end of the tube, either toward or away from the stimulus, there is no doubt that they exhibit a taxis. However, if only 6 out of 15 flies were at one end of the tube, does this represent a significant response or has the experiment failed to show a taxis?

Statistical tests can help us make up our minds in situations when the results are not clear-cut, when we need to make an unbiased judgment. Statistical tests are based on two premises: that you have a hypothesis to test and that you are willing to reject an answer that could be true.

Statistically Analyzing Simple Behaviors   **407**

Figure 32.2 Setup for studying phototaxis and geotaxis together.

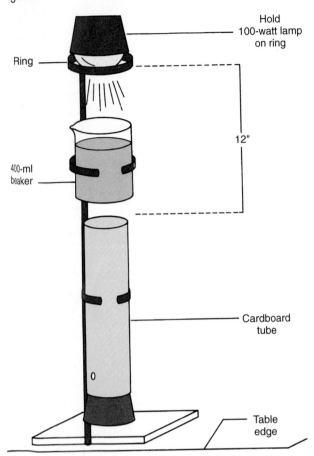

- Hold 100-watt lamp on ring
- Ring
- 400-ml beaker
- 12"
- Cardboard tube
- 0
- Table edge

Figure 32.3 Second setup for studying phototaxis and geotaxis.

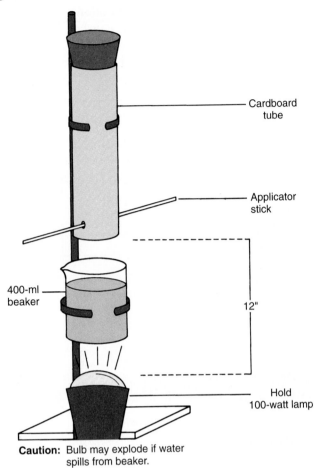

- Cardboard tube
- Applicator stick
- 400-ml beaker
- 12"
- Hold 100-watt lamp

Caution: Bulb may explode if water spills from beaker.

TABLE 32.6 Class data on combined phototaxis and geotaxis: light from the top

Fruit Fly Distribution

Group	ΔTemp (°C)	Top ⅓	Middle ⅓	Bottom ⅓
1				
2				
3				
4				
5				
6				
7				
8				
Sum across groups (observed)				
Expected by chance				
(obs-exp)				
(obs-exp)²				
(obs-exp)²/exp				

$\chi^2 = \dfrac{\Sigma(obs\text{-}exp)^2}{exp}$ across chamber areas = _____

Critical value for χ^2 from table C.4, p. 442 _____

d.f. = _____: Probability level _____

State null hypothesis for this experiment _____

_____

_____

_____

Total _____

State the alternative hypothesis _____

_____

_____

Accept or reject null hypothesis _____

Fruit Fly Distribution

Group	ΔTemp (°C)	Top ⅓	Middle ⅓	Bottom ⅓
1				
2				
3				
4				
5				
6				
7				
8				
Sum across groups (observed)				
Expected by chance				
(obs-exp)				
(obs-exp)²				
$\frac{(obs-exp)^2}{exp}$				

$\chi^2 = \dfrac{\Sigma (obs-exp)^2}{exp}$ across chamber areas = _____

Critical value for χ^2 from table C.4, p. 442 _____

d.f. = _____ : Probability level _____

State null hypothesis for this experiment _____

_____

_____

_____

Total _____

State an alternative hypothesis _____

_____

_____

Accept or reject null hypothesis _____

Statistical Testing of Hypotheses

The **chi-square (χ^2) goodness of fit test** can be used to test the null hypothesis (H_o), that is, whether the flies are equally distributed throughout the chamber within an acceptable level of deviation or whether they are clumped toward or away from the stimuli tested. The χ^2 (chi-square) test is discussed further in appendix C and you should read it before continuing.

The class data tables for each experiment have been set up with a χ^2 work table at the bottom. To determine the **observed** results, sum the entries in the columns. To determine the **expected** results, take the total number of flies (sum across observed results) and divide by 3. If flies randomly distribute in the chambers during the experiment, you would expect that one-third of the flies would be in each area. Enter only whole numbers as expected results.

The next three lines of the work tables are self-explanatory in terms of the arithmetic that needs to be done. What you are doing is developing a single number that is a measure of the amount of variation seen in the experiment compared to the expected random distribution proposed in the null hypothesis. If you sum across the last line of the table, you will have an estimation of the total variability in the results of the experiment. This is the calculated χ^2 for the experiment. Thus,

$$\chi^2 = \Sigma \frac{(obs-exp)^2}{exp}$$

When the observed and expected results are the same, χ^2 will equal zero: the greater the deviation of the observed results from those expected, the larger will be the value of χ^2.

How do you judge whether the χ^2 value you obtained is significant? You must refer to a table of critical χ^2 values, such as table C.4 on page 442. To use an χ^2 table, you must know the degrees of freedom and the acceptable confidence level.

The **degrees of freedom (d.f.)** is a number equal to the number of categories into which your data fall minus one. In your data, you categorized flies according to which one-third of the chamber they were found in. How many degrees of freedom should you use?_____

Students trained in statistics will realize that an assumption is being made here. To keep the analysis simple, we will assume that there is no interaction occurring between replication of the experiment by different laboratory groups and the location of the flies in a vial. If some groups consistently bump an upright vial knocking flies to the bottom, this assumption will be false and a contingency table analysis would have to be done. Students wanting to do such an analysis should consult a statistics textbook.

Each column in table C.4 contains **critical values** for χ^2 at several degrees of freedom and confidence levels. Scientists usually use the 95% confidence level, which means that 95% of the chance differences between observed and expected values will give an χ^2 value equal to or less than the value indicated for a specified degree of freedom. In other words, only 5% of the χ^2 values will be greater than this number due to chance.

If your calculated χ^2 value at your specified degrees of freedom is smaller than or equal to the value in the table, you cannot reject the null hypothesis because the agreement between the observed and the expected values is exceptionally good: fruit flies are randomly distributed in the vial. On the other hand, if your calculated χ^2 value is greater than the value in the table, you must reject the null hypothesis and infer the alternative: fruit flies respond to the variables being tested. You must look at the data to determine if they are positive or negative in their taxes.

For each experiment, compare your χ^2 value to that in the table. Must you accept or reject the null hypothesis in each experiment? What does that mean in terms of the model you proposed? Record your decisions below.

Experiment 1: Phototaxis

Experiment 2: Geotaxis

Experiment 3: Geotaxis with light above

Experiment 4: Geotaxis with light below

Experiment 5: Chemotaxis (if done)

Learning Biology by Writing

This lab is a well-defined experiment. The objectives and the hypotheses being tested are clear, and a lab report should be easy to write. In the introduction to your lab report, describe those purposes and hypotheses. In the methods section, describe briefly how you made your observations. Finally, in the data section, present your summary tables and anecdotal observations. The discussion should answer the questions in the introduction and generally discuss the significance of your conclusions. Be sure to include a discussion of sources of error. Finally, you might want to adopt an ethological approach to behavior and speculate on the survival value to the fruit fly of these taxes or lack of taxes. Recognize, however, that your statements are speculative and would have to be tested by other experiments before they could be accepted as facts.

As an alternative assignment, your instructor may ask you to turn in Lab Summary 32 at the end of the lab topic.

Internet Sources

In this lab topic, you studied, among other things, the response of fruit flies to gravity. Using **Alta Vista** search the World Wide Web for information about geotaxis in fruit flies. Several of the citations returned will be for a database called Flybase. It is a database for the genetics of fruit flies. Read the reports that are cited and determine why geotaxis was mentioned in the report.

Lab Summary Questions

1. What evidence do you have from your experiments that fruit flies are positively or negatively phototactic? Geotactic? Is the influence of phototaxis or geotaxis stronger?

2. In your own words, describe what a χ^2 (chi-square) goodness of fit test is used for.
3. What is a null hypothesis? What is an alternative hypothesis? Can both be true?
4. If your χ^2 value leads you to reject a null hypothesis, is the alternative hypothesis automatically true? Why, or why not?

Critical Thinking Questions

1. In the phototaxis experiments, how could you determine whether the flies are responding to light or to heat?

2. Speculate on the evolutionary advantage of a fruit fly being positively phototactic and negatively geotactic.

LAB TOPIC 33

Estimating Population Size and Growth

Supplies

Preparator's guide available on WWW at
  http://www.mhhe.com/dolphin

Equipment

Balances, with 0.01 g sensitivity

Materials

Wooden pegs or applicator sticks
Small plastic bags
4-m string
40-cm string
About 1,000 beads per student lab group (all one
    color, in coffee cans)
Beads of another color
Markers
Means of generating pairs of random numbers
If lab done outside
    Maps of quadrat sampling area
    Resource books on identification of common lawn
        weeds in area
    Identified samples of lawn plants in lab for reference
If lab done inside
    Table top covered by 1-m² paper ruled in 10-cm grid
    Beads to scatter on table
    Boards 1 m × 5 cm to edge grid
Microcomputer and population growth software

Prelab Preparation

Before doing this lab, you should read the
introduction and sections of the lab topic that have
been scheduled by the instructor.

You should use your textbook to review the
definitions of the following terms:

dispersion
exponential
logarithm
mark and recapture
population
population density
quadrat

You should be able describe in your own words
the following concepts:

Population growth curves
Exponential growth

Logistic growth
Carrying capacity

As a result of this review, you most likely have
questions about terms, concepts, or how you will do
the experiments included in this lab. Write these
questions in the space below or in the margins of the
pages of this lab topic. The lab experiments should
help you answer these questions, or you can ask your
instructor for help during the lab.

Objectives

1. To use quadrat sampling techniques to estimate the
   size of nonmobile populations
2. To simulate mark-and-recapture methods for
   estimating the size of large, mobile populations
3. To plot logistic growth curves, using prepared data
4. To use semilog graph paper for plotting population
   growth
5. To investigate factors influencing population
   growth, using computer simulations

Background

Organisms do not usually exist as isolated individuals in
nature but are parts of larger biological units called **populations.** These reproductive and evolutionary units consist of
the members of a single species residing in a defined geo-
graphical area, for example, in a small park, in a mountain
range, or on a continent. A population of clover in a lawn
may be easy to recognize because of its compactness,
whereas a widespread population of moose in northern
Minnesota may be less apparent. Organisms within each
population are interlinked by reproductive gene flow, and
natural selection operates on individuals within these populations to shape a population's adaptation to an environment over several generations. Populations are the fundamental units of species on which the mechanisms of
evolution operate.

Biologists study populations for two reasons: to understand gene flow, selection, and adaptation mechanisms and to manage the populations. Management addresses the problem of the relationship between humankind and the environment. In some cases, management involves controlling the size of populations of noxious organisms, such as rabid skunks, mosquitoes, or parasites of humans. In other cases, the goal may be to increase populations of beneficial organisms, such as lumber-producing trees, edible fish, or game species. To manage populations, it is necessary to know the basic biology of the organisms involved as well as to understand the physical and biological environment.

Studies of populations start with such basic questions as: How many individuals are in the population? How is the population distributed in the study area? Is the population increasing or decreasing in size? To answer these questions, biologists must use sampling procedures that allow them to estimate the number of individuals in the population distributed in a certain space, that is, the **population density.** Rarely is it practical to count the entire population. In this lab, you will use two often-employed techniques of estimation: **quadrat sampling** for nonmobile animals and plants and **mark-and-recapture techniques** for mobile animals. With the passage of the National Environmental Protection Act in 1970, these techniques have been and are, frequently used by biologists who are trying to determine the environmental impact of changes in ecosystems.

Both techniques are based on random sampling statistical procedures; care must be taken to assure that randomness occurs when these techniques are applied to an actual biological problem. Sampling plots must be randomly located and representative of the study areas, and animals must be randomly captured and released so that they can freely mix with the total population. If randomness is not realized, the sampling procedure will lead to erroneous estimates. Sampling procedures repeated over time allow the investigator to determine whether a population is growing or declining.

Estimation of population size alone, however, is not sufficient for making management decisions. Additional information is required, such as what resources are needed by and available to the population, how other organisms influence the population, and what the population age structure is like. The question of resources involves the physical environment, for example, water quality, soil type, temperature, or available nesting sites. The influence of other organisms on a population affects food availability, parasitism, and predation. Population structure refers to age distributions in the population. A stand of white pine trees could consist of all young, all old, or a mixture of young and old trees. Harvest practices and replanting programs would depend on the population structure in a particular timber tract. Similar problems exist in managing deer herds and coho salmon and in assessing the effects of agricultural pesticides on harmful insects.

Many biologists feel that population ecology is the most demanding and comprehensive field of biology because the investigator must understand and deal with environmental variables, metabolic efficiency, physiological reactions, hereditary mechanisms, and the interactions of organisms.

LAB INSTRUCTIONS

You will learn two estimation techniques used by population ecologists: quadrat sampling for plants and the mark-and-recapture technique for animals. You will also investigate the phenomenon of exponential growth in populations.

Quadrat Sampling of Vegetation

The structure and composition of nonmobile organisms, such as terrestrial plant communities, can be sampled by a number of techniques. The most frequently used procedure involves using randomly located plots of standard size called **quadrats.** Quadrat sampling varies in terms of the size, number, shape, and arrangement of the sample plots, all of which depend on the information sought and the nature of the populations or communities being studied.

Technique

The size of a plot is determined by the size and density of the plants being sampled. It should be large enough to include a number of individuals but small enough to allow easy separation, counting, and measuring of those individuals. Obviously, plots of different sizes would be used to estimate the number of trees in a forest in contrast to the number of cattails in a swamp.

For the sampling procedures to be statistically valid, quadrats must be chosen randomly within the study site. To do this, a baseline is laid out along one side of the study area. A pair of random numbers is chosen that corresponds (1) to distance along the baseline and (2) to perpendicular distance from the baseline to the center of the sampling quadrat. The units of length will depend on the size of the study area and may be in feet, meters, paces, or any other appropriate linear measure.

Inside Quadrats

Because weather might not allow you to perform quadrat sampling in the field, the technique can be simulated in the laboratory. If the indoor simulation is to be used, your instructor will have 1 m² of paper taped to a table top. It will be ruled into a 10-cm grid. How many 10-cm squares are in 1 m²?_____

One edge of the paper will have the 10-cm squares numbered 1 through 10. An adjacent edge will have them lettered A through J. The location of each 10-cm square is given by a letter-number combination. For example, H8 is found at the intersection of column 8 with row H.

Randomly scattered on the grid will be several hundred beads, some of one color and some of another. The colors represent juveniles and adults of the same species. Some squares will have boxes taped to them to exclude beads. These areas are uninhabitable by the species and should be subtracted from the total area available.

▷ Your task is to estimate the total population size in the study area and to estimate the juvenile population size by quadrat sampling.

Student pairs will each be assigned three random coordinates, such as A5, D2, and F7. Count the number of beads of each color in each assigned square and calculate an average per square. Record below.

▷ Place these estimates on the blackboard and calculate a grand average across all lab groups. Record below.

▷ Determine the total number of inhabitable squares in the study area and multiply your individual average estimates per square by this number to get your estimation of total population size and number of juveniles. Record below.

▷ Repeat these calculations, using the grand averages from the combined lab data.

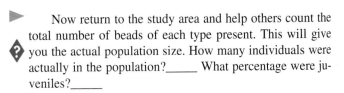

Now return to the study area and help others count the total number of beads of each type present. This will give you the actual population size. How many individuals were actually in the population?_____ What percentage were juveniles?_____

▷ Calculate the percentage of error for the estimations based on your data and those based on the class data, using the relationship:

$$\% \text{ error} = \frac{\text{estimate} - \text{real} \times 100}{\text{real}}$$

 Why are the % errors different for the estimations based on your data and the class data?

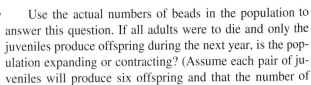 Use the actual numbers of beads in the population to answer this question. If all adults were to die and only the juveniles produce offspring during the next year, is the population expanding or contracting? (Assume each pair of juveniles will produce six offspring and that the number of males equals the number of females.)

Outside Quadrats

If you are to do this lab outside, your instructor will provide you with a map of a lawn on your campus that will serve as your study area. You will also be given a set of random coordinates, a piece of string 40 cm long and one 4 m long, and five pegs. Study the map so that you understand where the study area is and the orientation of the baseline.

▷ Take your lab manual, a pencil, and your strings to the study area. Using your coordinates, pace out the distances

TABLE 33.1   Number of nongrass plants in quadrat

Species	Number	Comments

along the baseline and perpendicular from it to locate the center of your sampling quadrat. Place a peg at this point. Tie the ends of the 4-m string together to form a loop. Using the other four pegs, stretch the string as a square around the center peg (the area inside the string will be 1 m²).

Count the number of nongrass plants contained within your quadrat. Keep separate counts for each species. If you cannot identify what kind of plants they are, make up a temporary name and then check resource books available in the laboratory. (Specimens can be collected and placed in plastic bags to prevent drying before later identification.) Record your counts in table 31.1.

If, by chance, you have a quadrat with thousands of clovers in it, you will not be able to accurately count them in the time provided. If this is the case, switch to the 40-cm string and lay out a smaller square around the center peg. Count the clover within this area and record the count separately.

To determine the density of grass in the lawn, use the 40-cm string to lay out the quadrat around the center peg or the upper-right corner. Count and record the number of grass plants in this quadrat. Consider any aboveground stem to be a grass "plant" but realize they will be connected to other plants by underground stems. Convert the count per small quadrat to the count per large quadrat by multiplying by 100 (= 1m² ÷ [0.1 m × 0.1 m]).

Note any unusual features in your study area. For example:

Are there any ant hills?

Is the quadrat located on a path? Under a tree? In the open?

Is there drainage into the area?

Are plants uniformly distributed or in clumps?

Remove one of each type of broadleaf plant from the lawn by cutting it off at the soil level. Collect several dozen grass plants in the same way.

Analysis

After you have completed your counts, notes, and collection, return to the lab. Put your data on the blackboard, including brief summarizations of your notes. After all lab groups have reported, the data should be discussed to see if any should be rejected because of atypical localized situations, such as chemical spills, trench excavation, or other isolated interferences. Combine the acceptable data and record them as a class summary in table 33.2.

Calculate the mean and standard deviation for the number of each type of plant per square meter. (Refer to lab

TABLE 33.2 Class results from quadrat analysis

Group	Area Size	No. of Grass Plants	No. of Other Plants (Specify)
1			
2			
3			
4			
5			
6			
7			
8			
9			
10			

topic 4 or appendix C for a description of how to calculate standard deviations.)

Using your measurements of grass and broadleaf weed density, how many of each type of plant would be found in a hectare? In an acre?

How many grass plants and other species are contained in the entire study area as defined by your instructor?

Weigh each type of plant you collected. Calculate the average weight of a grass plant. Since you know the number of plants in a hectare and the weight of each type of plant, calculate the aboveground wet biomass in a hectare of lawn.

A **hectare** is a unit in the metric system equivalent to 2.47 acres or 10,000 m^2. This would be equivalent to an area of 100 meters by 100 meters. How many square meters were contained in the total study area? Remember that length times width equals area. How many hectares are in the study area?

Mark and Recapture

In the field, it is difficult to measure the size of a population of randomly dispersed, mobile animals, but an estimation can be obtained by mark-and-recapture methods. In

TABLE 33.3 Results from sampling parameters assigned to you

Record conditions: no. to be marked = _____ no. to be recaught = _____

	Trial 1	Trial 2	Trial 3	Average
Marked				
Unmarked				

these methods, living individuals are trapped or collected, marked, and released at the site of capture. A record is kept of the number marked. At a later time that allows for dispersal, animals are again collected from the population. Some of these will be marked and some will be unmarked due to immigrations and emigrations of individuals to and from the collection sites.

Assuming the animals are randomly dispersed, the frequency of marked and unmarked animals in the second collection will allow you to estimate the total population size, using the following proportion (Lincoln-Peterson method): *The number of animals marked is to the total number of animals in the population as the number of marked animals recaptured is to the number of animals recaptured.* Or,

$$\frac{M}{N} = \frac{R}{C}$$

solving for N

$$N = \frac{MC}{R}$$

where
N = total number of individuals in population
M = number of animals marked and released (50, 100, or 150 depending on your assignment by the instructor)
C = total number of animals caught in second sample (60 or 120 depending on your assignment)
R = number of marked animals caught in second sample (recaptured)

Technique

In the lab are several cans containing 700 to 1,000 beads of one color. These beads are to simulate animal populations. Each lab group will be given a can and will be asked to estimate the population size, using mark-and-recapture methods. Each group will use a slightly different sampling regime. The results of the different groups will be compared.

Two groups should each mark 50 beads in their population, while two others should mark 100 and two others, 150. Do this by removing the assigned number of beads from the can and replacing them by adding the same number of different color beads to the can. By comparing results with other groups, you will determine if the size of the marked sample influences accuracy. One group from each of the above pairs should recapture 60 beads from its popu-

lation, while the other group will recapture 120. This will test the effect of the recapture sample size on accuracy.

Once you are sure of your group's assignment for the number of beads to be marked and the number to be recaptured, start the experiment. Add "marked" individuals to the population in the can. Close the can and shake well to assure random mixing. Without looking, withdraw a bead and tally its color on a piece of scrap paper. Return the bead to the can, stir the beads, and withdraw another, adding to the tally. Continue sampling, one bead at a time, until you withdraw the number of beads corresponding to your assigned recapture sample size. After every five draws put the top back on the can and shake it to assure random mixing.

Count how many beads in the recapture sample are marked and how many unmarked. Record the counts in table 33.3. Return the beads to the can and repeat the recapture sampling twice more. Record the results. Calculate an average value for your samples.

Analysis

Using the average values and the Lincoln-Peterson formula, estimate the population size in your can.

Now count all the beads (both colors) in your can and compare the real population size to the estimated population size.

Estimated number in population _____
Actual number in population   _____
Calculate the % error in your estimation:

$$\% \text{ error} = \frac{\text{estimate} - \text{real} \times 100}{\text{real}}$$

TABLE 33.4 Effects of sampling regime on % error

	Recaptured Sample Size	
	60	120
Size of marked sample		
50		
100		
150		

TABLE 33.5 Cell counts from *Acanthamoeba* cultures

Age (hrs)	Average Cell Density (Cells/ml)	Standard Deviation
0	1,000	± 50
10	2,100	± 120
30	10,000	± 450
50	44,000	± 3,000
70	160,000	± 10,000
90	330,000	± 15,000
110	600,000	± 40,000
150	1,050,000	± 40,000
210	1,400,000	± 30,000
250	1,430,000	± 40,000

Lab groups should now share their results by placing their % error values on the board in a facsimile of table 33.4. Record class values in table 33.4.

Take these class data and plot them on graph paper with the absolute value (no plus or minus sign) of % error (y-axis) as a function of marked sample size (x-axis). Two lines will be drawn on the graph: one for the recapture sample of 60 and the other for 120. If both of these lines are extrapolated to zero percent error, how many individuals must be marked before you will get correct estimates?

What fraction of the total population would this number be?

Describe the relationship between % error, marked sample size, and recapture sample size.

If you were optimizing effort and accuracy, what sampling strategy would be best? Why?

A Growth Curve Problem

Populations of organisms, whether they be oak trees, frogs, amoebas, or bacteria, have the inherent reproductive potential to increase exponentially in number over generations. Actual size of the population is limited by the available resources, such as food, water, and habitat.

Table 33.5 lists data about the growth of laboratory populations of the protozoan *Acanthamoeba,* a soil amoeba commonly used in research laboratories because it is easily grown in cultures. This amoeba can be grown on a peptone medium much like bacteria are in flasks. At low population

levels, the amoebas will grow and reproduce by mitosis, an asexual process. The data in the table come from 13 different cultures and have been combined to give an average value plus or minus one standard deviation.

You will plot this data to see what a population growth curve looks like. The same data will be plotted twice, once on arithmetic paper and once on semilog graph paper. You may not be familiar with semilog paper. Look at the sheet on page 423. Note that the spacing of vertical lines is equal, but the horizontal lines have unequal spacings numbered 1 through 9, and the cycle repeats four times on the page.

The spacing of these lines corresponds to the logarithm (base 10) of the number indicated on the line. Each cycle on the paper corresponds to a power of 10. Thus, if we designate the lowermost 1 as being 1,000, the 1 in the next cycle above is 10,000 and the 1's in the remaining cycles above are 100,000 and 1,000,000, respectively.

The numbers designated in each cycle take on the place value of the 1 below them. Thus 4 in the second cycle from the bottom in our example would be 40,000 because that is the place value of that cycle. The number 1,500,000 would be located on the heavy line between 1 and 2 in the uppermost cycle. Semilog paper is a convenient way of plotting the logarithm of a number without looking for the logarithm in a table.

At the end of the exercise, on both types of graph paper, plot as a function of time the cell densities from table 33.5.

Look at the two plots and note the differences. One is a straight line that plateaus at the end, and the other sigmoid. Biologists recognize three general areas on these curves: the exponential growth phase, the declining growth phase, and the stationary growth phase.

During **exponential growth,** every cell is dividing to give two cells, and they divide to give four, and then eight, and so on in an exponential progression. A measure of exponential growth is the **doubling time** of the population, the time that it takes the population to double. Doubling time represents the balance between birth and death rates in a population. The doubling time can be read directly from the graph you have drawn. What is the exponential doubling time on the following:

arithmetic paper _____
semilog paper   _____

Do the doubling times significantly differ from each other in the two types of plots? How can they be different if you used the same data?

The problem is that you probably read the curves in different regions. Note that on semilog paper, the population has a constant doubling time until it reaches 100,000 cells/ml and then the growth rate slows. On arithmetic paper, the slope of the line constantly changes, making it nearly impossible to define the period of exponential growth. This is one of the reasons semilog paper is used to plot growth curve data. The end of exponential growth is clearly shown as the place where linearity is lost. Exponential growth can be described by the exponential equation:
$N_T = N_0 e^{kt}$
where

N_T = number of cells at some time (t)
N_0 = number of cells to start
e = base of natural logarithms
t = elapsed time
k = a constant

This equation can be written as

$$\log N_T = \log N_0 + 2.3\,kt$$

which is the equation of a straight line. The reason that the growth curve is a straight line on semilog paper is that you have really plotted $\log N_T$ as a function of time (t). That is, you have plotted the above equation, where N_0 is the intercept and k is the slope of the line.

If it took approximately 52 hours for the cultures to reach a population density of 50,000 amoebas per ml, how many more hours will it take to reach 10^5 amoebas per ml? (Hint: look at graph you made.)

Now look at the growth curve and find the cell concentration at which the culture is in stationary growth, neither increasing nor decreasing in number. This is called the **carrying capacity** of the environment. In the culture, amoebas are dying at the same rate as they are dividing, so that the number stays constant.

Name at least three factors that would determine the carrying capacity of an environment.

Computer-Generated Growth Curves

Because the underlying mathematics of population growth curves are well understood, a number of computer programs are available to simulate population growth. Your lab instructor will describe the programs available in your lab and how to use the computers.

Most of the available programs allow you to vary the starting number of individuals in the population, the doubling time (generation time), the number of offspring produced per mating, and the carrying capacity of the environment. Use the program to understand the following situations.

Compare two populations in which population 1 starts with twice as many individuals as population 2, but population 1 produces twice as many offspring per mating as population 2. Test the following null hypothesis: No differences in the shape of the growth curves between these two populations are expected. What would be the alternative hypothesis?

Run the simulation program under the above conditions. Must you accept or reject the null hypothesis? Why?

Explore the other variables that can be changed in your program. Describe below what you have discovered.

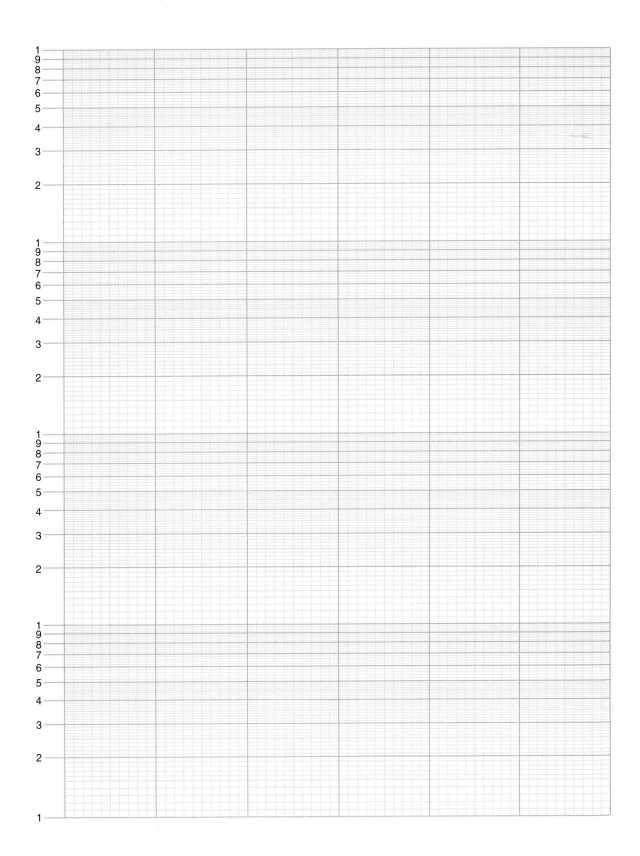

Estimating Population Size and Growth **423**

LAB TOPIC 34

Performing Standard Assays of Water Quality

Supplies

Preparator's guide available on WWW at
http://www.mhhe.com/dolphin

Equipment

Millipore Sterifil System (plus extra sterile filters and
culture pads)
Incubators set at 37°C
Hot plates
pH meter
Fluorescent light
Thermometers

Materials

Aspirators and vacuum tubing
Beaker tongs
1-liter (or larger) beaker or pan
Sterile petri plates (60 mm)
Forceps
5-ml syringes or pipettes
Pipettes or graduated cylinders for water samples
Depression plates
10-ml plastic syringes
18-gauge needle with end ground flat
0.2-ml pipettes
25-ml Erlenmeyer flask
Microburets (1 ml, in hundredths) (can use 4ml pipette
with hose and clip)
200-ml bottles (some clear; others covered with foil)
Duckweed
Elodea
Aluminum foil
Granite chips
Limestone chips
Safety goggles

Solutions

70% ethanol
Sterile water samples
Water samples from campus swimming pool, local
ponds or streams, or other sources (above and
below sewage plant discharge pipe would be
instructive)
Sterile MF-Endo broth
Escherichia coli slant culture

Winkler reagents (MnSO$_4$, KOH, H$_3$PO$_4$, starch, Na
thiosulfate, NaI)
0.01 M H$_2$SO$_4$

Prelab Preparation

Before doing this lab, you should read the introduction
and sections of the lab topic that have been scheduled
by the instructor.

You should use your textbook to review the
definitions of the following terms:

buffer
dissolved oxygen
E. coli
nomogram
water quality

You should be able to describe in your own words
the following concepts:

Sterile technique
Filtration

As a result of this review, you most likely have
questions about terms, concepts, or how you will do
the experiments included in this lab. Write these
questions in the space below or in the margins of the
pages of this lab topic. The lab experiments should
help you answer these questions, or you can ask your
instructor for help during the lab.

Objectives

1. To determine coliform bacterial content of water
   samples from various sources, using membrane
   filtration and culture techniques
2. To measure dissolved oxygen in water, showing
   differences in oxygen content due to respiratory
   demand and photosynthetic production
3. To demonstrate buffering capabilities of bedrock

Background

For some, the term **environmental quality** means simply the preservation of wilderness areas from such modern practices as natural resource harvesting, agricultural technology, and industrial or domestic development. This view conveys the idea that environmental quality should be maintained for recreational or aesthetic reasons. The total concerns for environmental quality, however, have a much broader base. The survival of humankind in the near future depends on our success in managing natural resources and in maintaining acceptable standards of environmental quality. Clearly, one requirement of civilization is water of high quality.

All societies require a plentiful supply of freshwater for drinking, food preparation, and cleanliness. Perhaps less obvious is the need for even larger amounts of good quality water in various industries, for cooling or washing processes. Also agriculture in many parts of the world would be impossible without abundant water for irrigation. For example, over 75% of the water used in California is used on crops. A major problem facing modern society is retaining water of sufficient quality and in sufficient amounts to supply these needs.

Freshwater is available from two sources. **Surface water** is the water contained in ponds, lakes, or rivers. It typically contains a large amount of organic matter, including suspended detritus, algae, and bacteria. Before this water can be made **potable** (drinkable) and usable in many industrial processes, it must be filtered and chemically treated to remove disease-causing organisms and organisms whose growth could clog water pipes and holding tanks. The second source of freshwater, **groundwater,** flows beneath the earth's surface in **aquifers,** layers of porous sand and pebbles. This water is typically low in organic content, usually containing no living organisms or detritus because of the filtering action of the aquifers. Groundwater often requires little or no treatment. It can be drawn from natural springs or human-made wells and is the preferred source for domestic and industrial use.

Heavy use of groundwater in areas of the United States has caused the water table to drop over 50 feet in the last 50 years. The result is that old wells have gone dry, new wells must be dug deeper, and the earth is collapsing in sinkholes where the soil is no longer supported by subterranean water. Before the end of this century, our demand for quality groundwater will exceed the available supply in many areas. The available supply of surface water is also limited. Water recycling will become more common, and water rights will assume a new importance.

Pollutants reduce the usability of both surface and groundwaters. The slow seepage of surface water into aquifers usually helps cleanse the water of impurities, but if surface waters contain many pollutants or if earth structure is disturbed through sinkhole collapse, contaminants may enter the groundwater. Seepages from chemical or domestic dumps, from cesspools, or from pesticides have contaminated water supplies throughout the country. When such water is used domestically, the chances of bacterial or viral diseases, possible birth defects, cancers, or toxic-waste poisoning are increased.

Water recycling is expensive because of the costs involved in removing wastes from water effluents. If polluted water is discharged downstream from an industry's water source, that industry's water quality is not affected. Instead, the cost is passed on to the consumers downstream who must treat the water before they use it. Though the ethics of this method of waste disposal are questionable, this has been and still continues to be the mode of operation in many watersheds. Fortunately, state and federal agencies enforce laws that require industries and municipalities to remove pollutants from their wastewater.

LAB INSTRUCTIONS

In recent years, easy-to-perform, reliable tests have been developed to check water samples for biological purity. If these indicator tests suggest possible problems, then more elaborate analyses are performed. In this lab, you will perform two tests: one for coliform bacteria content and one for dissolved oxygen content. Though samples will be provided, you could collect your own samples from local streams, ponds, or wells.

Coliform Bacteria Test

Drinking water and swimming areas are routinely monitored by public health authorities to determine if sewage bacteria are present. Raw sewage may contain disease-producing bacteria, such as the typhoid organism, *Salmonella typhosa,* or others that cause gastrointestinal distress. These organisms, when present in low numbers, are difficult to detect, so screening procedures based on other organisms are used. *Escherichia coli* is a very common, usually nonpathogenic inhabitant of the vertebrate intestine that is shed with the feces. *E. coli* is easily detected by a **coliform test,** an assay for nonspore-forming, rod-shaped bacteria that ferment the sugar lactose. Public health officials use the coliform test as an indicator for the presence of raw sewage. When coliform counts are high, they assume that pathogenic bacteria are present and preventive measures are initiated.

Sanitary engineers have developed a standard coliform assay. In this test, a water sample is drawn through a small-pore filter to trap bacteria. The filter with the bacteria on its surface is then placed on a specially formulated growth medium called **MF-Endo medium,** which encourages the growth of coliform bacteria while allowing other bacteria to grow only very slowly. The coliform bacteria break down the sugar lactose in the growth medium, forming a class of organic chemical compounds called **aldehydes.** The aldehydes react with a pink dye in the medium called **basic**

Figure 34.1 Millipore Corporation Sterifil apparatus.

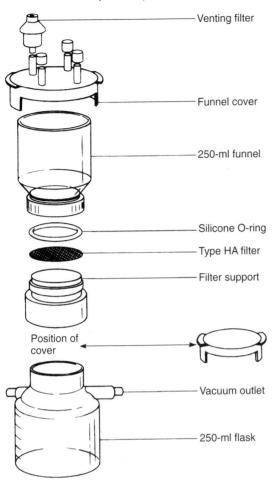

Figure 34.2 Method for adding filter to Millipore apparatus.

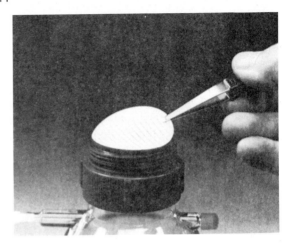

Remove the apparatus with tongs and assemble. Do not touch the inside of the funnel with your fingers.

Obtain a packet of presterilized membrane filters (with 0.45-micron pores) and a pair of forceps. The tips of the forceps should be sterilized by dipping them in alcohol and then quickly passing them through a flame. After they have cooled for a few seconds, use the forceps to pick up a membrane filter without puncturing it and lay it on the filter support. Center the filter and screw on the funnel with the cover on as in figure 34.2. Do not overtighten the funnel, or the filter may tear. If it does, install a new filter.

Your water samples should now be added to the funnel, and the bacteria collected by vacuum filtration. All glassware used to handle the water sample from the time the sample was taken until it is placed in the filter apparatus must be sterile; otherwise, the collection technique may add bacteria and give false indications of bacterial content.

The size of the sample required is variable depending on the estimated pollution of the water source. The standard sample volume is 100 ml. If the water source is considered clean—for example, a chlorinated pool or tap water—few bacteria will be present, and larger volumes could be filtered but this is not standard practice. On the other hand, if the water sample was taken from an obviously polluted stream or pond, a much smaller sample of 0.1 to 1.0 ml may be appropriate. If a volume of less than 25 ml is used, the sample should be mixed with sterile water to disperse the bacteria so that they will spread out on the filter and not be concentrated in one spot. To disperse the bacteria in small samples, first add 25 ml of sterile water to the funnel and then add the sample by pipette. Swirl the apparatus.

Once the sample is in the funnel and the cover is on, connect the receiver flask to a vacuum source such as an aspirator. This will draw the water through the membrane filter, but bacteria being about 1 micrometer in diameter will not pass through the 0.45-micrometer (10^{-6} meters) pores of the filter. When all water has passed through the filter, disconnect the vacuum line and shut off the water.

fuchsin, causing the developing coliform bacteria colonies to have a greenish, metallic sheen.

Since few other organisms make aldehydes from lactose, the appearance of colonies with a green sheen indicates that coliform bacteria are present and that raw sewage has at some time mixed with the water sample. However, this does not automatically mean that the water is unsafe. The number of coliform bacteria is a measure of the pollution level, and health officials have established minimum acceptability standards for different types of water. When these coliform standards are exceeded, however, disease-producing organisms are assumed to be present and consuming the water is a health hazard.

Filter Procedure

Several water samples are available in the lab, including sterilized water, water to which *E. coli* were purposely added, and samples from streams, pools, and so on. Your instructor will assign you a sample.

To collect and concentrate bacteria, you will use the apparatus shown in figure 34.1, consisting of a cover, funnel, filter, and receiver flask. The cover and funnel must be sterilized by immersion in boiling water for five minutes.

Though you cannot see them, any bacteria that were in the sample are now on the surface of the filter. If these bacteria are fed with a growth medium, each will develop into a visible colony consisting of several thousand bacterial cells. By counting the colonies, you can estimate the number of bacteria that were originally in the water sample.

Culture Procedure

▶To culture the bacteria, take a sterile petri dish and, using standard sterile technique, add 2 milliliters of sterile MF-Endo medium. Do this by lifting only one side of the lid and inserting a sterile, cotton-plugged pipette or by pouring the medium in from a sterile tube. Next add a sterile culture pad, using sterile forceps to handle it. After the medium has soaked into the pad, remove the filter from the filtration apparatus with sterile forceps and put it with the bacteria side up on top of the medium-soaked pad in the petri dish (fig. 34.3). Replace the petri dish top. Nutrients will diffuse through the membrane filter to the bacteria. If the plates are incubated for 24 hours at 37°C or 48 hours at room temperature, each bacterium will produce a colony on the surface of the filter.

Write your name, the date, and the water sample source on the lid of the petri dish. The petri plates should be incubated upside down; otherwise, water droplets that condense on the plate top will fall on the filter and wash away developing colonies.

Analysis

▶After the colonies have had time to develop (24 to 48 hours), store the plates in a refrigerator until the next lab.

If a sample contained any bacteria, colonies should be visible on the surface of the filter. Those with a green sheen to them will be coliform *(E. coli)* colonies. Count these colonies and record your counts in table 34.1.

▶Compare your results with those of classmates who had the same or different water samples. Did those who had the same sample obtain the same results that you did? Did the different samples show any water source to be highly contaminated with sewage?

In your town, what is the accepted coliform count for swimming water? For drinking water? The Environmental Protection Agency has set a standard of 1 coliform colony from 100 ml of water as being sufficient cause for concern, whereas 4 colonies per 100 ml requires direct action.

Figure 34.3   Sequence in which materials are added to petri dish. Add all materials using sterile technique; do not hold lid off the dish for long periods or lay it on the table. Poor technique will result in contamination and erroneous conclusions.

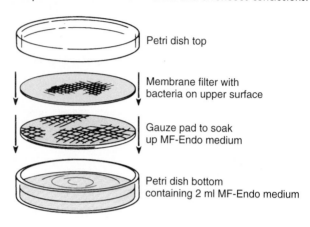

Petri dish top

Membrane filter with bacteria on upper surface

Gauze pad to soak up MF-Endo medium

Petri dish bottom containing 2 ml MF-Endo medium

Why were sterile water and known *E. coli* samples included in the set of samples?

Dissolved Oxygen

The suitability of aquatic habitats for fish and other organisms can be determined by monitoring the oxygen content. Fish require about 4 to 5 mg of O_2 per liter of water, whereas air-breathing aquatic insects and snails can inhabit waters with a much lower oxygen content. An adequate amount of oxygen must always be present. A *single day* of an algal or a bacterial bloom in a lake can decrease the oxygen content to zero and wipe out a fish population.

When organic wastes from municipal or industrial sources are added to a body of water, an oxygen demand is created as bacteria multiply and aerobically break down the organic compounds. Progress in the self-purification of water containing such wastes can be evaluated by accurate dissolved oxygen (DO) measurements.

Water will hold only a certain amount of dissolved oxygen, depending on temperature, salt content, and atmospheric pressure. When the photosynthetic rates of aquatic plants and algae are high, water may become saturated with oxygen and excess oxygen will escape into the atmosphere. If, on the other hand, the photosynthetic rates are low and the respiratory consumption of oxygen is high, oxygen dif-

TABLE 34.1 Coliform test results

	Sterile H₂O	E. coli (known)	Unknown Sample Source
Number of green colonies			
Volume of water sample			
Coliform bacteria per 100 ml of sample			

fuses from the atmosphere into water. Such exchanges are facilitated by the mixing actions of waves but often gradients develop with the highest levels near the surface. Have you ever seen fish close to the surface of a pond on a calm day? What do you suppose causes this behavior? Oxygen content measured in a particular body of water at noon on a sunny summer day will not be the same as that measured at night in midwinter under the ice. Why?

Winkler Method

Two methods are used for measuring dissolved oxygen: **polarographic,** or **electrode,** measurements and **chemical methods.** Both methods are in common use today. In this exercise, you will use the chemical method that was developed by Ludwig Wilhelm Winkler in Budapest in 1888. The **Winkler test** involves a sequence of chemical reactions that occur in proportion to the amount of oxygen in a water sample. These chemical reactions result in color changes that allow the investigator to quantify the amount of oxygen in the water sample.

In the test, manganous sulfate is added to the sample, along with a base to form manganese hydroxide, a flocculent whitish precipitate.

$$MnSO_4 + 2 KOH \rightarrow Mn(OH)_2 + K_2SO_4$$

If oxygen is dissolved in the sample, it reacts proportionately with the manganese hydroxide to form manganese dioxide, which is reddish brown:

$$2 Mn(OH)_2 + O_2 \rightarrow 2 MnO_2 + 2 H_2O$$

If this solution is acidified with a strong acid, such as phosphoric or sulfuric acid, and if sodium iodide is added, the floc will disappear and free iodine will be released in proportion to the amount of MnO_2 present, turning the solution yellow brown.

$$3 MnO_2 + 3 NaI + 4 H_3PO_4 \rightarrow I_3^- + 3 Na^+$$
$$+ 3 Mn_3(PO_4)_4 + 6 H_2O$$

When starch is added to this mixture, the solution will turn blue from the formation of a starch-iodine complex.

Adding sodium thiosulfate to the blue solution reduces free iodine to iodic acid, and the blue color disappears. If the thiosulfate concentration is known, the dissolved oxygen content of the sample can be calculated from the normality and the volume of thiosulfate used to reach the colorless end point.

Usually, the Winkler test is performed on 200-ml samples of water taken from streams or lakes. In 1962, Burke described a method that requires only a 10-ml sample. This method will be used here.

Micro-Winkler Procedure

1. Place the following volumes of reagents in separate depressions of a porcelain depression plate:
   a. 0.2 ml of manganous sulfate reagent
   b. 0.2 ml of alkaline-iodide-azide reagent
   c. 0.2 ml of concentrated H_3PO_4 acid (very corrosive; immediately wipe up spills or wash hands and clothing that contact acid)

> ### CAUTION
> These reagents are corrosive and will damage human tissue and clothing. Wear goggles to protect your eyes!

2. Obtain a 10-ml syringe without a needle and draw in slightly more than 10 ml of the water sample to be tested. Be careful not to draw in any air bubbles. If you do, start over; trapped air contains oxygen and will give false readings.

3. Measure the temperature of the water from which the sample was drawn and record._____

4. Now add an 18-gauge needle to the syringe and expel a small amount of fluid to get rid of air in the needle and to adjust the volume exactly to 10 ml.

5. Draw in the manganous sulfate from the depression plate, again being careful not to draw any air bubbles.

6. Mix the contents by rotating the syringe.

7. Now draw in the alkaline-iodide solution and wipe the tip of the syringe. Mix, allow the floc to settle, and remix.

Performing Standard Assays of Water Quality

Figure 34.4 Nomogram for calculating % saturation of dissolved oxygen in water at different temperatures. Use a ruler or other straightedge to find % saturation. Lay the straightedge on the nomogram so that its edge passes through the mg of oxygen you found for your sample and through the temperature of the water at the time you fixed your sample. The point where the straightedge crosses the diagonal line represents the % saturation of your sample.

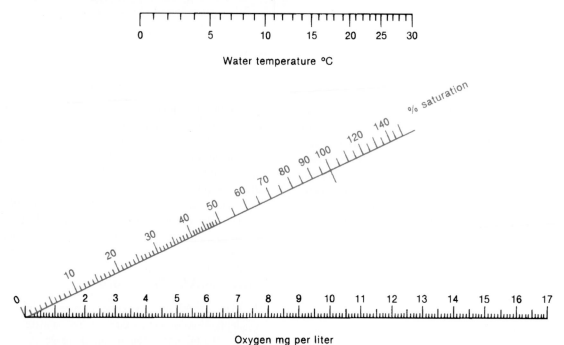

8. Carefully draw in the phosphoric acid, being sure to wipe the tip of the syringe. If the precipitate does not dissolve, add a little more acid.

9. When all the precipitate is dissolved, expel the solution from the syringe into a 25-ml Erlenmeyer flask and add a drop of starch solution. Mix. It should turn blue. Why?

10. Using a microburet or a pipette, very carefully titrate the sample with 0.025 M sodium thiosulfate solution until the blue color disappears. Placing a white card under the flask will help you see the end point. Add thiosulfate drop by drop, swirling the flask between drops.

Analysis

The concentration of thiosulfate was purposely chosen so that the milligrams of oxygen per liter of sample will equal the number of milliliters of thiosulfate used times 21.27. What is the oxygen content of your sample in mg of O_2 per liter?

A **nomogram**, as in figure 34.4, can be used to determine the percent saturation of the sample, that is, the ratio of the actual oxygen content to the maximum amount that can dissolve in water at that temperature. To use the nomogram, take a straightedge and line it up with the calculated oxygen concentration on the lower scale and with the temperature of the source of the sample on the upper scale. The percent saturation may be read where the straightedge crosses the middle line. For example, 9.5 mg of oxygen per liter of water from a source at 20°C would be 100% saturated.

Is your sample saturated? How much more oxygen (in mg per liter) could it hold?

If the mg of oxygen per liter of a sample remains constant, does the percent saturation increase or decrease as the temperature decreases?

TABLE 34.2   Oxygen content in light and dark bottles (mgO$_2$/$_L$)

| | \multicolumn{6}{c}{Sample Number} |
	1	2	3	4	5	Average
Light bottle						
Dark bottle						

Factors Affecting Oxygen Concentration

The amount of oxygen in a sample from a natural body of water is determined by the rates of photosynthesis and respiration of microscopic plankton and large plants and animals living in the water.

By using the Winkler test, it is possible to see the relative effects of respiration and photosynthesis. About 24 hours before the lab period, your instructor sealed duck weed and several sprigs of *Elodea* in bottles filled with water. No oxygen could enter from the atmosphere nor could any escape. One-half of the bottles were covered with aluminum foil and placed in a dark cabinet. The other bottles were placed in front of a bright fluorescent light. Your instructor will give you one of these light or dark bottles. Form a hypothesis that predicts whether the light or dark bottles will have a higher oxygen content.

H$_o$ =

H$_a$ =

You should open the bottle and quickly take a sample in your syringe for a Winkler test. Run the ten steps of the Winkler test as before. Measure the temperature of the water from which the sample was drawn and record._____

Put your value for dissolved oxygen on the blackboard in the lab and identify it as a light or dark sample.

Transfer these values from the board into table 34.2. When all students have reported their data, calculate an average. Under which conditions is respiration occurring?

Under which conditions is photosynthesis occurring?

 What would you expect to happen to the oxygen content if you placed the bottle from the light conditions into the dark for 12 hours?

Demonstration of Bedrock Buffering

Sulfur in various chemical forms is found in ores, coal, and petroleum products. When these materials are refined or burned, the sulfur is released as the gases sulfur dioxide or trioxide. In the atmosphere, the gases dissolve in water droplets and form sulfuric acid. When the droplets condense into raindrops, acid rain results.

Rain with a pH of 3 has been measured in some parts of the United States. Depending on the underlying geology of a region, the effects can be catastrophic. In New England, where granite is the bedrock, the pH of some streams, lakes, and soil has become so acidic that most plants and animals have been killed. In the Midwest, where the bedrock is limestone, the changes have not been as dramatic. Why do you think the effect is not as great in the Midwest?

Limestone regions have a buffering (acid-neutralizing) capacity. When acid rain falls on limestone, the following reactions occur:

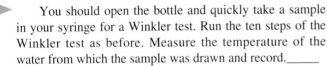

$$H_2SO_4 + CaCO_3 \rightarrow CaSO_4 + H_2CO_3$$

$$H_2CO_3 \rightleftharpoons H^+ + HCO_3 \rightleftharpoons H_2O + CO_2$$

The carbon dioxide formed in this way then escapes into the atmosphere. No similar buffering chemistry occurs in regions in which granite or other igneous rock formations underlie the soil.

To demonstrate the buffering capacity of various geological substrates, your instructor has prepared three beakers: one contains 0.01 M H_2SO_4; another, 0.01 M H_2SO_4 over granite chips; and the third, 0.01 M H_2SO_4 over limestone chips.

Measure the pH of all three beakers and record below.

0.01 M H_2SO_4 _____
0.01 M H_2SO_4 + granite _____
0.01 M H_2SO_4 + limestone _____

Based on your data, what do you conclude about the buffering capacity of granite versus limestone?

Learning Biology by Writing

In this laboratory, you have performed coliform tests that should allow you to decide the quality of the water in your samples. Would you consider the water potable? Suitable for swimming? Suitable as an environment for fish? Write a report describing where the samples came from, what tests were performed, and the results. Raise other questions about water quality not covered by the tests. Finally, briefly describe what it would take to bring the water up to drinking standards.

As an alternative assignment, your lab instructor may ask you to turn in answers to the Lab Summary and Critical Thinking Questions that follow.

Internet Journey

In this lab you performed a standard water test that determines the number of coliform bacteria in water samples. This test is widely used. Use a search engine to search the World Wide Web for information on coliform bacteria. Several of the URLs returned will give information on what levels of bacteria are acceptable for what uses of water. Determine the levels that are acceptable for (1) drinking, (2) swimming, (3) boating, and (4) industrial use. Record the URLs for later reference.

Lab Summary Questions

1. If most strains of *E. coli* are not pathogenic, why should you worry if they are in a water sample?

2. In your own words, describe how you isolated bacteria from a water sample and tested for coliforms.
3. What is a nomogram? Does the solubility of oxygen increase or decrease with increasing temperature? At the same temperature and with the same number of bacteria and algae in the samples, would you expect that saltwater or freshwater samples would have a greater oxygen content?
4. What natural processes add oxygen to natural bodies of water? Describe several sources. What natural processes remove oxygen from bodies of water? Describe how these processes of addition and removal can vary over time.

Critical Thinking Questions

1. *E. coli* is used as an indicator organism when assessing water for possible human fecal contamination. Why is this coliform count not a sole indicator of microbiological water quality (potability)?

2. If a pond is green in color, it probably has a large algae population. Under what conditions would you expect the oxygen concentration to be low, perhaps even zero?

3. Describe how you could use the light and dark bottle techniques to measure primary and secondary productivity in a pond.

4. Make a list of the factors that add dissolved oxygen to a natural body of water. Make a second list of those factors that remove dissolved oxgen. Describe how the measured dissolved oxygen content is the result of the balance between these processes.

APPENDIX A

Significant Figures and Rounding

In the laboratory, students often ask how precise they should be in recording measurements during experiments. The question of precision also arises when doing calculations to solve laboratory problems. A few simple explanations and rules are given here to guide you in these quantitative aspects of laboratory biology.

What Are Significant Figures?

Significant figures are defined as *the necessary number of figures required to express the result of a measurement so that only the last digit in the number is in doubt.* For example, if you have a ruler that is calibrated only in centimeters and find that a pine needle is between 9 and 10 cm long, how do you record the length?

The definition of significant figures tells you that you should estimate the additional fraction of a centimeter in tenths of a centimeter and add it to 9 cm, thus indicating that the last digit is only an estimate. You would never write this additional fraction as hundredths or thousandths of a centimeter because it would imply a precision that did not exist in your measuring instrument.

However, suppose you have a ruler calibrated in millimeters and measure the same pine needle, finding that the needle is between 93 and 94 mm long. You should then estimate the additional fraction in tenths of a millimeter and add it to 93 mm which would be 9.3 cm plus the estimate.

Memorize and use this rule throughout the course: *When recording measurements, include all of the digits you are sure of plus an estimate to the nearest tenth of the next smaller digit.*

Doing Arithmetic with Significant Figures

Other rules apply when doing calculations in the laboratory. Several situations you will encounter are discussed in the following paragraphs.

When converting measurements from one set of units to another in the metric system, be sure not to introduce greater precision than exists in the original number. For example, if you have estimated that something is 4.3 cm long and wish to convert it to millimeters, the correct answer is 43 mm, not 43.0 mm because the number of centimeters was known only with precision to a tenth of a centimeter and not a hundredth of a centimeter as 43.0 mm implies.

When performing multiplication or division involving numbers with different levels of significant figures, recognize that the answer should be expressed only with the precision of the number in the calculation that shows the least number of significant figures. For example, if you wish to calculate the weight of 10.1 ml of water and you are told the density of water is 0.9976 g/ml, you would multiply the density times the volume to obtain the weight. However, the correct answer would be 10.1 g, not 10.07576 g. Because the water volume measurement is known only to three significant figures, the latter number conveys a precision that is not justified given the uncertainty of the water volume measurement.

When performing additions or subtractions, the answer should contain no more decimal places than the number with the least number of digits following the decimal place. Thus, 7.2°C subtracted from 7.663°C yields a correct answer of 0.5°C not 0.463°C. If the first number had been known with a precision of 7.200°C, then the latter answer would have been correct.

What Is Rounding?

The last example introduces the concept of rounding to the appropriate number of significant figures. The rules governing this are straightforward. You should not change the value of the last significant digit if the digit following it is less than five. Therefore, 3.449 would round off to 3.4 if two significant figures were required. If the value of the following number is greater than five, increase the last significant digit by one. Therefore, 88.643 would round off to 89 if two significant figures were required.

There is some disagreement among scientists and statisticians as to what to do when the following number is exactly five, as in 724.5, and three significant numbers are required. Some will always round the last significant figure up (in this case 725), but others claim that this will introduce a significant bias to the work. To eliminate this problem, they would flip a coin (or use another random event generator) every time exactly five is encountered, rounding up when heads was obtained and leaving the last significant digit unchanged when tails was obtained. Recognize, however, that if the number were 724.51 or greater, the last significant digit would always be rounded up to 725.

Examples of Rounding

49.5149 rounded to 5 significant figures is 49.515
$(= 4.9515 \times 10)$

49.5149 rounded to 4 significant figures is 49.51
$(= 4.951 \times 10)$

49.5149 rounded to 3 significant figures is 49.5
$(= 4.95 \times 10)$

49.5149 rounded to 2 significant figures is 50
$(= 5.0 \times 10)$

49.5149 rounded to 1 significant figure is 50
$(= 5 \times 10)$

APPENDIX B

Making Graphs

Graphs are used to summarize data—to show the relationship between two variables. Graphs are easier to remember than are numbers in a table and are used extensively in science. You should get in the habit of making graphs of experimental data, and you should be able to interpret graphs quickly to grasp a scientific principle.

In using this lab manual, you will be asked to make two kinds of graphs—line graphs and histograms. **Line graphs** show the relationship between two variables, such as amount of oxygen consumed by a tadpole over an extended period of time (fig. B.1). **Histograms** are bar graphs and are usually used to represent frequency data, that is, data in which measurements are repeated and the counts are recorded, such as the values obtained when an object is weighed several times (fig. B.2).

Line Graphs

When you make line graphs, always follow these rules.

1. Decide which variable is the dependent variable and which is the independent variable. The **dependent variable** is the variable you know as a result of making experimental measurements. The **independent variable** is the information you know before you start the experiment. It does not change as a result of the dependent variable but changes independently of the other variable. In figure B.1, time does not change as a result of oxygen consumption. Therefore, time is the independent variable and the amount of oxygen consumed (which is dependent on time) is the dependent variable.

2. Always place the independent variable on the x-axis (the horizontal one), and the dependent variable on the y-axis (the vertical one).

3. *Always label* the axes with a few words describing the variable, and *always put the units* of the variable in parentheses after the variable description (fig. B.1).

4. Choose an appropriate *scale* for the dependent and independent variables so that the highest value of each will fit on the graph paper.

5. Plot the data set (the values of y for particular values of x). Make the plotted points dark enough to be seen. Always use pencil not pen in case you need to erase. If two or more data sets are to be plotted on the same coordinates, use different plotting symbols for each data set (·, ×, ⊙, ⊗, ▫, *etc.*).

6. Draw *smooth curves* or *straight lines* to fit the values plotted for any one data set. Do not connect the points

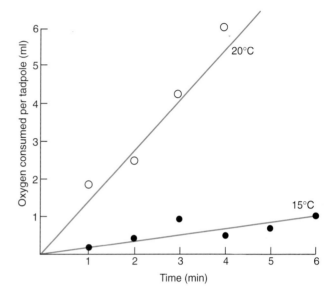

Figure B.1 Oxygen consumption by a tadpole at two temperatures.

with short lines. A smooth curve through a set of points is a visual way of averaging out chance variability in data. Do not extrapolate beyond a data set unless you are using it as a prediction technique, because you do not know from your experiment whether the relationship holds beyond the range tested.

7. Every graph should have a legend, a sentence, explaining what the graph is about.

Histograms

In making histograms, the count data are always on the y-axis. The categories in which the data fall are on the x-axis. For example, the data from which figure B.2 was drawn are as follows:

Class results from a series of weighings of the same sample (in grams)				
61.0	60.0	60.0	59.8	58.0
61.5	61.5	61.0	60.9	58.0
59.0	60.0	60.2	61.7	60.6
59.7	59.0	60.3	63.0	60.4
62.0	59.0	60.7	58.5	59.0

Figure B.2 Histogram of a hypothetical series of sample weights obtained by weighing the same sample several times. An average was calculated and is indicated by an arrow.

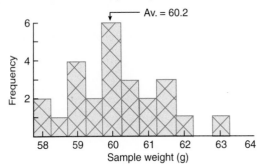

Distribution of sample weights

To make the histogram, the x-axis was laid out with a range of 58 to 64 so that all values would be included. The values were then marked on the graph as lightly penciled Xs, one in each square of the graph paper for each observation. After all data were plotted, bars were then drawn to show the frequencies of measurements. On bar graphs, it is a good idea to show the average value across all measurements with an arrow. In calculating an average, remember the significant figure rule (appendix A).

A final word should be said about how to draft graphs. Be neat! Remember most people do not trust sloppy work. Always print labels and use a sharp pencil, not a pen. Use a ruler to draw straight lines and a drafting template called a French curve to draw curved lines.

APPENDIX C

Simple Statistics

Quantitative data may be expressed in two forms, **count data** or **measurement data.** Count data are discontinuous variables that always consist of whole numbers. They are derived by counting how the results from an experiment fall into certain categories; for example, in a genetic cross between two heterozygotes, you expect to obtain a genotypic ratio of 1:2:1. Measurement data are continuous variables obtained by using some measuring instrument. The precision of the measurement depends on the fineness of the scale on the instrument; for example, a ruler calibrated in centimeters is not as precise as one calibrated in millimeters. Accuracy differs from precision in that it depends not on the scale used but on the accuracy of calibration and the proper reading of the scale.

Whenever measurements are made, there are potential sources of error. Instruments and the humans who read them make random errors that affect accuracy. If the instrument is properly calibrated and if the person who is reading it is careful, the percent error is small and will be randomly distributed around the true measurement. If several readings are taken, the true value will be closer to the mean than to any single measurement.

In some cases, another source of error may be introduced. Bias occurs when an instrument is improperly calibrated or when the operator makes a consistent error in reading or sampling. For example, if a watch that is five minutes slow is used to measure the time of sunset for several days, the data will reflect a consistent bias, showing sunset as occurring five minutes earlier than it really did. Similarly, if a balance, spectrophotometer, or pipette is improperly calibrated, it will consistently yield a biased estimation either over or under the true value; and if the operator of the device misreads the instrument, additional bias enters. For example, if one looks at a car speedometer from the passenger's seat, the speed of the car appears to be lower than it actually is because of the viewing angle. No statistical procedure can correct for bias; there is no substitute for proper calibration and care in making measurements.

Assuming that all sources of bias have been ruled out, there is still the problem of dealing with random fluctuations in measurement and in the properties (size, weight, color, and so on) of samples. In biological research, this is especially important since variation is the rule rather than the exception. For example, white pine needles are approximately 4 inches long at maturity, but in nature, the length of the needles varies due to genetic and environmental differences. A biologist must constantly be aware that biological variability and bias influence the results of experiments and any analysis should include procedures to minimize the effects.

Dealing with Measurement Data

When several measurements are made by an individual or by a class, most would agree that it is best to use the **mean** or **average** of those measurements to estimate the true value. However, determining the average alone may mean that important information concerning variability is ignored. Look at Table C.1, which contains two sets of data: the average temperature in degrees centigrade at 7:00 P.M. each day in September for two different geographic locations.

The average temperature for each location can be calculated by adding the readings for that location and dividing by the total number of readings:

$$\text{average temperature} = \frac{\Sigma \text{ Readings}}{N}$$

where Σ equals "sum of"

For both data sets, the averages are the same (15°C). Based on the averages alone, one might conclude that the two locations have similar climates. However, by simply scanning the table, one can see that location A has a more variable temperature than location B. Such temperature variations, especially those below 0°C, may have a tremendous effect on organisms; many plants and small insects may freeze at subzero temperatures. Therefore, reporting only the average temperature from this set of data does not convey crucial information on variability.

The **range** of values can convey some of this information. Location A had a mean temperature of 15°C with a range from –5 to 35, while location B's mean temperature was 15°C with a range of 10 to 20. Unfortunately, the range of values has a limited usefulness because it does not indicate how often a given temperature occurs. For example, if another location (location C) had 15 days at –5°C and 15 days at 35°C, it would have the same mean and range as location A, but the climate would be harsher.

To overcome some of the limitations of range, the data could be plotted in frequency histograms, with the x-axis showing the daily temperature and the y-axis showing the frequency of that temperature, or how often it occurs. The data for all three locations are plotted in Fig. C.1. It is now obvious, looking at the plotted data, that these three locations have quite different climates in September even though the mean temperatures are the same and the ranges overlap for two of the three. However, this method of reporting is cumbersome.

TABLE C.1 September temperatures in °C at 7:00 P.M. for two locations

Day	Location A	Location B
1	0	15
2	0	20
3	5	20
4	10	20
5	15	20
6	15	15
7	20	10
8	25	10
9	30	10
10	35	10
11	30	10
12	35	15
13	30	20
14	30	13
15	25	17
16	25	18
17	20	19
18	20	20
19	19	10
20	15	11
21	15	12
22	11	13
23	10	17
24	10	15
25	5	10
26	5	20
27	0	10
28	0	10
29	–5	20
30	–5	20

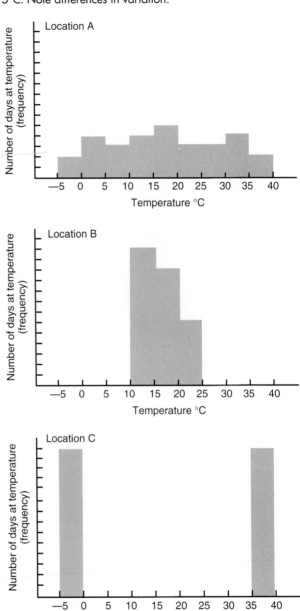

Figure C.1 Frequency histograms of temperatures at three locations each having the same mean temperature of 15°C. Note differences in variation.

An efficient way to report information about variability and its frequency is to calculate the **standard deviation** from the mean. This value is calculated by determining the difference between each observation and the average. If a group of measurements is distributed randomly and symmetrically about the mean, then the standard deviation defines a range of measurements in which 68% of the observed values fall. This is demonstrated in Fig. C.2.

Fig. C.3 shows that distributions can differ in three ways: the means may be different, the standard deviations may be different, or both the means and the standard deviations may be different. Obviously, when the standard deviation is large, the variability is great, and when the variability is small, the standard deviation is small.

Figure C.2   A frequency distribution representing a normal curve. On normal curves, the mean ($\bar{x}$) coincides with the mode, or the most frequently occurring value, and also with the median (the value at which 50% of all values are higher and 50% are lower). The shaded area corresponds to plus or minus one standard deviation about the mean, and will always include 68% of all the observations.

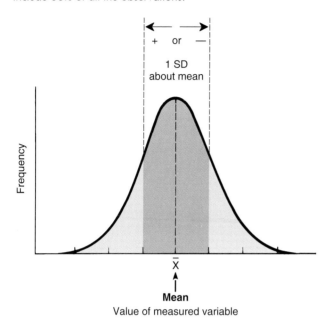

The following formula is used to calculate standard deviation:

$$\text{standard deviation }(\sigma) = \pm\sqrt{\frac{\Sigma\,(x_0 - \bar{x})^2}{n - 1}}$$

where

$\sqrt{}$ = square root

Σ = sum of

x_0 = an observed value

$\bar{x}$ = average of all observed values

n = number of observed values

The computation of standard deviation is best performed by setting up a calculation sheet as in Table C.2, which contains the data from location A in the earlier example. The value for x_0 is the temperature observation on each day, and $\bar{x}$ is the average temperature for the month, 15°C.

The mean and the standard deviation for this set of data are 15 ± 12°C. This means that the average temperature of 15°C plus or minus 12°C (3°C to 27°C) represents 68% of the temperatures you would expect to measure in this location in September. If the same figures are calculated for the data set from location B, the results are

Figure C.3   Three comparisons of normal curves that differ in their means and standard deviations. The dotted line represents the mean, and the shaded area represents plus or minus one standard deviation. In (a), the curves differ in their means but the standard deviations are the same. In (b), the means are the same but the standard deviations are different. In (c) both the means and the standard deviations differ.

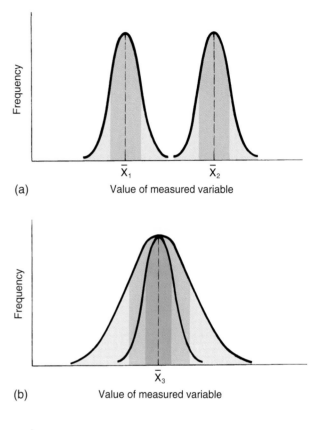

(a)   Value of measured variable

(b)   Value of measured variable

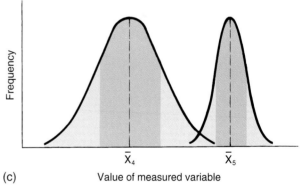

(c)   Value of measured variable

15 ± 4°C (not 4.2 because of significant figure rule; see appendix A). The mean and the standard deviation thus help convey an accurate impression of the similarities and differences between the two sets of data.

In teaching laboratories, each student usually makes only one or two measurements of a biological phenomenon.

TABLE C.2 Calculation sheet for standard deviation using Location A data

n	x_o	$(x_o - \bar{x})$	$(x_o - \bar{x})^2$
1	0	−15	225
2	0	−15	225
3	5	−10	100
4	10	−5	25
5	15	0	0
6	15	0	0
7	20	5	25
8	25	10	100
9	30	15	225
10	35	20	400
11	30	15	225
12	35	20	400
13	30	15	225
14	30	15	225
15	25	10	100
16	25	10	100
17	20	5	25
18	20	5	25
19	19	4	16
20	15	0	0
21	15	0	0
22	11	−4	16
23	10	−5	25
24	10	−5	25
25	5	−10	100
26	5	−10	100
27	0	−15	225
28	0	−15	225
29	−5	−20	400
30	−5	−20	400
	$\Sigma = 450$		$\Sigma = 4182$

$$\bar{X} = \frac{\Sigma X_0}{n} = \frac{450}{30} = 15$$

$$SD = \pm \sqrt{\frac{4182}{30-1}}$$

$$SD = \pm 12$$

The concept of standard deviation is not appropriate for so few readings; a range and average are more suitable. However, if the class combines its readings so that there are 20 or 30 measurements, it is better to calculate the average and the standard deviation. Form the habit of reporting averages with some estimate of the variability when presenting data.

Comparing Count Data: Dealing with Variability

Often in experiments, theories are used to predict the results of experiments and data are recorded as counts or frequencies in categories. For example, Mendelian genetics can be used to predict the progeny from a cross between two heterozygotes. In cases where there is clear dominance and recessiveness, we would use our understanding of genetics to predict that offspring would occur in a ratio of 3:1. What happens when the actual results from a genetic cross of this type do not exactly fit the predictions? First, we would review the data for any obvious errors of technique or arithmetic and eliminate the errors. If the data still do not conform exactly to our predictions, we must decide whether the variations are within the range expected by chance.

When variations from the expected are small, we usually just round off and say that there is a good fit between the data and the expected. However, this is somewhat arbitrary and can become a problem as deviations from the expected become large. When are deviations no longer small and insignificant in comparison to the expected results (i.e., the theory is not a good predictor)?

Statisticians confront this subjectivity by using a statistical test called the **chi-square** (χ^2) **goodness of fit test.** This test creates a number, the χ^2, which summarizes the differences between the data and what was expected. If that number is large, then the theory may be wrong. If it is small, the theory appears to be a reasonable predictor of the results. Rules and a standard table are used to make the decision as to whether this number is larger than we would expect by chance. The steps in using this test are outlined below.

Scientific Hypothesis

The χ^2 test is always used to test a hypothesis known as the **null hypothesis (H_o).** It is based on the scientific theory known as the **model,** which is the basis for predicting the results. A null hypothesis proposes that there is no difference between the results **actually obtained** and those **predicted** from a model. Stated in another way, differences between observed and expected are due to chance variations in results and not due to the action of any biological or physical law.

A null hypothesis is paired with an **alternative hypothesis (H_a),** which proposes an alternative explanation of the results not based on the same scientific model as was the null hypothesis. The acronym "HoHa" emphasizes the coupling of hypotheses in statistical testing.

If a statistical test indicates that the results of an experiment do not fit the expected, then the null hypothesis must be rejected and the alternative hypothesis is true by inference: the data variation is greater than that expected by chance, and a factor other than that tested is influencing the results, implying that the model is not correct.

TABLE C.3　Frequency table of observed and expected results from hypothetical fruit fly cross

	Dom-Dom	Dom-Rec	Rec-Dom	Rec-Rec	Total
Observed (O)	150	60	67	23	300
Expected (E)	169	56	56	19	300

Rejection of H_o does not prove H_a! This is a common mistake in the use of statistics. Rejection of H_o only rejects the model on which H_o is based, and any number of factors could cause nonagreement between the observed and the expected. For example, erroneous data could be included, arithmetic errors could have been made, or another variable(s) was not adequately controlled.

Let us suppose that you are conducting breeding experiments with fruit flies. You are looking at a dihybrid cross involving dominant and recessive traits located on autosomes. Mendel's principle of independent assortment predicts that you should obtain offspring in the phenotypic ratio of 9:3:3:1. When the breeding experiments were finished, the following results were obtained:

150 phenotypes dominant for both traits

60 phenotypes dominant for the first trait and recessive for the second

67 phenotypes recessive for the first trait and dominant for the second

23 phenotypes recessive for both traits

How would you determine whether these results were an acceptable variation from the expected 9:3:3:1 ratio? The actual ratio in this case is close to 7:3:3:1. Has the experiment failed to show Mendelian inheritance, or are the results simply within the limits of chance variability?

First, a null hypothesis should be formulated and then an alternative, mutually exclusive, hypothesis proposed. For the above experiment, these would be:

H_o　There is no significant difference between the results obtained and those predicted by the Mendelian principle of independent assortment.

H_a　Independent assortment does not predict the outcomes of this experiment.

Note how H_o and H_a are mutually exclusive and both cannot be true.

Testing the Null Hypothesis

Because we are dealing with count (frequency) data, which should conform to those predicted by a model, the chi-square goodness of fit test can be used to compare the actual and predicted results.

First, a frequency table is created (table C.3). The entries on the first line are the actual (observed) results from an experimental cross in the lab. Those on the second line are the expected results obtained by multiplying the total by the fractions expected in each category ($9/16 \times 300$, $3/16 \times 300$, etc.).

The summary chi-square statistic is calculated using the following formula:

$$\chi^2 = \Sigma^n \frac{(O_i - E_i)^2}{E_i}$$

where

Σ indicates "sum of"
O_i is the observed frequency in class i
E_i is the expected frequency in class i
n is the number of experimental classes (in this case 4)
χ^2 is the Greek letter *chi,* squared

It should be clear from inspecting this formula that the value of χ^2 will be 0 when there is perfect agreement between the observed and expected results, whereas the χ^2 value will be large when the difference between observed and expected results is large.

To calculate χ^2 for the data in table C.3, the following steps are required:

$$\chi^2 = \frac{(150 - 169)^2}{169} + \frac{(60 - 56)^2}{56} +$$

$$\frac{(67 - 56)^2}{56} + \frac{(23 - 19)^2}{19}$$

$$= \frac{(-19)^2}{169} + \frac{(+4)^2}{56} + \frac{(+11)^2}{56} + \frac{(+4)^2}{19}$$

$$= 2.14 + 0.29 + 2.16 + 0.84$$

$$\chi^2 = 5.43$$

Therefore, for this experiment, the variability from expected result is now summarized as a single number, $\chi^2 = 5.43$.

Making a Decision about the Null Hypothesis

To determine if a value of 5.43 is large and indicates poor agreement between the actual and predicted results, the calculated χ^2 value must be compared to a **critical χ^2 value** obtained from a table of standard critical values (table C.4). These values represent acceptable levels of variability obtained in random experiments of similar design.

TABLE C.4 Critical values for χ^2 at different confidence levels and degrees of freedom (d.f.)

d.f	$\chi^2.90$	$\chi^2.95$	$\chi^2.98$	$\chi^2.99$	$\chi^2.999$
1	2.7	3.8	5.4	6.6	10.8
2	4.6	6.0	7.8	9.2	13.8
3	6.3	7.8	9.8	11.3	16.3
4	7.8	9.5	11.7	13.3	18.5
5	9.2	11.1	13.4	15.1	20.5
6	10.6	12.6	15.0	16.8	22.5
7	12.0	14.1	16.6	18.5	24.3
8	13.4	15.5	18.2	20.1	26.1
9	14.7	16.9	19.7	21.7	27.9
10	16.0	18.3	21.2	23.2	29.6
15	22.3	25.0	28.3	30.6	37.7
20	28.4	31.4	35.0	37.6	45.3
30	40.3	43.8	48.0	50.9	59.7
40	51.8	55.8	60.4	63.7	73.5
60	74.4	79.1	84.6	88.4	99.7
100	118.5	124.3	131.1	135.8	149.5

Confidence Levels (column group header over the five χ^2 columns)

To use table C.4, you must know a parameter called the **degrees of freedom (d.f.)** for the experiment; it is always numerically equal to one less than the number of classes in the outcomes from the experiment. In our example, the experiment yielded data in four classes. Therefore, the degrees of freedom would be 3.

In addition to knowing the degrees of freedom for an experiment, you must also decide on a **confidence level,** a percentage between 0 and 99.99, which indicates the confidence that you wish to have in making a decision to reject the null hypothesis. Most scientists use a 95% confidence level. Any other confidence level can be used and several (but not all) are given in table C.4.

If you now read table C.4 by looking across the row corresponding to 3 degrees of freedom to the column for 95% confidence, you see the value 7.8. This critical value means that 95% of the χ^2 values for experiments with 3 degrees of freedom in which the H_o is true fall below 7.8 due to chance variation and that only 5% will be above that value due to chance.

The next step is to compare the calculated χ^2 value to the critical value for χ^2. When this is done, two outcomes are possible:

1. The calculated value is less than or equal to the critical value. This means that the variation in the results is of the type expected by chance in 95% of the experiments of a similar design; the null hypothesis cannot be rejected.

2. The calculated χ^2 value is greater than the critical value. This means that results would occur only 5% of the time by chance in similar experiments. Stated another way, these results should not be accepted as fitting the model; the null hypothesis should be rejected.

When these decision-making rules are applied to our calculated χ^2 value, 5.43 is obviously less than 7.8. Therefore, although our ratio of results was not 9:3:3:1, it is within an acceptable limit of variation and we cannot reject the null hypothesis. A model based on the Mendelian principle of independent assortment predicts the results of the experiment. Note, however, that you have not proven your H_o; you have simply failed to reject it with 95% confidence in your decision. This is more than a subtle difference because it indicates that scientific knowledge is probabilistic and not absolute.

A Hypothetical Alternative

For illustration, let us suppose that another group did the same experiment but got different results. When they calculated their χ^2 value, it was 9.4. If they went through the same decision making steps that we just did, they would have to reject the null hypothesis: Mendelian genetics did not have predictive power for their experiment. In rejecting the null hypothesis, they can have 95% confidence that they are not making a mistake by rejecting what is actually a true null hypothesis. Stated another way, there is a 5% chance that they are wrong when they reject H_o.

Those who have studied table C.4 might suggest a change in strategy here. If this second group changed its confidence level to 97.5%, the critical value becomes 9.8, which is greater than the calculated value of 9.4, thus keeping the group from rejecting the null hypothesis. This illustrates that by choosing a higher value for a confidence level at a constant number of degrees of freedom, the critical value will be larger, allowing one to accept almost any null hypothesis. Is this not arbitrary, the very situation we sought to avoid by invoking this statistical test?

The solution to this dilemma is found in what is really being tested by the chi-square goodness of fit test. The comparison of calculated and critical χ^2 values is done to attempt to falsify the null hypothesis, to reject it at a predetermined confidence level. If H_o cannot be rejected, then by default it is accepted. *You have not proven H_o, you have simply failed to disprove it.* The confidence level that is stated before each test represents the confidence you have in rejecting the null hypothesis, not the confidence you have in accepting it. As you increase the confidence level, you decrease the likelihood of rejecting H_o.

Statisticians often speak of type I and type II errors in statistical testing. A **type I error** is the probability that you will reject a true null hypothesis. A **type II error** is the probability that you will accept a false null hypothesis. As you increase the confidence level, you reduce the probability that you will reject a true null hypothesis (type I error), but you increase the probability that you will accept a false one (type II error). The confidence level of 95% is used by convention because it represents a compromise between the probabilities of making type I versus type II errors. The basis of this conservatism in science is the recognition that it is better to reject a true hypothesis than it is to accept a false one. Once a confidence level is stated for an experiment, it should not be changed according to the whim of the experimenter who wants a model to have predictive power. Those who do run the risk of accepting fiction as fact.

APPENDIX D

Writing Lab Reports and Scientific Papers

Verbal communication is temporal and easily forgotten, but written reports exist for long periods and yield long-term benefits for the author and others. Gregor Mendel's work is a perfect example. When he finished his research he gave a verbal presentation to a scientific meeting, but few understood it. That was in 1872. Fortunately, he also wrote a paper and published it. About 30 years later that paper was read by scientists who understood it and Mendel was given credit for founding the modern study of genetics.

Scientific research is a group activity. Individual scientists perform experiments to test hypotheses about biological phenomena. After experiments are completed and duplicated, researchers attempt to persuade others to accept or reject their hypotheses by presenting the data and their interpretations. The lab report or the scientific paper is the vehicle of persuasion; when it is published, it is available to other scientists for review. If the results stand up to criticism, they become part of the accepted body of scientific knowledge unless later disproved.

In some cases, a report may not be persuasive in nature but instead is an archival record for future generations. For example, data on the distribution and frequency of rabid skunks in a certain year may be of use to future epidemiologists in deciding whether the incidence of rabies is increasing. Regardless of whether a report is persuasive or archival, the following guidelines apply.

Format

A scientific report usually consists of the following:

1. Title
2. Abstract
3. Introduction
4. Materials and methods
5. Results
6. Discussion
7. Literature cited

There is general agreement among scientists that each section of the report should contain specific types of information.

Title

The title should be less than ten words and should reflect the factual content of the paper. Scientific titles are not designed to catch the reader's fancy. A good title is straightforward and uses keywords that researchers in a particular field will recognize.

Abstract

The purpose of an abstract is to allow the reader to judge whether it would serve his or her purposes to read the entire report. A good abstract is a concise (about 100 words) summary of the purpose of the report (hypotheses tested), the data obtained, and the author's major conclusions.

Introduction

The introduction defines the subject of the report. It must outline the scientific purpose(s) and hypotheses tested, giving the reader sufficient background to understand the rest of the report. Care should be taken to limit the background to whatever is pertinent to the experiment. A good introduction will answer several questions, including the following:

Why was this study performed?

Answers to this question may be derived from observations of nature or from the literature.

What knowledge already exists about this subject?

The answer to this question must review what is known about the topics, showing the historical development of an idea and including the conflicts and gaps in existing knowledge.

What is the purpose of the study?

The specific hypotheses being tested should be stated and the experimental design described.

Materials and Methods

As the name implies, the materials and methods used in the experiments should be reported in this section. The difficulty in writing this section is to provide enough detail for the reader to understand the experiment without overwhelming him or her. When procedures from a lab book or another report are followed exactly, simply cite the work, noting that details can be found there. However, it is still necessary to describe special pieces of equipment and the general theory of the assays used. This can usually be done in a short paragraph, possibly along with a drawing of the experimental apparatus. Generally, this section attempts to answer the following questions:

What materials were used?

How were they used?

Where and when was the work done? (This question is most important in field studies.)

Results

The results section should summarize the data from the experiments without discussing their implications. The data should be organized into tables, figures, graphs, photographs, and so on. But data included in a table should not be duplicated in a figure or graph.

All figures and tables should have descriptive titles and should include a legend explaining any symbols, abbreviations, or special methods used. Figures and tables should be numbered separately and should be referred to in the text by number, for example:

1. Figure 1 shows that the activity decreased after five minutes.
2. The activity decreased after five minutes (fig. 1).

Figures and tables should be self-explanatory; that is, the reader should be able to understand them without referring to the text. All columns and rows in tables and axes in figures should be labeled. See appendix B for graphing instructions.

This section of your report should concentrate on general trends and differences and not on trivial details. Many authors organize and write the results section before the rest of the report.

Discussion

In writing this section, you should explain the logic that allows you to accept or reject your original hypotheses. You should not just restate your results, but should emphasize interpretation of the data, relating them to existing theory and knowledge. Speculation is appropriate, if it is so identified. Suggestions for the improvement of techniques or experimental design may also be included here. You should also be able to suggest future experiments that might clarify areas of doubt in your results.

Literature Cited

This section lists all articles or books cited in your report. It is not the same as a bibliography, which simply lists references regardless of whether they were cited in the paper. The listing should be alphabetized by the last names of the authors. Different journals require different formats for citing literature. The format that includes the most information is given in the following examples:

For articles:

Fox, J.W. 1988. Nest-building behavior of the catbird, *Dumetella carolinensis. Journal of Ecology* 47: 113–17.

For books:

Bird, W.Z. 1990. *Ecological aspects of fox reproduction.* Berlin: Guttenberg Press.

For chapters in books:

Smith, C.J. 1989. Basal cell carcinomas. In *Histological aspects of cancer,* ed. C.D. Wilfred, pp.278–91. Boston: Boston Medical Press.

When citing references in the text, do not use footnotes; instead, refer to articles by the author's name and the date the paper was published. For example:

1. Fox in 1988 investigated the effects of hormones on the nest-building behavior of catbirds.
2. Hormones are known to influence the nest-building behavior of catbirds (Fox, 1988).

When citing papers that have two authors, both names must be listed. When three or more authors are involved, the Latin *et al. (et alia)* meaning "and others" may be used. A paper by Smith, Lynch, Merrill, and Beam published in 1989 would be cited in the text as:

Smith et al. (1989) have shown that . . .

This short form is for text use only. In the Literature Cited, all names would be listed, usually last name preceding initials.

There are a number of style manuals that provide detailed directions for writing scientific papers. Some are listed in further readings at the end of this section.

More and more students turn to the Word Wide Web (WWW) and search engines to locate background information for reports. This is certainly acceptable but you should be aware that mere publication on the WWW does not assure that something is true. Because electronic sources may disappear overnight or a source may be changed (updated) in the time between a first and second access, such sources must be cited differently from printed materials. Citation of an electronic source should include:

Author's last and first names with middle initial; Date of publication on the Internet, including revision dates; Title of the electronic document; The URL contained within angle brackets; and the date on which you accessed it. An example is:

Darwin, C. 1845 and 1997. The voyage of the Beagle. Project Guttenberg. <ftp://ibiblio.org/pub/docs/books/gutenberg/etext97/vbgle10.zip> Accessed 2001.

General Comments on Style

1. All scientific names (genus and species) must be italicized. (Underlining indicates italics in a typed paper.)
2. Use the metric system of measurements. Abbreviations of units are used without a following period.
3. Be aware that the word *data* is plural while *datum* is singular. This affects the choice of a correct verb. The word *species* is used both as a singular and as a plural.
4. Numbers should be written as numerals when they are greater than ten or when they are associated with

measurements; for example, 6 mm or 2 g but *two* explanations or *six* factors. When one list includes numbers over and under ten, all numbers in the list may be expressed as numerals; for example, 17 sunfish, 13 bass, and 2 trout. Never start a sentence with numerals. Spell all numbers beginning sentences.

5. Be sure to divide paragraphs correctly and to use starting and ending sentences that indicate the purpose of the paragraph. A report or a section of a report should not be one long paragraph.

6. Every sentence must have a subject and a verb.

7. Avoid using the first person, I or we, in writing. Keep your writing impersonal, in the third person. Instead of saying, "We weighed the frogs and put them in a glass jar," write, "The frogs were weighed and put in a glass jar."

8. Avoid the use of slang and the overuse of contractions.

9. Be consistent in the use of tense throughout a paragraph—do not switch between past and present. It is best to use past tense.

10. Be sure that pronouns refer to antecedents. For example, in the statement, "Sometimes cecropia caterpillars are in cherry trees but they are hard to find," does "they" refer to caterpillars or trees?

After writing a report, read it over, watching especially for lack of precision and for ambiguity. Each sentence should present a clear message. The following examples illustrate lack of precision:

1. "The sample was incubated in mixture A minus B plus C." Does the mixture lack both B and C or lack B and contain C?

2. The title "Protection against Carcinogenesis by Antioxidants" leaves the reader wondering whether antioxidants protect from or cause cancer.

The only way to prevent such errors is to read and think about what you write. Learn to reread and edit your work.

Further Readings

CBE Style Manual Committee. 1994. *Scientific style and format: CBE style manual for authors, editors, and publishers.* 6th ed. New York: Cambridge University Press.

McMillan, V.E. 1997. *Writing papers in the biological sciences.* 2d ed. New York: St. Martin's Press, Inc.

Pechenik, J.A. 1997. *A short guide to writing about biology.* 3d ed. White Plains: Longman Publishing Group.

CREDITS

Illustrations

Chapter 3

3.10b: From John W. Hole, Jr., *Human Anatomy and Physiology*, 5th ed. Copyright © 1990 The McGraw-Hill Companies, Inc. Reprinted by permission. All Rights Reserved.

Chapter 4

4.1: From David Shier, Jackie Butler, and Ricki Lewis, *Hole's Human Anatomy and Physiology*, 7th ed. Copyright © 1996 The McGraw-Hill Companies, Inc. Reprinted by permission. All Rights Reserved.

Chapter 21

21.7: From Charles F. Lytle, *General Zoology Laboratory Guide*, 12th ed. Copyright © 1996 The McGraw-Hill Companies, Inc. Reprinted by permission. All Rights Reserved.

Chapter 23

23.2b: From Randy Moore, W. Dennis Clark, and Darrell S. Vodopich, *Botany*, 2nd ed. Copyright © 1998 The McGraw-Hill Companies, Inc. Reprinted by permission. All Rights Reserved.

Chapter 25

25.4: From Kingsley R. Stern, *Introductory Plant Biology*, 6th ed. Copyright © 1994 The McGraw-Hill Companies. Reprinted by permission. All Rights Reserved.

Chapter 27

27.9: From David Shier, Jackie Butler, and Ricki Lewis, *Hole's Human Anatomy and Physiology*, 7th ed. Copyright © 1996 The McGraw-Hill Companies, Inc. Reprinted by permission. All Rights Reserved.

Chapter 28

28.1: From Kent M. Van De Graaff and Stuart Ira Fox, *Concepts of Human Anatomy and Physiology*, 4th ed. Copyright © 1995 The McGraw-Hill Companies, Inc. Reprinted by permission. All Rights Reserved; **28.7a:** From Kent M. Van De Graaff, *Synopsis of Human Anatomy and Physiology*. Copyright © 1997 The McGraw-Hill Companies, Inc. Reprinted by permission. All Rights Reserved; **28.7b:** From David Shier, Jackie Butler, and Ricki Lewis, *Hole's Human Anatomy and Physiology*, 7th ed. Copyright © 1996 The McGraw-Hill Companies, Inc. Reprinted by permission. All Rights Reserved; **28.8:** From Kent M. Van De Graaff, *Human Anatomy*, 4th ed. Copyright © 1995 The McGraw-Hill Companies, Inc. Reprinted by permission. All Rights Reserved.

Chapter 29

29.4: From Kent M. Van De Graaff and Stuart Ira Fox, *Concepts of Human Anatomy and Physiology*, 4th ed. Copyright © 1995 The McGraw-Hill Companies, Inc. Reprinted by permission. All Rights Reserved.

Chapter 30

30.1: From Kent M. Van De Graaff and Stuart Ira Fox, *Concepts of Human Anatomy and Physiology*, 4th ed. Copyright © 1995 The McGraw-Hill Companies, Inc. Reprinted by permission. All Rights Reserved; **30.8a:** From David Shier, Jackie Butler, and Ricki Lewis, *Hole's Human Anatomy and Physiology*, 7th ed. Copyright © 1996 The McGraw-Hill Companies, Inc. Reprinted by permission. All Rights Reserved; **30.11:** From Kent M. Van De Graaff, *Human Anatomy*, 4th ed. Copyright © 1995 The McGraw-Hill Companies, Inc. Reprinted by permission. All Rights Reserved.

Chapter 31

31.3: From David Shier, Jackie Butler, and Ricki Lewis, *Hole's Human Anatomy and Physiology*, 7th ed. Copyright © 1996 The McGraw-Hill Companies, Inc. Reprinted by permission. All Rights Reserved; **31.10:** From Kent M. Van De Graaff and Stuart Ira Fox, *Concepts of Human Anatomy and Physiology*, 4th ed. Copyright © 1995 The McGraw-Hill Companies, Inc. Reprinted by permission. All Rights Reserved.

Photos

Chapter 2

Fig 2.2: Courtesy of Leica, Inc.; **Fig 2.7:** © Cabisco/Phototake

Chapter 3

Fig 3.1a–c: © David M. Phillips/Visuals Unlimited; **Fig 3.2:** © John J. Cardamone, Jr./Biological Photo Service; **Fig 3.3:** © Biophoto Associates/Photo Researchers, Inc.; **Fig 3.8a:** © Ed Reschke; **Fig 3.9:** © MHHE/Dennis Strete, photographer; **Fig 3.10a** and **Fig 3.11:** © Ed Reschke

Chapter 4

Fig 4.6: © M. Abbey/Visuals Unlimited

Chapter 5

Fig 5.1: Courtesy Brinkman Instruments, Inc.; **Fig 5.5a–c:** Courtesy of Bausch & Lomb

Chapter 8

Fig 8.3a(1–4) and **Fig 8.3b:** © McGraw-Hill Higher Education, Kingsley Stern, photographer

Chapter 9

Fig 9.2: © B. John/Visuals Unlimited; **Fig 9.4a–e:** © Cabisco/Visuals Unlimited; **Fig 9.4f;** © John D. Cunningham/Visuals Unlimited; **Fig 9.4g:** © Cabisco/Visuals Unlimited; **Fig 9.5a:** © Fred E. Hossler/Visuals Unlimited; **Fig 9.6:** © Cabisco/Phototake.

Chapter 11

Fig 11.1: © K.G. Mutri/Visuals Unlimited; **Fig 11.2:** © David Dressler/Huntington Potter, Life Magazine, Time Inc.

Chapter 14

Fig 14.4a,b: © Bruce Russell/BioMedia Associates; **Fig 14.8a:** © James W. Richardson/Visuals Unlimited; **Fig 14.11a–d:** © Cabisco/Visuals Unlimited; **Fig 14.12a:** © Will Troyer/Visuals Unlimited; **Fig 14.13b:** © Eric Grave/Photo Researchers, Inc.

Chapter 15

Fig 15.3: © Runk/Schoenberger from Grant Heilman; **Fig 15.6a:** © David S. Addison/Visuals Unlimited; **Fig 15.6b:** © Runk/Schoenberger from Grant Heilman; **Fig 15.6c:** © Ed Reschke; **Fig 15.7a,b:** © John D. Cunningham/Visuals Unlimited

Chapter 16

Fig 16.1a: © George Loun/Visuals Unlimited; **Fig 16.1b:** © Runk/Schoenberger from Grant Heilman; **Fig 16.1c:** © Science VU/Visuals Unlimited; **Fig 16.2a:** © John D. Cunningham/Visuals Unlimited; **Fig 16.2b:** © Robert & Linda Mitchell; **Fig 16.3:** © Jack M. Bostrack/Visuals Unlimited; **Fig 16.4a:** © Doug Sokell/Visuals Unlimited; **Fig 16.4b:** © George J. Wilder/Visuals Unlimited

Chapter 17

Fig 17.1: © Biophoto Associates/Science Source/Photo Researchers, Inc.; **Fig 17.3b:** © James Richardson/Visuals Unlimited; **Fig 17.3c:** © Bruce Iverson/Visuals Unlimited; **Fig 17.5b:** © James Richardson/Visuals Unlimited; **Fig 17.5c:** © Ed Degginger/Color-Pic, Inc.; **Fig 17.5d:** © John Gerlach/Visuals Unlimited; **Fig 17.6a:** © William Ormerod/Visuals Unlimited; **Fig 17.6b:** © Richard Thom/Visuals Unlimited; **Fig 17.6c:** © Bill Keogh/Visuals Unlimited; **Fig 17.7b:** © Biophoto Associates; **Fig 17.8a:** © Stephen Kraseman/Peter Arnold, Inc.; **Fig 17.8b:** © Bob Ross/RARE Photographer; **Fig 17.8c:** © John Shaw/Tom Stack & Associates

Chapter 18

Fig 18.1a,b: © R. Calentine/Visuals Unlimited; **Fig 18.1c–h:** © Cabisco/Visuals Unlimited; **Fig 18.3a–d:** © Cabisco/Visuals Unlimited; **Fig 18.8:** © Cabisco/Visuals Unlimited; **Fig 18.9** © Peter Parks/Oxford Scientific Films

Chapter 19

Fig 19.12: © Fred Whittaker; **Fig 19.13b:** © Ed Reschke; **Fig 19.13c:** © Robert & Linda Mitchell; **Fig 19.15a,b:** © Stan W. Elms/Visuals Unlimited

Chapter 20

Fig 20.2a: © Runk/Schoenberg from Grant Heilman; **Fig 20.2b:** © Robert & Linda Mitchell; **Fig 20.8a:** © Ray Coleman/Photo Researchers, Inc.; **Fig 20.8b:** © Daniel W. Gotshall/Visuals Unlimited; **Fig 20.8c:** © Mindy E. Klarman/Photo Researchers, Inc.; **Fig 20.11:** © John D. Cunningham/Visuals Unlimited

Chapter 21

Fig 21.6: © Ken Taylor

Chapter 22

Fig 22.2a: Michael DiSpezio; **Fig 22.2b:** © Daniel Gotshall/Visuals Unlimited; **Fig 22.2c:** © Carl Toessler/Tom Stack & Associates; **Fig 22.2d:** © Daniel Gotshall/Visuals Unlimited; **Fig 22.2e:** © Bill Ober/Visuals Unlimited; **Fig 22.2f:** © Museum of New Zealand TePapa Tongarewa; **Fig. 22.3a:** © Michael DiSpezio; **Fig. 22.4a** and **d:** © John D. Cunningham/Visuals Unlimited

Chapter 23

Fig 23.6: © Cabisco/Visuals Unlimited

Chapter 24

Fig 24.3a: © Cabisco/Phototake; **Fig 24.4a:** © John Troughton and Lesley Donaldson; **Fig 24.4b:** © Cabisco/Phototake; **Fig 24.9b:** From Katherine Esau, *Anatomy of Seed Plants,* 2nd edition. Copyright © John Wiley & Sons, Inc. Reprinted by permission of John Wiley & Sons, Inc.

Chapter 25

Fig 25.1a: © W.P. Wergin & E.H. Newcomb/Biological Photo Service; **Fig 25.6:** © Nancy Vandersluis

Chapter 26

Fig 26.4: © Cabisco/Visuals Unlimited; **Fig 26.5:** © Ed Reschke; **Fig 26.6a, Fig 26.8a–c:** © Cabisco/Visuals Unlimited; **Fig 26.11a–c:** Courtesy of William H. Allen, Jr.; **Fig 26.12:** © John D. Cunningham/Visuals Unlimited

Chapter 27

Fig 27.6a: Copyright R.G. Kessel and R.H. Kardon, *Tissues and Organs: A Text-Atlas of Scanning Electron Microscopy,* 1979, W.H. Freeman and Company **Fig 27.6b:** © Keith Porter; **Fig 27.7c:** © Tom Eisner/Cornell University; **Fig 27.10a,b:** © Gregory J. Highison/Courtesy of Frank N. Low

Chapter 28

Fig 28.5: © Ken Taylor; **Fig 28.9:** © W. Rosenberg, Iona College/Biological Photo Service; **Fig 28.10** Copyright R.G. Kessel and R.H. Kardon, *Tissues and Organs: A Text-Atlas of Scanning Electron Microscopy,* 1979, W.H. Freeman and Company; **Fig 28.11a,b:** © Ed Reschke

Chapter 29

Fig 29.5: © Vincent Gattone and A. Evan; **Fig 29.7:** Photograph by John Vercoe; **Fig 29.10:** Courtesy of William Radke; **Fig 29.11:** From W. Bloom and D.S. Fawcett, *A Textbook of Histology,* 10th edition, Fig. 33.2, p. 859. Copyright © 1975 Chapman & Hall

Chapter 30

Fig 30.3: Courtesy of Dr. Michael K. Reedy; **Fig 30.4:** © Janzo Desaki, Ehime University School of Medicine, Shigenobu, Japan; **Fig 30.8b:** Courtesy of John W. Hole; **Fig 30.8c:** Copyright R.G. Kessel and R.H. Kardon, *Tissues and Organs: A Text-Atlas of Scanning Electron Microscopy,* 1979, W.H. Freeman and Company

Chapter 31

Fig 31.6 and **Fig 31.7:** © Theron O. Odlaug

Chapter 34

Fig 34.2: Courtesy of Millipore Corporation

Illustrators

Dean Biechler: 1.1, 1.2, 3.7, 3.8b, 4.3, 4.4, 7.1, 7.3, 8.1, 8.2, 14.2, 14.4c,d, 14.8b, 15.9, 16.9, 16.10, 17.2, 17.3a, 17.4, 17.5a, 17.7a, 18.2, 18.7, 19.3, 19.6, 19.7, 19.8, 19.9, 19.11, 19.13a, 19.14, 19.16, 20.5, 20.12, 21.1, 21.2, 21.5, 21.8, 22.3b, 22.4b, & c, 23.3, 24.10, 25.1b, 25.3, 25.6, I.1, 28.2, 28.3, 28.4, 28.6, 29.6, 30.6, 30.7, 31.4, 31.9, 31.11. Karla Burds: 27.1, 27.2, 29.2, 29.3, 30.5. Chris Peterson: 15.4, 15.8, 16.6, 22.1, 26.3. Meredith McLain: 11.4, 11.5, 12.2, 22.3c, 24.1. Ann Mackey-Weiss: 26.1, 26.2, 30.10, Margaret Hunter: 3.6, 19.5. Shawn Gould: 17.9, 31.5. Mike Gipple: 4.2. Dan Lanigan: 15.2.